Environmental Change and Agricultural Sustainability in the Mekong Delta

ADVANCES IN GLOBAL CHANGE RESEARCH

VOLUME 45

For further volumes:
http://www.springer.com/series/5588

Mart A. Stewart • Peter A. Coclanis
Editors

Environmental Change and Agricultural Sustainability in the Mekong Delta

Springer

Editors
Mart A. Stewart
Department of History
Western Washington University
Bond Hall 320
98225-9061 Bellingham
Washington, USA
Mart.Stewart@wwu.edu

Peter A. Coclanis
Department of History
University of North Carolina
Hamilton Hall CB# 3195
27599-3195 Chapel Hill
North Carolina, USA
coclanis@unc.edu

ISSN 1574-0919
ISBN 978-94-007-0933-1 e-ISBN 978-94-007-0934-8
DOI 10.1007/978-94-007-0934-8
Springer Dordrecht Heidelberg London New York

Library of Congress Control Number: 2011928080

Cover design: deblik

Photo credit: Cover photo of Tien Giang, Mekong Delta, by Mart Stewart

Printed on acid-free paper

Springer is part of Springer Science+Business Media (www.springer.com)

Acknowledgements

The idea for this volume grew out of a conference, "Environmental Change, Agricultural Sustainability, and Economic Development," held at Can Tho University in Can Tho, Vietnam in late March 2010. The conference focused on environmental change in the Mekong Delta and on adaptation and mediation strategies for responding to it, and included many deeply researched presentations; revised versions of some of these are included among the essays in this volume. The conference also included papers that were comparative or about issues elsewhere in the region that illuminated studies of the same issues in the Mekong Delta; several of these enriched the discussion as a whole, and revised versions are included here, also. We would like to thank all of the participants in the conference for their contributions and for making the conference a success. We are especially grateful to Professor Vo Tong Xuan of An Giang University, perhaps the leading expert in the world on agriculture in the Mekong Delta, for his expansive and stimulating keynote address. We also thank the many faculty, staff, and students of Can Tho University who helped us convene the conference. The conference co-organizers, Dr. Nguyen Hieu Trung, Dean of the College of the Environment and Natural Resources (CENRes), and Dr. Nguyen Be, Head of the Department of Environmental and Natural Resource Management in CENRes, played crucial roles in making the conference a success; the support of Can Tho University's rector, Dr. Nguyen Anh Tuan, was also essential. We also thank Dr. Tony Waldrop, erstwhile Vice Chancellor for Research and Economic Development at UNC-Chapel Hill, for the financial support provided by his office. UNC-Chapel Hill's newly launched Global Research Institute (GRI) provided strong financial support as well. One of the editors of this volume is Director of the GRI, and he is extremely pleased that the conference and now this volume benefitted from the funding provided. That this volume is the first book-length project supported by the GRI is something about which the University of North Carolina-Chapel Hill can be very proud. We would also like to thank Rachel Hynson for last-minute proofreading assistance and Margaret Deignan and Juno Martina at Springer Scientific for helping us move this project from manuscript to book.

Bellingham, Washington — Mart A. Stewart
Chapel Hill, North Carolina — Peter A. Coclanis

Contents

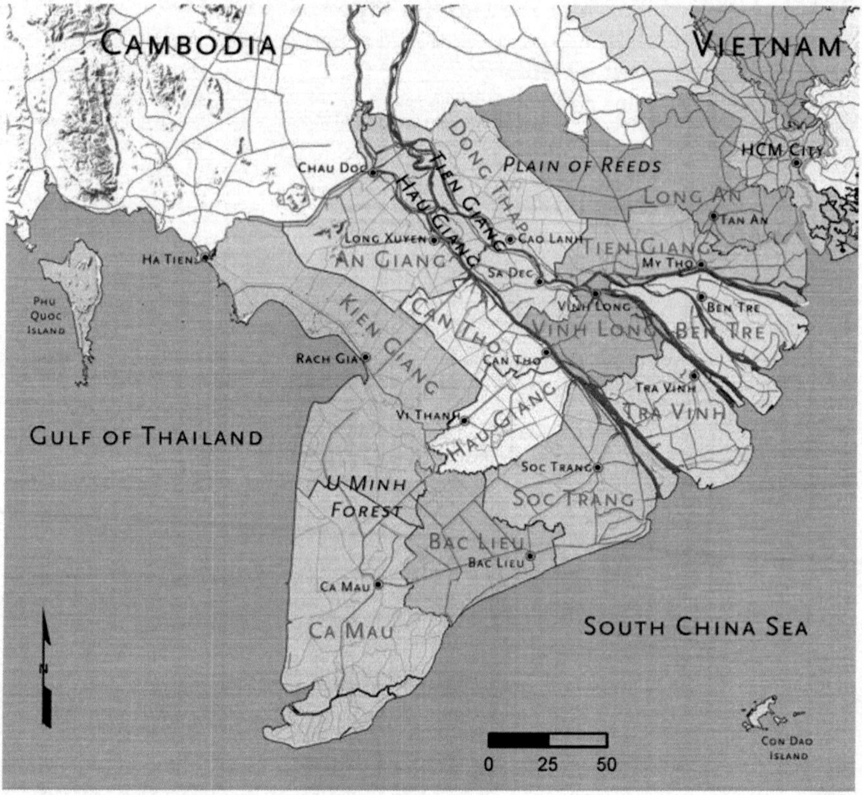

Mekong Delta of Vietnam: Provinces, cities, and major branches of the Mekong River (created by David Biggs)

Contributors

David Biggs
Department of History, University of California at Riverside, California, USA
dbiggs@ucr.edu

Jörn Birkmann
WISDOM Project, United Nations University, Institute for Environment and Human Security (UNU-EHS), Hermann-Ehlers-Str. 10, 53113 Bonn, Germany

Suppakorn Chinvanno
Southeast Asia START Regional Center, Chulalongkorn University, Rm. 508, 5th Floor, Old SWU Pathumwan Bld. N.5, Bangkok, Thailand
suppakorn@start.or.th

Peter A. Coclanis
Department of History, University of North Carolina, Hamilton Hall, CB# 3195, Chapel Hill, NC 27599-3195, USA
coclanis@unc.edu

Nguyen Dac Cu
College of Environment and Natural Resources, Can Tho University, Can Tho, Vietnam

Vo Thanh Danh
School of Economics and Business Administration, Cantho University of Vietnam, Can Tho, Vietnam
vtdanh@st.ctu.edu.vn

Boris Fabres
Centre For Marinelife Conservation and Community Development (MCD), Suite 3104, Level 31, Bldg. 34 T, Hoang Dao Thuy Street, Cau Giay District, Hanoi, Vietnam
bafabres@yahoo.com

Matthias Garschagen
WISDOM Project, United Nations University, Institute for Environment and Human Security (UNU-EHS), Hermann-Ehlers-Str. 10, 53113 Bonn, Germany
garschagen@ehs.unu.edu

Nguyen Lam Giang
Program for the Development of Organic and Fair Trade Cocoa in Vietnam, Helvetas Vietnam, Helvetas Swiss Intercooperation, Hanoi, Vietnam

N. V. Hao
Research Institute for Aquaculture No. 2, Ministry of Fisheries, 116 Nguyen Dinh Chieu St., District 1, Ho-Chi-Minh City, Vietnam

Amy M. Hay
Department of History and Philosophy, University of Texas – Pan American, Texas, USA
ahay@utpa.edu

Huynh Truong Huy
School of Economics and Business Administration, Can Tho University, No. 211, Street 30/4, Can Tho City, Vietnam
huyseba@gmail.com

Olivier M. Joffre
WorldFish Center, Greater Mekong Regional Office, #35 Street 71, Phnom Penh, Cambodia
o.joffre@cgiar.org

Thau Sok Kalyan
Research Team for ProVention Consortium Research and Action Grants, Asian Disaster Preparedness Center (ADPC), Bangkok, Thailand
tintin_thau@yahoo.com

Le Nguyen Doan Khoi
School of Economics and Business Administration, Can Tho University, Campus 1 – 30/4 Street, Hung Loi Ward, Ninh Kieu District, Can Tho City, Vietnam
khoihathanh@gmail.com

Ngo Thi Phuong Lan
Department of Anthropology, University of Social Sciences and Humanities, Vietnam National University – Ho Chi Minh City, 10 –12 Dinh Tien Hoang Street, District 1, Ho Chi Minh City, Vietnam
ngophuonglan@yahoo.com

Thi Thu Huong Le
Institute for Environment and Human Security, United Nations University, Hermann-Ehlers-Str.10, 53113 Bonn, Germany

Louis Lebel
Unit for Social and Environmental Research,
Chiang Mai University, Chiang Mai, Thailand
llebel@loxinfo.co.th

Yasuhiro Matsui
Graduate School of Environmental Science, Okayama University,
3-1-1 Tsushima-naka, 700-8530 Okayama, Japan

Shahbaz Mushtaq
Australian Centre for Sustainable Catchments,
University of Southern Queensland, Toowoomba 4350, Australia

H. H. Ngai
National Breeding Center for Southern Freshwater Aquaculture,
Research Institute for Aquaculture No. 2, An Thai Trung Commune,
Cai Be District, Tien Giang Province, Vietnam

Nguyen Vo Chau Ngan
College of Environment and Natural Resources,
Can Tho University, Can Tho, Vietnam
nvcngan@st.ctu.edu.vn

Ha T. Nguyen
Department of Mechanical and Materials Engineering,
School of Environmental Studies, Queen's University,
60 Union Street, Kingston, ON, Canada K7L 3N6
pearce@me.queensu.ca

Dang Kieu Nhan
Mekong Delta Development Research Institute,
Can Tho University, Can Tho, Vietnam
dknhan@ctu.edu.vn

Quan Minh Nhut
School of Economics and Business Administration, Can Tho University,
Campus 1 – 30/4 Street, Hung Loi Ward, Can Tho City, Ninh Kieu District, Vietnam
qmnhut@ctu.edu.vn

Joshua M. Pearce
Department of Mechanical and Materials Engineering, Queen's University,
60 Union Street, Kingston, ON, Canada, K7L 3N6
nguyenh@me.queensu.ca

Nguyen Huu Phong
VIE/020 Hyacinth Project, Can Tho University, Can Tho, Vietnam

Vo Van Phong
Program for the Development of Organic and Fair Trade Cocoa in Vietnam, Helvetas Vietnam, Helvetas Swiss Intercooperation, Hanoi, Vietnam
vo.phong@helvetas.org

Fabrice G. Renaud
Institute for Environment and Human Security, United Nations University, Hermann-Ehlers-Str.10, 53113 Bonn, Germany WISDOM Project, United Nations University, Institute for Environment and Human Security (UNU-EHS), Hermann-Ehlers-Str. 1053113, Bonn, Germany

Sela Samath
Research Team for ProVention Consortium Research and Action Grants, Asian Disaster Preparedness Center (ADPC), Bangkok, Thailand
samathsela@yahoo.com

NguyenVan Sanh
Mekong Delta Development Research Institute, Can Tho University, Can Tho, Vietnam
nvsanh@ctu.edu.vn

Zita Sebesvari
Institute for Environment and Human Security, United Nations University, Hermann-Ehlers-Str.10, 53113 Bonn, Germany
sebesvari@ehs.unu.edu

N. Sheriff
Policy, Economics and Social Sciences Discipline, WorldFish Center, Jalan Batu Maung, Batu Maung, Bayan Lepas, Penang 11960, Malaysia

Serey Sok
Research Office, Royal University of Phnom Penh, Phnom Penh, Cambodia
sereyhawaii@yahoo.com

Nguyen Phu Son
Mekong Delta Institution, Can Tho University, Can Tho City, Vietnam
npson@ctu.edu.vn

Tong Phuoc Hoang Son
Institute of Oceanography, 01 CauDa Street, Nha Trang City, Vietnam
tongphuochoangson@gmail.com

Mart A. Stewart
Department of History, Western Washington University, 516 High St., BH 320, Bellingham, WA 98225-9061, USA
Mart.Stewart@wwu.edu

Natedao Taotawin
Faculty of Social Science Department of Sociology and Anthropology, Chiang Mai University, Chiang Mai, Thailand
tnatedao@gmail.com

Olivier Tessier
École française d'Extrême-Orient (French School of Asian Studies),
5a xóm Hà Hôi, Hanoi, Vietnam
otessier2002@yahoo.fr; olivier.tessier@efeo.net

Nguyen Phuc Thanh
Graduate School of Environmental Science, Okayama University,
3-1-1 Tsushima-naka, 700-8530 Okayama, Japan
and
College of Environment and Natural Resources, Can Tho University,
Can Tho, Vietnam
npthanh@ctu.edu.vn; phucthanhn@gmail.com

Ram C. Bastakoti
Water Engineering and Management, Asian Institute of Technology,
Klong Luang, Pathumthani, Thailand
rcbastakoti@gmail.com

Nguyen Hieu Trung
College of Environment and Natural Resources,
Can Tho University, Can Tho, Vietnam
nhtrung@ctu.edu.vn

Le Anh Tuan
DRAGON Institute – Mekong – CanTho University,
Campus II, Street 3/2 CanTho City, Vietnam
latuan@ctu.edu.vn

Nguyen Thanh Tung
Institute of Economic and Fishery – Southern Branch of Vietnam,
731 Ly Thuong Kiet Street, Tan Binh District, Ho Chi Minh City, Vietnam
ngthanhtung198@yahoo.com

Daniel Valenghi
Program for the Development of Organic and Fair Trade Cocoa in Vietnam,
Helvetas Vietnam, Helvetas Swiss Intercooperation, Hanoi, Vietnam
daniel.valenghi@helvetas.org

Le Hoang Viet
College of Environment and Natural Resources, Can Tho University,
Can Tho, Vietnam

Charles Waugh
Department of English, Utah State University, Logan, Utah 84322-3200, USA
charles.waugh@usu.edu

Chapter 1
Introduction

Mart A. Stewart and Peter A. Coclanis

The Mekong Delta of Vietnam is one of the most productive agricultural areas in the world. The Mekong River fans out over an area of about 40,000 km^2, and over the course of many millennia has produced a region of fertile alluvial soils and constant flows of energy. Today, about a fourth of the Delta is under rice cultivation, making this area one of the premier rice granaries in the world. The Delta has always proven a difficult environment to manipulate, however, and because of population pressures, increasing acidification of soils, and changes in the Mekong's flow, environmental problems have intensified. The confluence of agriculture and economy in the region with larger flows of commodities and capital over time has also had an impact on the region: For example, its reemergence in recent decades as a major rice-exporting area has linked it inextricably to global markets and their vicissitudes. And most recently, the potential for sea level increases because of global warming has added a new threat, one that makes the Delta a place where local, regional, and global environmental changes are dramatically converging.

Because most of the region is on average only a few meters above sea level and because any increase of sea level will change the complex relationship between tides and down-river water flow, the Mekong Delta is one of the areas in the world which is most vulnerable to the effects of climate change. A meter increase in sea level could displace millions of people and wreak havoc on the productive capacity of agricultural lands in the Delta – and would at the same time, according to a recent Oxfam report, severely set back Vietnam's overall development goals. Moreover, larger environmental justice issues are at stake as well: The Mekong Delta will likely bear a severe burden from climate change despite the fact that as

M.A. Stewart (✉)
Department of History, Western Washington University, 516 High St., BH 320, Bellingham, WA 98225-9061, USA
e-mail: Mart.Stewart@wwu.edu

P.A. Coclanis
Department of History, University of North Carolina, Hamilton Hall, CB# 3195, Chapel Hill, NC 27599-3195, USA
e-mail: coclanis@unc.edu

M.A. Stewart and P.A. Coclanis (eds.), *Environmental Change and Agricultural Sustainability in the Mekong Delta*, Advances in Global Change Research 45,
DOI 10.1007/978-94-007-0934-8_1, © Springer Science+Business Media B.V. 2011

recently as 2000 Vietnam as a whole produced only 0.35% of the world's greenhouse gasses, one of the lowest contributions in the world.

Climate change presents a particular, and a particularly recent, challenge to agriculture and livelihoods in the Delta. Other challenges continue to shape the Delta environment and in turn the capacity for agricultural sustainability in the region. Though Delta residents have always known how to live with floods, changes in flooding patterns pose a particular challenge, especially when they transform patterns of water flow that are essential for rice agriculture. The modernization of Delta agriculture has increased productivity but has created a growing problem with pesticide pollution. The pests that thrive in monoculture continue to plague – in spite of massive and increasingly sophisticated control programs – rice agriculture in the region. Long-term problems with soil fertility, saline intrusions, and soil acidification shape and modify and sometimes challenge substantially different strategies for wresting a living out of the Delta soils. In some locales adjacent to major branches of the Mekong, changes in water dynamics and in shorelines because of massive excavations of sand for sale to developers in places such as Singapore have also caused problems. Moreover, wetlands and forests that are important buffers to agricultural lands have shrunk as rice cultivation has expanded, and biodiversity in general has suffered the same fate in the Delta as it has everywhere that monocultures have expanded. Perhaps the biggest challenge to Delta agriculture comes from far upstream, and is as much a political and technological change as an environmental one: Dams constructed or soon to be constructed on the upper reaches of the Mekong, mostly by the Chinese, have already begun to make it a different river altogether – one with less water, reduced fisheries, and fewer seasonal rhythms and enriching sediments from upstream. The new highly managed Mekong will impoverish environments on the lower Mekong, and make the Delta and Delta farmers more vulnerable to water shortages and droughts as well.

Environmental challenges and changes are amplified by social, economic, and political ones. Livelihoods in the Delta are currently shaped as much by the wrenching changes of modernization and urbanization as by increasingly erratic flows of water and nutrients from upstream and changing tidal flows from down. Many Delta households have a member or more who contribute nonfarm income to the household by working in the new urban industrial zones of Ho Chi Minh City or Can Tho. The flow of resources, commodities, capital, and labor between the Delta and the burgeoning metropolis of Ho Chi Minh City has grown to such intensity and has acquired such complexity that as one of the chapters in this volume argues, it no longer makes sense to think about parts of the Delta as "rural" at all, but as peri-urban. If links to global commodity markets drive even the smallest rice farmers to produce one crop and not another, the powerful market middlemen who represent their personal contacts with these markets compel them to produce certain varieties of rice and do it in certain ways. Rapidly changing institutional arrangements have also had an impact on Delta agriculture and those who practice it: the decline of social welfare networks in the last 20 years, the changing relationship between private and public sectors and the consequent modulation of government authority, the development of agricultural research programs and outreach programs for farmers, and the growing influence of donor economies and of nongovernmental organizations in agricultural research and reform programs have all shaped Delta agriculture – and efforts to develop new strategies of adaptation to environmental change.

This collection represents altogether a mosaic of investigations into environmental change and agriculture in the Mekong Delta. While many of the authors attempt interpretive contributions, mainly in the interest of informing or connecting to emerging agricultural policy discussions, most of them report research about discrete topics that nonetheless are deeply related to the larger problem of mitigation of and adaptation to environmental change. Much of the interpretive writing about the Delta and about agriculture in dynamic environments in the developing world in general attempts to develop new insights into ideas about development trajectories, the roles of different actors within national or regional contexts, the history and larger context of human and environmental relationships in the site of study, and are more focused on engaging with other scholars and scientists to develop analytical and interpretive frameworks than in solving real-world problems. They are, in a word, academic. To be sure, most of the chapters in this collection depend on analytical models that come out of traditional academic disciplines, but they focus on strategies for understanding environmental change in the Delta that might yield real policy and/or real solutions to the challenges that millions in the Delta currently face. Rather than swooping down over the Mekong Delta terrain in an interpretive flyover, they wade around in it, and do so in a way that attempts to tell us what to do about problems at the same time that they describe them. Several of the chapters in this volume do provide us with valuable discussions of the larger historical context of environmental change in the Mekong Delta, or link current changes comparatively with similar developments elsewhere in an effort to illuminate those changes, or talk about cultural values or the political culture of NGOs and other important actors. But most of the chapters focus on discrete research questions that are relevant to understanding environmental and related changes in the Delta and what might be done about them. Many of the scholars also live and work in the area as well, and understand just what the stakes are for residents of the place – that their work is rooted in the Mekong Delta in more ways than one makes their research contributions especially valuable. Much of the work on environmental and related problems in the developing world also seeks to make an ideological argument beyond the problem itself, to persuade as well as explain. The value of problem-based research is that it focuses on the problem; the value of a collection of problem-based research papers is that collectively something can be made of them that addresses the problem, and not simply the discursive world of those who are thinking about it. There is a profound need, given the seriousness of the problems at hand in the Delta and the noisy distractions that prevent us from seeing them clearly, for such "shovel-ready" scholarship, written by people with boots on the ground.

Part I
Environmental Change in the Mekong Delta – Actions and Agencies

Chapter 2
Think Global, Act Global in the Mekong Delta? Environmental Change, Civil Society, and NGOs

Boris Fabres

Abstract The Mekong delta is experiencing rapid growth in foreign investment and trade, urban development, industry, agriculture, and fisheries. With these come externalities of river-flow interruptions, habitat destruction, and pollution interacting with climate change to weaken environmental services. Results can present unprecedented livelihood challenges to communities in one of the most environmentally vulnerable areas of Vietnam. Overcoming the increasing stress and its consequences requires new skills, a forward-looking governance framework, and engagement of millions of delta households. Questions therefore arise about how to address the magnitude and geographic scale of such challenges by developing appropriate institutional and human resources how and to ascertain society's proper role in this process. Civil-society studies have mostly examined the political and societal context of nongovernmental organizations (NGOs) in Vietnam, but NGO initiatives in environmental restoration are at an early stage and less frequently studied. Defining the urgent environmental pressures on the Mekong delta, especially in aquatic systems, this chapter outlines the need for an expanded vision in identifying solutions and widening society's contributions. This chapter also details the institutional constraints to society and NGO action. The chapter argues that more visible global engagement in restoration and management is needed given Vietnam's position as an emerging, natural resource-dependent society and its increasingly international-linked delta economy.

Keywords Civil Society • Environment • Mekong • NGO • Vietnam

---

B. Fabres (✉)
Centre For Marinelife Conservation and Community Development (MCD), Suite 3104, Level 31, Bldg. 34 T, Hoang Dao Thuy Street, Cau Giay District, Hanoi, Vietnam
e-mail: bafabres@yahoo.com

M.A. Stewart and P.A. Coclanis (eds.), *Environmental Change and Agricultural Sustainability in the Mekong Delta*, Advances in Global Change Research 45,
DOI 10.1007/978-94-007-0934-8_2,

2.1 Introduction

The impressive record by Vietnam in reducing poverty and unemployment, improving health, attracting investment, expanding exports in the production of food as well as manufactured goods and services has been termed a model of development and a success story (Thoburn 2009). After more than a decade of continuous high growth rates and progressive integration with the world economy, Vietnam's development has, however, been challenged by multiple international crises in food, finance, health, and internal macroeconomic instability (Riedel 2009). Both national and international agencies have commented on the need to reexamine development strategy and to focus rather more on the quality of growth as Vietnam approaches middle-income status (Nguyen Anh Ngoc et al. 2009; Pincus 2010).

Development has come at an environmental cost, and nowhere have the impacts of development been more pronounced than in Vietnam's coastal zone. Most industrial and commercial activity occurs in these areas, and most of Vietnam's major cities, population, and people of low-income status are located in the coastal zone and delta areas. Large-scale finance and infrastructure investments are focused mainly along the coastline and in its expanding urban centers. The Mekong delta (including 12 provinces, Ho Chi Minh, and Can Tho cities) has been a leading driver of national development and international trade. About 27% of the country's manufacturing firms and 29% of manufacturing workers are located in Ho Chi Minh City (Dore et al. 2008). The 2009 Provincial Competitiveness Index indicates that 10 of the 12 delta provinces, along with Ho Chi Minh and Can Tho cities, are ranked in the top 22 positions (Malesky 2009), and national production statistics (Table 2.1) identify the delta as the main business center, rice and fisheries producer, and revenue earner in Vietnam.

Ho Chi Minh City and environs captured almost 30% of foreign direct investment (FDI) in Vietnam from 1988 to 2007 (Athukorala and Tien Quang Tran 2008). The region is also the major contributor to foreign-exchange earnings through

Table 2.1 Contribution of Mekong delta production to Vietnam's economy (Source: GSO 2009)

Industry, agriculture, and fishery statistics	Mekong Delta[a]	Percent in country
Gross industry output (Billion VND)[b]	460,993	31
Net business turnover (Billion VND)[c]	1,194,230	35
Planted area of paddy rice ('000 Ha)	3,889	52
Production of paddy rice ('000 tons)	20,788	53
Number of farms	58,896	49
Number of fish farms	25,770	74
Aquaculture water surface ('000 Ha)	762	72
Aquaculture shrimp production (tons)	315,691	81
Aquaculture fish production (tons)	1,428,972	77
Gross output of fisheries (Billion VND)	34,407	69
Production of fisheries (tons)	2,744,145	60

[a]Includes Ho Chi Minh city, Can Tho city, and 12 provinces
[b]Current prices
[c]For 2007

exports, including 90% of national rice exports and aquatic products that have driven national fisheries expansion, making this sector the fourth highest earner of foreign exchange for Vietnam. By 2008, catfish (*Pangasius* and *Pangasionodon* sp.) aquaculture exports (mainly from the delta) reached 117 countries and territories and were valued at US$ 1.45 billion. At the same time, shrimp (mainly *Penaeus monodon*) exports were valued at US$ 1.63 billion. Catfish and shrimp exports combined accounted for 68% of Vietnamese aquatic-product exports. These national contributions and Ho Chi Minh City's revenue-sharing with other provinces have led Malesky (2008) to argue that the city has special national leverage. Vind and Fold (2010) have proposed that the city is emerging in international importance as part of a "Global Commodity Chain" with extended impact on the countryside through the attraction of migrant labor and international connectivity.

Coastal development pressures will intensify as Vietnam plans to increase the contribution of the coastal and maritime economy to more than 50% of national GDP by 2020 (CPV 2007). As the southeast region and delta continue to attract investments (Dinh Thi Thanh Binh 2010) and international exposure, it is also likely that this region will increasingly become even more of a public environmental "hotspot." The future brings both great opportunities and worrying challenges. No longer considered only a biodiversity or conservation issue, environmental threats are a growing and critical social and economic concern, especially for many communities whose livelihoods depend directly on living resources. The need to accelerate action to reduce environmental damage in Vietnam and to restore ecosystems is being increasingly voiced by government, citizens, international nongovernmental organizations (INGOs), and donors (Anon 2008; MPI 2009; Vietnam Development Report 2007).

In the Mekong delta, despite the growing evidence of environmental damage, top-down management action has been slow and inadequate; and water-based environmental threats are increasing. Broadening civic engagement in policy development, decision-making, and implementation has been suggested as much-needed, new development pathways are emerging (Friend 2009; Hirsch 2001; Lebel et al. 2005). While local-level action is essential to improve the quality of economic development in the delta and to overcome environmental threats, it is proposed that much greater emphasis than presently exists must be given to the reality of the delta's growing international development drivers, national leverage, and linkages between the two. Local initiatives should take advantage of international conservation trends and develop effective advocacy and diplomacy to influence and mitigate effects that arise outside the delta. To do this effectively, given the social complexity, wider and more organized community engagement is also needed.

2.2 The Environmental Situation in the Delta

The Mekong delta spans 39,000 km^2 in Vietnam with 61% in agriculture and aquaculture use (Deltares 2009). In terms of fish species, the delta is reported to be the most biodiverse region of the Mekong River, with 73 families and 481 species.

Its ecosystems are, however, vulnerable due to agriculture, urban and industrial pollution, eutrophication, sediment fluctuation, dredging, water diversion, habitat damage, and over-harvesting (ICEM 2010). At basin levels, Vietnam has lost more than 60% of the mangrove forests in wetlands and tidal floodplains, especially in deltaic areas, with more than 2,000 km^2 lost in the last 20 years. Only 21% of existing mangrove areas are original forests, the rest have been replanted (VEPA 2005). Sub-tidal coastal and marine wetlands, which form more than 25% of Vietnam's delta wetlands, are under stress (Snidvongs et al. 2003). A wide diversity of farming, rice cropping cycles, and aquaculture systems have displaced delta wetlands (Phan Minh Thu and Populus 2007; Joffre and Bosma 2009; Sakamoto et al. 2009), and urban and industrial development have also led to a reduction in cultivable land (Tran Thi Van and Ha Duong Xuan Bao 2007). Clearing of coastal forests for agriculture and aquaculture has contributed to ecosystem fragmentation, the intensified energy of coastal waves and winds reaching inland, increased flood risk, and facilitated extension of saline regimes inland (Mazda et al. 2004). Globally, the Mekong coastal ecosystem is ranked third in terms of threats from cumulative land-based impacts of nutrients, organic and inorganic pollution, and the coastal population (Halpern et al. 2009).

Pollution loads in the Mekong delta are significant. Ho Chi Minh City is ranked as the first and Can Tho City as the fourteenth most polluting municipality/province in Vietnam (Dore et al. 2008). Pollution is a mix of agriculture and aquaculture chemicals and waste, as well as industrial and urban solid, liquid, and gaseous discharges. Monitoring of the delta has identified extremely high concentrations of veterinary antibiotics linked to livestock farming (Managaki et al. 2007), and pesticide use is estimated to cost US$ 251 million in terms of the loss of use of water resources (Dang Minh Phuong and Gopalakrishnan 2003). Health costs of pesticides are estimated at US$ 650/farmer/ km^2 (Nguyen Huu Dung and Tran Thi Thanh Dung 1999). A bio-monitoring program initiated by the Mekong River Commission (Dao Huy Giap et al. 2010) showed that in 2008 only 40% of sites sampled in the Mekong delta of Vietnam were rated as Class "B" (good).

Perhaps, however, the most contentious trans-boundary issue is the development of dams. Up to 153 major dams are in various stages of development throughout the Mekong basin. These barriers will result to changes in hydrology, sediment load, water quality, and flow interruptions. Kummu et al. (2010) provide estimates of the amount of sediment trapped by existing and planned reservoirs on the mainstream and subbasins of the Mekong River, concluding that more than 50% of the total basin sediment load could be trapped, with potentially severe impacts on the floodplains and delta area in Cambodia and Vietnam.

With 40–70% of the total fish catch in the Mekong reported to depend on migrating species (Barlow et al. 2008; Baran and Myschowoda 2008), the future of fisheries' livelihoods in the world's largest inland fishery looks bleak. Vietnam's delta fisheries are considered as "High Risk" due to biodiversity changes and barriers to completion of life cycles and migrations. Poor households that are the most dependent on capture fisheries will be the most affected (Baran and Myschowoda 2009; Kang et al. 2009).

Delta-capture fisheries' indicators already show downward trends. Declines in catches and sizes of large migratory fish and an increase in low-value species in Mekong fisheries are documented by Baran and Myschowda (2008), and coastal fishery surveys of Soc Trang, Bac Lieu, and Ca Mau provinces indicate that 70% of catches are composed of young and immature fish (Tran Van Viet and Tran Xuan Loi 2007). Catch rates of the inshore trawler fleets in Ca Mau and Bac Lieu provinces from 1996 to 2002 also show reductions of 30–40% in shrimp catches (Christensen and Dang Van 2008).

The Mekong River may not only have near-coastal impacts. Delta coastal fisheries' productivity was earlier linked by Chevey (1933) to annual flooding of the Tonle Sap Lake in Cambodia. Oceanographic studies suggest that the Mekong River plume stimulates phytoplankton growth and oceanic primary production in the western South China Sea (Bombar et al. 2010; Voss et al. 2006), with about 77% of the surface water in the central part diluted by Mekong River flow (Chen et al. 2009). Impact on the Gulf of Thailand is less studied; however, the river's plume is known to reach the middle part of the Gulf (Wolanski and Nguyen Huu Nhan 2005). The downstream ecosystem linkages of the plume, including possible effects on offshore fisheries, remain to be determined.

2.3 Climate Change in the Delta

Superimposed on national development changes are the environmental, social, and economic impacts of climate change. The planning scenario for Vietnam is an average sea level rise (SLR) of 1 m by 2100, modeled to affect 37.8% of the Mekong delta and 23% of the greater Ho Chi Minh City area. Expected results are inundation and salinization of land, coastal infrastructure, industrial enterprises, critical biodiversity, and protected areas. By 2050, about one million persons are estimated to be threatened with displacement (Carew-Reid 2008; UNDP 2009). Ho Chi Minh City is projected to be among the ten most globally exposed cities by the 2070s (Nicholls et al. 2008). Indirect effects and economic and social impacts are less well understood, though intra-regional migration is expected to increase. It has also been suggested that nearer-term climate challenges may be translated into international trade pressure through tariffs, taxes, restrictions to "carbon neutral" products, or the purchase of carbon credits being applied to exporting countries that do not cut emissions (Baumüller 2010; Kirkpatrick et al. 2010).

Of special concern in the delta are climate-change synergies with the multiple stressors of ongoing economic development. These include (1) Coastal subsidence caused by excessive groundwater extraction and deforestation (Pham Thi Viet Nga 2008) leading to increased areas and depth of salt-water intrusion; (2) Coastal deforestation – for example, of mangroves – that accelerates erosion, intensifies wind energy reaching the coast, increasing tidal heights, flooding, salt-water intrusion, and greenhouse-gas levels (Mazda et al. 2002, 2004); (3) Water-flow interruptions through improper barriers, dikes, or dams that may increase the risk of flooding (Le Thi Viet Hoa 2007); (4) Urban and industrial development practices that amplify surface-temperature increases due to urban heat-island effects, already documented up to 9° higher in Ho Chi Minh City (Tran Thi Van and Ha Duong Xuan Bao 2010); (5) Overfishing that makes recruitment in fish populations more variable and more vulnerable to environmental and climate extremes (Anderson et al. 2008). How these combine to shape ecosystem change and their dependent delta communities warrants much greater attention, especially the need to avoid adaptive responses that focus only on selected effects, for example, sea level rise, but in doing so create other environmental and social stresses.

2.4 Environmental Restoration: Developmental Issues

Economic progress in the delta depends on restoring environmental capital while advancing economic development. This is a scientific and social challenge it, and directs development planning to value the delta's environment and ecosystems in ways not limited to marketable commodities or restricted to present value. These research initiatives are still at an early stage in Vietnam but are delivering valuable information for planning. For example, Do Nam Thang and Bennett (2008) demonstrate that a biodiversity conservation program in Tram Chim National Park generates a net social benefit of US$ 0.15–0.96 million. Timber extraction is forecast to increase at an annual rate of 6% beginning in 2010, the Mekong River Delta being one of the most important timber-producing regions (World Bank 2010). However, Vu Xuan Nguyet Hong et al. (2007), in comparing forest-use value in Vietnam, estimated conservation-oriented forests to generate higher values than production/timber use. Development externalities in the delta (pollution, deforestation, habitat damage, overfishing, river-flow interruptions) and incentives also need to be quantified in financial and social terms and used in planning decisions, management, and enforcement.

Administratively, at the country level in Vietnam, there are a range of implementing and collaborating institutions that make environmental governance more challenging. Key institutions and organizations relating to the United Nations Framework Convention on Climate Change (UNFCCC) include 7 ministry and ministry-level institutions and 13 national institutes (MonRE 2005), and for water resources management-9 ministries and 18 departments (Do Thi Nham 2008). In addition, there are academic institutes, provincial- and municipality-level departments, and university-based institutions. In terms of legislation, there are national laws, regulations, decrees, directives, decisions, ordinances, and circulars at country, provincial, and local levels. From 1987 to 2003, over 100 of these legal instruments have been issued in the fisheries sector alone (Pomeroy et al. 2009). Despite this profusion, there may be overlaps, inconsistencies, loopholes, or other inadequacies – for example, a low level of fines, inconsistent monitoring and enforcement, and rent capture – leading to minimal compliance even by serial offenders.

Settlement of citizens' administrative complaints also strains existing systems. Hoang Ngoc Giao et al. (2009) concluded that there is an increasing trend of formal complaints in diverse areas, including land management and the environment, and that complaints were becoming more complicated and diversified. Citizens, however, had difficulty in organizing, presenting, and settling their claims, state organs were overloaded in handling them, and inefficiency in existing mechanisms may in fact prolong and increase complaints. These conditions suggest that there are opportunities for third-party perspectives and engagement through national and international civil-society organizations (CSOs) to strengthen citizen capacities to manage their claims and to assist the government in improving the systems that process them.

The challenges of competing geographic scales of interventions are also significant. Major projects, planning, and research programs in the Mekong operate at scales defined by boundaries as the "Greater Mekong Sub-Region, Mekong Basin, Mekong River, Lower Mekong, and Mekong Delta." These provide a "top-down" framework and a "big picture" context that promotes cooperation at country and regional levels to address trans-boundary challenges. However, this scaling tends to ignore the local, where production activities occur and also where most of the initiatives aimed at reversing environmental damage and its social impacts take place. Some have asked what impact does this "rescaling" of the challenge, this larger-scale emphasis and region-wide institutional effort, have on development of civil-society groups' efforts to come to terms with these same environmental issues (Hirsch 2001)?

In parallel, community-based levels of management are often proposed as the major form of intervention for a number of reasons: perceived inadequacies of top-down or central government management; a preconceived belief in grass-roots solutions; trust in the effectiveness of traditional and culturally embedded approaches; inadequate analysis of local situations and drivers; practical limits imposed by short-term and underfinanced interventions to handle "manageable" situations with a visible result; and limitations of implementing agencies. Often, site-specific reasons for the apparent failure of "top-down, central government" approaches (the commonly used description) are not examined. The result is that scattered community-based projects are becoming a dominant development strategy, and they continue to be favored by donors to civil-society groups and formal NGO programs. Past World Bank portfolios for funding this approach approximated US$ 7 billion (Mansuri and Rao 2003). Interestingly, in a study of four project sites in Vietnam with a view to involving intermediaries in introducing Payment for Environmental Services (PES) programs, all interviewees claimed that local organizations had the least power among possible intermediaries (Pham Thu Thuy et al. 2010).

Criticism has also been raised of the validity of exclusive dependence on participatory and community-based derived knowledge and research (Rambo 2009), and there have also been difficulties in identifying achievement of success in community-based management (Shackelton et al. 2010). Nevertheless, it has been argued that in community intervention, "there is no management 'end-point' to be achieved, only continual adjustments and capacity-building efforts," using an experimentive and adaptive management approach, but still requiring higher-order governmental support (Truong Van Tuyen et al. 2010). Where threats, drivers, and control originate only locally, in most cases solutions can be found locally. In the cases of national, trans-boundary, regional, and international origins of physical disturbances or economic-demand-driven local actions, as in the Mekong delta, permanent solutions need multilevel management, as well as wider stakeholder and institutional involvement (Berkes 2007). Identifying such boundaries that need to be addressed, Cash et al. (2006) propose spatial, temporal, jurisdictional, institutional, management, network, and knowledge components. Berkes (2006) cautions that "overemphasis on community-based management runs the risk of defining issues at one level instead of many. Communities may be complex and embedded in further complexity due to intervening layers and external drivers, and because of the problem of fit between institution and ecosystem boundaries." Cudney-Bueno and Basurto (2009) show that failure to obtain higher-level, cross-scale governance support for community-based management efforts can lead to their demise. These perspectives place greater demands on civil-society groups and NGOs to develop strategies that rigorously examine each situation, develop criteria and forms of entry and exit, work at different scales, and justifying these in their development plans.

Two main development challenges emerge from these tensions. First, there is the challenge of establishing connectivity and support between the wider framework across countries and the local groundwork in communities. Establishing this connectivity continues to elude donors, development agencies, government planners, civil-society organizations operating locally, and researchers. Secondly, there is the challenge of integrating local project interactions, developing a typology of success factors and upscaling, and learning from successes and failures in a way that fits the complexity of the environmental challenge. A common shortcoming is interpreting success as temporary progress at one or a few sites without knowing or covering the wider scope of interventions or the mechanism to extend initial achievements.

Traditional development has been mainly direct and based on expansion of organizations to service a wider base, for example, larger government and nongovernmental organizations (NGOs). Other options that need to be explored with greater effort are the indirect, multiplying impact of creating new and autonomous organizations and changing behavior in existing organizations (Uvin et al. 2000). Progress can be made by being open to combinations of participatory approaches, not prejudging solutions, and involving those whose actions either contribute to environmental threats, are affected by them, or whose participation is needed to enable success. Not surprising, this is challenging. Different sectors, levels and participation capacities are involved. The inertia of habits, historical and asymmetric power relationships have to be overcome, where local groups engage more organized, technically prepared and better funded institutions.

Of additional concern in incorporating community individuals and groups and exercising devolved management authority are potential abuses such as local elite capture and rent-seeking by wealthier or politically connected community members, empowering themselves by rearranging local authority. Case studies show elite capture to be pervasive in community-based projects. Civil-society organizations have a role to play in working toward establishing systems that allocate rights transparently, are accountable, have legitimacy and trust, and provide opportunities for community members to voice their concerns (Béné et al. 2009; Fritzen 2007; Platteau 2004). Reviewing international NGO (INGO) activity in Vietnam from 1990 to 1999, caution was given to ensure that the selection process for local partners did not inadvertently create new local elites with special advantages (Nguyen Kim Ha 2001).

There has been a significant amount of enabling legislation in Vietnam by the government to promote expanded public participation and new approaches, but implementation has been slow. Recent attempts in Cat Tien National Park to develop PES partnerships with local communities have revealed a number of fundamental challenges. These include limitations in the communities' capacities to organize, negotiate, and sustain agreements, the need for a deeper understanding of the diversity of community attitudes, their past government relations, and costs of engagement (Petheram and Campbell 2010). Government institutional arrangements also add transaction costs for communities and NGOs through the spreading of governance responsibilities for similar functions among different agencies. Pham Thu Thuy et al. (2008) note that "four biodiversity monitoring systems belong to different departments in MARD (Ministry of Agriculture and Rural Development) and MoNRE (Ministry of Natural Resources and Environment). Each of these agencies uses its own approaches, indicators, and standards, with no shared systems or agreements."

The international biodiversity importance of the Mekong delta also warrants greater management emphasis. There are 30 protected areas with 19 declared of national importance and 4 areas of international biodiversity significance (FPD/MARD, UNDP and IUCN 2006). The delta also hosts three international biosphere reserves (Kien Giang, Can Gio, and Mui Ca Mau) as part of the World Network of Biosphere Reserves coordinated by the United Nations Educational, Scientific, and Cultural Organization. However, the design of most of the delta's protected areas covers relatively small areas, and encroachment, fragmentation, and sea level rise threaten their future (Pilgrim 2007). The management framework of the biosphere reserves needs strengthening, especially with regard to increasing local community involvement beyond statutory "consultations" and devolving greater authority to local management boards. Of the more than US$ 325 million of financing for major protected areas in Vietnam allocated from 1995 to 2005, most (63%) came from Official Development Assistance (ODA) but with inadequate funding for conservation activities. On average, funding was equivalent to 0.1–0.2% of total ODA and 0.1% of annual government expenditure (FPD/MARD, UNDP and IUCN 2006). It would seem reasonable that biodiversity areas of national and international significance and world biosphere reserves merit higher levels of management and conservation support.

NGOs have an important role in advocating for improved protected area management systems, reducing further ecosystem fragmentation, and working toward active and informed community involvement, improved livelihoods, and financial support for protected areas. In the context of protected-area ecotourism in Vietnam, García-Herrera (2009) identified a range of possible NGO roles: as a bridge between stakeholders, for example, communities, government, and the private sector; acting as an objective intermediary and facilitator; providing training, capacity development, and conservation awareness; and contributing to the implementation of eco-certification. Ecotourism development in Vietnam is a special case needing action and rigorous standards. Often used as a marketing tool in mass tourism, many ecotourism ventures pose serious threats for protected areas. The Institute for Tourism Development Research reported that 95% of Vietnam's tourism sites are advertised as offering ecotourism and that 90% of ecotourism guides lack environmental knowledge (Ngoc Anh 2007).

2.5 Challenges in Fishery Management

The Mekong delta is the center of Vietnam's fisheries industry and aquatic-product exports. As Vietnam's economy moves from a production orientation to an international-market orientation, fishery management requirements have increased to include a mix of traditional and novel approaches involving government, the private sector, and the public, with both obligatory and voluntary options. In particular, a set of globalized conservation and management values continues to emerge, embedded in multinational treaties and conventions. Regulatory action is becoming increasingly demanding as international law and "Soft Law" instruments for environment-related issues are numerous. The CIESIN database[1] lists 62 international environment-related and fishery agreements to which Vietnam is signatory. These carry obligations and related implementation costs, and they are managed by semiautonomous international agencies with different styles and rates of implementation and different scales, for example, from global environmental change to site-based programs and sector-focused goals (Cochrane and Doulman 2005; Schipper and Pelling 2006).

Fishery-management approaches in Vietnam, however, have not yet moved toward recovery-oriented plans (Pitcher 2001). Rather than rebuilding and restoring ecosystems, the current approach prioritizes the sustaining of existing levels of degraded ecosystems or present numbers of wildlife. Recovery planning will require new skills to evaluate and apply new analytical and planning methods; for example, the transition from single-species management based on surplus-production models to ecosystem-based analysis and management which emerged in the 1970s in Southeast Asian fisheries (Pauly and Chuenpagdee 2003). Some are suggesting the integration of physical, biological, and social sciences into the "Earth System Science" and "Global Change Science" categories; this proposal reflects the scale and interdependencies of environmental systems and human actions (Lehodey et al. 2006; NRC 2007). New forms of collaborative research and management also need to be explored using mobile technologies, the Internet, and

---

[1] ENTRI – Environment Treaties and Resource Indicators. Center for International Earth Science Information Network (CIESIN), Columbia University, New York, USA. http://sedac.ciesin.columbia.edu/entri/countryProfile.jsp?ISO=VNM

community volunteers as citizen scientists in environmental data collection, monitoring, and cooperating in knowledge creation.

One of the results of moving toward a market economy is the increased export focus of fishery products. The fishery sector has become of major export importance with a value of US$ 4.25 billion in 2009 (VASEP 2010), and there are plans for it to develop as the "leading sector for the course of industrialization and modernization of the agriculture sector" (Prime Minister of Government 2006). Investment and development need to be managed to minimize environmental damage. The main export products of farmed catfish (*Pangasius* and *Pangasionodon* sp.) and shrimp (mainly *Penaeus monodon*), with activities centered in the Mekong delta, are still heavily dependent on imported feed. Edwards et al. (2004) report that high-quality fish meal for fish fingerlings (mainly *Pangasius* sp.) and crustaceans (mainly shrimp) in aquaculture is mainly imported, representing about 90% of the total fish meal used. Annual fish meal imports have increased from 14,000 tons in 1999 to about 400,000 tons, with imports reported from Thailand, Hong Kong, Taiwan, and the USA and fish oil from Korea (Le Nguyen Doan Khoi 2007; Le Thang Hung and Huynh Pham Viet Thuy 2007). The environmental impacts of export-oriented aquaculture also affect local-capture fisheries. Estimates are that by 2004, about 100,000–120,000 tons of unsustainably harvested marine fish were being used for *Pangasius* catfish feed (APFIC 2005).

It has been argued (Deutsch et al. 2007) that there is a need to expand the discussion and management of aquaculture production from the local farm site to include the global marine-production system supporting the farm in order to ensure traceability of aquaculture inputs, including knowledge of any impacts on source ecosystems. This applies to fish-processing enterprises in Vietnam as well, as there is increasing demand for stocks of imported fish to process due to local plants operating at about 50% of capacity (Anon 2010). Resolving these environmental challenges is complicated by the "stretching" of the international aquatic-production chain with alternative and intermediate suppliers obscuring original ecosystem impacts (Lebel et al. 2002).

There is recognition that while progress is being made in conservation, the number, types, and scale of fisheries-management challenges have increased; and traditional input and output management controls need to be expanded to adopt additional, flexible, and more novel approaches that also commercially benefit producers. In 2009, a community-level capture fishery for clams (*Meretrix lyrata*) in Ben Tre province received international certification from the Marine Stewardship Council (MSC) – the first MSC fishery certification in Vietnam and Southeast Asia (Tran Truong Luu et al. 2009). Market-based and voluntary interventions are increasingly being considered, including certification, and life cycle and value chain approaches for shrimp and fish aquaculture after being given the export focus of many Mekong delta products (Vo Thi Thanh Loc et al. 2010). A number of consultations and trials have been initiated towards meeting international certification standards for shrimp (mainly *Penaeus monodon*) and finfish aquaculture products (Anon 2009a; Vu Dzung Tien and Griffiths 2009). The selection, application, and monitoring of certification schemes, regulations, and standards can, however, be confusing; and guidance is especially needed for small producers, processors, and exporters (Ha Trieu 2010). There are schemes promoted for varied objectives (e.g., conservation, consumer health, animal and social welfare) by industry, governments, NGOs, "organic" and "fair-trade" labels, and codes (Corsin et al. 2007). This diversity, the continuing evolution of the fisheries and regulation schemes, and technical understanding challenge government agencies and NGOs as service providers and agents of change in the industry, especially small producers who have greater difficulty in compliance.

2.6 Civil Society, Vietnamese NGOs, and the Environment

In situations where environmental resources are overexploited, other nonmarket environmental values are not accounted for. Dependent communities are poor and have limited voice, rights, and access to influence policy or enforcement; corruption influences decision-making more than knowledge. In such situations, attempts to restore environmental benefits are demanding. These parameters, in essence, define the work for emerging civil-society organizations and NGOs, whose leadership, management style, and governance arrangements also greatly influence program success.

Vietnam has a record of informal environmental regulation or "community-driven regulation" as described by O'Rourke (2002), where local communities supported by local social organizations respond to local environmental events. It was reported, for example, in 2009 that farmers' associations from Ho Chi Minh City, Dong Nai, and Ba Ria Vung Tau provinces received nearly 11,000 letters demanding legal action against an industrial river polluter. Cases are also reported in the national media and sometimes involve public responses by academics and government agencies (Phung Thuy Phuong and Mol 2001; Vietnam Net 2009). However, such community actions tend to be ad hoc, local, and reactive, often based on immediate and visible economic and health threats, not institutionalized precautionary approaches to avoid long-term environmental damage.

In the study of civil society and its organizations in Vietnam, it is useful to recollect the history of a society emerging through a long colonial period, national conflict, and centrally planned socialist governance toward a market economy, as well as from an agriculture-based economy to an industrial economy. The study both of civil society and of civil-society-linked organizations and NGOs in Vietnam has been undertaken in particular by foreign observers concerned mainly with the evolution, societal, regulatory, and state relations of organizations and only minimally on sectoral issues (though agriculture is better analyzed than others) and on environmental issues. Following the transition of organizations from their origins to the present day, it is important to identify their situational constraints and their developing capacities as partners in dealing with environmental change.

Initial attempts to develop Vietnamese NGOs (VNGOs) and operations of international NGOs (INGOs) in Vietnam are reported as hesitant and challenging

(Hannah 2007). Early analyses in Vietnam commented on the closeness of the first organizations to the government, in some cases, with lateral transfers of individuals from government agencies reconstituting themselves as separate or connected organizations, with a minimal of action-oriented activities (Pednekar 1995). As late as 2001, it was reported that in Vietnam independent local or national environmental NGOs that performed the role of mobilizing communities and an environmental movement were absent (Hirsch 2001; Phung PT and Mol 2001). But later reviews (Lux and Straussman 2004) conclude it is "increasingly clear that the forms, functions, and behavior of Vietnamese NGOs are more and more similar to non-state counterparts in the rest of the world." Sabharwal and Than Thi Thien Huong (2005) also identified a broad trend in that the "growth of development NGOs [is] essentially different from the organizations that emerged as a product of the retrenchment process during the late 1980's." More recently, however, the annual report presented to the government by international donors led by the World Bank still noted, "the processes of creating new civil society organizations are burdensome, and not all are treated equally …. in Vietnam's devolving system, the devolution of civil society has been approached cautiously" (Vietnam Development Report 2010). Recent legislation covering the process to establish and operate associations remains highly legalistic and associations tightly controlled (Prime Minister of Government 2010).

A formal national role for NGOs in economic development and management was recognized in the Government of Vietnam's Five-Year Socioeconomic Development Plan (SEDP) 2006–2010. In this plan, NGOs would act as service providers in humanitarian activities, addressing poverty and environmental management of pollution and sanitation. In addition, the SEDP encourages civil society to "engage in managing and monitoring some public fields" and planning, implementing, and monitoring the SEDP from central to local levels (GOV 2006). There also seems to be a rapid growth of organizations calling themselves NGOs or sometimes VNGOs, whether or not they fit Western conceptions and categories. Nørlund et al. (2006) list a typology of six main types of national organizations with a wide range of functions (mass organizations, umbrella organizations, professional associations, faith-based organizations, VNGOs, and informal groups). Kerkvliet (2003) and Nørlund (2007) suggest that useful evaluation criteria to consider are an organization's origin, affiliation with and dependency on the state, funding source, voluntary membership, governance, work program, and nonprofit orientation. The numbers in each category are publicly un-quantified, but an estimated 65–70 million persons are members.

Organizations may register under a variety of decrees and regulations with varying rights, responsibilities, and geographic jurisdictions. The Vietnam Union of Science and Technology Associations (VUSTA) is estimated to have 1.15 million members across the country (Nørlund et al. 2006). Citizen surveys in Vietnam indicate that 7.6% of persons in 1999 identified themselves as belonging to an environment/conservation organization (Dalton 2005), a higher percentage than in the 17 European countries surveyed. While the diversity of registration procedures and mechanisms facilitates growth in numbers of organizations, it challenges the development of effective governance, performance standards, and accountability.

The debate on VNGO performance in Vietnam is ongoing but not yet informed by comprehensive assessment or public access to information on outcomes and results. There is some skepticism of aid-imposed "civil society" and "participatory" solutions (Molenaers and Renard 2009). Concerns are also expressed regarding whether NGOs are compromised by being aid-driven or because of government restrictions, and questions have arisen as to whether or not their results are really superior to those of government institutions (Gray 1999). Similarly, it has been argued that many NGOs place emphasis on delivering short-term, project-specific outputs rather than meaningful community engagement (Pham Thu Thuy et al. 2010). Organizational limitations in VNGOs have been identified in management, leadership, technical and language skills, networking, communication with government agencies, narrow geographic focus, and staff turnover. On the other hand, noted assets include high motivation levels of their predominantly young staffs, a more egalitarian approach to working with communities, ability to mobilize voluntary resources, lower costs than commercial enterprises, and ability to minimize operational overheads (ADB 1999; AusAID 2000; TAF 2008).

In Vietnam, Nørlund (2007) identified civil-society organizations (CSOs) among the least influential organizations yet notes "all types reach down to the grassroots level better than similar government programs and policies. In that regard, CSOs have had an impact." Similarly, Wischermann (2003) notes "there is a widening gap between what society is in a desperate need of and what state and/or the economy can deliver. Civic Organizations are filling this gap and they are filling it in a very specific way." The space for civil-society action in climate-change action seems particularly open. A recent poll in Vietnam (World Bank 2009) records 93% of respondents indicating that the country has a responsibility to take steps to deal with climate change even if other countries did not, with 77% reporting that the government was not doing enough.

In terms of socioeconomic impact, an external evaluation for Vietnam commissioned by the European Commission found community-based organizations in rural areas to be "highly effective in poverty reduction, working with vulnerable groups, etc. Their work is not as visible as it deserves to be" (Particip GmbH 2009). An even less publicized role for NGOs (especially VNGOs) in Vietnam is their acceptance of volunteers to participate in fieldwork, projects, and events, and the opportunities they provide for student projects and theses. While such collaborations are often temporary, they provide informal education through "on-the-job" exposure, allowing many citizens (especially youth) to gain a deeper knowledge of society and local communities, and in the case of environmental-focused NGOs, of the issues facing biodiversity and natural resources in Vietnam.

2.7 International NGOs in the Mekong Delta

International NGOs (INGOs) have been increasingly active in Vietnam, with development assistance growing from an estimated US$ 20 million in 1990 to US$ 260 million in 2008, with about 650 INGOs estimated to operate now in Vietnam

(Anon 2009b). Together with 29 bilateral and 19 multilateral agencies, including United Nations Organizations, they form the international ODA and development community in Vietnam (DFID 2008), in addition to other INGOs not in Vietnam which also provide technical and financial support. According to Fforde (2005), INGOs and donors mainly channeled their efforts directly to and through state and mass organizations, a situation indicated as operating outside the mainstream of social development, compromising both their understanding of Vietnamese society and their effectiveness.

INGOs, however, played and still play significant roles, by: (1) increasingly funding local organizations including promoting grant programs reserved for VNGOs, the level of funding disbursed often limited by the local organization's technical and management capacity; (2) acting as a bridge between international and local organizations and government agencies, particularly beneficial when the concept of VNGOs was new in Vietnam; (3) facilitating capacity development of VNGOs by providing training and mentoring; and (4) contributing to VNGO leadership through transfers of Vietnamese professionals who previously worked in INGOs.

Records[2] indicate that between 2004 and 2008 at least 99 INGOs operated in the Mekong delta (including Ho Chi Minh and Can Tho cities) with an apparent increasing trend from 74 to 81 over this period. Of these, only 12–18 list themselves as having an Environment, Conservation, Wildlife, or Integrated Development focus. Most INGOs (76–81%) worked in Ho Chi Minh City, and while all provinces had INGO activity, 90–95% of INGOs worked only in four provinces or fewer. This pattern presents a case to have strong institutional coordination and information exchange so that knowledge of the delta's environment as a whole can be obtained and consolidated.

2.8 Information and Communication Challenges and Advocacy

The delta's environmental challenges, origins, scale, and locations require an effective exchange of experiences, successes, failures, knowledge, and open communication. However, development-oriented organizations operating in Vietnam do not tend to share information either at professional or at local community levels. There are some systems in place, however. For example, the VUFO-NGO Resource Centre based in Hanoi facilitates 17 sectoral working groups and moderates 28 Internet-based mailing lists with over 9,000 subscribers, but its operations are precarious "experiencing a financial deficit – to various degrees – since its establishment in 1993" (VUFO-NGO Resource Centre 2010). These initiatives are important to avoid "NGO fatigue," where a number of NGOs converge on the same

[2] Viet Nam INGO Directories, 2004-2005, 2006, 2007, 2008. Vietnam Union of Friendship Organizations (VUFO) NGO Resource Centre, Hanoi, Vietnam.

government authorities or local communities with different and sometimes conflicting messages or similar surveys.

While information on VNGO and INGO activities is reported to supervisory agencies and donors as operating requirements, for members of the public, accessing information on VNGO and INGO activities in Vietnam is time consuming, and information on achievements is not easily obtainable and is of variable quality. Institutional and project evaluation reports are rarely made public on an on-going basis or with outcome information. In a review of comanagement experiences, Swan (2010) comments "what is required of protected area comanagement in Vietnam is to learn and share lessons learnt, from both international and national experiences. At present, this is not occurring to a sufficient degree; the same mistakes are being repeatedly made, and few projects are critically evaluating, documenting, and disseminating achievements and failures." There are benefits to INGOs and VNGOs in improving transparency and the communication of information on their programs, not only to be accountable but also to extend their effectiveness. For example, taking advantage of the strong connections of VNGOs with local communities and INGOs with international communities will extend the geographic impact and advocacy effectiveness of their collaboration. The potential remains great even for basic information technology use in information-sharing. In a survey by the Asia Foundation, only 35% of VNGOs had Web sites (TAF 2008), and, according to another source, 45% of INGO's working in the Mekong delta listed Web sites (VUFO-NGO Resource Centre 2009). Given the estimated 23 million Internet users in Vietnam (VNNIC 2010), the limited Web site development of NGOs seems a continuing lost opportunity.

In terms of technical information, the greatest need in Vietnam relates to the fact that national research is not being adequately covered and made available by national databases, library systems, and national institutions, thereby limiting peer review and making research results unavailable for national development. There is very limited circulation of many institution and university journals, published proceedings, results of workshop and conferences, project reports, and especially theses and dissertations from national universities. Paradoxically, publications produced by Mekong basin country institutions or by researchers and students of these countries in foreign institutions or (published in foreign journals), are sometimes more easily accessed through international search engines and databases than national databases or library systems in the Mekong basin countries that produce them. There are some documentation initiatives in Vietnam such as the National Center for Scientific & Technological Information – *NACESTI* (*www.vista.gov.vn*) and *Vietnam Journals Online* (*www.vjol.info/index.php*) that show great promise but are still limited in coverage and need technical and financial support to expand because of the rate of knowledge creation.

Foreign journals, commercial publications, and aggregation services such as the *Web of Science* still remain the main source of peer-reviewed information, but they are increasingly expensive and in languages not usable by all Mekong basin scientists. General Internet search engines such as *Google Scholar* are helpful and often competitive with commercial services in identifying new knowledge on Vietnam, as are other Internet-based thematic aggregating services, for example, the *Social*

Science Research Network – SSRN (*http://ssrn.com*) and *EconPapers* (*http://econpapers.repec.org/*) in social sciences. But not all information can be obtained without cost. International Open Access journals, Open Archive initiatives, and sponsored access to online journals through *AGORA*, *HINARI, and OARE*[3] are a significant help to life science researchers in Vietnam. But these sponsored-access services are not widely known in Vietnam, even within institutions that have free access to them. Very few NGOs seem to recognize the need, and even fewer pay attention to data and information-sharing and facilitating access as priority needs. Nor does this seem to be a development area favored by donors.

Obtaining information on the Mekong environment at a basin-wide scale is also challenging. There are, for example, development and donor agency networks and Web sites (Table 2.2). Most are traditional Web sites hosting information on publications, news, research updates, meetings, links, and some with national pages. Web site maintenance also needs to be improved, as many Websites have broken links and inadequate updates. Mekong information is also embedded in national and international organizations with Southeast Asia programs, including NGOs and global databases. While all these sources help with identifying information on completed and current research and issues, Internet technology is still not being used to share and take advantage of individual site content; nor is it being customized for the benefit of local users. Each Web site has to be browsed separately, and, as new information is added, the task becomes more time consuming. Importantly, in many cases the information contained is not in the main languages of the peoples of Mekong basin countries.

In terms of natural resource information, the Mekong River Commission (Campbell 2005) describes the Mekong as "one of the more poorly studied rivers … there are very few studies on the geomorphology of the river, very few published studies on any aspect of the ecology of the Mekong … we are limited by the poor knowledge of the biota." This reflects underrepresentation of riparian countries in research and in leading research, "Gray Literature" not being known even within countries generating it. There is a limited sharing of information. An analysis of 4579 freshwater biology papers published from 1992 to 2001 in 9 international journals showed only 1.6% were authored or coauthored by scientists based in Asia (Dudgeon 2003). Regionally, because of political implications, the disclosure of impact assessments at Mekong basin and national levels also tend to remain sequestered by regional agencies or countries requesting the studies (Cronin and Hamlin 2010).

Taking advantage of and contributing to international information systems is becoming increasingly important. Riparian countries in the Mekong basin share a common biodiversity. In the case of fish species for example, there are species, in the Mekong basin ecosystem that Vietnam has in common with its neighbors (from 20 species with Myanmar to 205 species with Cambodia). Table 2.3 shows the common occurrence of freshwater fish species for Mekong ecosystems in riparian countries.

[3] AGORA (Access to Global Online Research in Agriculture), HINARI (Health InterNetwork Access to Research Initiative), OARE (Online Access to Research in the Environment).

Table 2.2 Organizations, programs, and networks in the Mekong River Basin

Name	Type of activity	Organization	Web site
Greater Mekong Sub-Region Program (GMS)	Program	Asian Development Bank	www.eppo.go.th/inter/GMS/GMS.html
Mekong River Commission (MRC)	Program	Intergovernmental organization of Cambodia, Laos, Thailand, Vietnam	www.mrcmekong.org
International Rivers Network	Network	International Rivers Network (Nongovernmental)	www.internationalrivers.org/en/southeast-asia
Mekong Program on Water Environment and Resilience (M Power)	Network	Hosted by Chiang Mai University, Thailand	www.mpowernet.org/
Greater Mekong Sub-Region Academic and Research Network (GMSARN)	Network	Academic and Research institutions in Cambodia, China, Laos, Myanmar, Thailand, Vietnam	http://mpuhost04.ait.ac.th/gmsarn/sitegmsarn2009/home.php
Greater Mekong Sub-Region Energy and Environment Network (GMSEENet)	Network	Project of GMSARN	www.gmseenet.org/
Sustainable Mekong Research Network (Sumernet)	Network	Hosted by Stockholm Environment Institute (Sweden)	www.sumernet.org
Mekong Migration Network (MMN)	Network	Civil society and government organizations in Cambodia, China, Laos, Thailand, Vietnam	www.mekongmigration.org
University Network for Wetland Research and Training in the Mekong Region	Network	Mahidol University, Thailand	www.en.mahidol.ac.th/wetland/index.html#bg
Greater Mekong Sub-Region Tertiary Education Consortium Trust (GMSTEC)	Consortium	Universities in Australia, Cambodia, China, Laos, New Zealand, Thailand, Vietnam	www.gmstec.org/
Mekong Delta Development Research Institute (MDI)	Institute	Can Tho University, Vietnam	www.ctu.edu.vn/institutes/mdi/english/
Delta Research and Global Observation Network (DRAGON) Institute	Institute	Hosted by Can Tho University, Vietnam	http://deltas.usgs.gov/rivers.aspx?river=mekong
Mekong Institute (MI)	Institute	Hosted at Khon Kaen University, Thailand. Inter-Governmental Organization of Cambodia, China, Laos, Myanmar, Thailand, Vietnam	www.mekonginstitute.org
Australian Mekong Resource Centre (AMRC)	Center	University of Sydney, Australia	www.usyd.edu.au/mekong
Mekong Sub-Region Social Research Center (MSSRC)	Center	Ubon Ratchatani University, Thailand	www.mssrc.la.ubu.ac.th
MekongInfo	Project	Hosted by the Mekong River Commission	www.mekonginfo.org

Table 2.3 Common freshwater fish species occurring in the Mekong Basin (Data provided by E. Capuli [Pers. Comm., WorldFish Center, Los Baños, Philippines] on March 1, 2010)

Country[a]	Myanmar	Thailand	Laos	Cambodia	Vietnam
China	12	79	104	64	45
Myanmar		30	26	19	20
Thailand			331	294	169
Laos				325	192
Cambodia					205

[a]Species identified for the Mekong basin ecosystem in each country

Identifying the commonality of Mekong biodiversity helps to prioritize research and management and save costs by countries using each other's knowledge and experiences. As these countries are highly biodiverse, it is unlikely that even the medium-term countries will have the human and financial resources to research every species and local ecosystem. Using credible international information systems like FishBase (www.fishbase.org) that are available to the public, allows countries to take advantage of global knowledge of local importance. The value of this is even more important when countries that share natural resources contribute information to such global information systems: When any country contributes, all countries benefit. The urgency and importance of keeping track of new knowledge to manage the Mekong basin is underscored by the pace of knowledge creation. In 2009, the Mekong River Commission reported 19 regional initiatives in climate change (MRC 2009); and since 1993, published literature on climate-change impacts on marine systems alone is reported to have increased exponentially (Harley et al. 2006).

The application of new knowledge and the identification of priority areas for advocacy, both in developing policy and in addressing on-the-ground environmental issues, should be a logical part of an NGO's program to make change happen. This is, however, only slowly developing in Vietnam. While demand to extend activities to advocacy may exist, with a survey showing 68.3% of organizations to be interested, the reasons offered for inaction include lack of training (70% of staff indicated not attending any course or training in advocacy) and the legal framework for NGOs to participate in policy-making not being clear (TAF 2008).

Most VNGOs still operate with a service-delivery orientation: With a site-specific and local focus, "advocacy" is not considered a main function and usually limited to recommendations to local authorities and communities (Norlund et al. 2006). Some VNGOs do contribute to consultations, drafting of national legislation, and communicating the opinions and experiences of local communities. There are, however, generally not active and dedicated investigative actions by civil-society organizations or VNGOs to hold individuals, institutions, or the private sector accountable. Serial investigative reporting does occur through a few Vietnamese newspapers on polluting enterprises, and illegal use of wildlife, contributing to building public and political pressure for enforcement (Mol 2008). A novel environmental initiative is the work of the VNGO "Environment for Nature" which established a "Wildlife Crime Unit" or a national telephone hotline and database to track occurrences of illegal wildlife use, consumption, and trade

(Sumrall 2009). The Vietnam Forum of Environmental Journalists (www.vfej.vn/en), founded in 1998 with a mission to "motivate and assist Vietnamese environmental journalists to actively contribute to environmental protection for sustainable development through journalism activities," is also an important organization in aggregating and sharing environmental information and impacts in Vietnam. In general, however, VNGOs have not used either the local or international media effectively in highlighting successful environmental interventions, situations needing attention, or in offering solutions. There does however seem to be a positive trend in some areas. For example, a 2006–2009 survey of print and broadcast media in Vietnam (Pham Huy Dung 2009) shows VNGOs to be an increasing source, as well as the second most important source of media information for climate change.

The limited access to local and international information, weak thrust in advocacy, and limited media use significantly limit the impact of VNGO work and represent a serious imbalance in VNGO activity and impact in Vietnam. Most effort continues to be limited to local fieldwork, with much less on using communication tools to spread success stories and solutions and to document practices that have not worked. Opportunities to provide guidance, inspire communities that have overcome environmental threats, motivate others that need similar benefits, and facilitate future collaboration are being lost.

2.9 Conclusion

Objective conditions require that environmental solutions in the Mekong delta incorporate international action. The driving forces include increasing foreign investment and markets for products that drive the delta's economy; legal obligations under international agreements on the environment; the presence of sites of international biodiversity significance in the delta; Vietnam's position downstream from riparian country use of the Mekong River; and the need for international technologies to restore the delta's environment. The situation Vietnam faces in the delta due to the changing global climate also requires international advocacy and diplomacy in order to secure support for a climate-change mitigation and adaptation campaign.

The rate of environmental change and the limited effectiveness of centralized and top-down restoration point to the need for new forms of collaboration and partnerships. Despite incomplete and variable information quality, there are enough indications in Vietnam of a trajectory of progress and the increasing potential of organized civil society to be effective and work with the government and private sector. VNGOs and local community organizations can lead, partner, and support actions to address environmental challenges in the Mekong delta. INGOs can continue to play important roles in bringing international linkages, perspective and advocacy; facilitating application of advances in science; and capacity building of individuals and organizations. At the same time, there is much work to be done in improving transparency, accountability, staff technical capacities, and the administrative framework under which VNGOs and INGOs operate. In terms of capacities, there is a widening range of technical skills needed to match new forms of knowledge and technology, as well as soft skills in relationship building, communication, and advocacy. It is unlikely that any one organization will have all the skills, so the need for active networks, partnerships, and flexible use of services will increase.

Effective programs will need a combination of traditional tools and new approaches that match the scales and stresses, span governments and the international private sector, and are both voluntary and regulatory. Despite the progress and potential of VNGOs to participate effectively from the central government to the community household level, VNGOs have human, financial, and logistical limits. Goals need to be redefined so that communities are not permanently dependent on organizations based in the main urban areas and so that local communities can be guided to self-organize, learn from, and guide other communities. Lastly, "success" cannot be viewed as temporary environmental restoration in individual local sites, no matter how spectacular results may appear, but as permanent, environmental restoration embraced by affected communities across Vietnam and supported by institutional and management change and regulatory enforcement.

References

ADB (1999) A study of NGOs – Vietnam. Asian Development Bank (ADB), Manila

Anderson CNK, Hsieh C, Sandin SA, Hewitt R, Hollowed A, Beddington J, May RM, Sugihara GA (2008) Why fishing magnifies fluctuations in fish abundance. Nature 452(17 Apr):835–839

Anon (2008) International Non-governmental Organisation (INGO) statement for the Vietnam Consultative Group meeting, Hanoi, 4–5 Dec 2008

Anon (2009a) Better management practices (BMPs) for striped catfish (Tra Catfish) farming practices in the Mekong Delta, Vietnam. Version [2] Post-Stakeholder Consultations Fisheries Victoria. Department of Primary Industries, Australia, Network of Aquaculture Centres in Asia-Pacific, Research Institute for Aquaculture No. 2, Can Tho University, Vietnam

Anon (2009b) NGOs: neglect at your own risk. Vietnam Financ Rev 39(July 2009):24

Anon (2010) Vietnam to expand fisheries sector. www.worldfishing.net/features/new-horizons/vietnam-to-expand-fisheries-sector. Accessed 20 May 2010

APFIC (2005) APFIC Regional workshop on Low Value and "Trash Fish" in the Asia-Pacific region. Food and Agriculture Organization of the United Nations, Regional Office for Asia and the Pacific, Asia-Pacific Fisheries Commission

Athukorala P, Tran TQ (2008) Foreign direct investment in industrial transition: the experience of Vietnam. In: Paper prepared for the research workshop, emerging trends and patterns of trade and investment in Asia, 9th Global Development Network conference, Brisbane, 1–2 Jan 2008

AusAID (2000) Assisting local communities: evaluation of government funded NGO projects in Vietnam. QualityAssurance Series No. 18, Commonwealth of Australia

Baran E, Myschowoda C (2008) Have fish catches been declining in the Mekong River Basin? In: Kummu M, Keskinen M, Varis O (eds) Modern myths of the Mekong. A critical review of water and development concepts, principles and policies. Helsinki University of Technology, Helsinki, pp 55–64

Baran E, Myschowoda C (2009) Dams and fisheries in the Mekong Basin. Aquat Ecosyst Health Manage 12:227–234

Barlow C, Baran E, Halls A, Kshatriya M (2008) How much of the Mekong fish catch is at risk from mainstream dam development? Catch Cult 14:16–21

Baumüller H (2010) Building a low carbon future for Vietnam. Technological and other needs for climate change, mitigation and adaptation. Energy, environment and development programme paper 09/02. Chatham House, London

Béné C, Belal E, Ousman Baba M, Ovie S, Raji A, Malasha I, Njaya F, Mamane AN, Russell A, Neiland A (2009) Power struggle, dispute and alliance over local resources: analyzing 'Democratic' decentralization of natural resources through the lenses of Africa Inland Fisheries. World Dev 37:1935–1950

Berkes F (2006) From community-based resource management to complex systems. Ecol Soc 1145. www.ecologyandsociety.org/vol11/iss1/art45/ 15p. Accessed 15 July 2010

Berkes F (2007) Community-based conservation in a globalized world. Proc Nat Acad Sci 104:15188–15193

Bombar D, Dippner JW, Doan Hai Nhu, Nguyen Lam Ngoc, Liskow I, Loick-Wilde N, Voss M (2010) Sources of new nitrogen in the Vietnamese upwelling region of the South China Sea. J Geophys Res. doi:10.1029/2008JC005154

Campbell I (2005) Status of the lower Mekong River and future research priorities. In: Claire-Ashton E, Nguyen Thi Bich Nhi, Nielsen T (eds) Workshop report "research priorities for the Mekong delta – environmental status and future requirements". Can Tho University, Vietnam, pp 23–26

Carew-Reid J (2008) Rapid assessment of the extent and impact of sea level rise in Vietnam. Climate change discussion Paper 1, ICEM – International Centre for Environmental Management, Brisbane

Cash DW, Adger W, Berkes F, Garden P, Lebel L, Olsson P, Pritchard L, Young O (2006) Scale and cross-scale dynamics: governance and information in a multilevel world. Ecol Soc 118. www.ecologyandsociety.org/vol11/iss2/art8/. Accessed 12 Mar 2010

Chen W, Dai M, Liu Q, Huh C, Miao Y (2009) Signature of Mekong River plume in the western South China Sea revealed from 228Ra and 226Ra. American Geophysical Union, Fall Meeting 2009, Abstract #OS23A-1183. http://adsabs.harvard.edu//abs/2009AGUFMOS23A1183C. Accessed 14 May 2010

Chevey P (1933) The Great Lake of Cambodia. The underlying causes of its richness in fish. In: Proceedings of the fifth pacific science congress, Victoria, 1–4 June 1933 and 5–15 June 1933, Vancouver, pp 3809–3816

Christensen S, Dang Van Thi (2008) Catch rates as indicators of ecosystem health and exploitation status in the shrimp fishery in the South China Sea. Aquat Ecosyst Health Manage 11:212–224

Cochrane KL, Doulman DJ (2005) The rising tide of fisheries instruments and the struggle to keep afloat. Philos Trans R Soc Biol Sci 360:77–94

Corsin F, Funge-Smith S, Clausen J (2007) A qualitative assessment of standards and certification schemes applicable to aquaculture in the Asia–Pacific region, RAP Publication 2007/25. Food and Agriculture Organization, Bangkok

CPV (2007) Vietnam to become a strong marine based economy by 2020. Communist Party of Vietnam website. www.cpv.org.vn/details_e.asp?id=BT1850735493. Accessed 18 May 2007

Cronin RP, T Hamlin (2010) Mekong tipping point: Hydropower Dams, human security and regional stability. www.stimson.org/pub.cfm?ID=980. Accessed 27 Aug 2010

Cudney-Bueno R, Basurto X (2009) Lack of cross-scale linkages reduces robustness of community-based fisheries management. PLoS One 4:e6253

Dalton RJ (2005) The greening of the Globe? Cross-national levels of environmental group membership. Environ Polit 14:441–459

Dang Minh Phuong, Gopalakrishnan C (2003) An application of the contingent valuation method to estimate the loss of value of water resources due to pesticide contamination: the case of the Mekong Delta, Vietnam. Int J Water Resour Dev 19:617–633

Dao GHuy, Kunpradid T, Vongsobath C, Do Thi Bich Loc, Prum S (2010) Report on the 2008 biomonitoring survey of the lower Mekong River and selected tributaries. MRC Technical Paper No. 27, Mekong River Commission, Vientiane

Deltares (2009) Towards sustainable development of deltas, estuaries and coastal zones. Description of Eight Selected Deltas, Delft

Deutsch L, Gräslund S, Folke C, Troell M, Huitric M, Kautsky N, Lebel L (2007) Feeding aquaculture growth through globalization: exploitation of marine ecosystems for fishmeal. Global Environ Change 17:238–249

DFID (2008) Vietnam: country assistance plan 2007–2011. Department for International Development, London

Dinh BTT (2010) Agglomeration economies and location choices by foreign firms in Vietnam. FIW Working Paper. No. 45

Do Nam Thang, Bennett J (2008) Would wetland biodiversity conservation improve social welfare? A case study in Vietnam's Mekong River Delta. GMSARN Int J 2:101–108

Do NT (2008) Water related ecosystems. Project on Water Sector Review, Vietnam. Asian Development Bank (ADB.TA.4903-VIE) Consultancy Report, ADB, Manila

Dore G, Brylski P, Nygard J, Tran Thi Thanh Phuong (2008) Review and analysis of the pollution impacts from Vietnamese manufacturing sectors. Rural development, natural resources and environment unit, East Asia and Pacific region. The World Bank, Washington, DC

Dudgeon D (2003) The contribution of scientific information to the conservation and management of freshwater biodiversity in Tropical Asia. Hydrobiologia 500:295–314

Edwards P, Le Anh Tuan, Allan GL (2004) A survey of marine trash fish and fish meal as aquaculture feed ingredients in Vietnam. ACIAR Working Paper No. 57, Australian Centre for International Agricultural Research

Fforde A (2005) Instability, the causes of development success and the need for strategic rethinking. School of Development Studies, Melbourne University Private. http://www.aduki.com.au/Adam%20Fforde's% 20Research%20output.htm. Accessed 16 Feb 2010

FPD/MARD, UNDP, IUCN (2006) Protected areas policy study draft technical report. Forestry Protection Department Ministry of Agriculture and Rural Development, United Nations Development Programme and IUCN – The World Conservation Union, Hanoi

Friend RM (2009) Fishing for influence: fisheries science and evidence in water resources development in the Mekong Basin. Water Altern 2:167–182

Fritzen SA (2007) Can the design of community-driven development reduce the risk of elite capture? Evidence from Indonesia. World Dev 35:1359–1375

García-Herrera JJ (2009) Handbook of ecotourism in protected areas of Vietnam. Fundación Desarrollo Sostenido (FUNDESO), Madrid

GOV (2006) The five-year socioeconomic development plan 2006–2010. Government of the Socialist Republic of Vietnam, Hanoi

Gray M (1999) Creating civil society? The emergence of NGOs in Vietnam. Dev Change 30:693–713

GSO (2009) Statistical yearbook of Vietnam 2008. General Statistics Office (GSO), Statistical Publishing House, Hanoi

Ha Trieu (2010) Global GAP and ISO 22000 – "Passport" for seafood from Mekong Delta. Vietfish 7:54–57

Halpern BS, Ebert CM, Kappel CV, Madin E, Micheli F, Perry M, Selkoe KA, Walbridge S (2009) Global priority areas for incorporating land–sea connections in Marine conservation. Conserv Lett 2:189–196

Hannah J (2007) Local non-government organizations in Vietnam: development, civil society and state-society relations. PhD dissertation, Department of Geography, University of Washington, Seattle

Harley CDG, Hughes AR, Hultgren KM, Miner B, Sorte CJB, Thornber CS, Rodriguez LF, Tomanek L, Williams SL (2006) The impacts of climate change in coastal marine systems. Ecol Lett 9:228–241

Hirsch P (2001) Globalisation, regionalisation and local voices: the Asian Development Bank and rescaled politics of environment in the Mekong region. Singapore J Trop Geo 22:237–251

Hoang Ngoc Giao, Tran Van Son, Nguyen Thi Thu Van, Dinh Van Minh, Nguyen Van Kim, Nguyen Thi Hanh (2009) The mechanism to settle administrative complaints in Vietnam.

Challenges and solutions. Summary report. The Asia Foundation and the Research Institute for Policy, Law and Development, Hanoi

ICEM (2010) Aquatic systems baseline assessment working paper. Prepared by the International Centre for Environmental Management (ICEM) for the Mekong River Commission Secretariat (MRCS) Strategic Environmental Assessment, Vol II, 8 Mar 2010

Joffre OM, Bosma RH (2009) Typology of shrimp farming in Bac Lieu Province, Mekong Delta, using multivariate statistics. Agric Ecosyst Environ 132:153–159

Kam SP, Badjeck M-C, Teh L, Be Năm VT, Hien TT, Hue NT, Phillips M, Pomeroy R, Sinh LX (2010) Economics of adaptation to climate change in Vietnam's aquaculture sector: a case study. Report to the World Bank. WorldFish Center, in collaboration with the Sub-National Institute for Agricultural Planning and Projection, Can Tho University, Centre for Marinelife Conservation and Community Development, Can Tho

Kang B, Perret L, He D, Li Y (2009) Are the fish of the upper and lower Mekong interconnected? Chin J Oceanol Limnol 27:400–407

Kerkvliet BJT (2003) Introduction: grappling with organizations and the state in contemporary Vietnam. In: Kerkvliet BJT, Heng RHK, Koh DWH (eds) Getting organized in Vietnam. Moving in and around the socialist state. Institute of Southeast Asian Studies, Singapore, pp 1–24

Kirkpatrick C, Freeman NJ, Kim N, Ninh B (2010) Managing risk and attaining equitable growth. The Asia Foundation, Hanoi

Kummu M, Lu XX, Wang JJ, Varis O (2010) Basin-wide sediment trapping efficiency of emerging reservoirs along the Mekong. Geomorphology 119:181–197

Le KDN (2007) Description of the *Pangasius* value chain in Vietnam. CAS Discussion paper No 56, Oct 2007, Centre for ASEAN Studies, University of Antwerp, Antwerp

Le Thang Hung, Huynh Pham Viet Thuy (2007) Analysis of feeds and fertilizers for sustainable aquaculture development in Vietnam. FAO Fish Tech Pap 597:331–361

Le Thi Viet Hoa, Nguyen Huu Nhan, Wolanski E, Tran ThanhCong, Haruyama S (2007) The combined impact on the flooding in Vietnam's Mekong River delta of local man-made structures, sea level rise, and dams upstream in the river catchment. Estuar Coast Shelf Sci 71:110–116

Lebel L, Nguyen Hoang Tri, Saengnoree A, Pasong S, Buatama U, Le Kim Thoa (2002) Industrial transformation and shrimp aquaculture in Thailand and Vietnam: pathways to ecological, social, and economic sustainability? Ambio 31:311–323

Lebel L, Garden P, Imamura M (2005) The politics of scale, position, and place in the governance of water resources in the Mekong region. Ecol Soc 1018. www.ecologyandsociety.org/vol10/iss2/art18/. Accessed 10 May 2010

Lehodey P, Alheit J, Barange M, Baumgartner T, Beaugrand G, Drinkwater K, Fromentin J-M, Hare SR, Ottersen G, Perry RI, Roy C, van der Lingen CD, Werner F (2006) Climate variability, fish and fisheries. J Climate 19:5009–5030

Lux SJ, Straussman JD (2004) Searching for a balance: Vietnamese NGOs operating in a State-Led Civil Society. Public Admin Dev 24:173–181

Malesky E (2008) Straight ahead on red: how foreign direct investment empowers subnational leaders. J Polit 70:97–119

Malesky E (2009) The Vietnam provincial competitiveness index: measuring economic governance for private sector development. 2009 Final Report, Vietnam Competitiveness Initiative Policy Paper #14, Vietnam Chamber of Commerce and Industry and USAID Vietnam Competitiveness Initiative, Hanoi

Managaki S, Murata A, Takada H, Bui Cach Tuyen, Chiem NH (2007) Distribution of macrolides sulfonamides, and trimethoprim in tropical waters: ubiquitous occurrence of veterinary antibiotics in the Mekong delta. Environ Sci Technol 41:8004–8010

Mansuri G, Rao V (2003) Community based (and Driven) development: a critical review. Development Research Group, The World Bank, Washington, DC

Mazda Y, Magi M, Nanao H, Kogo M, Miyagi T, Nobuyuki Kanazawa N, Kobashi D (2002) Coastal erosion due to long-term human impact on Mangrove Forests. Wetlands Ecol Manage 10:1–9

Mazda Y, Magi M, Kogo M, Phan Nguyen Hong (2004) Mangroves as a coastal protection in the Tonkin Delta, Vietnam. In: Phan Nguyen Hong (ed) Mangrove ecosystem in the Red River Coastal Zone. Agricultural Publishing House, Hanoi, pp 315–326

Mol APJ (2008) Environmental governance through information: China and Vietnam. Singapore J Trop Geo 30:114–129

Molenaers N, Renard R (2009) The trouble with participation: assessing the new aid paradigm. In: Kremer M, van Lieshout MP, Went R (eds) Doing good or doing better. Development policies in a globalizing world. Scientific Council for Government Policy, Amsterdam University Press, The Hague, pp 255–278

MoNRE (2005) Stocktaking assessment on the United Nations framework convention on climate change and Kyoto protocol. Department of Environment, Ministry of Natural Resources and Environment, Hanoi

MPI (2009) Results-based mid-term report for implementation of the five-year socioeconomic development plan 2006-2010. Project 00040722 "Support to Socioeconomic Development Monitoring". Ministry of Planning and Investment, Hanoi

MRC (2009) Adaptation to climate change in the countries of the Lower Mekong Basin: regional synthesis report. MRC Technical Paper No. 24, Mekong River Commission, Vientiane

Ngoc A (2007) Ecotourism in name only? Vietnam Investment Review News, No 796. Release date 15 Jan 2007. www.vir.com.vn/Client/VIR/index.asp?url=content.asp&doc=11438#. Accessed 27 Jan 2007

Nguyen NA, Nhat Duc Nguyen, Chuc Dinh Nguyen, Nguyen Thang (2009) The global crisis and medium term growth prospects for developing countries: the case of Vietnam. DEPOCEN Working Paper Series No. 2010/08

Nguyen DH, Tran DTT (1999) Economic and health consequences of pesticide use in paddy production in the Mekong Delta, Vietnam. EEPSEA Research Reports 1999–2002

Nguyen HK (2001) Lessons learned from a decade of experience. A strategic analysis of INGO methods and activities in Vietnam 1990–1999. Vietnam Union of Friendship Organizations (VUFO) NGO Resource Centre, Hanoi

Nicholls RJ, Hanson S, Herweijer C, Patmore N, Hallegate S, Corfee-Morlot J, Chatreau J, Muir-Wood R (2008) Ranking port cities with high exposure and vulnerability to climate extremes. Exposure estimates. Environment Working Papers No.1, Organisation for Economic Cooperation and Development (OECD), Paris

Nørlund I, Dang DN, Bach ST, Chu D, Dang QN, Do Bich CMN, Tang CT, Vu MC (2006) The emerging civil society. An initial assessment of civil society in Vietnam. Vietnam Institute of Development Studies (VIDS), UNDP Vietnam, SNV Vietnam, CIVICUS Civil Society Index, Hanoi

Nørlund I (2007) Civil society in Vietnam. Social organizations and approaches to new concepts. The German Journal on Contemporary Asia (ASIEN), 105:68–90. http://www.asienkunde.de/english/index.php?folder=journal_asien&file=journal_asien.htm

Nørlund I, Dang Ngoc Dinh, Bach Tan Sinh, Chu Dung, Dang Ngoc Quang, Do Bich Diem, Nguyen Manh Cuong, Tang The Cuong, Vu Chi Mai (2006) The emerging civil society. An initial assessment of civil society in Vietnam. CIVICUS, VIDS, SNV, UNDP, Hanoi. www.civicus.org/new/ media/CSI_ Vietnam_ report%20.pdf,,160p. Accessed 14 Mar 2010

NRC (2007) Understanding multiple environmental stresses. Report of a workshop. National Research Council/National Academies, Washington, DC

O'Rourke D (2002) Motivating a conflicted environmental state: community-driven regulation in Vietnam. In: Buttel FH, Mol APJ (eds) The environmental state under pressure, vol 10. Elsevier Science, Amsterdam, pp 221–244

Particip GmbH (2009) Special focus 4: civil society in Vietnam 1: overview. evaluation of European Commission's Cooperation with Vietnam. Final report, Vol II, Annexes. Ref.: EuropeAid/122888/C/SER/ Multi, Particip GmbH, Freiburg, pp 180–182

Pauly D, Chuenpagdee R (2003) Development of fisheries in the gulf of Thailand Large Marine Ecosystem: analysis of an unplanned experiment. In: Hempel G, Sherman K (eds) Large marine ecosystems of the world. Elsevier, Amsterdam, pp 337–354

Pednekar SS (1995) NGOs and natural resource management in mainland Southeast Asia. Thai Dev Res Inst Q Rev 10:21–7

Petheram L, Campbell BM (2010) Listening to locals on payments for environmental services. J Environ Manage 91:1139–1149

Pham DH (2009) climate change coverage by the Vietnamese media: four-year trends 2006–2009. Institute of Health, Environment and Development, Hanoi, in association with the Vietnam Forum of Environmental Journalists and Internews' Earth Journalism Network, 5 Dec 2009. 8p. Source: www.earthjournalism.net/.../Vietnam%20media_Executive%20Summmary_2009_Final.pdf

Pham NVT (2008) Study on land subsidence due to groundwater exploitation in Ho Chi Minh city. M. Eng. thesis, Asian Institute of Technology, Pathum Thani

Pham Thu Thuy, Hoang Minh Ha, Campbell BM (2008) Pro-poor payments for environmental services: challenges for the government and administrative agencies in Vietnam. Public Admin Dev 28:363–373

Pham Thu Thuy, Campbell BM, Garnet S, Aslin H, Hoang Minh Ha (2010) Importance and impacts of intermediary boundary organizations in facilitating payment for environmental services in Vietnam. Environ Conserv 37:64–72

Phan Minh Thu, Populus J (2007) Status and changes of Mangrove Forest in Mekong Delta: case study in Tra Vinh, Vietnam. Estuar Coast Shelf Sci 71:98–109

Phung PT, Mol APJ (2001) Communities and informal regulators. Local protests against industrial pollution in Vietnam. In: Sustainability at the millennium: globalization, competitiveness and the public trust: 9th international conference of greening of industry network, Bangkok, 21–25 Jan 2001

Pilgrim J (2007) Effects of sea level rise on critical natural habitats in Vietnam. In: Proceedings of the international symposium on biodiversity and climate change – links with poverty and sustainable development, Hanoi, 22–23 May 2007, German Development Cooperation (GTZ), pp 17–20

Pincus J (2010) Moving in step with the world. Vietnam Invest Rev 956, 957 (Feb 8–21): 6–7.

Pitcher TJ (2001) Fisheries managed to rebuild ecosystems? Reconstructing the past to salvage the future. Ecol Appl 11:601–617

Platteau J-P (2004) Monitoring elite capture in community-driven development. Dev Change 35:223–246

Pomeroy R, Nguyen Kim Anh Thi, Ha Xuan Thong (2009) Small-scale marine fisheries policy in Vietnam. Mar Policy 33:419–428

Prime Minister of Government (2006) Decision No.242/2006/QD-TTg of 25 Oct 2006, on approval of the Fisheries Export Development Program by 2010 and its orientation by 2020 (Unofficial English Translation)

Prime Minister of Government (2010) Decree No. 45/2010/ND-CP of 21st April 2010 on the Organization, Operation and Management of Associations (Unofficial English Translation)

Rambo AT (2009) Are the farmers always right? Rethinking assumptions guiding agricultural and environmental research in Southeast Asia. Asia Pacific Issues No. 88 (May 2009), East-West Center, Honolulu

VUFO-NGO Resource Centre (2009) VUFO-NGO Resource Centre INGO Directory 2008. Vietnam Union of Friendship Organizations (VUFO) NGO Resource Centre, Hanoi

VUFO-NGO Resource Centre (2010) Financial overview and next Steps. Presentation Summary. INGO Forum 22 April 2010, Vietnam Union of Friendship Organizations (VUFO) NGO Resource Centre, Hanoi, Vietnam

Riedel J (2009) The global economic crisis and its long-run implications for Vietnam. School of Advanced International Studies, The Johns Hopkins University, USA. Paper prepared for the United Nations Development Programme (UNDP), Hanoi

Sabharwal G, Than Thi Thien Huong (2005) Civil society in Vietnam: moving from the margins to the mainstream. Department for International Development (UK), Hanoi

Sakamoto T, Phung Cao Van, Kotera A, Khang Nguyen Duy, Yokozawa M (2009) Detection of yearly change in farming systems in the Vietnamese Mekong Delta from MODIS time-series imagery. Jpn Agric Res Q 43:173–185

Schipper L, Pelling M (2006) Disaster risk, climate change and international development: scope for and challenges to integration. Disasters 30:19–38

Shackleton CM, Willis TJ, Brown K, Polunin NVC (2010) Reflecting on the next generation of models for community-based natural resources management. Environ Conserv 37:1–4

Snidvongs A, Choowaew S, Chinvanno S (2003) Impact of climate change on water and wetland resources in Mekong River Basin: directions for preparedness and action. Southeast Asia START Regional Center Report Number 12

Sumrall KA (2009) Confronting illegal wildlife trade in Vietnam: the experience of education for nature-Vietnam. Practicum submitted in partial fulfillment of the requirements for the degree of Master of Science (Environmental Policy), University of Michigan, Apr 2009

Swan S (2010) Co-management: concepts and practices in Vietnam. Deutsche Gesellschaft für Technische Zusammenarbeit (GTZ), Hanoi

TAF (2008) Training needs assessment of civil society organizations in Vietnam: organizational development and community mobilization in the policy and law-making process civil society empowerment and participation in the policy and law-making process in Vietnam project. The Asia Foundation, Hanoi

Thoburn J (2009) Vietnam as a role model for development. World Institute for Development Economics Research, United Nations University, Research Paper No. 2009/30

Tran VT, Ha BX (2007) Urban land cover change through development of imperviousness in Ho Chi Minh City, Vietnam. In: Asian Association on Remote Sensing (AARS) (ed) Proceedings of 28th Asian conference on remote sensing, Kuala Lumpur

Tran Thi Van, Ha Duong Xuan Bao (2010) Study of the impact of urban development on surface temperature using remote sensing in Ho Chi Minh City, Northern Vietnam. Geogr Res 48:86–96

Tran LT, Holt T, Hough A (2009) Final report for Ben Tre clam fishery. Ref: 82036 v4, Moody Marine Ltd., Moody International Certification, Wyvern Business Park, Derby

Tran VV, Tran LX (2007) The status of inshore fishing on aquatic resources on the coastal area of the Mekong of Vietnam. In: Proceedings of the 11th international symposium on efficient application and preservation of marine biological resources, Ministry of Education and Training, Vietnam Fishery & Aquaculture Institution Network, Nha Trang University, Vietnam, 1–2 Nov 2007, pp 174–178

Truong Van Tuyen, Armitage D, Marschke M (2010) Livelihoods and co-management in the Tam Giang lagoon, Vietnam. Ocean Coast Manage 53:327–335

UNDP (2009) Climate change fact sheet: the effects of climate change in Vietnam and the UN's response. United Nations Development Programme, Hanoi, 19 Oct 2009

Uvin P, Jain PS, Brown LD (2000) Think large and act small: towards a new paradigm for NGO scaling. World Dev 28:1409–1419

VASEP (2010) Vietnamese seafood exports Jan–Dec 2009. Vietnam Association of Seafood Exporters and Producers (VASEP). www.vasep.com.vn/vasep/eCustomer.nsf/A1 D1 EB BE 97F143BC4 72 56F1C00304D3F/ CD7A73E292 EBE9704725 76C600 142 E37. Accessed 8 Mar 2010

VEPA (2005) Overview of wetlands status in Vietnam following 15 years of Ramsar Convention implementation. Vietnam Environment Protection Agency, Ministry of Environment and Natural Resources, Hanoi

Vietnam Development Report (2007) Vietnam: aiming high. Joint donor report to the Vietnam consultative group meeting, Hanoi, 14–15 Dec 2006

Vietnam Development Report (2010) Modern institutions. Joint donor report to the Vietnam consultative group meeting, Hanoi, 3–4 Dec 2009

Vietnam Net (2009) Pollution Report ordered to push VEDAN to Compensate. http://english.vietnamnet.vn/tech/ 2009/06/853694/, 1p.. Accessed 19 Dec 2009

Vind I, Fold N (2010) City networks and commodity chains: identifying global flows and local connections in Ho Chi Minh City. Glob Netw 10:54–74

VNNIC (2010) Report on internet statistics of Vietnam (Jan 2010). Vietnam Internet Network Information Center, Hanoi. www.thongkeinternet.vn /jsp/trangchu/index.jsp. Downloaded 3 Feb 2010

Vo LTT, Bush SR, Le Xuan Sinh, Nguyen Tri Khiem (2010) High and low value fish chains in the Mekong Delta: challenges for livelihoods and governance. Environ Dev Sustain, 12:889–908

Voss MD, Bombar LN, Dippner JW (2006) Riverine influence on nitrogen fixation in the upwelling region off Vietnam, South China Sea. Geophys Res Lett 33 (L07604). doi: 10.1029/2005GL025569

Vu Dzung Tien, Griffiths D (2009) Shrimp GAP and BMP in Vietnam: policy, current status and future direction, fisheries sector programme support (FSPS II) project. Ministry of Agriculture and Rural Development, Hanoi

Vu Xuan Nguyet Hong, Nguyen Manh Hai, Ho Cong Hoa P, Silva FT, Mortensen JB (2007) Accounting for Vietnam's wealth: the role of natural resources. Central Institute for Economic Management, Ministry of Planning and Investment, Hanoi

Wischermann J (2003) Vietnam in the era of Doi Moi: issue-oriented organizations and their relationship to the government. Asian Surv 43:867–889

Wolanski E, Nguyen NH (2005) Oceanography of the Mekong River Estuary. In: Extended abstract presentation, International conference on Deltas, Ho Chi Minh City, 10–16 Jan 2005, 5p. www.megadelta.ecnu.edu.cn/

World Bank (2009) Public attitudes toward climate change: findings from a multi-country poll. World Development Report 2010 Background Paper Report 52066, The World Bank, Washington, DC

World Bank (2010) Socialist Republic of Vietnam – aligning public spending with strategic priorities in the forestry sector. Report 53101, The World Bank, Washington, DC

Chapter 3
Fixing the Delta: History and the Politics of Hydraulic Infrastructure Development and Conservation in the Mekong Delta

David Biggs

Historians and many other writers have widely acknowledged the political importance of flood control and land reclamation in river deltas. Whether one is considering ancient dikes on the Red River near Hanoi – works first planned 1,000 years ago by Emperor Ly Thai To following the hoof prints of a magical white horse – or modern levees on the Mississippi River in New Orleans – works built by U.S. Army Corps of Engineers following a mantra of rationalization and efficiency – the importance of these works to the states that maintain them is clear. The fortunes of Vietnamese emperors and American presidents have risen and fallen with the successes and failures of flood control. With the global spread of environmentalist movements and increasing concerns, especially in river deltas, about rising sea levels, political debates about building or maintaining infrastructure have become more complicated with concerns about environmental impacts.

There is another factor that has gained limited acknowledgment in these debates about the future of hydraulic infrastructure projects, and that is the historical one. Engineers and policymakers, even when faced with opportunities to start fresh and abandon troubled or failing projects, often choose to rebuild and patch up the old structures. American historian of technology Theodore Porter (1995) might suggest that powerful engineering bureaucracies such as the Army Corps of Engineers are keen to protect their status in government by continuing the same kinds of expensive, high-tech solutions that require their continued services. Engineers often do think historically to the extent that they spend considerable effort studying past surveys and historic data accumulated at a project site. Some have even become so fascinated by their sites that they take up the pen to write history. For example, the Vietnamese writer Nguyen Hien Le (1971), who wrote one of the first modern histories of Đồng Tháp (Plain of Reeds), traveled the Mekong Delta as a surveyor with the colonial-era Department of Public Works. Historic infrastructure, even failing infrastructure, often carries with it powerful cultural associations to the current state – thus prohibiting abandonment. River deltas, with their tendency to

D. Biggs (✉)
Department of History, University of California at Riverside
e-mail: dbiggs@ucr.edu

M.A. Stewart and P.A. Coclanis (eds.), *Environmental Change and Agricultural Sustainability in the Mekong Delta*, Advances in Global Change Research 45,
DOI 10.1007/978-94-007-0934-8_3, © Springer Science+Business Media B.V. 2011

flood, continually threaten to undo such works. Like palimpsests, the seemingly blank, muddy surfaces of a delta often hide all but a faint trace of older canals, roads, and levees that have almost vanished. States often spend great sums to prevent their roads, canals, bridges, and monuments from disappearing, too.

With major changes in governance of the Mekong Delta region over the past 100 years, decision-makers continue to struggle not only with the economic and political dimensions of flood control but also with the lack of historical continuity in governance over 20 years. This chapter argues that fixing such problems as degraded waterways or disappearing wetlands requires not only solving technical issues at hand but also adapting fixes to fit the historical reality of a given place. How might historians contribute to sustainable development? They can play a role in expanding knowledge of the environmental and engineering past surrounding specific projects and places. By providing more access for readers and contemporary development specialists into the engineering archives, local histories, and state records, historians can aid in contextualizing contemporary issues, possibly even shaping the debate. The following three sections each examine a site in the Mekong Delta where different debates have emerged about how to fix them. There is some play, too, with the word "fix" to mean both an action to "repair or improve" a site and an action to "place or anchor" something more permanently.

3.1 Fixing the Border: Vinh Te Canal

One of the challenges, especially for Vietnam, has been managing the long international border with Cambodia that stretches a few hundred kilometers across the Dong Thap and Long Xuyen flood plains. In the fall when the Mekong floods, much of this territory is under a few meters of water; and since the border was first established in 1757, many people have crossed it with little regard for issues of sovereignty. Roughly, a fifth of the delta's area, about a million hectares, lies inside Cambodia; and every year tens of thousands of fishermen, merchants, Theravadan Buddhists, ethnic Vietnamese, ethnic Cambodians, and ethnic Chinese make pilgrimages to see family in both directions. The permeable nature of this invisible border and the ties that bind people across it have plagued governments intent on controlling smugglers and others who cross it. Waterways and levees constructed along this border thus represent an effort to more permanently fix the geopolitical boundaries between Vietnam and Cambodia.

In 1818, the Vietnamese King Gia Long made the first effort to fix this border in the landscape with a massive border canal project. He ordered General Nguyen Van Thoai to build a canal connecting the Hau Giang River with the Gulf of Siam to move Nguyen forces quickly to the gulf coast if needed and seal the border from potential Siamese incursions. The general assembled a work force of some 5,000 laborers and spent about 5 years building the waterway (Nguyen 1972, p. 159). The construction was repeatedly challenged, by Khmer workers who worried about losing land to Vietnamese settlers and by stagnant water that contributed to about 800 deaths from disease. The general, upon completion of the project, received a promotion to “Jade Marquis” (*ngoc hau*); and his wife, Ms. Chau Thi Te, was honored for her aid to the sick by having the canal named after her.

From the beginning, as a hydraulic and defensive work, the Vinh Te Canal was troubled. Le Van Khoi, the adopted son of the southern viceroy Le Van Duyet, started a secessionist rebellion in 1832 and invited Siamese forces to help him fight the Nguyen. Siamese naval ships entered Vinh Te Canal from the Gulf and quickly built camps along the Vinh Te Canal. They overtook Vietnamese forces garrisoned at Phnom Penh and much of the western delta region until 1845 (Son Nam 1988, p. 9). One of the vestiges from the Siamese occupation, so locals claim, is a species of Cambodian coconut tree called *thot not* (Khmer *th'not*) that produces a fruit fermented into a kind of beer that is still popular in the Cambodian countryside.

Not long after the Siamese left, the French arrived with steam-powered, ironclad gunboats. Their first attacks in the delta commenced in 1860 with major naval battles in the eastern delta in 1862. By 1868, the western region around Vinh Te Canal had turned into a dangerous harbor for groups organizing raids on French outposts, some led by veterans of the Nguyen Army. As the admiral-governors consolidated their hold on the Mekong Delta in the 1870s, they commenced surveys to map out canals, roads, and even railroads. However, they left Vinh Te Canal alone. The canal's course continued to deteriorate through the last decades of the nineteenth century, and the colonial regime focused more on building roads and railroads in Cambodia and Vietnam than fixing waterways that tended to be clogged with local traffic (Edwards 2006; Biggs 2011, 35–37). A naval hydrographer, Jacques Renaud, surveyed the canal in 1879 and made a strong case for re-dredging it to bring that region of the delta back economically to the regional trading center it had been in the late 1700s (Rénaud 1879). However, with Cambodia as a French protectorate in 1884, the colonial government had much less concern with sealing the border as had the Nguyen. Renaud argued unpersuasively for the government to “continue that work of civilization begun by the Vietnamese.” (Rénaud 1879) Rather than spend great sums to build a more permanent marker of the political border, the colonial government left its topographers to choose any available solid points – not always easy to find in a river delta – and draft invisible lines across the flooded plains to signify an intercolonial border with little immediate importance.

This problem of the delta's border, however, returned violently in the postcolonial era. Fighting from 1945 to 1954 produced many conflicts on the border as not only the Viet Minh and the French forces clashed, but also over a million followers of a millenarian Buddhist political-religious group, Hoa Hao, organized their zones of control in the border area. With Cambodian King Sihanouk's determination of Cambodian neutrality in 1958, the Viet Minh found more sanctuaries on the Cambodian side of the border while groups loyal to the Republic of Vietnam (RVN) sought refuge inside Vietnamese territory. American advisors tried unsuccessfully to prevent smuggling and movement of resistance fighters across the invisible border. Nicholas Sellers, an American army captain serving in Ha Tien in 1967–1968, notes in the introduction of his historical work the frustration he found in patrolling this porous area (Sellers 1983). The Vinh Te Canal by 1968 had become another important crossing for land-based and water-based transport of goods through the port at Kampot or from the mountains north of Tay Ninh. After the 1968 Tet Offensive, the top American naval commander in Vietnam, Admiral Zumwalt, resolved to halt the traffic across the Cambodian frontier. Operation SEALORDS (Southeast Asia Lakes, Oceans, Rivers, Deltas Strategy) commenced with intensive aerial bombing of National Liberation Front bases as well as interdiction missions along the borders, including Vinh Te Canal. It was relatively short-lived on the border, and NLF traffic resumed shortly after the campaign ended (McQuilkin 1997).

After the American war ended in 1975, violent incidents soon erupted again on the Vinh Te Canal. The Khmer Rouge forces, in 1977, massacred over 3,000 ethnic Vietnamese villagers at Ba Chuc living several kilometers from the canal. This action, combined with attacks across the border in other locations prompted the Vietnamese to seize Phnom Penh and occupy Cambodia in 1978–1979. That occupation lasted until 1989 with the Paris Agreements commencing in September and concluding in October 1991.

With sustained peace after 1989 and Vietnam's economic resurgence after the *doi moi* reforms set in 1986, work re-dredging Vinh Te Canal commenced in 1998 as part of the national government's region-wide plan to improve irrigation, transport, and agriculture in the delta (Socialist Republic of Vietnam 1996). I first traveled along the banks of the Vinh Te Canal in 2000 and found the waterway and more so its embankments buzzing with activity. The old road that ran from Chau Doc to Tinh Bien Border Crossing was being elevated on a much higher levee as well as paved and widened to accept container trucks and tourist buses. Barges with diggers mounted on them proceeded to excavate the canal channel and repair the banks with the fill. Anyone traveling in the Mekong Delta today will likely confirm that these roadways have in most cases overtaken the canals as the prime conduits for passenger traffic and commerce in consumer goods. Meanwhile, waterways such as Vinh Te Canal still look much the same as they likely did in 1823 or 1879. Boat traffic is limited to shallow-bottom junks, fishing boats, and barges.

With the roadwork and renewed attention on the border in peacetime, the Southern Institute of Social Sciences and Humanities convened a conference on Vinh Te in 1999 that drew some press attention. Reports and news articles on this gathering of eminent national historians highlighted the canal's important role in the nineteenth century for protecting the border. Old historical sites including the tomb of General Thoai and his wife Ms. Te were visited, and historical sources describing Vietnamese claims to the area were explained.

Thus the history of the Vinh Te Canal offers a good example of fixing not so much hydraulic problems in the landscape but claims to sovereignty. A series of high floods that struck the delta in 2000 and 2001 have likely filled in much of the re-dredged channel, but the roadway remains a vital link for international commerce. The historical conference and more intense investment in the border region since 2000 suggests that in the long run, solving hydrological problems here may be less important than in other, more agriculturally productive areas.

3.2 Fixing Colonial Mistakes: Phung Hiep

Today, the seven-way canal intersection of Phung Hiep is a destination for boatloads of tourists who come to see its floating market, a sprawling cluster of boats selling all manner of vegetables and fruits from their decks. Like many towns in the region south of the Hau Giang River, Phung Hiep sprang up in response to the colonial-era dredging campaigns that brought several million migrants into newly drained flood plains. One of the main canals leading here, Lai Hieu Canal, was started in 1906 and allegedly followed along an "elephant road" (*duong tuong*) cut by migrating herds of elephants over the centuries. It was a transverse canal, meaning that it traversed a bowl-shaped floodplain and connected the strong-flowing Hau Giang River with the smaller Cai Lon River that drains into the Gulf of Siam rather than the South China Sea. Such canals aided the cutting of vast freshwater mangrove forests and the establishment of rice paddy, but they also complicated the tidal flux so essential for rice cultivation by connecting two different tidal influences coming from the South China Sea and the Gulf of Siam.

Several earlier projects further upstream on the Hau Giang River (Thot Not and Xa No Canals) had met with relatively little local resistance, but when colonial engineers connected the Lai Hieu Canal to an existing canal intersection at Phung Hiep they met a very different situation. Measurements of both the volume of water passing through this intersection and the relative heights of the different waterways were inaccurate. Instead of bringing water from the Hau Giang through the new lands and draining through the Cai Lon River, the new canal caused water to fill up the shallow bowl-like depression of lands that had already been cultivated. One of the largest landowners in this area, a Vietnamese woman who had inherited a plantation when her French husband died, brought a massive lawsuit against the colony seeking damages. Her *metis* son Henri, educated in France and living in Saigon, sought the help of a Paris lawyer, and the case reached the high court in Hanoi (Vietnam National Archives #2 [VNA2] 1911). After a lengthy court case and much race-based criticism of the Vietnamese heiress and her mixed-race son, the courts eventually sided with the Department of Public Works (DPW). The chief engineer of the DPW argued that local authorities had not carried out additional works correctly, thus causing the flooding (VNA2 1912).

In an attempt to open new areas to colonization and remedy the localized flooding, the colonial government initiated one of the longest transverse canals running from Phung Hiep to Ca Mau, the Quan Lo – Phung Hiep Canal. This canal, completed in 1918, produced the seventh and final opening at Phung Hiep and did mitigate some of the flooding on other canals. However, it also shifted the flooding problem to new areas, especially the terminus in the Ca Mau Peninsula. Agricultural engineers, writing in 1942 about the colonial canal-building efforts, noted huge pools of stagnant water that formed around Ca Mau, perfect breeding grounds for mosquitoes and moths (VNA2 1949).

Much like other modern canalization schemes underway in the world at that time, engineers had a lot of sway in government and with business leaders, but they often struggled to get the hydrology right. Recognition of problems in surveying altitude in the Mekong Delta led to a new survey in 1911. The military topographer assigned to the job noted in his introduction:

> In Cochinchina, where there are immense areas without a single solid structure, without masonry, without any rock, it is not possible to produce benchmarks even close to the network of Europe (Régnier 1911, pp. 12–13).

In many areas, surveyors used the corners of stone tombs and the brick foundations of village communal houses for reference points, knowing full well that often the heavy structures could sink into the soft clay foundation below.

Perhaps the greatest evidence that public works engineers had miscalculated flows and heights in the Phung Hiep area came after the Great Depression and during the Japanese Occupation (1941–1945) when the budgets for waterway maintenance was severely limited. Many areas silted up and even prevented deeper draft French boats from passing. With the French War commencing in 1946 and a blockade placed on the Transbassac (Hau Giang) in 1950, substantial changes to these canals were impossible in some areas for another 30 years.

Rather than redesign the waterway system, the colonial government, starting the late 1930s, embarked upon a "Dutch dike strategy" for flood control that has had long-term consequences extending to the present day. Rather than figure out ways to drain water from agricultural areas, the government instead experimented with building encircling flood dikes or polders (*casiers*) that would at least protect lands inside. Doing this then require mechanical pumping equipment to insure adequate water levels. This strategy was especially popular in the western areas of Hau Giang near Long Xuyen (Service du Génie Rural 1943). Since the 1950s, this strategy has resulted in widespread adoption of both large-scale water pumping stations by provincial and state authorities as well as individually operated motor pumps owned by farmers. During the American War, big contracts went to Asian firms such as Nippon Koei to construct saltwater intrusion barriers and maintain these flood dikes (Nippon Koei 1966).

After 1975, the Socialist Republic of Vietnam, while free to improve the delta's hydraulic infrastructure without the threat of war, has been limited in its options for straying from the "Dutch dike strategy" in areas such as Phung Hiep and Ca Mau. Electricity and diesel as well as pumping machinery are cheap enough to permit total reliance on mechanical means to maintain water levels in local sections of the water grid. Also, since the 1990s the central government in Hanoi has sought to divest itself of fiscal responsibility for maintaining many secondary and tertiary waterways by shifting the burden onto local municipalities and rural districts. Thus while there is steady talk about a master plan for developing the delta, there is little evidence that any significant concerted efforts at managing water regionally have succeeded (World Bank 1999).

Thus, the seven-way intersection (*nga bay*) at Phung Hiep, while largely a result of French engineering mistakes in the early twentieth century, shows little sign of disappearing today. Today, the crossroads is a thriving town (now called Nga Bay) and tourist site in the lower delta landscape. As with other urban or industrial sites, it depends upon mechanical pumping to prevent floods, saltwater intrusion, and to keep water flowing during droughts. Fixing colonial-era mistakes since 1945 has involved not only bringing in large-scale technological solutions such as polders and pumping stations but also small-scale technology such as motorized pumps.

3.3 Fixing Forests: U Minh

Finally, a very different kind of hydraulic fix is involved for the remaining bits of peat swamp and mangrove forest located in the Ca Mau Peninsula at a site called U Minh. Since the 1930s, this dense, swampy forest was famous as a refuge for Vietnamese nationalists fleeing the French secret police. Such leaders as Vo Van Kiet (prime minister of Vietnam in the 1980s–1990s) helped establish Communist Party cells in the forest; and during the French War, the Viet Minh's Ninth Military Region was headquartered here. Throughout most of the colonial era, the government mainly considered plans to excavate the peat and clear the land of the twisted, gnarled trunks of cajeput (*Malaleuca cajuputi*). Cajeput had little value as furniture wood, but it provided the main source of charcoal used to stoke the boilers of steamboats, trains, and electric power plants. These "charcoal forests" also supplied delta inhabitants with cooking fuel long before gas stoves were available. (Even today, one can visit delta markets and find cajeput charcoal sold for use in cooking.) While the colonial forest service did mount concern about rapid deforestation of upland forests, it was mainly interested in cajeput forests for the lost taxes from unregistered cutting. The volume of cut wood from the delta averaged 2.5 million cubic meters per year from 1880 to 1904 with approximately 5,000 ha of forest land cleared each year (VNA2 1904). The distant forests of the Ca Mau Peninsula were the last remaining stands in the Mekong Delta, and by 1940 they had become base areas for Vietnamese nationalists.

Concern with preserving the mangrove forests only appeared in government reports in the 1940s as agricultural engineers (not civil engineers), seeing the havoc brought by canals and unregulated cutting, argued for preserving the forests to sustain freshwater levels and thus the nearby fields. The deep layers of peat acted as a sponge and held water during the dry season. However, it is difficult to separate the sudden colonial concern in the 1940s for preserving the forest with concerns emerging at the same time in Saigon about a growing Vietnamese resistance based there. After the end of the French War in 1954, the Republic of Vietnam (RVN) commenced projects to resettle thousands of army veterans around the forest to wrest control of the region from the Viet Minh who had controlled it since 1946. The RVN initiated plans to create forest preserves from the remaining freshwater and saltwater mangroves. A former colonial rural engineer, Jammé, visited the site as a consultant and wrote a report imploring the government to stop digging canals around the forest as they were causing deterioration of the peat soils and sure to lead to salt intrusion in the dry season (VNA2 1957).

However, as the American War reached U Minh in late 1968, any concern for the ecological or hydrological stability of this forest was sacrificed to military operations. After the 1968 Tet Offensive, the U.S. Military Assistance Command ramped up its efforts to destroy sanctuaries for the National Liberation Front including the bases in U Minh. The SEALORDS strategy mentioned above included several amphibious invasions into the forest as well a series of bombing strikes from B52s. Defoliants such as Agent Orange were also dropped on select sites within the forest, and these chemicals did what repeat incendiary bombing could not by killing large tracts of the forest (McQuilkin 1997; Biggs 2005).

This dual political and environmental history of U Minh as a "cradle of the revolution" (*noi cach mang*) and the last vestige of a freshwater mangrove and peat swamp ecosystem (severely compromised by war) has presented many challenges to conservationists in the postwar era. Immediately after the war, one of the base areas adjacent to the forest was converted into a prison while many adjacent parcels of land were distributed to revolutionary war veterans, many of whom had suffered through the terrible fighting and sought to establish new lives in Ca Mau after 1975 rather than return home. By 1990, the pressure of settlers cutting down remaining stands of trees and establishing farms or shrimp ponds led Prime Minister Vo Van Kiet, a veteran of the area, to decree the upper portion of U Minh as a forest reserve (BirdLife 1999, p. 3). The National Assembly elevated it to a National Park several years later. However, in the spring dry season of 2002, approximately 2,700 ha of the park burned down in a fire that ignited the peat soils and burned for several weeks. Several thousand firefighters had to pump in salt water from the coast to finally douse the flames.

In the postmortem reports and news articles, the blame for the fires spread to almost every group. The first to receive it were the predominantly ethnic Khmer inhabitants who continued to hunt for honey, using low smoldering fires to smoke the hives. International NGOs tended to blame farmers whose use of motor pumps and ground wells had caused the water table to lower. While the water table in areas surrounding U Minh was certainly low, the park was protected by an encircling dike that maintained higher water levels inside – a twist on the polder scheme mentioned above. Its largely speculation what was the main cause for the peat to dry out. Perhaps park officials intentionally lowered the water for some reason? Perhaps water leached out naturally through the porous peat to lower levels outside? Most likely, some combination of these causes helped the fire to ignite then to spread (Center for International Disaster Information 2002).

State forestry agencies have managed to reforest large areas of the burned forest, but really "fixing" this ecosystem as a peat swamp, freshwater mangrove forest remains an elusive goal. Besides the economic pressures on the land and the competing uses of it, one of the biggest problems remains that until perhaps the last few years popular views of cajeput remain focused on it as a common commodity, a rot-proof construction timber, and a source of charcoal for cooking fuel. Only since the mid-1990s has anyone suggested anything different. Furthermore, with over a century of colonial-era neglect and the wars that followed, there is no significant historical precedent for wetland forest conservation. More so than most wetlands, the forests in the Mekong Delta were subjected not only to burning and cutting but also bombing and chemical defoliation.

3.4 Conclusion

Each of the three examples given above demonstrates one thing in common: Agencies responsible for repairing, upgrading, or conserving a given landscape rarely work only to maximize environmental gains. Instead, the motives that often prompt officials to act may have their roots in symbolic gestures (declaring sovereignty), constructive gestures (creating technologically enhanced, built environments), or commemorative gestures (recognizing the sacrifices of a war-torn past). This is not only the case in the Mekong River but in most wetland environments today from the Mississippi to the Red River (Vietnam) to the Colorado, the Yangtze, and the Nile. It is much more the norm that political or business concerns trump environmental or conservationist ones. However, the threat of rising sea levels, prolonged droughts, and major modifications to water flow in the rivers that drive the creation and sustenance of river deltas means that stakeholders may have to act differently in the future. Failure to respond to environmental challenges may mean disastrous flooding, as in the case of New Orleans, significantly higher costs to maintain infrastructure, and continued loss of biodiversity in protected areas. Taking the historical perspective into greater account and recognizing *why* certain interests top environmental ones is an important first step in the work of developing new ways to fix past problems in the Mekong Delta.

References

Biggs D (2005) Managing a rebel landscape: conservation, pioneers, and the revolutionary past in the U Minh forest, Vietnam. Environ Hist 10(3):448–476

Biggs D (2011) Quagmire: Nation-building and nature in the Mekong Delta. University of Washington Press, Seattle

BirdLife International Vietnam and Institute of Ecology and Biological Resources (1999) The conservation of key wetland sites in the Mekong Delta. Royal Netherlands Embassy, Hanoi

Center for International Disaster Information (2002) OCHA situation report No. 1 – Vietnam – Forest fire 9 April 2002. http://www.cidi.org/disaster/02a/ixl119.html

Edwards P (2006) The tyranny of proximity: power and mobility in colonial Cambodia 1863–1954. J Southeast Asian Stud 37:421–443

McQuilkin WC (1997) Operation SEALORDS: a front in a frontless war, an analysis of the brown-water navy in Vietnam. MA. Thesis, U.S. Army Command and General Staff College, Fort Leavenworth, pp 44–46

Nguyễn HL (1971) Bảy ngày trong Đồng-Tháp-Mười: Du kí và biên khảo. Trí Đăng, Saigon

Nguyễn VH (1972) Thoại Ngọc Hầu và Những Cuộc Khai Phá Miền Hậu Giang. Hoa Sen, Saigon

Nippon Koei Company Ltd (1966) Design report on Ngan Ro Barrage project, Oct 1966, NL402, VNA2

Porter TM (1995) Trust in numbers: the pursuit of objectivity in science and public life. Princeton University Press, Princeton

Régnier P (1911) Nivellement général de la Cochinchine: Rapport du chef du brigade [General leveling of Cochinchina: report of the chief of the Brigade]. Imprimerie commerciale M. Rey, Saigon

Rénaud J (1879) Étude sur l'approfondissement du canal de Vinh-té et l'amélioration du port d'Hatien. Excursions et reconnaissances 1:66–73

Sellers N (1983) The princes of Hà-tiên (1682–1867). Thanh-Long, Brussels
Service du Génie Rural (1943) Compte Rendu d'Enquête dans la Province du Chau Doc. File H62/9, Goucoch divers, VNA2
Socialist Republic of Vietnam (1996) Prime Minister Decree No. 99, Feb 9, 1996, On long-term orientation and the five-year plan of 1996–2000 for development of irrigation, transport and construction in rural areas of the Mekong River Delta. http://www.asianlii.org/vn/legis/laws/oloatfpo19962000fdoitaciraotmrd1368/. Accessed 23 Sept 2010
Sơn N (1988) Lịch sử An Giang. Long Xuyên: nhà xuất bản Tổng hợp An Giang
Vietnam National Archives Ctr #2 [hereafter VNA2] (1904) Nombre de stères de bois à brûler vérifies, 5 Mar 1904, file IA 13/308(12), Fonds Goucoch
VNA2 (1911) Le gouverneur generale de l'Indochine à lt-gouv de la Cochinchine Letter no. 1832, 14 June 1911, file IB 25/124, Fonds Goucoch
VNA2 (1912) Chief engineer, Direction générale des travaux publics, 1912, file IB 25/124, Fonds Goucoch
VNA2 (1949) M. Combot and Paul Emery, Étude sur les travaux d'hydraulique dans le Trans-Bassac, 20 Oct 1942, file H.62/3, Toa Dai Bieu Chinh Phu Nam Viet
VNA2 (1957) Tờ Trình Của Trưởng Khu Thủy-Nông, 19 Nov 1957, file C18/1, GTCC, TTLTQG2
World Bank (1999) Project appraisal document on a proposed credit of SDR 72.8 million to the Socialist Republic of Vietnam for the Mekong Delta water resources project, 15 Mar 1999, World Bank, Washington, DC

Chapter 4
Outline of the Process of Red River Hydraulics Development During the Nguyễn Dynasty (Nineteenth Century)*

Olivier Tessier

Abstract The history of water management in the Red River Delta presents a stark contrast with the Mekong Delta. While the latter was an ancient site of contact between the Khmer, Siamese, Malay, Chinese, Chăm, and Vietnamese populations, its large-scale hydraulics planning are of relatively recent date, having been initiated during the seventeenth century under the Nguyễn lords. In the Red River Delta, the conquest and development of the territory is intimately related to the history of water management, as well as to the hydrographical network which shaped it through a long process of alluvial deposit. The basic need for self-protection against the river's violent floods is a constant feature of the ancient past and the modern period in Vietnam's history and largely contributed to the structuring of relations between the state and the peasantry.

This chapter's aim is to present several important features of the history of hydraulics policy implemented in the northern delta during the nineteenth century by the Nguyễn dynasty, drawn from one principal documentary source, the imperial annals. Indeed, it was in the nineteenth century that the most concerted efforts were made, to such an extent that on the eve of the colonial intervention, the construction of the Red River Delta's dike system was complete. This unprecedented investment should not mask, however, the instability and ambiguity of the hydraulics policy implemented by this dynasty's successive emperors; this was illustrated by discontinuity in the administration and management of the dikes and alternating periods of state involvement and withdrawal, whereby part of its prerogatives was abandoned to peasant communities. Looking beyond this contrasting conclusion, we may consider that the Nguyễn dynasty played a key role in the area of water management, establishing the foundations for a modern and rational system in the delta.

---

*I warmly thank Nicholas Stedman who translated this article.

O. Tessier (✉)
École française d'Extrême-Orient (French School of Asian Studies), 5a xóm Hà Hôi, Hanoi, Vietnam
e-mail: otessier2002@yahoo.fr; olivier.tessier@efeo.net

M.A. Stewart and P.A. Coclanis (eds.), *Environmental Change and Agricultural Sustainability in the Mekong Delta*, Advances in Global Change Research 45,
DOI 10.1007/978-94-007-0934-8_4, © Springer Science+Business Media B.V. 2011

The history of development of the Red River Delta contrasts sharply with equivalent development in the Mekong Delta. Although the Southern Delta has long been a meeting place for the Khmer, Siamese, Malay, Chinese, Chăm, and Vietnamese people and was home to the former kingdom of Funan (site of *Óc Eo*), famed for its skill in water management; the process of large-scale hydraulics development is relatively recent and was instigated in the seventeenth century under the leadership of the Nguyễn lords. With the establishment of the dynasty of the same name at the beginning of the nineteenth century, colonization intensified through the foundation of villages and citadels and the digging of canals for purposes of irrigation, trade, and defense, thus asserting the irrevocable status of the southern regions as part of the newly-reunified Vietnamese national territory. This continued until the country's division into two independent entities (1954). In fact, the ultimate stage of the conquest of the Mekong basin was organized by the French colonizers. In counterpoint to the absolute priority given to building dykes in the North, land reclamation in the South depended upon draining and desalinating vast tracts of land by digging a dense network of canals drawing off water into the sea. It is therefore this latticework of primary canals, linked directly to the rivers, that shapes the contours of the "hydraulic land subdivisions" (*cashiers hydrauliques*) into as many independent units, from the perspectives of irrigation and drainage. Finally, in the Mekong Delta human activity is entirely focused on the river or the canals, and historically evolved along the waterways, a distinctive feature of the regional culture that inspired the writer Sơn Nam's expression "river civilization" (*văn minh sông nước*) to describe this multiethnic society (Sơn Nam 1970).

In the Northern Delta, dyke-building was the response adopted since the beginning of the Christian era by the Viet people to protect themselves from violent river floods, particularly those from the Red River. This uninterrupted process of building and strengthening a network of ever-more complex dykes had two main consequences. First, it determined how rural areas were organized. Villages grew up on the occasional outcrops rising above the Delta, protected by dykes, in a closed system where almost the only people having direct dealings with the rivers were specialized guilds, in particular those of boatmen (river transport) and fishermen. Second, this process contributed greatly toward structuring the relationship between the state and the rural population, if only because in this almost exclusively rural country, agricultural production represented the largest source of the imperial state's revenue through the taxing of peasant farmers and land. For centuries, the state was therefore obliged to devote efforts toward protecting this source of income, one that was indispensable to its very existence. This occurred in a region noted for its chronic agricultural insecurity, where the threat of drought hanging over crops harvested in the 5th month gave way to the risk of rapid rises in water levels and floods spelling the destruction of crops normally harvested in the 10th month. This accounts for a central characteristic of the Delta. It is a vast alluvial plain (14,700 km^2), long paralyzed in its development by human intervention. The dyking up of the two main fluvial systems, the Thái Bình and Red River (Map 4.1), has for centuries immobilized irregularities in the topographical relief that were only temporary and that the natural spreading action of floodwaters would have progressively smoothed out if the rivers were not contained within a network of more than 2,000 km of dykes.

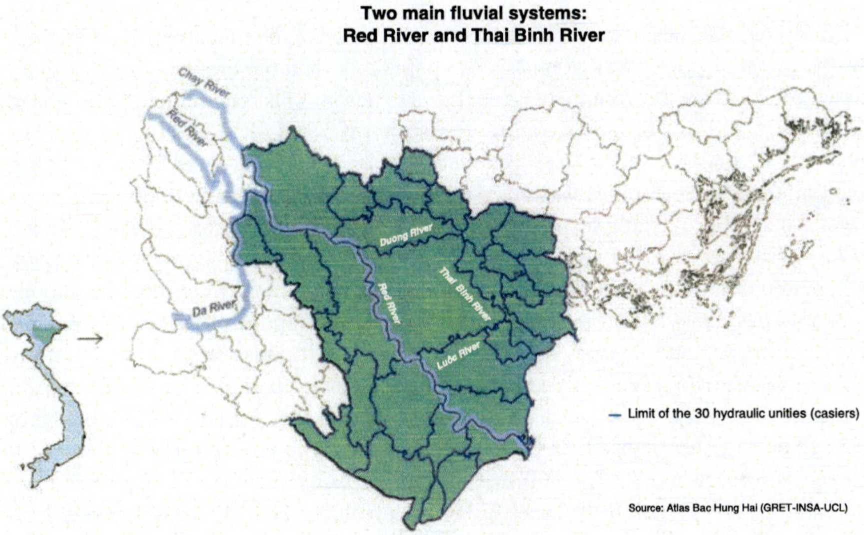

Map 4.1 Two main fluvial systems: Red River and Thái Bình River (Source: Atlas Bac Hung Hai, GRET-INSA-UCL)

The Thái Bình River, which flows into the northeast of the Delta, does so at a powerful but regular rate. The system it comprises with its tributaries and distributaries typically experiences manageable changes in water level, easily contained by its network of dykes. The Red River system is quite the opposite. When it bursts its banks, swollen by the waters of its major confluents (Sông Đà and Sông Lô), it does so with unusual violence and causes devastating flooding. But the river has also been the principal architect of the Delta and its prime provider of fertility, thanks to the considerable mass of sediment deposited over the centuries. The depth of the accumulated layer reaches several tens of meters.[1] Its rate of flow, which can rise to 28,000 m^3/s in times of flooding, elevates the river's water level enormously, leaving it literally perched up above the plain. In this configuration, the dykes that confine the river normally prevent the floodwaters from spreading beyond the elevated riverbanks. In concrete terms, during the dry season in winter, the river's average low-water level is about 2.50 m above sea level. As soon as the rainy season begins (June–October), the level rises rapidly and can reach 10 m in a few days. The elevation of the capital's historic center is 5 m; Hanoi can therefore be at the river's mercy for the entire rainy season. This seasonal threat is a reality for the majority of human settlements and land in the Delta, with the exception of the higher-standing area in the northwest.

[1] The quantity of sediment carried by the Red River, estimated at 130 million metric tons per year, puts this river in eighth place worldwide in terms of solid matter carried, coupled with a rate of flow that does not make it a major river (Béthemont 2000).

It follows that in this exclusively rural country, one of the fundamental obligations of the imperial power was to protect the population and the crops in order to guarantee and increase the volume of agricultural taxation. This requirement necessitated a ceaseless struggle against flooding. Between 1803 and 1861, the imperial annals recorded 27 major floods in the Red River Delta or almost one every 2 years, bearing in mind that every year breaches in dykes occurred locally, causing flooding of lesser importance. As an example, during the flood of 1830, the dykes protecting Thăng Long – Hanoi were breached over several hundreds of meters, letting the river flow so far into the town that it flooded the citadel. 1,000 soldiers were mobilized to try and hold back the waters.[2] Although imperial historiographical sources record crops being destroyed and famines brought about by these floods, they do not offer any statistical evaluation. Records kept during the colonial period make it possible to measure their scale a posteriori. During the floods of 1913, about 30 local breaches caused damage to about 100,000 ha of paddy fields where the crop was completely ruined, a loss of at least 150,000 metric tons of paddy (Brénier 1914). The severe flooding of 1915 caused 48 breaches in dykes alone. Half of the waters flowing down the Red River poured out onto the plain, flooding 365,000 ha or a quarter of the Delta's total area (Figs. 4.1 and 4.2).

Fig. 4.1 Flood in Hanoi, 1926 (Source: Photography Library EFEO, funds Vietnam, ref. VIE04098)

[2] *Đại Nam Thực Lục*, t. III, q. 70 (2007: 112–113).

Fig. 4.2 Flood in Bắc Ninh province, 1923 (Source: Photography Library EFEO, funds Vietnam, ref. VIE04108)

In this context, the aim of this chapter is to examine the hydraulic policy implemented in the Red River Delta by successive emperors of the Nguyễn dynasty (1802–1945) until 1882, the beginning of French colonization in Tonkin. Despite doubts raised by the continuation of costly dyking efforts that did not even guarantee satisfactory protection against floods and the changes of direction that were a characteristic feature of imperial policy, we will attempt to demonstrate that the emperors of this dynasty were in certain aspects precursors in the domain of water management and established the basis for the integrated development of the Red River Delta.

4.1 The Nguyễn Dynasty: Working Toward a Fully-Dyked Red River Delta

In Vietnam, as in China, building and maintaining dykes was one of the Emperor's exclusive privileges. The celestial mandate that bestowed "supreme authority," that is to say, the legitimacy to command his fellows, called in return for the benevolent protection by the sovereign of his subjects. This political and symbolic dimension was reasserted during the *Nam Giao* sacrifice, a ceremony for worshipping the heavens, where "The Emperor seems to have made himself the representative and the messenger of his people: in the name of all, he prostrates himself, presents offerings, gives thanks, makes requests" (Cadière 1992, 85). As one who can intercede between the earthly and the celestial worlds, he could therefore be held responsible for a breach in a dyke causing serious flooding, the people seeing in it a sign of Heaven's disapproval regarding some of the sovereign's actions. In parallel with this request for divine assistance, realized every year during the "furrow-opening ceremony" (*lễ tịch điền* or *lễ hạ điền*) that inaugurated the season of work in the fields, the need to protect the people and their crops also obliged the imperial power to get directly involved in hydraulic systems in an attempt to shield the countryside from the river's wrath.

Before the Nguyễn dynasty came to power, considerable efforts at hydraulic development of the Delta had already been made by the preceding dynasties, in particular the Trần and the Lê.[3] The Red River had thus been completely dyked from its entry into the Delta to its mouth since the thirteenth century. Every year, however, breached dykes and flooding affected entire provinces, frustrating the constant efforts of the rural population to ensure its subsistence. As for the other component of water management, irrigation, it must be acknowledged that initiatives taken since the fifteenth century by the imperial power had not brought about any notable improvements. No irrigation works drawing directly from the rivers were carried out before the nineteenth century.

Although the capital of the newly reunified country had been transferred to Huế in 1802, the first emperors of the Nguyễn dynasty proceeded to lavish special attention upon the socioeconomic recovery of the North of the country, on its knees after several decades of war and devastation. In order to demonstrate their power, they set themselves up as the instigators of major construction works, building roads, bridges and ports, as well as powerful citadels with Vauban-style fortifications in order to contain peasant revolts (Le Thanh Khoi 1992, 348–349). This was an ironic consequence of Nguyễn dynasty hegemony in the first half of the nineteenth century: While the establishment of an absolute monarchy enabled the modernization of infrastructure in the Red River Delta, the increased control exerted over the population by a widely-corrupt bureaucratic and mandarin administration bred growing discontent in the rural population that led, on several occasions, to rebellion.

In the field of hydraulics, the first Emperor Gia Long (1802–1819), then Emperors Minh Mạng (1820–1840), Thiệu Trị (1841–1847), and Tự Dức (1848–1883) ordered major works aimed principally at managing the rises in the Red River's water levels.

---

[3]The first mention of dyke-building work following the long period of Chinese domination dates from the end of the eleventh century and is inscribed in *A complete history of the Việt*. In the 2nd month of 1108, King Nhân Tông of the Lý dynasty had a dyke built on the river bank at Cơ Xá Village near where Long Biên Bridge now stands (*Việt Sử Cương mục Tiết Yếu* 2000: 118). This first explicit reference does not necessarily mean that dykes had not been built in still more ancient times. Protective earthworks were present in the former provinces of Sơn Tây and Hưng Hóa since the beginning of the Christian era and were apparently the result of technology borrowed from China (Gourou 1936, 84).

According to the figures of the inspection carried out in 1829 by the chief mandarin responsible for dykes (*Đê chính*), Lê Đại Cương; the total length of principal dykes in the Red River Delta (counted in 739 communes (*xã*) belonging to 38 subprefectures (*huyện*) of 5 provinces (*tỉnh*)) was 952 km (238,660 *trượng*), of which 144.5 km (36,127 *trượng*) were built in 26 years by the first two emperors of the dynasty.[4] The efforts expended were such that P. Gourou judged that on the eve of colonization, the Red River Delta was completely enclosed, meaning that the network of dykes was almost as dense as that which he could observe at the beginning of the 1930s (1936, 85). It stretched to nearly 2,000 km of principal dykes and 2,000 km of secondary dykes (Map 4.2).

In spite of this impressive overall appraisal, the Nguyễn's hydraulic policy involved several U-turns and discontinuities that highlighted the weight of unilateral decisions taken by each of the emperors and the severe dysfunctions of the mandarin administration. Two aspects worthy of mention were at once causes and consequences of this situation.

4.1.1 First Aspect: Instability of the Administration of Dykes

In 1809, Emperor Gia Long reformed the administration of dykes and created a specialized central department placed under the authority of a mandarin of the Court, quan Đê chính ở Bắc Thành.[5] Simultaneously, he enacted regulations structured into eight articles that dealt with the execution of works and their inspection, with the protection of constructions, and with the planning of future works associated with an assessment of their costs.[6] Finally, just as with the code of the Lê emperors, Gia Long's code made provisions for sentences up to and including the death penalty for any person guilty of "secretly breaching river dykes" – (Art. 395) and of "not acting in a timely manner and not repairing dykes" – (Art. 396),[7] this last article echoing the main elements laid out in the code of the Lê emperors.

---

[4] *Đại Nam Thực Lục*, t. II, q. 61 (2007, 892–893).

[5] The Imperial City of Thăng Long was renamed Bắc Thành (City of the North) by Nguyễn Huệ (Quang Trung), first sovereign of the short-lived Tây Sơn dynasty (1788–1802); the city then took the name Hà Nội in 1831 within the context of the great nationwide administrative reform implemented by Minh Mạng.

[6] "Regulations on dykes for Bắc Thành are decreed. The King, placing great importance in the prevention of river floodwaters, requests city dignitaries to take counsel from previous experiences and Court dignitaries to consider the matter again and decides to decree regulations [articles follow]" *Đại Nam Thực Lục*, t. I, q. 39 (2007, 764–765).

[7] Art. 395: "Those who secretly breach river dykes (built by the state), will be punished with one hundred strokes of the trượng [rattan cane]; those who secretly breach pond dykes or levees (belonging to individuals), will be punished with eighty strokes of the trượng […]." Art. 396: "On any occasion where river dykes have not been (built up and) repaired (before damage occurs), or if (although repaired) they have not been restored in a timely manner, the officials and employees responsible for managing this service will be punished with fifty strokes of the trượng […]." (Philastre 1876, 742–745).

Map 4.2 Map of dykes in Red River delta, 1905 (Source: Gauthier, 1930:172)

Despite this set of measures aimed at rationalizing the technical and human management of the dyke network, breaches and flooding continued apace under the reigns of Gia Long and Minh Mạng, with their share of deadly floods, famine, and peasant revolts. Every year or almost, from the sixth month onward, the *Đại Nam Thực Lục* made a grim inventory of more or less extreme calamities brought about by the violence of flooding or by drought, with the emperor granting emergency aid in the form of rice and money to the victims, along with total or partial exemptions from income tax. This was why, after the catastrophic floods of 1827, the Emperor gave orders as follows:

> Bắc Thành [*the North of the country*] has experienced serious flooding, the dykes have given way in Sơn Tây, Sơn Nam, and Nam Định Provinces; houses and paddy fields are

> under water, and many people have drowned. The city dignitaries have sent people out all over the countryside to distribute aid to those affected and have reported back to me. [...] I have read these reports of the deplorable situation of the flood victims; the assistance they have received is insufficient. I therefore command provincial mandarins to distribute further aid. For every fatality, man, woman, child, or elderly person, there will be granted 3 strings of sapeques (small denomination coins: quan tiền) to his or her family; if the person is very poor, he will be granted 2 strings of sapeques and 1 bushel of rice (phương gạo); if the person is poor, he will be granted 1 string of sapeques and 1 bushel of rice.[8]

Following these catastrophic floods, Minh Mạng issued, in the 8th year of his reign (1828), an order that removed responsibility for construction and upkeep of dykes from provincial mandarins accused of negligence and incompetence, bestowing these tasks upon a specialized body of public servants. This text specified with precision the dimensions of the various categories of dykes, which had to be built up compared to the standard heights under Gia Long and indicated where they must pass; it also ordered that bamboos be planted at their bases and that each year before the floods, provision be made of baskets and bamboo in order to facilitate repairs; finally, it made provision for the construction at the confluence of the Red River with the Claire River (Sông Lô) of a temple to the water spirit. The following year, the Bắc Thành Dykes Service was founded, headed by a military mandarin, the Commander of provincial regiments, and a civil mandarin, the former Governor of Nam Định Province.[9]

But all this was to no avail. So Minh Mạng decided once again in 1833 to completely reform the administration of dykes. He disbanded the specialized department that he had created 6 years earlier and returned the administration of dykes to provincial mandarins who henceforth were responsible for the portion of river that crossed their respective territories.

> Provinces are henceforth placed under the responsibility of a governor and mandarins of the prefecture charged with serving the interests of the people. Order is given to entrust public and private dykes to their competence [...]. Since management of dykes is carried out by local mandarins, the abolition of the Dykes Service is under consideration. The Sovereign has consulted the Court authorities on this subject. The latter being unanimously in favor of abolition, the Emperor has therefore decided that this ruling should be applied.[10]

This complete reversal of policy was motivated by the following analysis. Evaluation showed that the specialized mandarins concentrated solely on construction of and repairs to the works, without concern for agricultural activities and, in particular, for the possibilities of setting up drainage or irrigation ditches.[11]

---

[8] *Đại Nam Thực Lục*, t. II, q. 47 (2007, 648).

[9] *Đại Nam Thực Lục*, t. II, q. 57 (2007, 812).

[10] *Đại Nam Thực Lục*, t. III, q. 92 (2007, 536–537).

[11] Extract from the text: "Abolition of the Bắc Kỳ dykes administration [...] Building of dykes is aimed above all at protecting agricultural activities. For a long time, during the steep autumnal rises in water levels, the mandarins responsible for dykes concentrate solely on building and repairing earthworks without concern for agricultural activities [...]. Despite this, local mandarins take the liberty of not reporting directly; a delay is therefore inevitable. Normally bad weather is enough to lead to material damage." *Đại Nam Thực Lục*, t III, q. 92 (2007, 536).

The floods of 1856 and 1857, which totally swamped 6 of the 11 Red River Delta provinces, called for a reaction from the Sovereign. In 1857 Emperor Tự Đức restored the dykes but without any noticeable improvement to the upkeep of untended hydraulic infrastructures. At the beginning of the second half of the nineteenth century, the country was in a permanent state of war, caused by an influx of thousands of Chinese mercenaries and pirates (the famous *Black Flag*, *Red Flag*, and *Yellow Flag*) making incursions as far as the center of the Delta and by numerous rebellious movements rising up against imperial authority. The North of Vietnam was plunged into chaos, and its starving inhabitants deserted the countryside, as Tự Đức noted bitterly in a royal decree in 1859:

> Our country has only indifferently fertile land. Only Hà Tĩnh, Nghệ An, Thanh Hóa and Ninh Bình Provinces can be considered relatively affluent. As for the people of the North, their loyalty leaves much to be desired. Sơn Tây, Bắc Ninth, Thái Nguyên, Lạng Sơn, Hưng Hóa, and Tuyên Quang Provinces, continually harassed by bandits and pirates, constantly echo to the clash of weapons... Our resources are now finished, the people are exhausted, finances are lacking, and the state treasury is empty.

From 1858 onward, the situation deteriorated still further with the French colonial intervention, first in the center, then in the south of the country, which occupied all the attention of the Court in Huế. Consequently, the imperial annals barely make any more references to the question of hydraulic management in the North. All we know is that dyke servicing was once again dissolved in 1862 due to budgetary constraints. The position of "Special Envoy to the Emperor for the Inspection of Dykes" was created in 1876, but the position was eliminated two years later.

4.1.2 Second Aspect: Debate Over the Usefulness of Maintaining and Strengthening Dykes

This administrative and technical reorganization must be placed within the context of a fundamental debate, whose terms were laid out as early as 1803 by Emperor Gia Long. He ordered the mandarins and inhabitants to discuss the usefulness of keeping the dykes or of leveling them off.[12] As the network of dykes and counter-dykes became ever denser, farmers and mandarins began to realize the hindrance that could

[12] Second year in the reign of Gia Long [1803], Autumn, 8th month: "The King orders mandarins and inhabitants of Bắc Thành to discuss the usefulness of earthworks. His orders are these: favoring beneficial actions and abandoning harmful ones is the politician's primary vocation. Previous and current experiences must be taken into judicious consideration. Districts located along the rivers in your locality have built dykes to ward off flooding. Yet the dykes are breached every time the water level rises; the paddy fields are swamped and human and animal losses occur. You, who were born or who work in these places, you understand the land and its people well. You are therefore called upon to express your opinion as to the usefulness of maintaining the dykes or of leveling them off. Any practicable initiative may be rewarded." *Đại Nam Thực Lục*, t. I, q. 22 (2007, 572–573).

be caused to the irrigation and drainage of Delta paddy fields.[13] They witnessed the distressing sight of the Red River flowing in full spate while the rice wilted away in the paddy fields. So it was that *The Descriptive Geography of the Emperor Đồng Khánh* (1888–1889) related, concerning Sơn Tây Province:

> The earth and waters of low-lying districts are healthier, similar to those of the southeastern provinces. Heavy rains coming from the high ground in neighboring Hưng Hóa and Tuyên Quang Provinces arrive during the few weeks between summer and autumn; the rivers begin to flow very rapidly; the water level can reach 18 or 19 feet, while the crops, planted out on the other side of the dyke, are hit by drought.[14]

People began to wonder if it would not be wiser to simply remove these costly and dangerous constructions to let the waters of the major rivers spread out freely during the summer months over the whole Delta, like a slow and progressive flood and not a cataclysm. This would result in a high level of moisture that should solve the irrigation problem while increasing the fertility of the soil with alluvium deposits. In other words, the question was to determine whether the dykes were not a cure worse than the sickness they were supposed to treat.

It was certainly the case that after each major flood, the inhabitants of the inundated regions demanded the removal of the dykes, for example, in 1804, 1825, 1833, 1835, 1847, 1872, and 1879. The main reason motivating such demands was that when floodwaters breached an upriver stretch of dykes, the downriver sections remained intact and prevented the waters from receding when the river water level went down, thus eliminating any hope of a harvest (Pouyanne 1931, 20). Although no emperor of the Nguyễn dynasty ever decided to order a Delta-wide destruction of dykes, this insistent debate between supporters and opponents of dyke destruction persisted throughout the nineteenth century and gave rise to alternative solutions aimed at lessening the impact of imperfections in the dyke system.[15]

This explains why, when he reformed the dykes administration in 1833, Minh Mạng called for the mandarins of the capital and the Delta provinces to make propositions to improve water management and flood control. Some of these officials, drawing upon the opinions of local authorities and notables, considered it pointless and costly to seek perpetually to strengthen the dykes. They believed that a better course of action would be to widen the fresh arroyos opened up by the river's current, where dykes had been breached, in order to improve runoff. Although sympathetic

[13]Widespread dyke building compartmentalizes the Delta into hydraulic units called *casiers* (hydraulic land subdivisions), independent from each other, and used on the one hand to supply irrigation water from the river and on the other, to drain excess rainwater back toward the river. Today the Delta is split up into 30 primary *casiers* of sizes varying between 5,000 and 180,000 ha, a series of separate and unconnected hydraulic systems that must be drained throughout the rainy season.

[14]*Đồng Khánh địa dư chí* (2004, 1650).

[15]For an overview of the question and, in particular, of the evolution of the arguments developed by the supporters and opponents of a total or partial abolition of dykes in the nineteenth century, an evolution that shows that the various positions of those involved were not fixed but rather informed by past experiences, see the article by E. Poisson (2009) listed in the bibliography.

to this argument, the Emperor worried about the consequences of a solution that would inevitably cause flooding of low-lying areas and basins located below elevated riverbanks. For the famous mandarin and man of letters Nguyễn Công Trứ, the alternative solution to an imperfect dyke system was to dredge the Red River's natural distributaries and to dig artificial ones, in order to lower the water level of the river by increasing drainage capacity in its outlets. Although Minh Mang initially prioritized the deepening of the Sông Đuống River; in 1833, on the advice of two special mandarins, he decided to dig and clear out the Cửu An River, which had to serve both as a distributary to the Red River at Hưng Yên and as an irrigation canal.

From 1835 to 1836, 20 km were dug in order to make the link between the Red River and the Cửu An River, of which the mean bed was also dredged and widened over more than 40 km (Đỗ Đức Hùng 1998, 44). In tandem with this excavation work, the dykes of the low-lying areas of Hưng Yên Province were either greatly lowered or simply flattened down to ground level.

The impact of this major construction project was the complete opposite of what was hoped for. Four months after the completion of work, the channel from the Cửu An River into the Red River was swept away in three places, Hưng Yên and Hải Dương provinces were submerged under 2 m of water, and the town of Hửng Yên was completely flooded. It was the most disastrous flooding ever experienced in these two provinces (Đỗ Đức Hùng 1998, 45). It earned Nguyễn Công Trứ virulent criticism from the Emperor, who named him responsible for the situation and ordered him to repair the damage. Early in 1837, large-scale reinforcement work began on the Cửu An River. In response to the pleadings of the inhabitants, the upriver channel of the Cửu An River was completely sealed off, and its only remaining function was as a drainage canal to evacuate water from low-lying areas in Hưng Yên Province toward the downstream part of the Sông Luộc – Bamboo Canal (Pouyanne 1931, 21). There was no other option but to rebuild and strengthen the dykes in Hưng Yên, Hải Dương, and Nam Định provinces, a task to which Emperor Thiệu Trị set himself during the 6 years of his reign (1841–1847). The latter not only had the dykes rebuilt in Hưng Yên Province on the left bank of the Cửu An River, but he also had new ones built on the right bank during the winter of 1841–1842.

Yet despite this disastrous experiment, the debate was rekindled in 1847,[16] then in 1852 by Emperor Tự Đức who made no secret of his misgivings over the

---

[16]The governor of Hà Ninh Province, Nguyễn Đăng Giai, one of those in favor of destroying the dykes, wrote one of the most thorough reports on the question in 1847. He offers no less than 11 separate arguments to support his position. He dwelt on the exorbitant costs entailed by dyke building, maintenance and inspection; he recalled the human and economic disasters regularly brought about by breached dykes; he highlighted the unfairness generated by the construction of these large-scale earthworks that ate up agricultural land and caused conflicts between people living within and without the dykes, etc. Although the Emperor was in agreement with Nguyễn Đăng Giai on one point, that the construction of dykes under the Trần dynasty had been a mistake, he considered it too late to go back and that there was no alternative solution available exclusive of dyke building, since dredging the Red River tributaries alone could not guarantee a satisfactory evacuation of floodwaters (Poisson 2009, 88–90).

continued dyking up of the Delta, as witnessed by an annotation written in his own hand in the margin to a passage of *Đại việt sử ký* concerning the construction of the Đinh Nhĩ Dyke: "That was a poorly conceived undertaking responsible for countless disasters."[17] Nevertheless, after 5 years of procrastination, it was finally those in favor of maintaining and strengthening the dyke system who carried the day; most dignitaries at Court recognizing that attempts at abandoning certain dykes had led over the previous decade to flooding with a frequency and on a scale never seen before.

Once the matter had been decided, the dykes department was reestablished in 1857. The new mandarin in charge of the department (Đê chính) made a series of ten specific propositions consisting of as many priorities in terms of hydraulic and regulatory projects to control rises in water levels (Đỗ Đức Hùng 1979, 52). On this basis and in tandem with new dykes and counter-dykes built along the Red River to keep it within its banks, dredging work was undertaken with the use of "harrows" (submerged booms) pulled by ships to deepen the river's mouth and thus enable a more rapid runoff of its tumultuous waters. Efforts were also made to divert part of its flow toward the Thái Bình River in order to lower the water level of the river and thus limit the violence of its spates. This was how, in 1858, Tự Đức came to dig a new entry to the Sông Đuống, downstream from the former one, now completely blocked by alluvial deposits (Chassigneux 1914, 101). But in 1862, before the work on the Sông Đuống could be completed, the dykes department was once again dissolved and the policy of major hydraulic-development work placed on hold. The Emperor justified this decision by citing financial problems and especially the serious disruption occasioned by French colonizing incursions into the North and the South of the country.

Finally, in 1871, the debate was once again activated by a petition from a mandarin, Nguyễn Thành, who argued in favor of the removal of all dykes from Bắc Kỳ. The results of the survey conducted in 1872 by the Emperor among the various provincial governors showed a real diversity of understanding of the problem that actually reflected quite faithfully the diversity of hydrographical and geomorphologic situations in each of the Delta's provinces. A pragmatic approach, personified by Phạm Thận Duật, seems to have gained the upper hand. In 19 petitions addressed to the Emperor between 1876 and 1878, he defended a logic of dyke consolidation, combined with dredging of the Red River's distributaries. This was because the problem of runoff waters in the Upper and Middle Delta could not be solely attributed to the dyke system, as it also derived from the topographical configuration (Poisson 2009, 94).

[17] *Chronique du Bulletin de l'École française d'Extrême-Orient* (July–December 1905, 482–483). The (anonymous) author explains that on the occasion of the meeting of the Dykes Commission held after the 1905 floods, he had looked up "the oldest mentions made in indigenous texts of dyke-building work on the Red River."

4.2 Public Dykes, Private Dykes

The coexistence of two categories of dykes (public and private) is an historical reality intrinsic to the very process of dyke building in the Red River Delta. In 1665, the *Cương mục* established when dykes should be repaired and set out the different kinds of work necessary depending on their importance:

> In the tenth month of every year, the administration service (*thừa ty*) of each province (*xứ*) should order the authorities of each sub-prefecture (*huyện*) under their command to make a survey of the places on the dykes needing repair work. Minor work should be carried out by the people of the communes (*xã*) endangered by the strength of the current, under the surveillance of the prefectoral authorities; as for major works, a mandarin will be ordered to direct them.
>
> (Quoted by Langlet 1970, 119)

However, for the first time during the Nguyễn dynasty, these two categories were explicitly dissociated. In his survey carried out in 1829 in five Delta provinces, the chief mandarin in charge of dykes (*Đê chính*) Lê Đại Cương, counted 698 km of private secondary dykes and 16 sluice gates in the same category.[18] The proportion of private dykes therefore represented at this period more than 40% of the total 1,650 km dykes surveyed. On what criteria was this typology based?

Public dykes were built by order of the imperial state under the direction of either the provincial or specialized mandarin administration, depending on the period. They were considered strategic in the sense that they guaranteed the protection of large geographical and human units, since their breaching could cause flooding which affected district administrative centers, provincial capitals, and whole provinces. These were the most significant earthworks that lined either side of the Red River's flood plain and the banks of its most unruly tributaries and whose construction and upkeep had scrupulously to follow the standard norms defined by the state in terms of size and compactness. In order to periodically mobilize the large workforce needed to carry out such infrastructure projects, the state had two possible forms of recourse at its disposal. The first was the mobilization of villagers through the corvée system (set by the Gia Long code at 60 days per year, per person enrolled). For major construction projects, thousands or even tens of thousands of rural inhabitants were mobilized in this way, often assisted by soldiers (Fig. 4.3). In 1835 (10th month), in his report to the Emperor concerning the dredging of the sông Cửu An, Nguyễn Công Trứ gave some impressive figures:

> Dredging rivers and building dykes are hard work. The hiring of 20,000 coolies is requested (6,000 coolies from Nam Định, 4,000 from Hải Dương, 3,000 from Hưng Yên, and 3,500 from each of the provinces surrounding Hà Nội and Bắc Ninh). Work will begin in the 1st month of next year.[19]

---

[18] *Đại Nam Thực Lục*, t. II, q. 61 (2007, 648).

[19] *Đại Nam Thực Lục*, t. IV, q. 160 (2007, 784).

Fig. 4.3 Reconstruction of Gia Lâm's dyke, 1926 (Source: Photography Library EFEO, funds Vietnam, ref. VIE06801)

Repairing sections of dykes that had been damaged or carried away by floodwaters also consumed large quantities of workers, as shown by the report of the governor (*Tổng đốc*) of Hà Nội – Ninh Bình Province, Đoàn Văn Trường, presented to the Emperor in the 10th month of 1833: "4,000 coolies have been mobilized to repair broken stretches of dyke in Chương Đức, Hoài An, and Thanh Liêm Districts of the Province."[20]

The second recourse was to moblize a paid workforce (money and rice) in the form of piecework. This paid workforce was made indispensable by the considerable number of men and women needed during the few months of the dry season that separated the 10th-month harvest from the first spring floods.

It is easy to appreciate that the combination of these two possible methods of mobilizing a workforce represented a source of tension between the financial means that the state was ready to invest in building and repairing dykes and their effective use in the field. While some reports presented to the Emperor made explicit mention of the reticence of rural subjects to participate in the works, it also appeared that local notables embezzled money allocated by the state for paying workers by imposing upon inhabitants a financial contribution or a share of the work through the corvée system. In 1809, for the construction of new dykes in the

[20]*Đại Nam Thực Lục*, t. III, q. 110 (2007, 858).

North *(Bắc Thành)*, the mandarins Đặng Trần Thường and Nguyễn Khắc Thiệu asked permission from Emperor Gia Long:

> The dykes in Sơn Tây, Kinh Bắc, and Sơn Nam Provinces have collapsed. We plan to have three fresh sections built and two existing stretches strengthened by using a paid workforce [piecework]. To rebuild other damaged stretches of dyke where the current [of the river] is not too strong, the local workforce will be mobilized. The King has given his consent.[21]

Although the financial efforts made in the nineteenth century varied greatly from one emperor to the next and from one policy on major hydraulic works to another, investment in this field was never questioned, in spite of the continual destruction of constructions that fed the debate concerning their usefulness. In the Annals of the reign of Minh Mạng, the Emperor recalls the importance of the state's financial contribution and, using this as a starting point, levels bitter reproaches at mandarins of Sơn Tây, Sơn Nam, and Nam Định provinces following a series of breached dykes and serious floods:

> The dykes are of great importance to the inhabitants of your provinces. You are well aware that We are far from miserly, since every year We place at your disposal a credit of 100,000 ligatures (*strings of coins*), with payments in kind to watch over the proper maintenance of the dykes. [...] Why did you not take the necessary precautions in advance to avoid such catastrophes? Why did you not, when the violence of the waters increased, take fresh precautions? [...] This proves your negligence.
>
> (quoted by Chassigneux 1914, 100)

Although some reports presented to the Emperor explicitly state the reticence of farmers to take part in the work, it also appears that local dignitaries embezzled the money allocated by the state to pay workers. They did this by imposing upon inhabitants a financial contribution or forced-work participation by means of the corvée system. In 1838, the Emperor called the Service of Public Works to order:

> Preferring leisure to hard work is something common to all. But only arduous labor gives the people a prosperous and peaceful life. With the constant concern to prevent flooding and encourage agricultural activities to feed the people, we spare neither our efforts nor our means. [...] For dyke-building work, we have renounced the former practice consisting of making use of outside labor in the form of piecework and prefer to mobilize local inhabitants to strengthen dykes. [...] But the common people have a bad habit of lazily hiring others to do their work. In addition, dishonest local dignitaries make up stories through which they profit by forcing ordinary people to give them money. This situation not only occasions unnecessary losses for the state treasury but also causes the people distress, a state of affairs we deem unacceptable.[22]

By default, the private dykes were considered to be secondary and less strategic because they were built along portions of watercourses that were more stable and regular and because potential rises in water level and overflow only affected human settlements and restricted territories. The state therefore made an

[21] *Đại Nam Thực Lục*, t. I, q. 37 (2007, 744).

[22] *Đại Nam Thực Lục*, t. XX, q. 189 (1968, 47–48).

empirical distinction between public and private domains, namely a sharing of tasks and responsibilities governed by its financial capacities at any one time, its political priorities for the governance of the kingdom in general, and for hydraulics management in particular. Private works projects were carried out at the initiative of peasant-farmer communities and self-financed by them after they had secured, and this was a crucial point, the Emperor's agreement.[23] Such agreement was necessary since, according to the land laws, the state was the sole legitimate and permanent landowner; peasant farmers being only franchise-holders whose rights to land use were conditional upon employing it to good effect and upon payment of taxes.

Work carried out locally could be on a minor scale and only involve the inhabitants of a single isolated village or commune, as recorded on some village stelae. The stela of Hồ Khẩu phường preserved in Chúc Thánh Pagoda and dated 1858 explains that repairs following a breached dyke required large sums of money and that an appeal had to be made for contributions from private donors.

> The stretch of dyke located in the Quảng Bố neighborhood gave way, water flooded the Hồ Khẩu neighborhood, and its inhabitants had to repair it. As this work was very costly, it was necessary to appeal to the people's generosity. Mrs. Nguyễn Thị Vạn gave 30 quan tiền [*strings of sapeques*] and asked that the anniversary of her death be celebrated in Chúc Thánh Pagoda. The local people put up a stela to pay homage to her and instigated the tradition of celebrating the anniversary of her death.[24]

But structures built at the initiative of villages were sometimes on such a scale that their completion undeniably required public involvement. Their existence exposed the state's shortcomings and its struggle to manage Delta-wide hydraulic planning and, by the same token, the people's ability to organize themselves locally to carry out public works in the common interest. This state of affairs was prevalent at the very beginning of the Nguyễn dynasty, as illustrated by the case of construction in the Mĩ Lương – Yên Sơn region (today Chương Mỹ and Quốc Ai Districts, Hà Tây tỉnh) of the Thập Cửu Dyke. The construction had a total length of about 50 km (3.1 m wide at the base and 2.3 m high), in order to protect 8,000 *mẫu* (2,890 ha) of paddy fields. Remarkably, it was built from 1808 to 1812 solely by the inhabitants of a group of 19 communes, and it gave rise to an extremely detailed village agreement (*bản khoán ước*) that clearly laid out the responsibilities incumbent on each commune in terms of monitoring, protection, upkeep of the earthworks (buffalo grazing, bamboo plantations, etc.), financial contributions, and penalties in case of breach of contract and intentional damage. Given its scale, once work was completed, the installation was placed under state responsibility by royal decree; thus integrating it into the public domain without the outlay of a single sapeque! (Huy Vu 1978).

---

[23] Article n°389 (Item I) of the Gia Long Code entitled "Building without permission," states that works can only be undertaken after approval by the higher authority that sets the rate of pay due those required to fulfill the corvée, *op. cit.*

[24] *Thư mục thác bản văn khắc Hán Nôm Việt Nam*, t. 1 (2007, 94).

It must however be emphasized that the passivity and self-interested caution displayed by the central power in this instance of building the Thập Cửu Dyke was not always the rule, as witnessed by the numerous requests for construction of earthworks recorded in the annals and whose approval was accompanied by state financial support:

> A new dyke has been built in Mai Xá commune, Nam Định Province, more than 200 trượng in length. [...] The commune's inhabitants all proved themselves willing. The provincial authorities reported this to the King who made known his pleasure and gave his permission [to build]. When work was completed, a reward of 5,000 strings of sapeques (*quan tiền*) was made to the inhabitants.[25]

In other words, even during these periods of disengagement, in order to encourage local initiatives, the Emperor could grant total or partial exemptions from taxes for up to 3 years, mobilize the army to assist the people, bestow rewards and honorary decorations to deserving dignitaries and commoners, and compensate those whose paddy fields were covered over by earthworks.

The relative significance of public intervention compared to local initiatives followed the changes in the allocation of responsibility for dykes. This reflected quite closely the fickleness of the relationship that successive emperors of the Nguyễn dynasty maintained with the dyke-building system for the main Delta rivers. For instance, the dissolution of the dykes department in 1833 and the search for alternative solutions to dyke building, particularly through efforts to discharge the Red River into the Cửu An River, coincided with a clear disengagement by the state with regards to its sovereign prerogatives in matters of hydraulic management that it temporarily entrusted to peasant-farmer communities.

4.3 Foreshore Reclamation and Irrigation: the First Planned Development

Although successive emperors of the Nguyễn dynasty concerned themselves above all with works to protect against flooding from the Red River, they also brought about technical innovations in the field of agricultural hydraulics.

The first involved the reclamation of foreshores (the foreshore being that part of the shore between the high- and low-water marks) and of offshore bars. Strictly speaking, this is not a nineteenth-century innovation, since historical sources testify to foreshore reclamation since the thirteenth century, a development that may even have been the reason behind the first embankment works being built in the Delta. The novelty derives more from the rationalization of foreshore reclamation through dyking up land temporarily exposed by the sea, thus making large-scale land recovery possible.

[25] *Đại Nam Thực Lục*, t. XVIII, q. 176 (1967, 360).

The first and largest undertaking of this kind was begun in 1828 and involved the foreshore of Thái Bình and Ninh Bình provinces. The works carried out under the leadership of Nguyễn Công Trứ enabled the creation in 1829 of two fresh coastal districts, Tiền Hải and Kim Sơn:

> Kim Sơn District is created and attached to Yên Khánh District, Ninh Bình. Nguyễn Công Trứ is nominated Dinh điền sứ [*mandarin in charge of putting fallow land to good use*]. Outside Hồng Lĩnh Dyke, the area of fallow land is judged to be 14,620 mẫu (*5,280 hectares*) that are allocated to more than 1,260 needy inhabitants. In this area there are 3 villages, 22 estates (ấp), 24 farms (*trại*), and 4 hamlet subdivisions (*giáp*), divided into 5 cantons (*tổng*). Permission has been requested to create a district there to be called Kim Sơn.[26]

At the turn of the twentieth century, these two districts totaled 20,000 ha of arable land and fed 120,000 inhabitants.

Once again, in conjunction with the involvement of the central power, fresh-land reclamation projects were also the result of local initiatives. For example, a stela dated the 10th day of the 6th month of the 32nd year of Tự Đức's reign (1880), tells us that Hải Yên Commune (today in Quảng Ninh Province) sealed off 50 *mẫu* of new land reclaimed from the sea; thanks to a network of dykes and mini-dams that preserved this land from salty waters.

But the most significant innovation concerned irrigation. The possibility of obtaining a second annual rice harvest on low-lying and mid-altitude Delta land, namely two-third of the total arable area, depended exclusively on climatic conditions during the winter, in particular sufficient rains. Recorded episodes of drought along with food shortage and famine are a constant of national history in the imperial annals. For example, it is noted for the year 1343: "this year no rains, famine, many peasants became thieves - especially servants-slaves in noble families."[27] Or again, for 1835 (5th month), "The paddy fields of all the districts in Sơn Tây Province have dried up, the price of rice has spiraled. Given the absence of a crop from this year and the lack of rice from last year [also due to drought], I [Governor of the Province] request a reduction by half of taxes payable by the people."[28]

Not until the beginning of the nineteenth century did a technique exist for drawing water from the river. Until this period, it seemed inconceivable that one might technically control an opening in a dyke without it degenerating into a full-scale breach synonymous with flooding. The principle consisted of procuring the water directly from the river by using underground conduits set solidly within the dykes and easy-to-close pipes or sluice gates. This system, aimed at improving upon the only palliative to drought available until this time, namely stored water. It was designed, using only gravity, to solve problems linked to irrigation of paddy fields and drainage of excess water accumulated accidentally, following a breach in a

---

[26] *Đại Nam Thực Lục*, t. II, q. 58 (2007, 843).

[27] *Đại Việt sử ký toàn thư*, bản kỷ toàn thư, quyển VII, kỷ nhà Trần, f° 12a, (1993, t. II, 128).

[28] *Đại Nam Thực Lục*, t. III, q. 152 (2007, 13).

dyke, or normally in low-lying areas of hydraulic land subdivisions (*casiers hydrauliques*) after heavy rains.

The first written records alluding to work carried out directly within the structure of dykes date it back to the first years of Gia Long's reign. The first constructions were of a rudimentary design, made from hollowed tree trunks that sometimes compromised the watertightness of dykes. Later, much more sophisticated installations, such as vaulted aqueducts, were built of bricks using a very strong, sticky, and durable mortar. *The Descriptive Geography* of the Emperor Đồng Khánh (1888–1889) mentions the existence of many of these mechanisms:

> The Kim Ngưu River [Bắc Ninh Province], which flows in Văn Giang District, formerly called Tế Giang, is a distributary of the Red River. In this province, it flows out of Sơn Hô Châu Commune, passes through the Phụng Công and Công Luận sluice gates – now very narrow, like the river itself –, loops round Từ Hồ Commune then enters Hưng Yên Province.[29]

In 1829, the inventory of hydrological installations lists 50 main sluice gates and 16 private sluice gates that all served the dual purposes of drainage and irrigation. Note, too, that the report made to Emperor Minh Mạng in 1833 by the three mandarins who initiated the reform of the dykes administration encouraged the wider use of such devices:

> [...] Concerning stretches of dykes that run alongside rivers, if the installation of sluice gates promotes agricultural activities, this must be rapidly noted in order to request permission to build them. The sluice gates will be opened in the event of drought or flooding, while the crops of the fifth or tenth lunar months are growing, to ease irrigation or drainage. They will be closed during sharp rises in water levels. For tributaries located within the dyke, particularly for shallow and dried-up sections, dredging work will occur to help water drain away. Along the stretches where sluice gates are installed, piles of earth and equipment used for protecting dykes must be provided. All decisions about dredging or dyking must be taken only after a field inspection by local mandarins, as the common people should not be made responsible for undertaking the entire task. [...][30]

How effective were these sluice gates? They gave satisfactory results in the areas near the sea where the insecurity of some crops was due much more to floodwaters from the Red River than to the excess of rainwater. On the other hand, as soon as one moved away from these coastal areas, it must be admitted that they brought fewer improvements than expected. The first reason for this was of a technical nature. Their insufficient dimensions when initially designed, aggravated by silting phenomena when water levels rose, did not allow these offtakes to let pass a sufficient volume of water. But above all, during the high-water season, in the higher and intermediary areas of the Delta, the level of the dyked rivers in spate was much higher than the level in low-lying areas of the hydraulic land subdivisions (*casiers hydrauliques*), thus preventing the evacuation through the sluice gates of excess waters that continued to form vast lakes. As for the question of irrigation, the fall in the rivers' water levels left most of the pipes out of the water for the winter season. In short, although the emperors of

[29] *Đồng Khánh địa dư chí* (2004, 1612).

[30] *Đại Nam Thực Lục*, t. III, q. 92 (2007, 536–537).

the Nguyễn dynasty understood very well that the Delta's rivers represented an inexhaustible reserve of water that agriculture must use, the topography of the Delta itself did not allow for irrigation by simple gravitational force. The absence of mechanical means of pumping prevented them from overcoming this difficulty (Chassigneux 1912, 96).

For this reason, irrigation using arroyos and ponds filled by the natural flow of the rivers remained the norm until the construction of gravity-driven networks. In this system, each peasant farmer was responsible for the supply of water to his plots of land and bailed from a specific place located on the side of an arroyo or a pond, called a bailing station. These stations being private, the owner had priority over any other villager wishing to irrigate his paddy fields. The bailed water was then either directly poured into the field or went into a canal that led to the farmer's land.[31]

All the emperors enacted incentive measures to increase water reserves available for the 5th-month rice crop, whether they consisted of clearing arroyos, digging irrigation canals, or raising the height of *talwegs* (mini-dyke-dams). But even more than the construction of dykes, the fluctuations in investment granted by the imperial state and a certain powerlessness to propose efficient solutions for irrigation were two factors that impelled peasant-farmer communities to design and make their own installations, as evidenced by this stela of Đắc Sở Commune (Đan Phượng District, Hà Đông Province), engraved in 1854:

> Hydraulic systems are at the service of agricultural activities. Some mandarins, venerable village elders, gave money to finance the installation of a stone sluice gate in the commune with the aim of improving irrigation and drainage. On the stela are written the names of those having made their contribution such as Nguyễn Văn Uyên, deputy head of the canton, Nguyễn Kim Nguyên, deputy head of the village....[32]

Although it appears impossible to establish a quantitative assessment of these local initiatives, to obtain a general idea let us note that at the end of the nineteenth century, combined public and private installations for irrigation and/or drainage enabled a double annual rice crop on one-third of the total surface of the Red River Delta.

Helpless when confronted with the droughts that regularly afflicted the country, the Nguyễn dynasty emperors, like previous dynasties, sought divine intervention to hasten the long-awaited rains. In several passages of the annals devoted to the quest for such divine assistance, the Emperor ponders the reasons that may have brought about such celestial wrath and what action should be taken to assuage it. Such action was of two kinds.

First, the Sovereign made endless gestures of propitiatory piety in order to counterbalance any lapses and excessive harshness attributable to his government, these being deemed to be the causes of natural disasters. In practice, he granted amnesties to large numbers of prisoners and imposed upon the Court and upon himself a simple and austere lifestyle, similar to that of the people, outlawing feasts,

---

[31] Bailing itself was mainly accomplished using one of two cheap and simple tools, chosen depending on the height of the water being bailed, which are still in use to this day: the tripod bailer (*cái gầu sòng*) and the roped basket (*cái gầu giai*).

[32] *Thư mục thác bản văn khắc Hán Nôm Việt Nam*, t. I (2007, 802–803).

music, and other pleasures pertaining to his rank. In the 3rd month of 1824, the Sovereign's birthday, "having noted that many areas are suffering from drought, the Emperor cancels all the usual customs associated with this feast day, such as displays of singing and music, as well as fireworks and torchlight dances."[33]

Similar measures were taken in the event of flooding, as illustrated by this order from 1833:

> The North (*Bắc Kỳ*) is suffering from flooding, the inhabitants there are enduring dismal times. I am their father, how can I live in joy and peace! From today onwards, the dishes prepared for the Court must be reduced by half. Displays of singing and dancing must be cancelled. All the birds in the Royal Garden must be freed. All those whose place is in the Inner Palace (*nội cung*) must dress according to their rank and in a sober manner; the royal servant women are forbidden from excessive use of fabrics and foodstuffs.[34]

Second, he addressed prayers and presented offerings to mountain and river gods in order to restore harmony between heaven and earth. He also ordered that religious ceremonies be celebrated in the temples and communal houses located in the regions affected by the drought, even if he sometimes voiced doubts as to the efficacy of such methods. This passage from 1826 documents the Sovereign's incredulity:

> […] For some weeks, as it has not rained, the weather is so hot that a man feels it burning in his stomach, and the plants in the paddy fields are wilting away. In the palace, not a single night have I slept peacefully. Yesterday evening, seated at Court, my eyes lifted to the heavens. I saw a great black cloud appear, but it was quickly scattered by the southeasterly wind, a sign that rains will not fall easily here. The Emperor turned to Phan Huy Thực and said to him: I wish to call upon the local deities. If someone can make the rains come, he will be rewarded; otherwise, we will suspend our prayers. But rain and wind are the product of nature; the gods can do nothing to change them. […][35]

Finally, lacking an efficient technical mastery of the irrigation process, the King granted partial or total tax exemptions and had rice from the royal granaries distributed to somewhat alleviate the hardships suffered by the people. In 1825, the drought having damaged the 5th-month rice crop in five districts of Hải Dương and Nam Định Provinces in the North (Bắc Thành), the Emperor ordered:

> […] In all the districts struck down by this blight and who have lost their crops for the year, the people cannot escape famine. For this reason, the regions that could not plant rice are exempted from taxes; for those who have lost 4/10th of their crops, taxes are reduced by 2/10th, for 5/10th of losses the reduction is 3/10th, for 6/10th the reduction is 4/10th, for 7/10th the reduction is 5/10th, and for 8/10th of losses the exemption is total. Repayment is waived on loans of rice granted last year by the state [*royal granaries*].[36]

The Nguyễn dynasty allocated a massive financial investment to the domain of hydraulics that made it possible to dyke up the whole Delta. This occurred despite

---

[33] *Đại Nam Thực Lục*, t. II, q. 26 (2007, 347).

[34] *Đại Nam Thực Lục*, t. III, q. 100 (2007, 665–666).

[35] *Đại Nam Thực Lục*, t. II, q. 40 (2007, 524–525).

[36] *Đại Nam Thực Lục*, t. II, q. 34 (2007, 438–439).

the fact that the country's capital was transferred to Huế, after centuries of being embedded in the heart of the Red River Delta. This unprecedented effort should not however mask the instability and ambiguity of the hydraulic policy pursued by successive emperors. Examples of these problems include: Discontinuity in the administration and management of dykes alternately entrusted to a specialized department or relegated to the responsibility of provincial mandarins; bureaucratic unwieldiness linked in part to the distance from the center of power and to failures of the mandarin administration; alternate periods of involvement followed by disengagement by the state giving up part of its prerogatives to peasant-farmer communities, the latter finding themselves obliged to take sole responsibility for the construction and upkeep of a growing number of installations.

The inability of the emperors to fulfill the terms of the "celestial contract" that assigned to them the responsibility of protecting their subjects from the natural elements and in particular from flooding, contributed without any doubt to the popular discontent evidenced by the frequency of successive peasant revolts in the nineteenth century: 4 revolts a year under Gia Long; 11 revolts a year under Minh Mạng; 8 revolts a year under Thiệu Trị; 3 revolts a year under Tự Đức; in total, more than 400 rebellions in about 60 years.

But over and above this checkered record, it can still be said that the Nguyễn dynasty played a key part in the domain of hydraulics by creating the foundations for modern and planned development of the Red River Delta. In addition to widespread dyke building, Emperors Gia Long and Ming Mạng were the first to have tried to resolve the crucial problem of irrigation by tapping directly into the resource constituted by the practically inexhaustible reserves of water supplied by the rivers. Admittedly, the results obtained were mediocre; in a deltaic region such as this one, the installation of hydraulic networks can only be carried out on a large scale in order to enable not only irrigation but also drainage of excess water during the monsoon. Such efforts call for significant mechanical means for pumping water and for the digging of robust drainage and irrigation canals. For these precursors, limits were above all of a technical nature, whatever the opinion of Western observers. Indeed, it was not until the 1920s that the colonial engineers and experts come up with effective and realistic irrigation projects for the middle and lower Delta, thanks to an increased reliance on thermal, then electrical-pumping stations. Despite this admission of powerlessness, the emperors of this dynasty nevertheless had an overall vision of the complexity of the hydraulic workings of the Delta and the sometimes harmful consequences of its development, as proved by the debate initiated as far back as 1803 by Gia Long about the possibility of either strengthening or leveling of the existing network of dykes.

References

Béthemont J (2000) Les grands fleuves: entre nature et société, Deuxième édition. Armand Colin, Paris

Brénier H (1914) Essai d'Atlas Statistique de l'Indochine Française. IDEO, Hanoi-Haiphong

Cadière L (1992) Croyances et pratiques religieuses des Vietnamiens, Tome I. EFEO, Paris (re-edition 1922)

Chassigneux E (1912) L'irrigation dans le delta du Tonkin. Revue de Géographie annuelle VI:1–121

Chassigneux E (1914) Les inondations et les digues du Tonkin. Conférence faite à l'école coloniale, le 13 Mars 1914, Imprimerie Chaix, Paris

Đại Nam Thực Lục [True Account of Đại Nam] (1967–68) tập XVIII & XX. nhà xuất bản Khoa Học Xã Hội, Hà Nội

Đại Nam Thực Lục [True Account of Đại Nam] (2007) tập I, II, III, IV. nhà xuất bản Giáo Dục, Hà Nội

Đại Việt sử ký toàn thư [Book of the complete historical memoirs of Đại Việt] (1993) t. II. nhà xuất bản Khoa Học Xã Hội, Hà Nội

Đỗ Đức Hùng (1979) Về trị thủy – thủy lợi ở nước ta nửa đầu thế kỷ XIX [Concerning the harnessing of rivers and irrigation in our country at the beginning of the 19th century]. Tập chí Nghiên Cứu Lịch Sử 4(187):46–56

Đỗ Đức Hùng (1998) Vấn đề trị thủy ở Hưng Yên dưới thời Nguyễn (thế kỷ XIX) [Problems with hydraulic engineering works in Hưng Yên Province during the Nguyễn dynasty (19th century)]. Tập chí Nghiên Cứu Lịch Sử 5(270):42–47

Fontenelle JP et al Atlas vùng Bắc Hưng Hải (Việt Nam) (2000) [Atlas of the Bac Hung Hai Polder]. nhà xuất bản Nông Nghiệp, Hà Nội

Gauthier J (1930) Digues du Tonkin, Exposition coloniale internationale – Paris 1931. IDEO, Hanoi-Haiphong

Gourou P (1936) Les paysans du delta tonkinois, publications de l'École française d'Extrême-Orient. Les Éditions d'art et d'histoire, Paris

Huy Vu (1978) Vài nét về đê điều, thủy lợi ở làng xã Việt Nam thời trước, [Some aspects of dyke-building and irrigation in Vietnamese villages long ago]. Tập chí Nghiên Cứu Lịch Sử 3(180):46–53

Langlet P (1970) La tradition vietnamienne, un État national au sein de la civilisation chinoise. BSEI XLV:1–395

Lê Thành Khoi (1992) Histoire du Viet Nam des origines à 1858. Sudestasie, Paris

Ngô Đức Thọ, Nguyễn Văn Nguyên, Papin P (eds) (2004a) Đồng Khánh địa dư chí [The descriptive geography of the emperor Đồng Khánh]. EFEO/Purple Ink, Hà Nội

Ngô Đức Thọ, Nguyên NV, Papin P (eds) (2004b) *Đồng Khánh địa dư chí* [The Descriptive Geography of the Emperor Đồng Khánh]. EFEO – nhà xuất bản Thế Giới, Hanoi

Philastre P-L-F (1876) Le Code Annamite. Ernest Leroux, Tome II, Paris

Poisson E (2009) Détruire ou consolider les digues du fleuve Rouge. Aséanie 23:77–96

Pouyanne A (1931) L'hydraulique agricole au Tonkin, Exposition coloniale internationale – Paris 1931. IDEO, Hanoi-Haiphong

Sơn Nam (1970) Đồng bằng sông Cửu Long hay là Văn minh miệt vườn [The Mekong delta or the Orchard civilisation]. An Tiên, Saigon

Thư mục thác bản văn khắc Hán Nôm Việt Nam [Catalogue of inscriptions from Vietnam] (2007) t. I & III.Viện Nghiên Cứu Hán Nôm – EFEO – EPHE, Hanoi

Việt Sử Cương mục Tiết Yếu [A complete history of the Việt] (2000) nhà xuất bản Khoa Hoạc Xã Hội, Ha Noi

Chapter 5
"A Kind of Mylai … Against the Indochinese Countryside": American Scientists, Herbicides, and South Vietnamese Mangrove Forests

Amy M. Hay

In a 1972 book review of *Harvest of Death*, a collection of essays examining chemical warfare in South Vietnam, plant biologist Arthur Galston compared the US herbicide campaign to the "willful destruction of an entire people and its culture" that had happened during World War II and had been condemned at the Nuremberg trials (Galston 1972). Developing his analogy further, Galston noted that a crime of such magnitude against humanity was labeled genocide. In the case where "the willful and permanent destruction of an environment in which a people can live in a manner of their own choosing ought," according to Galston, "to be designated by the term ecocide." *Harvest of Death*, then, became "a document of a kind of Mylai perpetrated against the Indochinese countryside." Galston represented one side of a sometimes bitterly divided group of American scientists as they sought to assess the damage done to the South Vietnamese countryside, most particularly the coastal mangrove forests of the Mekong Delta.

As the employment and scale of chemical herbicides increased during the Vietnam conflict, American scientists began questioning the effects the massive defoliation campaign might be having on the Vietnam countryside. As early as 1967, biologists and plant scientists challenged US government claims that the herbicides being used, which included a 50/50 mixture of (2,4-dichlorophenoxy) acetic acid and (2,4,5-trichlorophenoxy) acetic acid (hereafter 2,4-D and 2,4,5-T), better known as Agent Orange, were safe for humans and animals. As spraying operations continued, some members of the American Association for the Advancement of Science (AAAS) fought to get the association involved in evaluating the potential ecological damage being done by the massive amounts of chemical herbicides being sprayed over the South Vietnamese countryside. This effort succeeded, and a study was commissioned. This study, produced by the AAAS Herbicide Assessment Commission (HAC), sent American scientists to the South Vietnamese jungle in 1970 to assess the ecological effects of herbicide spraying. Four years later, the National Academy

A.M. Hay (✉)
Department of History and Philosophy, University of Texas – Pan American
e-mail: ahay@utpa.edu

M.A. Stewart and P.A. Coclanis (eds.), *Environmental Change and Agricultural Sustainability in the Mekong Delta*, Advances in Global Change Research 45,
DOI 10.1007/978-94-007-0934-8_5, © Springer Science+Business Media B.V. 2011

of Sciences (NAS) released its own massive report on ecological conditions, *The Effects of Herbicides in South Vietnam.* Taken together, these two studies have become the standard by which the damage done by herbicides in the Mekong Delta in particular, and South Vietnam more broadly, is measured (Tran Triet et al. 2004; Young 2009). That this is the case is somewhat ironic because the scientists involved fell on both sides of the debate regarding the lasting harm done to the Vietnamese ecosystem. Contemporary scientists' unquestioned acceptance of the 1970 and 1974 studies may be even more ironic, given the controversy that surrounded both reports at the time they were done. The dwindling area sustaining South Vietnam's mangrove forests suggests that history and science may need to be read differently to stop the disappearance of a vanishing ecosystem.

This chapter examines the report produced by the AAAS' Herbicide Assessment Commission in 1970 and the NAS report in 1974, as each tried to assess the ecological harm caused by the herbicide campaign in South Vietnam. It considers the broader history of the mangrove forests in South Vietnam, the research and production of the reports themselves, the subsequent controversies, and the reports' legacies. These two reports represent the official involvement of two prestigious bodies of American scientists; they also demonstrate the contested nature of scientific knowledge, objectivity, and opinions about the ecological damage being done in South Vietnam. The US herbicide campaign in South Vietnam plays an important, if unfortunate, part in the history of environmental change in the Delta. The American scientists involved in these reports, acting not only in their roles as scientists but also as human beings concerned about the natural environment and people of Vietnam, tried to address the harm they thought was being done. At the same time, cultural, economic, and political factors all played important roles not only in the production of scientific knowledge in the 1970s, but also in the ecological fate of the mangrove forests of the Mekong Delta, right down to the present day.

In contextualizing the effects of wartime herbicide sprayings, the French colonial period represents a good starting point when considering the ways this natural environment had been understood and used. French scientists helped colonial administrators evaluate and intervene in the forests of Vietnam, and the mangrove forests in particular in South Vietnam. Many saw the mangrove swamps as impediments to progress, untamed wilderness that as late as 1930 had been the sole preserve of wild elephants. Engineers offered technical expertise that promised to make such unused and wasted land more productive. While the French were not the first to settle, or even try and improve this region, their efforts marked the advent of the particularly colonial enterprises of "modernization" and "progress" (Biggs 2003; Käkönen 2008). Entrusted to the Public Works Department, French engineers embarked on an improvement campaign centered on canal construction, an enterprise which ultimately resulted in 165 million cubic meters of earth moved over a 50-year period, from 1880 to 1930, and changed the natural environment of the Delta. French attention to the various canal projects waxed and waned, and seemingly achieved a sense of a partially domesticated wilderness (Käkönen 2008). But short-term success in improvements in transportation or rice production had hidden, and often more negative effects, on the overall ecology of the "mangrove swamps"

(Biggs 2003). Along with the Public Works Department's canal-building modernization projects, the French colonial government also sold the land at cheap prices to wealthy landlords, who then rented the property to thousands of poor migrants (Biggs 2005). During the 1940s and 1950s, French scientists continued their efforts to manage the natural landscape and resources of Vietnam. Paul Maurand, a French forester, completed one of the early vegetative maps of Indochina in 1943, while Claude Moquillon reported specifically on the mangrove forests of Ca Mau in 1950. These studies showed 400,000 total hectares of mangrove forests, with 200,000 ha in the Ca Mau peninsula (Hong, San 1993). They also set a baseline for forest growth in the area for later scientific investigations. Foreign attempts to modernize and control the mangrove forests of the Mekong Delta continued after World War II, although a new set of actors entered the picture.

As part of a wave of global anticolonial revolts following the end of World War II, the French Indochina War (later known as the First Indochina War) saw the incursion of American engineers and technology into the swamps of South Vietnam. Following international irrigation programs in India and China, the United States Bureau of Reclamation attempted to duplicate in Vietnam the modernizing success of the Franklin D. Roosevelt administration's Tennessee Valley Authority engineering projects. Intrinsically tied to Cold War politics, the Bureau initially sponsored training programs for foreign diplomats and engineers, but by 1952 sent scientific advisors (Biggs 2006). Within a decade they would be followed by experts of another sort, ones focused on using American technology to support military operations. These activities sought to destroy the unruly and resistant Mekong Delta mangrove swamps that resisted other forms of enlightened colonial scientific and military control.

Even as French and American engineers tried to tame the swampy lands of the Mekong Delta, Vietnamese rebels used the undesirable spaces in their efforts to resist first French colonial rule, and eventually American troops. Filled with landless migrants, the Mekong Delta became the site of various rebellions against colonial rule and wealthy landowners. By 1940, Ho Chi Minh's Viet Minh used Mekong Delta forests as bases of operations that offered not only refuge, but also cells that produced weapons and propaganda. At the end of World War II, disaffected youth flocked to the U Minh as one of the two centers of the liberation movement. Rebels appeared out of the forest to conduct guerilla warfare against French colonial troops. In the process of defying French colonial authorities, the National Liberation Front rebels (NLF) redistributed the swamp wasteland to over 30,000 of their supporters, formerly landless migrants. Eventually the NLF "liberated" the entire region, giving more than 500,000 families slightly more than 564,000 ha of land (Biggs 2005).

For many scientists in the 1960s, the production of scientific knowledge demanded well-defined research agendas, valid research methods and reproducible results, and a stance of neutrality and rationality. While American scientists showed a long history of engagement in political causes, for both progressive and less than exemplary motives and results, scientists in the postwar period were confronted with an especially vexing problem: How to reconcile the relationship between knowledge of the natural world and related activities, state sponsorship of such endeavors, and scientists' societal obligations (Moore 2008). In the postwar period, the majority of scientific funding came from the Department of Defense, understandable considering Cold War priorities. However, this reliance on government funding of individual and university research agendas, along with Cold War political repression, put scientists in difficult positions when it came to speaking out against programs they thought harmful or immoral. In the postwar period, these tensions increased as scientists began questioning their professional roles in the production of knowledge and societal obligations with respect to what knowledge was produced, and the ways such knowledge would be used. American involvement in Vietnam and the use of chemical and biological warfare presented an especially daunting problem for scientists, as some individuals pondered the role scientists should play in evaluating and advocating not just for human beings, but human and natural environments.

Little research has been done on postwar scientific activism, much less scientists' involvement in protesting the Vietnam War. Domestic Cold War concerns had led to political repression generally among scientists, as individuals were "disciplined" for poor support, much less outright protest, of government scientific programs, as the case of Robert Oppenheimer demonstrated (Wang 2002; Thorpe 2002). Scientists persisted, however, in questioning what their role and obligations within society should be, first with respect to atomic energy and nuclear weapons, and then regarding what kinds of research should be receiving state priority: military weapons or humane endeavors, such as disease and hunger relief. Postwar social movements like civil rights, or especially the student democracy movement and its critique of the American university system, intensified the pressure on scientists to engage the social issues around them. These same progressive movements led to greater awareness of environmental concerns, helped by the 1962 publication of Rachel Carson's *Silent Spring*. Most historical examinations have focused on scientific activism with respect to atomic energy and nuclear weapons (Wang 1999). As part of a larger project examining the development, use, and protests against Agent Orange, this chapter focuses on the role scientists played in challenging the use of chemical herbicides to defoliate the jungles of South Vietnam. Scientists represent an important, and mostly overlooked, group that opposed the use of chemical and biological weapons, an opposition that led many to oppose the war itself. While many other factors brought about the end of the war, scientific concerns, investigations, and protests put both the Johnson and Nixon administrations on the defensive, and such activities were the major factors in the Nixon administration's ending of defoliation operations in 1970. Scientific expertise led these individuals to question Agent Orange, and these questions eventually led these same scientists to protest the use of Agent Orange in Vietnam.

Ironically, the chemical war on Vietnam, which started just shortly before the publication of *Silent Spring* in 1962, prompted scientists in particular, but Americans generally, to examine the safety of chemical compounds and their use in the natural world. While Carson's famous investigation focused on DDT, an insecticide, *Silent*

Spring ignited a firestorm over the use of chemicals in agriculture and drew vociferous, often personal, attacks from the chemical industry. Just a year before its publication, the United States had begun a spraying program in South Vietnam designed to destroy the heavy jungle vegetative growth in hopes at once of exposing Viet Cong enemy movement, supply lines and caches, and of increasing the effectiveness of air attacks (Buckingham 1982). The herbicides, a mix of two phenoxy compounds that inhibited plant growth, were developed at the end of World War II. After the war, private industry began production of the herbicides for agricultural use, which saw rapid growth from 1950 to 1970, as farmers increased their purchases of pesticides from 25 cents per 0.4 ha to $3.65 for the same area (Galston 1971). When efforts to use fires to clear jungle growth in South Vietnam failed, the phenoxy herbicides once again drew military interest. What would eventually be called Operation Ranch Hand gained Secretary of Defense Robert McNamara's support in 1960, and defoliation operations began in January 1961.

As American defoliation operations intensified, so did scientific concern. Best known for its slogan, "Only We Can Stop Forests," Operation Ranch Hand used millions of liters of herbicides in less than a decade, spraying thousands of acres of the Vietnamese countryside. Defoliation operations began slowly, with 250,000 gal sprayed between 1962 and 1964, with activities peaking in 1967. By 1965, 400,000 gal had been ordered. Over 1.6 million acres were sprayed with approximately 3 gal per acre. Some of these spraying operations were intended to destroy crops by employing cacodylic acid and picloram, Agents Blue and White. Anticipated military demands of 5.6 million gallons in 1967 and 11.9 million gallons in 1968 strained the ability of the industry to produce enough herbicides. Spraying operations required specially equipped planes and helicopters staffed with a pilot, copilot, and technical specialist. When making a spraying run, these aircraft flew as close as possible to foliage at low speeds. Wind, temperature, and other weather conditions affected how the herbicide spray settled over the plant growth, which meant the best conditions for spraying were in the early morning when there is little air turbulence. Eighty percent of the herbicide spray settled on the top levels of the forest canopy, with as little as 6% reaching the ground-level vegetation. Parallel to the intensification of spraying operations, the defoliation campaign in Vietnam increasingly attracted the attention of biological scientists concerned about the use of chemical and biological weapons in the war. Almost from the beginning, the operation attracted censure. Critics charged that spraying herbicides represented an act of chemical warfare and violated the 1925 Geneva Convention banning the use of such agents. Military scientists responded that Agent Orange posed no harm to animals or humans. Other critics spoke out harshly against the use of Agents Blue and White, which were deliberately used to destroy crops with the hope that such destruction would pressure Vietcong forces. While biological scientists did not put aside their concerns for animal or human safety, they increasingly voiced concerns regarding the potential ecological damage defoliation might be causing to the Vietnamese ecosystem. As America increased its involvement with South Vietnam over the 1960s, accompanied by an intensification of its defoliation campaign, scientists began speaking out.

The AAAS investigation into the use of herbicides in Vietnam came about through the efforts of several scientists, but the efforts of E.W., or "Bert" Pfeiffer, stand out in particular. A wildlife biologist, Pfeiffer worked at the University of Montana, and had previous experience with the potential harm caused by the military in the form of nuclear fallout from atomic bomb tests. When Pfeiffer became aware of Operation Ranch Hand, he brought the matter of herbicide sprayings before the AAAS executive council, framing the matter as one of concern over the possible harm being done to the South Vietnamese ecosystem (Zwierler 2008). In the same year, Arthur Galston, the plant physiologist whose book review began this chapter, first attempted to address the potential harm of herbicide chemicals when he tried to put the issue on the American Association of Plant Physiology meeting agenda in the form of a letter to President Lyndon B. Johnson. While his efforts were thwarted, another letter sent in 1966 and signed by 22 scientists called upon President Johnson to stop the use of chemical anticrop chemical agents in Vietnam. This petition had received the support of the Council of the Federation of American Scientists, and included "some of the most distinguished chemists, biochemists, bacteriologists, physicists, biologists, and men of medical science in the American scientific community."[1] The letter made no comment on the war itself, and came about because of informal discussions held by Harvard chemists, biochemists, and bacteriologists. Less than 5 months later, a petition signed by over 5,000 scientists was sent to the White House protesting chemical warfare, and showed the dramatic increase in concern among scientists. Among the signers of both petitions was another scientist who would become closely associated with scientific protests against chemical warfare: Matthew S. Meselson, a geneticist from Harvard University. Meselson's involvement with military research and arms control dated back to 1963, when he worked as a consultant to the U.S. Arms Control and Disarmament Agency, and eventually asked to be assigned to review the government's chemical and biological warfare programs. Meselson had spent the subsequent years involved in changing government policy on chemical and germ warfare. Along with Pfeiffer and Arthur Westing, a former graduate student of Galston's, Meselson and Galston represented some of the most outspoken and active scientific critics of American defoliation activities.

After several revisions, Pfeiffer's resolution was passed by the AAAS in 1966. The resolution read in part: "modern science and technology [had] now given man unprecedented power to alter his environment and affect the ecological balance of this planet …" It went on to express the AAAS' concern over the long-range consequences of biological and chemical agents which modified the environment and recommended the establishment of a committee to study the potential effects of such chemical and biological warfare agents. Finally, the AAAS would work in cooperation with other public agencies and the government in "the task of ascertaining scientifically and objectively the full implications of major programs and activities which modify the environment and affect the ecological balance on a large scale" (Wolfle 1967). Adoption of the resolution resulted in the formation of an ad hoc committee at the 1967 March meeting.

[1]"22 Scientists Bid Johnson Bar Chemical Weapons in Vietnam," New York Times, September 20, 1966, 1.

Rene Dubos, a later skeptic of the benefits of modern scientific medicine, chaired the committee. It recommended that the AAAS establish a permanent, continuing committee dedicated to examining "environmental alteration." Eventually, the AAAS created a permanent committee to investigate environmental alteration and a specific inquiry into chemical and biological warfare in Vietnam. Over the next several months, the AAAS worked with the Department of Defense in trying to identify an appropriate means to gather the information needed to evaluate the potential harmful effects of chemical and biological warfare. At the same time these discussions were going on between the AAAS and the Department of Defense, arrangements had been made to gather information from the war zone in Vietnam through a contract with the Midwest Research Institute in Kansas City, Missouri. The MRI report was reviewed by the National Academy of Sciences, which noted the continuing difficulty of assessing the ecological consequences of herbicide use.

By the time of the 1969 Dallas meeting of the AAAS, it had become clear that the military's assessment of its herbicide operations would not be enough, although the issue remained a difficult problem for the AAAS to address. By 1969, the AAAS recommended that it undertake its own investigation of the ecological effects of the US herbicide campaign, in part because "Department of Defense officials do not have scientific assurance that herbicides cause no seriously adverse ecological consequences in Vietnam" (Nelson 1969). Although it was clear that not all members opposed the war in Vietnam, or even the US herbicide campaign, "most scientists find it difficult to oppose a purely scientific study." The action was viewed by some members as the "quickening conscience of the American scientific community." Margaret Mead expressed the opinion that other issues of environmental alteration through technology, such as the Aswan Dam in Africa, dwarfed the use of herbicides in Vietnam. She suggested that "The issue here [was] warfare, not defoliation." A panel presentation on the use of herbicides in Vietnam had generated some heated exchanges between Boysie E. Day of the University of California, Riverside, and Galston and Barry Commoner, as they argued over the findings of the MRI report. Day questioned the usefulness of a report "prepared in 60 days by people who do not understand herbicides," while Commoner pointed out that Day had been a consultant to the MRI Report.

At the same panel presentation, Fred Tschirley of the U.S. Agricultural Research Service presented a paper on a 1-month study of defoliation in South Vietnam. Tschirley reported that herbicide spraying had affected the ecology of the mangrove forests, although he expressed the belief that such changes were not irreversible. He recommended intensive study, after the war, of the Vietnamese ecology, the need for further assessment of ongoing herbicide operations as they affected forests and waterways, and changes in patterns of defoliation that might cause less long-term harm to the ecosystem. Tschirley's experience in gathering information captured the attention of AAAS members as well, as his three forays to defoliated forest areas had required him "to carry a carbine and a revolver for protection." As noted by in the press report, "obviously, it is much easier to authorize a study of the use of herbicides in Vietnam than it will be to carry it out in the near future."

American scientists concerned about the harm caused by Agent Orange in Vietnam succeeded in gaining professional support for investigations of the ecological conditions in South Vietnam, and these scientific investigations and the resulting reports represented another avenue by which they used their professional expertise to protest Agent Orange. The AAAS appointed the Herbicide Assessment Commission in December of 1969 with Meselson serving as the Commission leader. Meselson's appointment reflected his previous involvement with chemical and biological weapons and political activities rather than any environmental biological or chemical expertise. It would be his personal political skills that ensured the success of the committee and the study (Zierler 2008). Operating with a budget of $80,000, the team "reviewed the pertinent literature, consulted with numerous experts, and made a 5-week inspection tour of South Vietnam …" (Boffey 1971). The Commission presented its finding at the 1970 AAAS general meeting. The Commission issued a statement that said in part that US defoliation operations had resulted in "extremely serious harm." This judgment was supported by a number of assertions, among them that up to one-fifth to one-half of the South Vietnamese mangrove forests had been destroyed; that hardwood mangrove forests near Saigon had been killed and the area was being invaded by bamboo growth that would prevent reforestation by the hardwoods; that the Army's crop-destruction program had destroyed civilian food supplies rather than denying food to enemy soldiers; and that while there was no definitive evidence of adverse human health effects, further study was warranted. The AAAS study ignited a firestorm as various scientific and government agencies began questioning its findings.

The HAC examined several areas that defoliation operations might have affected. The commission also assessed the Army's crop-destruction program (primarily Agent White). With over 2,000 acres sprayed, the commission estimated the crops destroyed were sufficient to feed over 600,000 people a year. Along with its negative judgment of the crop-destruction program, the AAAS team undertook an assessment of the food chain itself, taking biological samples from fauna, flora, and human beings to try and evaluate whether or not toxic chemicals had entered the food chain. Coastal mangrove forests proved to be especially sensitive environment, with a single spraying of herbicides effectively defoliating and killing the forests. Equally problematic, the devastated landscape seemed resistant to the regrowth of new plants. As noted in the *Science* report of the HAC trip: "The only sign of new life was a few worthless ferns and shrubs … This lack of new plant life was deemed one of the most significant findings of the AAAS study …" Team member Arthur Westing expressed concern over the state of the mangrove forests as they served as the home for thousands of people, provided firewood and charcoal, and were an integral part of the coastal ecosystem, considered to be one of the most diverse in the world. Westing estimated the damage to the tropical forests as severe, with 35% of South Vietnam's 14 million acres sprayed, and a timber loss of $500 million. The forests might take decades to recover (Boffey 1971).

The popular press, for the most part, accepted the AAAS team's findings. In December 1970, the *New York Times* ran an article entitled, "Sprays in Vietnam Said to Level Fifth of Mangrove Area." The damage done by defoliation was described as catastrophic. The *Times* article linked the AAAS report with the Nixon administration's decision to cease herbicide sprayings (Sullivan 1970). A *Washington Star* news release quoted Arthur Westing's opinion that it might take a generation of growth to restore coastal mangrove swamps (Randal 1970). A piece in *Newsweek* focused on the issue of crop destruction, but still noted the destruction of mangrove trees in the Mekong Delta (Blight 1971). If nothing else, the press coverage reflected the changing popular view of government attitudes toward the usefulness of herbicide spraying in South Vietnam. The next scientific study of herbicide effects in Vietnam would receive very different reactions.

The National Academy of Sciences (NAS) became involved in Vietnam due to a 1970 Congressional order for an independent study of herbicide-spraying operations and their effects in South Vietnam. While the military had produced some assessments, government officials recognized that these reports were not being considered objective assessments of conditions. The Department of Defense (DOD) contracted the scientific investigation out to the NAS. The NAS sought to avoid controversy in this investigation by omitting scientists opposed to defoliation, even if some of them had previous experience on the AAAS Herbicide Assessment Commission or scientific training that gave them necessary expertise. Its efforts were not successful. The NAS proposal still attracted controversy when anthropologists from the American Anthropological Association refused to participate in the study because of DOD funding. The NAS negotiated this position by using other funds for the anthropological portion of the study. The NAS scientists were focused on producing verifiable scientific results, which meant that investigators limited themselves to a narrow range of issues (Phuong-Lan). As NAS President Philip Handler acknowledged in his preface to the final report, a variety of factors limited the team's abilities to fully assess herbicide effects, including the uncontrolled nature of the war itself, which affected accurate records, the ability to find or interview qualified observers, separate the effects of herbicide sprayings from other aspects of the war, or the dangers of trying to assess conditions during the wartime conditions and dangers in the very areas being assessed (Handler 1974).

The extent of the damage to the inland forests represented the most contested part of the study, even as it acknowledged the sensitivity of the mangrove forests to herbicide sprayings. In fact, the NAS study reaffirmed the earlier assessment expressed by the HAC's Westing when it reported that it might take up to a century for the mangrove forests to recover. Despite this judgment, the NAS study concluded that no herbicides remained in South Vietnamese soils, and scientists appeared to disregard the disruption of the South Vietnamese economic and social ways of life tied to the mangrove ecosystem. Although the NAS study was limited because of the lack of control areas with which gathered data could be compared, the report acknowledged that defoliation in the coastal mangrove forests resulted in increased mosquito and rat infestations. This in turn meant greater chances for malaria, which already had a long and unfortunate history in Southeast Asia (Handler 1974). Upon the NAS study's release, a former NAS chair publicly questioned the NAS study's conclusion regarding the effects of herbicides. At least three scientists – British biologist Paul W. Richards, South Vietnamese botanist Pham

Hoang Ho, and Harvard public health specialist Alexander Leighton – refused to endorse the final report (Morgan 1974). The *New York Times*' initial coverage of the NAS report emphasized the time needed for the South Vietnamese ecosystem to recover (Finney 1974) and discussed the controversy that the report's findings were generating. Accuracy in Media (AIM) immediately challenged the *Times* coverage in an article entitled "New York Times Readers Deceived on Defoliation Report." The AIM piece characterized the *Times* article as "rushed" and one that "grossly misrepresented the findings of the scientific study" (Accuracy in Media 1974). In his review of the NAS study, Don Willarejo singled out the damage done to the coastal mangrove forests as representative of the damage done more broadly throughout South Vietnam (Willarejo 1975). The NAS report attracted more controversy when team members protested the inclusion of Matthew Meselson on the scientific review board convened to evaluate the study's findings. Some recently uncovered evidence from the data gathered by NAS scientific investigators shows that the South Vietnamese people and countryside were exposed to almost seven million liters more of dioxin-laden herbicides than original numbers showed (Phuong-Lan 2003). It is important to note that neither the 1970 AAAS report nor the 1974 NAS study adequately addressed the possible harmful human health effects that the spraying of herbicides might have caused. Each study made attempts, but neither could gather verifiable scientific evidence of harm in the midst of war.

Even as some American scientists raised concerns about herbicide spraying in Vietnam, or participated in studies to assess potential harm, others staunchly defended the safety of herbicides and questioned their colleagues' motives. In his 1968 presidential address to the newly named Weed Science Society of America (WSSA), president Richard Behrens noted one cause for concern. Behrens worried that recent publicity over herbicide use in Vietnam came from scientists not trained in weed science. "I am deeply concerned that the AAAS Committee [to investigate herbicide use in Vietnam] will consist entirely of men not knowledgeable in weed science, men … who appear to be attacking herbicidal usage in Vietnam as a means of expressing their disapproval of American action in Vietnam" (Behrens 1968). A paper by an unknown author published in the 1973 Montana State Weed Conference proceedings directly addressed this dilemma. The paper began by noting the extensive media investigations and scientific studies examining the herbicide operations in Vietnam. Farmers paid little attention at first, but grew increasingly concerned about the negative coverage. "After all, it is the military and it is in a land on the other side of the world…" (Montana Weed Association 1973). Or sometimes, American scientists just did not care about the effects of herbicide sprayings in Vietnam. In another WSSA presidential address, E.G. Rodgers challenged the ban on 2,4,5-T based on findings taken from Vietnam. "A proposal was made in 1973 that 2,4,5-T be completely banned because of its alleged toxic effect on the food chain. The basis of that proposal apparently was an identification of toxic residues in fish in the Mekong Delta of South Vietnam after application of very high rates of this herbicide in military operations" (Rodgers 1974). Rodgers dismissed such concerns because the amounts sprayed in the United States were much lower, and actually risk was much lower in the United States. In 1975, the Council for Agricultural Science and Technology, composed of chemical industry organizations, devoted an entire report to the question of herbicide effects in Vietnam "and their relation to herbicide use in the United States," concluding that the "atypical" use of herbicides in Vietnam showed no connection with peacetime use. This "atypical" use of herbicides significantly destroyed the forests, and worse, damaged the ecosystem (Council for Agricultural Science and Technology, 1975). The war also intensified the status of the Delta's mangrove swamps as rebellious and resistant space, this time in regards to efforts to rehabilitate the environment.

Concerns over the health of the coastal mangrove forests in the Mekong Delta have continued to this day. Worldwide, mangrove deforestation rates are twice that of the rain forests, and mangrove forests in Southeast Asia show the highest rate of deforestation. Most scientists identify the Second Indochina War as one of the major factors in the escalating devastation of the southern mangrove forests, and current scientific studies of the mangrove forests rely on the scientific reports produced by the AAAS and NAS (Hong 1993; Stellman et al. 2003; Mayaux et al. 2005). Contextualizing these reports within the history of the region suggests a major consideration that eluded American scientists' conclusions about the ecocide happening in Vietnam and continues to be obscured in current-day studies of the area: The resistant nature of human beings embodied within a rebellious landscape, a stubborn defiance that risks losing the very space that formed and nurtures them. Contemporary scientists note the problems of aquatic farming, but fail to realize the ways the mangrove swamps' role as a base for armed resistance and liberated land have contributed to the problem. Former NLF army members settled in the Mekong Delta after the end of war with the United States. Many veterans continued to farm shrimp and refuse to stop even today, and scientists and conservationists have excused such behavior as either economic (food, wood, land), or try to use forestry models used elsewhere in resolving the problem (Biggs 2005). Ignoring the cultural history of resistance and pioneer ethos of the region, as both the AAAS and NAS studies did, prevents current scientists from identifying effective strategies of intervention.

Given the difficulties of assessing the human and ecological effects of herbicide sprayings in Vietnam, the one clear theme that emerges is the uncertainty of the scientific studies done in Vietnam during the early 1970s, an uncertainty that has continued until today. Wartime conditions made it extremely difficult to properly gather data and assess conditions in South Vietnam. Plant botanists clashed with chemists as scientists struggled with how to conduct a proper scientific investigation of ecological conditions in South Vietnam. Yet despite the confusion, controversy, and questions about scientific validity, American scientists examining South Vietnam agreed that some measure of harm had been done to the countryside, even if it lay beyond scientific efforts to evaluate harm done to the health of individuals. Concerned scientists pressured the American Association for the Advancement of Science to involve itself in the massive use of chemical herbicides in Vietnam. The Herbicide Assessment Commission produced results that supported concerns about the environmental harm being done by herbicide-spraying operations, even if those results remained unclear in forecasting the long-term consequences of herbicide use. While the National Academy of Science may have produced a more conservative assessment of harm, its scientists too agreed that the forests of South Vietnam had been harmed. Both reports identified the fragility of mangrove forests in the Mekong Delta to herbicide spraying, and both failed to fully assess the human damage that resulted from the ecological harm done.

Arthur Galston's imagery of ecocide, the willful destruction of an ecosystem akin to the willful destruction of a people, appeared to be a dramatic but not incorrect characterization of what the American military herbicide-spraying operations had wrought in the South Vietnamese countryside. Unlike the military atrocity of MyLai, which eventually attracted the attention and condemnation of the American people, the ecocide of South Vietnam remained mostly unresolved by American scientists. Some historians have argued that the advocacy of American scientists was in part the reason for the Nixon administration's decision to quit aerial-spraying operations in 1970 (Zierler 2008). Cultural considerations relating both to the swamp as refuge and American scientific clashes appear to be clear, however. Human action continues to harm the forests, and until science better incorporates the cultural factors along with the ecological, the mangrove forests will continue to disappear.

References

Accuracy in Media (1974) New York Times readers deceived on defoliation report. AIM Rep 3:1–3

Behrens R (1968) WSSA: progress and challenges. Weed Sci 16:411–413

Biggs D (2003) Problematic progress: reading environmental and social change in the Mekong Delta. J Southeast Asian Stud 34:77–96

Biggs D (2005) Managing a rebel landscape: conservation, pioneers, and the revolutionary past in the U Minh Forest, Vietnam. Environ Hist 10(3):448–476, Retrieved from http://www.jstor.org

Biggs D (2006) Reclamation nations: The U.S. Bureau of Reclamation's role in building water management and nation building in the Mekong valley, 1945–1975. Comp Technol Transf Soc 4:225–246

Boffey P (1971) Herbicides in Vietnam: AAAS study finds widespread devastation. Science 171:43

Buckingham W Jr (1982) Operation Ranch Hand: The Air Force and herbicides in Southeast Asia 1961–1971. Office of Air Force History, Washington, DC

Council for Agricultural Science and Technology (1975) Effects of herbicides in Vietnam and their relation to herbicide use in the United States. CAST Report, No. 46, Ames, IA

Finney J (1974) Vietnam defoliants study sees effects of 100 years. 22 Feb 1974, Folder 16, Box 02, Douglas Pike Collection: Unit 03 – Technology, The Virtual Vietnam Archive, Texas Tech University. Retrieved from http://www.vietnam.ttu.edu/virtualarchive/

Galston A (1971) Some implications of the widespread use of herbicides. Bioscience 21:891–892

Galston A (1972) Book review: aroused consciences. Science 176:154–156

Handler P (1974) The effects of herbicides in South Vietnam. Folder 15, Box 02, Douglas Pike Collection: Unit 03 – Technology, The Virtual Vietnam Archive, Texas Tech University. Retrieved from http://www.vietnam.ttu.edu/virtualarchive/

Hong PN, San HT (1993) Mangroves of Vietnam. The International Union for the Conservation of Nature and Natural Resources, Bangkok

Käkönen M (2008) Mekong Delta at the crossroads: more control or adaptation? Ambio 37:205–212

Mayaux P, Holmgren P, Achard F, Eva H, Stibig H, Branthomme A (2005) Tropical forest cover change in the 1990s and options for future monitoring. Philos Trans Biol Sci 360:373–384

Montana Weed Control Association (1973) The Orange controversy and its implications to weed control programs. In: State Weed Conference, 15–16 Nov 1973, Miles City, Montana, Folder 3814, Box 133, Alvin L. Young Collection. National Agricultural Library, Beltsville, MD

Moore K (2008) Disrupting science: social movements, American scientists, and the politics of the military, 1945–1975. Princeton University Press, Princeton

Morgan D (1974) Viet defoliation damage held serious. 23 Feb 1974, Folder 16, Box 02, Douglas Pike Collection: Unit 03 – Technology, The Virtual Vietnam Archive, Texas Tech University. Retrieved from http://www.vietnam.ttu.edu/virtualarchive/

Nelson B (1969) Herbicides in Vietnam: AAAS board seeks field study. Science 163:58

Phuong-Lan, Bui Thi (2003) When the forest became the enemy and the legacy American herbicidal warfare in Vietnam. Ph.D. dissertation, Harvard University, Cambridge

Randal J (1970) Vietnam Forest. 29 Dec 1970, Folder 13, Box 02, Douglas Pike Collection: Unit 03 – Technology, The Virtual Vietnam Archive, Texas Tech University. Retrieved from http://www.vietnam.ttu.edu/virtualarchive/

Rodgers EG (1974) Weed science today. Weed Sci 22:464–468

Stellman JM, Stellman S, Weber T, Tomasallo C, Stellman A, Christian R (2003) A geographic information system for characterizing exposure to Agent Orange and other herbicides in Vietnam. Environ Health Perspect 111:321–328

Sullivan W (1970) Sprays in Vietnam: mangrove area. Folder 13, Box 02, Douglas Pike Collection: Unit 03 – Technology, The Virtual Vietnam Archive, Texas Tech University. Retrieved from http://www.vietnam.ttu.edu/virtualarchive/

The Blight That Failed (01 January 1971) Folder 16, Box 02, Douglas Pike Collection: Unit 03 – Technology, The Virtual Vietnam Archive, Texas Tech University, Lubbock. Retrieved from http://www.vietnam.ttu.edu/virtualarchive/

Thorpe C (2002) Disciplining experts: scientific authority and liberal democracy in the Oppenheimer case. Soc Stud Sci 4:525–562

Tran T, Barzen JA, Le CK, Moore D (2004) Wartime herbicides in the Mekong Delta and their implications for post-war wetland conservation. In: Furukawa Hisao, Nishibuchi Mituaki, Kono Yasuyuki, Kaida Yoshihiro (eds) Ecological destruction, health, and development: advancing Asian paradigms. Trans Pacific Press/Kyoto University Press, Victoria/Kyoto, pp 199–211

Wang J (1999) American scientists in an age of anxiety: scientists, anticommunism, and the cold war. University of North Carolina Press, Chapel Hill

Wang J (2002) Scientists and the problem of the public in cold war America, 1945–1960. Osiris 17:323–347

Willarejo D (1975) Only we can prevent forests. Vietnam Courier, February, no. 33: 19–20. Effects on forest. Folder 04, Box 03, Douglas Pike Collection: Unit 03 – Technology, The Vietnam Archive, Texas Tech University. Retrieved from http://www.vietnam.ttu.edu/virtualarchive/

Wolfe D (1967) AAAS council meeting, 1966. Science 155:855–856

Young AL (2009) The history, use, disposition and environmental fate of Agent Orange. Springer, New York

Zierler, D (2008) Inventing ecocide: Agent Orange, antiwar protest, and environmental destruction in Vietnam. Ph.D. dissertation, Temple University, Philadelphia

Chapter 6
The Politics and Culture of Climate Change: US Actors and Global Implications

Charles Waugh

Abstract Despite the scientific consensus on global warming, many people in the USA,—both ordinary citizens and elected leaders alike—remain skeptical of the need to act, and in fact remain skeptical of the idea that humans are contributing to global warming at all. Thus, environmental justice arguments based on United States carbon emissions and the disproportionate impact of rising temperatures and rising sea levels on tropical developing nations such as Vietnam frequently fall on deaf ears. This chapter explores the political and cultural construction of this deafness, seeking a better understanding of how and why so many Americans refuse to act to address global warming. The two main sources of this deafness that this chapter address are (1) the politics of carbon-intensive energy producers such as the coal and oil industries, demonstrating the ways in which those industries have distorted the debate over global warming, have found eager allies in political candidates willing to accept large campaign contributions, and—with the help of other industries such as automobile manufacturing and home construction—have encouraged the second main source of denial: (2) a culture of aggrandized individualism that places greater value on personal identity construction than on the national and global common good. Once these sources are established, the chapter recommends strategies for using narrative to overcome cultural and political resistance to climate change mitigation that may be effective not only in the United States, but in Vietnam as well.

Keywords Climate change • Culture • Individualism • Narrative

Since the election of Barack Obama in the United States, much has been made by the American news media of the new green economy, with the president himself calling for an investment in renewable energy and climate change mitigation strategies that he believes will have the twofold effect of resuscitating the American

---

C. Waugh (✉)
Department of English, Utah State University, Logan, Utah, 84322-3200, USA
e-mail: charles.waugh@usu.edu

M.A. Stewart and P.A. Coclanis (eds.), *Environmental Change and Agricultural Sustainability in the Mekong Delta*, Advances in Global Change Research 45,
DOI 10.1007/978-94-007-0934-8_6, © Springer Science+Business Media B.V. 2011

economy and preparing the nation for a hotter world. But so far, the overall impact of many of these strategies has fallen short of their goals. For example, one plan, the Consumer Assistance to Recycle and Save Act of 2009 (better known by its nickname "Cash for Clunkers"), sought to reduce carbon emissions and to revitalize automobile sales by encouraging owners of inefficient and highly polluting vehicles to trade them in for newer, more efficient, and less polluting models. The government subsidized the program by supplying the money to buy the old vehicles in order to ensure they would be taken off the road and destroyed. Americans traded in nearly 680,000 cars and trucks for more efficient replacements, saving an annual average of $32,947,886 in fuel, and preventing an estimated 379,000 metric tons of CO_2 from being released annually (US Department of Transportation 2009). While any reduction of CO_2 emissions is a good thing, at just 0.15% of the United States' annual carbon emissions, such results fail to be inspiring, and in fact, raise the question of why more people did not take advantage of the program. After all, 680,000 may initially seem like a lot of cars, but it represents just 0.27% of the total number of registered vehicles in the United States, or the participation of just 0.22% of the US population.

Setting aside the economic reasons for not participating, from figures such as the United States' total annual carbon emissions as reported by the United Nations' Millennium Development Goals Indicators, we know that the United States is the second largest CO_2 emitter in the world, and for a long time was the number one emitter. As individuals, Americans live a more carbon-intensive lifestyle than all but a few nationalities (UN Statistics 2010). Why did not more of them take advantage of a program that would not only benefit themselves, but also help reduce this enormous carbon footprint?

The full response to that question has many complicated parts, and this chapter addresses several of them, but the short answer is that many people in the United States—both ordinary citizens and elected leaders alike—remain skeptical of core climate change assumptions. Some look at their local weather and refuse to believe the Earth is warming in any significantly different way than in other decades-long trends, and some, when they finally do confront the overwhelming data that the Earth is indeed growing warmer, quickly attribute that warming to the Earth's natural long-term cycles of warming and cooling, and in effect deny that human activity has anything to do with it. In both scenarios, these skeptics implicitly claim that nothing humans can do will make any difference. But if humans want to continue to enjoy an earth that still resembles the earth that we all know now, and if we wish to avert the catastrophic loss of life, society and capital that unchecked greenhouse gas emissions will cause, we must find a way to break through enough skepticism to make a difference. This chapter first analyzes American climate change skepticism to find its political and cultural origins, and then suggest ways to reframe or repurpose those origins to inspire people to act.

It might be easier to understand US climate change skepticism if seen through an example. In 2010 in the state where I live, Utah, the conservative state legislature responded to the US Environmental Protection Agency's "carbon reduction policies" by passing House Joint Resolution 12, which "urges the United States

Environmental Protection Agency to immediately halt its carbon dioxide reduction policies and programs and withdraw its 'Endangerment Finding' and related regulations until a full and independent investigation of the climate data conspiracy and global warming science can be substantiated" (Utah 2010). It follows with a litany of unsubstantiated reasons why climate change science is flawed or otherwise cannot be trusted, how climate change action will cost governments and tax payers billions of dollars that will simply line the pockets of climate change researchers, prevent the use of millions of acres of agricultural land, starve the people now eating food produced on that land, and "ultimately lock billions of human beings into long-term poverty" (Utah 2010). The paranoid hyperbole the bill expresses arises in part as a response to recent revelations that climate scientists altered findings to create a sense of urgency in the public, but it is also representative of the way many Utahns (though not all, obviously!) and many Americans feel about climate change. Letters to the editor published in my local newspaper often "debate" climate change, with typical entries often sounding like this one: "[Climate change science] is purely unadulterated 'horse puckey.' These clowns are in it for the money at yours and my tax money expense [sic]....With the recent revealing reports of false documents of global warming, Al Gore and his pals should be 'cut off' from our taxpayers' funding....Climate change is a scam at our expense. Believe it. You far-out loons are crazy" (Larsen 2010).

Neither the writer of this letter nor the sponsor of House Joint Resolution 12 is a scientist, nor are they apparently consumers of information from leading scholarly or even popular science journals or even mainstream media. So where do they get such strong feelings about climate change, and why would they believe climate change mitigation strategies are just some farfetched and elaborate conspiracy to steal their money?

One part of the answer is political: As early as 1998, the biggest energy producers in the United States, corporations such as Exxon, Shell, and BP, along with organizations that they support such as the American Petroleum Institute and the Cato Institute, began a systematic campaign designed precisely to create public skepticism on global warming. Taking the offensive on what they believed would be a downturn in demand for their products, or a tax on the carbon their products released, they began to funnel huge sums of money to organizations and scientists who would provide research to support the contentions that global warming did not exist and that human activity could not be definitively linked to climate change. Exxon alone ultimately paid more than $16 million between 1998 and 2004 to 40 different groups, who in turn paid researchers to produce reports to support their claims. Perhaps the most cynical part is that the oil corporations never had any intention of ever proving the validity of their paid scientists' work. In a memo leaked to the New York Times, the American Petroleum Institute declared that their campaign would be successful just by creating the *appearance* of doubt about climate change science. They knew as long as the media presented the issue of climate change as a *debate*, at least some of the public would believe the science was inconclusive and there would be no real motivation to limit fossil fuel exploitation or use. Since then, Exxon has claimed to "soften its stance" on global warming, and yet they have continued to support climate denial institutes with millions of dollars every year since making that claim (Union of Concerned Scientists 2007).[1]

The energy corporations have also made large donations to political campaigns, giving candidates sympathetic to energy exploration and development vast sums to spend during elections. Since 1990, oil and gas companies have contributed more than $238.7 million to federal level campaigns, with 75% going to Republicans (Center for Responsive Politics 2010a, b). Candidates claim campaign donations do not necessarily mean they do whatever the donors tell them to. But in practice, it seems this is often the case.[2] For example, in January 2010, the Alton Coal Development company made a $10,000 contribution to the reelection campaign of the energy sector-friendly governor of Utah, Gary Herbert. That same day, at the governor's request, Alton's application to open a mine in an environmentally sensitive area was fast-tracked for approval. When pressed, the governor claimed the two incidents were unrelated, and as long as no record exists of his asking for the money or the coal company asking for his help with the approval process, corruption like this goes unpunished (Foy 2010).

In the case of the George W. Bush administration, scores of energy sector personnel moved into the government itself. Both President Bush and Vice President Cheney had been energy corporation executives, and as they filled positions within their administration, they relied heavily on former industry lobbyists, lawyers, managers, and board members. For example, the chief of staff for the White House Council on Environmental Quality from 2001 to 2005 was Philip Cooney, who had spent the previous decade as a lawyer for the American Petroleum Institute. As CEQ Chief of Staff, Cooney was able to act as a filter for all environmental information distributed by the federal government, at times changing or deleting parts of reports that contained data or predictions contrary to industry opinion. When US government scientists in the Climate Change Science Program submitted their 2002 Strategic Plan, Cooney made 650 changes, removing estimations of threats such as reductions in glacial ice and available melt water, and adding qualifying words to projected outcomes and impacts to make them seem less likely. Another tactic used by the Bush administration to thwart the dissemination of science contrary to their energy-intensive political goals included requiring scientists nominated for advisory boards to pass interviews that tested them not for their expertise but for their political allegiance (Shulman 2006, pp. 117–123).

---

[1] The climate change denial actions of ExxonMobil and the institutes it supports are well documented. For example, see Cushman (1998) and Harkinson (2009).

[2] A prominent example at the federal level comes from James Inhofe, Republican Senator from Oklahoma and the most vocal skeptic of global warming in the Senate, who received $432,950 between 2005 and 2010 in campaign contributions from the oil and gas industry, $206,654 from electric utility political action committees, and $176,983 from lobbyists. It is hard not to imagine that he is not saying what the energy industries want him to say. Figures from Open Secrets.org, the Web site of the Center for Responsive Politics. Retrieved from http://www.opensecrets.org/politicians/summary.php?cid=N00005582&cycle=Career

And while the Bush administration is an extreme case of oil industry influence over US policy and opinions, that influence did not appear over night. Rather, as Naomi Oreskes and Erik Conway have convincingly shown (2010), it is the culmination of a long, concerted effort by conservatives to shift policy discussions away from scientific facts and the common, global good toward short-term economic interests. Following the lead of economists such as Thomas Schelling and William Nordhaus, whose reports on climate change in the early 1980s discounted the science and leading scientists' predictions of environmental disaster in favor of economic uncertainty and a "wait-and-see" attitude, groups such as the Marshall Institute have been churning out "information" about global warming that has never withstood scientific scrutiny and whose purpose is simply to cloud the issues and to encourage more delay and inaction (Oreskes and Conway 2010, pp. 177–180). One such Marshall Institute report, *Global Warming: What Does the Science Tell Us?* (Jastrow et al. 1989), employed tactics such as presenting only the data that supported their conclusions and misrepresenting peer-reviewed research to make the report look like a credible refutation of climate change science. Because the report came from a policy think tank, it had never been peer reviewed, but because it appeared to use real scientific data (albeit in an incorrect way), many policy makers were duped. Similarly, releasing such reports to the public through conservative media outlets such as the *Wall Street Journal* has allowed climate change deniers to distribute their ideas widely, but when leading scientists have written letters to the editor of the *Journal* to refute the science, the newspaper has either refused to print the rebuttals or edited them heavily to reduce their impact. Journals such as *Science* and *Nature* have eventually published full rebuttals, but to much smaller readerships and with the damage already done (Oreskes and Conway 2010, pp. 209–213).

What the Marshall Institute and the energy industries they serve have always realized is that what they write are policy suggestions, and that their audience is composed of policy makers. They are not scientists and they are not writing scientific articles that aim to have their science compete in the realm of science with other scientific claims. Instead, they have exploited the less rigorous standards of mainstream publishing to make their claims and have provided policy arguments and supporting materials to like-minded policy makers that say that the science is not the main thing to consider when it comes to climate change. And their story appeals to so many people in part because it preserves the status quo—no one has to do anything different in their version of the future—but in even larger measure because they often package these policy suggestions to take advantage of powerful elements in American culture.

Given such an overwhelming effort to mislead the public, it is no wonder that ordinary citizens such as my letter writing neighbor are convinced that global warming is not what the scientists say it is. In fact, the effort to deceive Americans about climate change has been so monumental; it would be easy simply to blame the politicians and the corporations for the denial of climate science and the perpetuation of a carbon-intensive way of life. But the situation is more complicated than that; in fact, it goes to the very core of American cultural identity. After all, what those corrupt corporations and politicians are really defending is the uninhibited freedom for individuals to emit as much carbon as they can personally afford, to buy as many things and to use as much energy as they can pay for. (It is no coincidence that several of the climate change denial groups funded by Exxon have names like the Frontiers of Freedom Institute and the Free Enterprise Institute). And as selfish as that sounds, it is a privilege that resonates with vast numbers of Americans. Of course, the corporations are still intimately involved, since they are selling what the American people are buying, but a deep complicity runs between the average American citizen and the corporations that sell them the things they want.

A good way to see this complicity comes from the automobile industry, which by nature involves both a carbon-intensive production process and product. Since the establishment of fuel efficiency standards in the United States in 1975, the American automobile industry has argued consistently for the more lenient standards that have allowed them to cater to base consumer demands: more passenger and cargo space, more powerful engines, and more luxurious interiors. At the same time, these more lenient standards have allowed them to focus less on efficiency, and more on the idea that a car is not just a car, but an extension of the owner's personality. A recent BMW commercial aired in the United States demonstrates this idea quite clearly. The commercial depicts beautiful cars being driven by people with big smiles on their faces, and the narrator says, "We realized a long time ago, that what you make people feel is just as important as what you make, and at BMW, we make joy" (BMW 2010). In the United States, big trucks are sold on the idea they make the owner tougher, more rugged. Minivans are sold by suggesting that parents are better parents if they buy the van that keeps their children safer, or now, better entertained. The car's efficiency is usually treated as an afterthought, even in ads for the most efficient cars. For example, commercials for the popular brand of hybrid electric and gas car, the Toyota Prius, mainly pitch a certain kind of way of being in the world, a personality that makes the world a happier place to live in.

In their "Harmony" TV commercial, Prius promises the buyer "more power and more space," two of the things many Americans want in their car, and that "the world gets fewer smog forming emissions." But the pitch for the environment is understated in comparison to the ad's music and visual effects. The song in the background is Loretta Lynn's, "Let Your Love Flow," and as we watch the Prius drive along a highway, the landscape is transformed from a bland white to a vibrant green and blue made up of smiling children in blossoming plant costumes. Presumably because in the United States so many people are still in denial about global warming, the ad does not even mention the reduction in carbon emissions, suggesting instead the reduction in smog (Toyota 2010). Apparently, the important thing to take away from the ad is, as BMW put it, how it makes you feel. But this sort of identity construction at least is relatively benign in comparison to what happens in Ford's "Built Ford Tough" commercials. In hundreds of different ads, various aspects of their trucks' designs are touted as being tougher than their competition, with the implication always being, as one ad asks the viewer directly at the end, "Are *you* tough enough?" (Ford 2007). In other words, the overdesign of the truck, adding weight and losing efficiency, is justified by a consumer who also wants to be seen as stronger and more capable of physical work than anyone else.

One way to explain this emphasis on individual identity construction, as well as the toughness message from Ford, comes from the United States' frontier history, whose mythology has been perpetuated by popular culture since the late eighteenth century. In brief, these stories feature individuals who go out into the wilderness with just the right tools and supplies and turn that wilderness into civilization. These are the stories of early explorers like Daniel Boone and Davy Crockett, but also of western cowboys like Wyatt Earp and Wild Bill Hickok who are portrayed as individuals who struggled heroically to tame the frontier. Of course the legends around these historical figures have exaggerated their accomplishments, and like all myths, some things are emphasized over others. It does not matter that Earp had brothers and friends, or that Crockett had a wife and children and neighbors helping him, the myth reinforces that these men as self-reliant individuals played an enormous role in making the United States what it is today.

American studies scholars have rigorously debated the origins and mythic quality of this rugged American individualism, and not all of that scholarship can be revisited here, but one common explanation suggests that, despite the truly *inter*dependent nature of early American life, "throughout the Revolutionary period and beyond, most Americans made their livings in agriculturally-based hamlets, or in free-standing farmer's homesteads of one or a few families. These material circumstances seem likely to have encouraged a clear sense of *economic self-reliance* or *independence* in much of the population, which, in turn, may have eventually promoted a generally more individualistic outlook among many settlers" (Grabb et al. 1999, p. 527). This emphasis on economic self-reliance runs through a great deal of American thought, even through the writing of the United States' greatest champions of the environment such as Ralph Waldo Emerson, Henry David Thoreau, and Wendell Berry, though they each in their own way also accompanied their celebration of independence with warnings not to become "the tools of their tools" (Thoreau 1854, p.132). A second explanation for American individualism comes from what political scientist Barry Shain calls a conflation of "local communal hostility toward the provincial or the imperial (later national) center…with a common 20th century hostility toward all authority" (Shain, 1994, p. 86).[3] Combining these two explanations gives us a competent portrait of the consumer that Ford truck commercials are trying to reach: Someone who earns his living by himself, with his hands, and who does not like being told what to do by anyone, and if not that person, someone who fantasizes about being him. In short, my letter writing neighbor.

[3] Such feelings are not exclusively American; many cultures and peoples feel similarly, even in Vietnam, where the saying goes, *Phép vua thua lệ làng* (the emperor's law yields to village custom). But unlike in Vietnam, where there is also a saying, *Biết thì thưa thốt, không biết thì dựa cột mà nghe* (if you know, speak up, if not lean against a pillar and listen), America's infatuation with individualism has led many to believe that all opinions have equal value.

The values this myth and these advertisements perpetuate are precisely the ones he expressed in his letter to my local paper quoted above. The value of rugged economic self-reliance and distrust of outside authority combine to express the fear that government agencies, scientists, and representatives are using climate change as a way to tell him what to do and to take his money. The nexus of corporations, politicians, and conservative media make the situation worse by presenting irrational opinions about individual rights and misleading pseudoscientific information as rational counterarguments to the scientific certainty of global warming, but to an extent, they are successful because they play on a distrust of authority that has been enshrined in a culture and history that many Americans believe began with the Declaration of Independence. At the same time, Americans are inundated with corporate messages to indulge their sense of individual importance, to enjoy what they think they have earned, and ignore requests made for the common, global good.

Because the only information some Americans get is what they hear on politically motivated radio and television programs, these citizens never fully confront the legitimacy of climate change science. But even if these Americans did engage with clearly articulated, sound scientific proof of climate change, their feelings of skepticism and distrust, and ultimately the fear of losing their privileged access to an overabundance of material goods would most likely prevent them from being convinced to do anything. Climate scholar Mike Hulme has applied planning theorist Horst Rittel's term "wicked problem" to these entrenched rejections of climate change science because they so intractably defy reasonable problem-solving strategies. But Hulme refuses to give up on the idea that it will take everyone in the world, especially climate change skeptics, to change our world for the better. Coincidentally, his answer to this "wicked problem" employs the exact same methods of myth-making, identity construction, and even a distrust of authoritative solutions that helped make the problem so intractable to begin with.

Hulme's book, *Why We Disagree About Climate Change* (2009), explores the many ways that people around the world perceive climate change, explaining how different values, beliefs, and forms of government combine to make "solving" climate change as a problem impossible. Because "solving" climate change involves "uncertainty; inconsistent and ill-defined needs, preferences and values; unclear understanding of the means, consequences or cumulative impacts of collective actions; and fluid participation in which multiple, partisan participants vary in the amount of resources they invest" (Carley and Christie 2001, p. 156), no authoritative, top-down solution will ever satisfy everyone, even if one could be devised. Instead, Hulme argues, "we need to approach climate change as an imaginative idea, an idea that we develop and employ to fulfill a variety of tasks for us" (p. 329).

Treating climate change as a source of inspiration frees us to go more directly to the forms of identity construction that will allow us to find common ground and take action that will help us all. As the Alliance of Religions and Conservation suggests, "Without…[narrative, myth, and metaphor], policies will have very few real roots…. Without narrative, few people are ever moved to change or adapt" (Alliance 2007), and Hulme agrees: "In a world where the globalizing powers of capital, trade and consumption separate us from the local stories that give meaning not just to climate, but also frequently to our lives, this [strategy] also offers a way of reconnecting ourselves with climate through the telling of stories" (p. 356). In other words, to mobilize the world to act on climate change, we have to "inspire," not to "convince." We have to construct stories about ourselves and climate that allow us to better examine who we are, who we want to be, and what we want from life.

Such a strategy seems especially well suited to American climate change skepticism. It allows us to build from our mythic past when we struggled to make a better life out of an unpredictable wilderness. It allows us to turn that distrust of authoritative, top-down decision-making into individualistic, everyday solutions that still allow for a high quality of life. And it allows us to construct narratives that connect the values of the past with the choices we must make in the present. As Hulme suggests, "We can use the idea of global climate change to tell ourselves new stories about … the consequences of our collective behaviours. We can use the idea of climate change to move beyond the separated categories of the physical and the cultural, beyond the framing of climate change that uses the language of problem and solution" (p. 357). Thus, we must express the idea of climate change in stories that can inspire the world, one individual at a time, to create a new way of understanding the choices we make and how they affect people who live far away from us.

Based on his analysis of the arguments that have obstructed climate change mitigation strategies in the past, Hulme proposes a taxonomy of four myths that represent four human impulses or emotions as a way of understanding how stories we tell about climate change might motivate us: (1) Eden, which arises out of our sense of nostalgia, (2) Apocalypse, which arises from our fears for the future, (3) Babel, which expresses our pride for our sense of mastery and control over the environment and other human beings, and (4) Jubilee, which engages with our sense of justice, of doing the right thing by others affected more dramatically by climate change (p. 358). But in order to know what will motivate a particular group best—as scholars of rhetoric and narrative from Aristotle to Kenneth Burke, Wayne Booth and others have shown—we first have to know something about the audience. A story can only do the work Hulme suggests when you know which story to tell.

For example, the climate story an individual American might tell another individual American, over the phone, between father and son or sisters or close friends, will be different from the story told by director Roland Emmerich to the American mainstream public in the climate change thriller *The Day After Tomorrow*. The story I would tell to my Vietnamese colleagues while working in Vietnam might not be the same story I would tell my neighbor when chatting about our gardens and the weather. And even within the larger community of climate change skeptics, the story I might tell to inspire the writer of the letter to my local newspaper would not necessarily be the story I would tell to Jeff Kueter, president of the Marshall Institute.

So what would it take to inspire a skeptic like my letter writing neighbor? As Shain and Grabb have suggested above, and as the American Tea Party movement has confirmed, the idea of America that comes from its revolutionary period is very much alive in the contemporary mainstream United States. The sense of individualism, of disliking being told what to do by some far off distant ruler, of this irascibility somehow defining a quintessential American spirit are all common features not only of Ford truck commercials but also of contemporary conservative discourse. An extreme variant, the Appleseed movement, goes one step further by connecting those same characteristics to the type of marksmanship conjured up by recollections of minute men militia. When pressed to explain why every American should own an AR-15 rifle and know how to shoot it, Appleseed founder Jack Dailey expressed the sentiment common to all these groups: "Because they [unspecified antecedent, suggesting anyone] want to tell us what to do. And we don't want them to tell us what to do" (Schwartz 2010, p. 41). Here, Hulme's use of four myths can really help make clear what is going on. Fear for the future, combined with pride for the mastery over nature and other human beings and a sense of justice, have all been wrapped together with nostalgia for a time when problems could be solved by shooting at them. But it is the nostalgia that is the most important part. When a Tea Partier reaches for that tricorne hat, or an Appleseeder reaches for a gun, he or she is reaching for the past.

T. Boone Pickens, the oil tycoon turned wind farm proponent, does the same in his attempts to convince conservatives of the need to turn away from imported oil. Consider the rhetoric from his Web site for the Pickens Plan: "Lend your voice to the call for American energy independence….We are organizing the New Energy Army in every Congressional District" (Pickens 2010). This is not about reducing our carbon footprint—it is a revolutionary call to arms!

But there is more from the American Revolution than just guns and independence that will resonate with Americans today. Stories of imperial imposition were commonplace in late eighteenth century colonial America, stories that told of unfair taxes yes, but also local stories of being taken advantage of by someone far away, stories of the inherent injustice in a colonial system that featured a distant king telling people what to do with the land they lived on. These stories involved real people who were forced to quarter soldiers in their homes, real people whose access to their local forests and its resources was limited by the Broad Arrow policy, real people whose potential was limited by the Royal Proclamation of 1763 barring them from westward migration.[4] The Revolution was the means through which they expressed their independent spirits, refused to be hobbled any longer, sacrificed their time and revenue, and rose to their full potential.

Stories that equate corporate greed, excess, and callousness with imperial rule can be marshaled in the same way, and have been already, perhaps most notably in the case of BP's former American CEO Tony Hayward's yachting vacation during the massive 2010 oil spill in the Gulf of Mexico. The same could be said for the public outcry in 2008 when gasoline prices went through the roof while the oil companies posted record profits and their CEOs raked in outrageous salaries and benefits.

---

[4] For examples of these stories, see Roberts (2010) and Calloway (2006).

But even these have not been enough to inspire my letter writing neighbors to give up their big trucks, ATVs, and snowmobiles. What else can be done? According to Hulme, it is "the local stories that give meaning not just to climate, but also frequently to our lives" (p. 358) that need to be told, so perhaps the stories my neighbors need to hear are the ones from local scientists predicting what climate change will mean for folks right here in my little mountain valley. Will climate change mean less snowfall? Will less snowfall mean fewer opportunities for skiing and snowmobiling? Will less meltwater mean the end of agriculture in our valley? Will droughts leave our forests more susceptible to mountain pine beetles and eventually wipe them out? Stories like these—intertwined with stories of family outings, camping and fishing and hunting stories, stories of local adventure—may finally touch the things my neighbors care about, help them imagine how their own lives will change in a warmer world, and make them nostalgic for the important role climate plays in the places and activities they hold dear. Motivating them to do something might then rely on appeals to the injustice of corporate indifference to climate, to their own frontier spirit, or to doing something locally before the federal government imposes a solution from outside. But the rub will always be to get their attention first with something that already interests them, and the more local the origin of the story, the science, the activities, and the impacts, so much the better.

Of course, this is not how new science tends to be disseminated. Instead, the bigger the news, the more important the national journal it comes out in. And most scientists probably do not want to spend all their time writing different versions of their work for different audiences. In these cases, perhaps collaboration with writers from various locations likely to be impacted by what the scientists have found would help get the job of adapting their work to local audiences done. Greater partnerships between scientists doing cutting-edge climate research and local writers who can publish in the types of newspapers, magazines, and Web sites that skeptics already read stand a much better chance of inspiring the presently uninspired than even the best-written article in top journals like *Science* and *Nature*.

But not everyone in the United States who still remains somewhat skeptical of climate change science need so much frontier, Daniel Boone-type inspiration. Many others, still less than literate in science, still unsure what to make of the conflicting claims, but with good hearts can be inspired by stories arousing their compassion and sense of justice. Moderately conservative "soccer moms" and dads around the country have already shown how issues that concern the health of their children can bring them to support environmental legislation and more climate friendly local and organic agriculture. Stories about their children's relationship with climate in the future could be effective, but so too could stories about the suffering of others. As writers such as Donald Brown have argued, ethical arguments about climate change need to be made more prominently in the United States (Brown 2010), and this group in particular seems likely to be a receptive audience to stories about the unequal distribution of negative effects of climate change and how children in other places around the world will have to bear a much greater burden than their own.

The clarity of this injustice will not be lost on such an audience. The biggest carbon emitters have well-developed economies and mainly occupy temperate zones with greater elevations that will allow them to better weather negative impacts. Meanwhile, developing countries like Vietnam that have small carbon footprints, tropical climates, and low-lying coastal geography are predicted by the World Bank to suffer the worst of the consequences (Dasgupta et al. 2007). Put succinctly, Vietnam and countries like it are being forced to pay the true cost of the United States' culture of consumption. With a compassionate audience, a story about the loss of ancestral lands, cultural customs, or displacement can resonate and make a difference, as long as it is centered around people and life experience, not just a vehicle for the science of sea level rise. This is the type of audience who can hear arguments about the injustice of the United States' carbon culture and how the overconsumption of goods and energy in the United States causes devastation in other parts of the world and be inspired to act. But because these stories have to some extent already begun to be told, perhaps the problem is less of content and audience but of distribution, and the solution may again lie with the need for help from more local writers, with more of these stories in local publications linking the rest of the world to even the smallest communities in the United States.

And while the focus of this chapter so far has been on the political and cultural origins of US climate change skepticism and the kinds of narratives that might inspire those skeptics into action, it is important to recognize that this process of adapting global science to local audiences must occur everywhere in the world, using narrative to link each and every locality to the political, cultural, and individual choices that impact and will be impacted by climate change, because the story of the injustice of climate change does not simply end with lifestyles in rich nations causing a disproportionate amount of climate affects in developing nations. In fact, it has a historical dimension and a contemporary quality of life dimension that makes the story even more complicated. The culture of consumption[5] and aggrandized individualism that has made the United States the leading carbon producer for decades has now become a global phenomenon, and the carbon emissions for developing nations have begun to rise, most dramatically in China, whose recent exponential growth in manufacturing and electricity production and consumption has brought it to the top of the list of yearly emissions, but also in Vietnam, whose carbon emissions increased by more than 300% between 1995 and 2005 (CAIT 2010). China's climate change ambassador Yu Qingtai has made strong arguments for considering the historical record of CO_2 emissions as the basis for future emissions reductions, arguments that attempt to reconcile the injustice of the West's carbon-intensive colonial period with the victims of that period's need to emit carbon to catch up with the West's infrastructure and standard of living (Vaughan 2009). But while this makes perfect sense historically and in terms of righting a wrong from the imperial past, to say that developing nations should emit as much

[5]For an historical analysis of how Americans came to associate the accumulation of material things with status and personal identity, see, for example, Cohen (2003).

carbon as have the developed nations does not make a lot of sense for the planet. What is needed is green development, development that gets the same social and economic results but without emitting the same amount of greenhouse gases.

Unfortunately, recent changes in Vietnam's Mekong Delta do not appear to be the kind of development that ultimately will be sustainable. The Intergovernmental Panel on Climate Change has predicted a 1–3 m rise in sea levels over this next century, which means that a significant portion of Vietnam's arable land will be submerged. With just a 1 m rise, 5% of Vietnam will be under water, and with the IPCC's worst-case scenario, a 5 m rise, 16% of Vietnam disappears beneath the waves. In all cases, the areas hit first and worst are the fertile river deltas where much of the nation's crops are grown, and where most of the people live. In the 5 m scenario, nearly 40% of the population—over 34 million people—will be displaced (Cruz et al. 2007). This is why the World Bank has predicted that Vietnam will be one of the five countries most impacted by climate change, with the world's worst impacts in lost GDP, population displacement, and urban areas destroyed (Dasgupta et al. 2007). And yet, despite these dire warnings, many Vietnamese are rushing headlong toward a far greater carbon-intensive way of life.

One of the ways Vietnam has traditionally managed to keep a low carbon profile is local, sustainable production of food. For millennia, crops were grown for local consumption, rice paddy was rotated with mung beans and other fallow crops for nitrogen fixing, and grazed by buffalo and ducks for fertilizer. But now most of the rice grown in Vietnam is done so for export with the aid of chemical fertilizers, and the rice preferred by the Vietnamese to eat comes from Thailand (Vietnam News 2010). Many farmers in the Mekong Delta have given up traditional farming altogether, and have turned to raising shrimp because it is perceived to be the only way to get ahead financially. But whether such entrepreneurs can actually get ahead is uncertain, and converting a farm from traditional, sustainable agriculture to shrimp aquaculture suddenly turns formerly carbon neutral land into one of the many cogs in a global greenhouse gas making machine.

Mr. Bich, a shrimp farmer in the Mekong Delta's Vinh Chau District, explained his experience making this transition. An international shrimp food producer gave him the loans to have his paddy excavated for shrimp ponds and to buy his initial stock of fry, the necessary equipment such as feeders and water agitators, and shrimp food. The food, which he had to sign a long-term contract for, comes from "trash fish" harvested from fisheries off the Pacific coast of South America. Since Mr. Bich converted all his land to shrimp production, the earnings he makes from the shrimp now have to provide enough to pay back the loan and continue to buy the contract's worth of shrimp food, plus all his other costs of living including the food he and his family eat, which may now come from as far away as Thailand, China, or even the United States. The land has been stripped of vegetation to increase the temperature of the ponds, which is good for producing shrimp, but not so good for keeping the house cool, preventing erosion, or growing fruits or vegetables for family consumption. Maybe if Mr. Bich were the only one who has done this, the impact would be minimal, but of course he is not alone. With the encouragement of both the provincial as well as the national government, many Vietnamese farmers have converted their farms completely to aquaculture—up to 98% of all farmers in some Mekong Delta districts (Ngo 2010).

When asked why he decided to switch to aquaculture, unsurprisingly Mr. Bich answered that he had heard a story about someone else in his district who had made a fortune that way. Here too narratives are playing an important role. But what other narratives, what "local stories that give meaning not just to climate, but also frequently to our lives" might be told in the Mekong Delta to raise awareness of impending climate changes and consequences?

Of course, the Vietnamese will know the best answers to this question. As an outside observer of Vietnamese culture, though, some possibilities come to mind. Ho Chi Minh's famous remark, "Không có gì quý hơn độc lập, tự do," (Nothing is more precious than independence and freedom) can be found on posters in public spaces throughout Vietnam. Stories that highlight sustainable agriculture could certainly draw on this sentiment, especially in contrast to the situation of having to follow the dictates of contracts with foreign companies to which one is perpetually in debt.

More generally, care for the environment can frequently be found in Vietnamese folk literature. Stories from folk tradition can do the same kind of work that myths of origination do to tap into deeply held and often unquestioned beliefs, feelings, and motivations. For example, most Vietnamese will already know the story of Chu Cuoi, the boy in the moon. In brief, the story involves a boy who finds a magic tree whose leaves heal all wounds and illnesses. He knows he must care for the tree diligently, and that as long as he does so, it will provide for him, but when he finally does allow it to be mistreated or neglected, the tree pulls up its roots and begins to rise into the sky. The boy quickly latches on, trying to weigh it down. But the tree continues upward, stranding the boy in the moon, where he sits to this day, playing his flute and pining for his terrestrial home. Different versions of the story change some of the details, but the essence of the story is always the same: the boy has an obligation to the tree and there are sad consequences when that obligation is not met.

To better highlight positive, proactive behavior, one might harness the story of Mr. Thirty, a tiger who takes care of an old woman after killing her son. In her grief, the woman roams the forest until she finds the tiger and then scolds him for preventing her son from caring for her in old age, shaming him deeply. The tiger begins to make up for her loss by delivering wild game to her door, and supports her for the rest of her life. Even after her death, the tiger continues to bring an offering of food to her grave on the thirtieth day of each lunar month. The story reminds us of the dangers the environment holds, as well as its ability to care for us when we act dutifully and with respect. It is also a story about obligations and the need to think about the larger ramifications of our actions.

But as before, every audience will be different, and so these stories may or may not resonate. For young, urban Vietnamese, perhaps the stories they most need to hear are ones that take what they may already know—climate change is real and will have huge negative impacts on Vietnam—with what they may not put together—that the endless pursuit and accumulation of material things such as the latest, most fashionable motorbike, car, or cell phone, when multiplied around the globe makes climate change's worst-case scenarios all the more likely. Vietnamese youth are also great candidates for telling stories of their own, for connecting with the rest of the world and making people in the industrial West understand how their lifestyles affect one another.

Telling stories about climate change that intertwine the deepest roots of our cultures with real people making real decisions in the world today can help us achieve what Hulme suggests:

> We should use climate change both as a magnifying glass and as a mirror. As a magnifier, climate change allows us to conduct examinations—both more forensic and more honest than we have been used to—of each of our human projects: whether they be of projects of personal well-being, self-determination, liberated or localized trade, poverty reduction, community building, demographic management, or social and psychological health. Climate change demands that we focus on the long-term implications of short-term choices, that we recognize the global reach of our actions, and that we are alert both to material realities and to cultural values. And as a mirror, climate change teaches us to attend more closely to what we really want to achieve for ourselves and for humanity. (pp. 362–363)

Stories can connect us with our past and inspire us to make good decisions for the future, all while providing a better understanding of the global dimensions of our actions in the present. Whether inspiring climate skeptics in the United States to act for the greater good, or helping farmers in Vietnam to imagine other ways to lift themselves out of poverty, scientists must become storytellers too, or enlist the help of local storytellers to adapt facts and predictions to plots and narrative, to give the science flesh and bone and heart.

Neither Hulme nor I mean to suggest that narrative alone can help us address global climate change. Only through science can we begin to know the full range of physical effects a warmer world will create. But as the interdisciplinary nature of this volume demonstrates, it will take approaching climate change from every conceivable intellectual direction to know what climate change will *mean*, and it will take stories that reflect each of our own cultures' values and aspirations to inspire the world, one individual at a time, to *act*.

References

Alliance of Religions and Conservation (2007 Dec 7) ARC/UNDP programme statement on climate change. Retrieved from http://www.arcworld.org/news.asp?pageID=207

BMW (2010) Joy commercial. Retrieved from http://www.youtube.com/watch?v=jl1rDlt4m60

Brown D (2010) Are ethical arguments for climate change action weaker than self-interest based arguments? why taking ethical arguments off the table is like a soccer team unilaterally taking the goalie out of the net. Penn State Rock Ethics Institute Climate Ethics. Retrieved from http://rockblogs.psu.edu/climate/2010/08/are-ethical-arguments-for-climate-change-action-weaker-than-self-interest-based-arguments-why-taking.html

Calloway CG (2006) The scratch of a pen: 1763 and the transformation of North America. Oxford UP, Oxford

Carley M, Christie I (2001) Managing sustainable development. Earthscan, London

Center for Responsive Politics (2010) James M Inhofe: campaign finance/money – summary – senator 2010. Retrieved from http://www.opensecrets.org/politicians/summary.php?cid=N00005582

Center for Responsive Politics (2010) Oil & Gas. Retrieved from http://www.opensecrets.org/industries/indus.php?ind=E01

Climate Analysis Indicators Tool (CAIT) (2010) Version 7.0. World Resources Institute, Washington, DC. Retrieved from http://cait.wri.org/

Cohen L (2003) Consumers' republic: the politics of mass consumption in postwar America. Vintage Books, New York

Cruz RV, Harasawa H, Lal M, Wu S, Anokhin Y, Punsalmaa B, Honda Y, Jafari M, Li C, Huu Ninh N (2007) Asia climate change 2007: impacts, adaptation and vulnerability. In: Parry ML, Canziani OF, Palutikof JP, van der Linden PJ, Hanson CE (eds) Contribution of Working Group II to the fourth assessment report of the Intergovernmental Panel on Climate Change, Cambridge University Press, Cambridge, pp 469–506. Retrieved from http://www.ipcc.ch/pdf/assessment-report/ar4/wg2/ar4-wg2-chapter10.pdf

Cushman J (1998 April 26) Industrial group plans to battle climate treaty. New York Times, p 1

Dasgupta S, Laplante B, Meisner C, Wheeler D, Jianping Yan (2007) The impact of sea level rise on developing countries: a comparative analysis. World Bank Policy Research Working Paper 4136, February 2007. Retrieved from http://www-wds.worldbank.org/external/default/WDSContentServer/IW3P/IB/2007/02/09/000016406_20070209161430/Rendered/PDF/wps4136.pdf

Ford (2007) Are you tough enough? Commercial. Retrieved from http://www.youtube.com/watch?v=iK-neU5OnsI

Foy P (2010, Feb 3) Utah Gov. Gary Herbert given $10 K by Alton coal development. Deseret News. Retrieved from http://www.deseretnews.com/article/print/700006916/Utah-Gov-Gary-Herbert-given-10K-by-Alton-Coal-Development.html

Grabb E, Baer D, Curtis J (1999) The origins of American individualism: reconsidering the historical evidence. Can J Sociol 24(4):511–533

Harkinson J (2009 Jul 1) Despite pledge, ExxonMobil still funding climate change deniers. *Mother Jones*. Retrieved from http://motherjones.com/blue-marble/2009/07/despite-pledge-exxonmobil-still-funding-climate-change-deniers

Hulme M (2009) Why we disagree about climate change: understanding controversy, inaction and opportunity. Cambridge University Press, Cambridge

Jastrow R, Nierenberg W, Seitz F (1989) Global warming: what does the science tell us? George C. Marshall Institute, Washington, D.C

Larsen R (2010 Feb 6) Gore & Co. Distorting facts. In Letters to the Editor. The Herald Journal. Retrieved from http://news.hjnews.com/opinion/letters_to_the_editor/article_1ea53ac8-ad80-5c99-a9db-fcd23e3f747e.html

Ngo Thi Phuong Lan (2010) From rice to shrimp: ecological change and human adaptation in the Mekong Delta. Paper given at the international conference on environmental change, agricultural sustainability, and economic development in the Mekong Delta of Vietnam, Can Tho University, Can Tho, 25–27 March 2010

Oreskes N, Conway EM (2010) Merchants of doubt: how a handful of scientists obscured the truth on issues from tobacco smoke to global warming. Bloomsbury Press, New York

Pickens T (2010) The Pickens Plan. Retrieved from www.pickensplan.com

Roberts SE (2010) Pines, profits, and popular politics: responses to the White Pine Acts in the colonial Connecticut River Valley. New Engl Quart 83(1):73–101

Schwartz M (2010) Firing line. The New York Times Magazine. 38–43, 1 Aug 2010

Shain B (1994) The myth of American individualism. Princeton University Press, Princeton

Shulman Seth (2006) Undermining science: suppression and distortion in the Bush administration. University of California Press, Berkeley

Thoreau HD (1854, 1981) Walden and other writings by Henry David Thoreau. Bantam, New York

Toyota (2010) Harmony commercial. Retrieved from http://www.youtube.com/watch?v=Tq4nrmnqY9o

Union of Concerned Scientists (2007 Jan 3) Scientists' report documents ExxonMobil's tobacco-like disinformation campaign on global warming science. Retrieved from http://www.ucsusa.org/news/press_release/ExxonMobil-GlobalWarming-tobacco.html?print=t

United Nations Statistics Division (2010) Millennium development goals indicators. Retrieved from http://mdgs.un.org/unsd/mdg/SeriesDetail.aspx?srid=749&crid=

US Department of Transportation. (2009) Consumer Assistance to Recycle and Save Act of 2009 Report to the House Committee on Energy and Commerce, the Senate Committee on Commerce, Science, and Transportation and the House and Senate Committees on Appropriations. Retrieved from http://www.cars.gov/files/official-information/CARS-Report-to-Congress.pdf

Utah State Legislature (2010 Feb 9) Climate change joint resolution. Retrieved from http://le.utah.gov/~2010/bills/hbillamd/hjr012.htm

Vaughan A (2009 Sept 2) A history of CO_2 emissions. Guardian.co.uk. Retrieved from http://www.guardian.co.uk/environment/datablog/2009/sep/02/co2-emissions-historical

Vietnam News Service (2010 Jan 7) Imported rice brands dominate markets in the Mekong Delta. Viet Nam News. Retrieved from http://vietnamnews.vnanet.vn/Economy/195638/Imported-rice-brands-dominate-markets-in-the-Mekong-Delta-.html

Part II
Social and Economic Dynamics

Chapter 7
Precarious Paddies: The Uncertain, Unstable, and Insecure Lives of Rice Farmers in the Mekong Delta

Peter A. Coclanis and Mart A. Stewart

Abstract In recent decades, the concept "precarious work" has gained great currency in the literature of the social sciences. Generally speaking, "precarious work" is defined as work that is uncertain, unstable, and insecure, wherein the preponderance of risks is borne by workers rather than by employers or government. Up until now, the concept has almost always been used with reference to wageworkers employed, whether formally or informally, in the manufacturing or service sectors. In this chapter, we broaden this valuable concept by extending it to the agricultural sector and by including various types of "insecure" agriculturalists – whether wage laborers, tenants, or owner operators – in its embrace. More specifically, we apply the concept to farmers and farming in the Mekong Delta. Rice farming in the Mekong Delta has never been easy, seldom been secure, and never been risk-free. Year-to-year fluctuations in weather conditions alone are enough to render farm life in the region so. Over the past two decades, though, uncertainty, instability, and insecurity have risen for many Delta farmers as a result of market reforms in Vietnam beginning in 1986, the vagaries of world commodity prices, changes in health care, disability, and pension schemes in the country, a variety of environmental changes associated with the modernization of rice agriculture and the increase of poorly monitored industry, and now, the threat of significant, if not drastic climate change, which has the potential to increase uncertainty, instability, and insecurity exponentially. We have used a variety of sources, including field investigations, to demonstrate the relevance of the "precarious work" concept to the Mekong Delta's rice cultivators.

P.A. Coclanis (✉)
Department of History, University of North Carolina, Hamilton Hall CB# 3195, Chapel Hill North Carolina, NC 27599-3195, USA
e-mail: coclanis@unc.edu

M.A. Stewart
Department of History, Western Washington University, 516 High Street, BH 320, Bellingham, WA 98225-9061, USA
e-mail: Mart.Stewart@wwu.edu

M.A. Stewart and P.A. Coclanis (eds.), *Environmental Change and Agricultural Sustainability in the Mekong Delta*, Advances in Global Change Research 45,
DOI 10.1007/978-94-007-0934-8_7, © Springer Science+Business Media B.V. 2011

Keywords Precarious work • Rice agriculture • Climate change • Mekong Delta

In the heat of mid-day, I plough my field
My sweat falls drop by drop like rain on the ploughed earth
Oh, you who hold a rice-bowl in your hands
Remember how much burning bitterness there is
In each tender and fragrant grain in your mouth

–traditional Vietnamese song

In recent decades, the concept "precarious work" has gained great currency both in policy circles and in the literature of the social sciences. Generally speaking, "precarious work" is defined as work that is uncertain, unstable, and insecure, wherein the preponderance of risks is borne by workers rather than by employers or government (Kalleberg 2009; Ross 2009). Up until now, the concept has almost always been used with reference to wageworkers employed, whether formally or informally, in the manufacturing or service sectors. Here we propose to broaden this valuable concept by extending it to the agricultural sector and by including various types of "insecure" agriculturalists – whether wage laborers, tenants, or owner operators – in its embrace. More specifically, we shall apply the concept to farmers and farming in the Mekong Delta, and, in so doing, utilize written records and interviews as well as our own fieldwork conducted in 2009.

Farming in general – and rice farming in the densely populated Mekong Delta in particular – has rarely been easy, seldom been secure, and never been risk-free. Deltas in general are fluid and dynamic environments that are both rich in ecological energy and highly unstable. The conditions that make the Mekong Delta and other deltas richly productive areas for agriculture also make for instability and precariousness. Floods that deposit upstream topsoil sediments can also sweep away efforts to cultivate them. The tidal currents that create dynamic swirls of energy at river mouths can also create conditions too saline for most agricultural crops. The hugely productive agricultural zone (about 40,000–45,000 km^2) of the Mekong Delta – the rice basket of Vietnam – is also a region of often unpredictable environmental change.

Year-to-year fluctuations in growing conditions are alone sufficient to render farm life precarious in the Delta today, and the effects of historical power asymmetries of various kinds – exploitative landlord–tenant relations, most notably – have left much of the population of the Delta without many safeguards against one or another possible calamity (Robequain 1939, 201–270; Sansom 1970, 18–52; Brocheux 1995; Biggs 2003; Cooke and Tana 2004; Biggs 2008; Biggs et al. 2010). Before going any further, let us be clear, though: Precariousness is not new to the area's inhabitants, and current problems in some ways seem less imposing in comparison to those caused in the past by heavy flood years, famines, and decades of war. It is, however, probably unproductive in any case to spend too much time assessing whether life is more precarious for the Delta's population today than was the case in earlier periods.[1] Reworking Tolstoy for our own purposes, it seems

[1] On the history of rice-growing in the Delta, see, for example, Xuan and Matsui, eds. (1998) and Biggs and Cronon (2011).

fair to say that life has always been precarious, if not unhappy in the Delta, but precarious in its own way. And over the past two decades, uncertainty, instability, and insecurity have taken new forms in the region, due in part to the changing political economy of Vietnam, but also to the changing environment, and to upstream engineering projects on the Mekong River itself.

In most explanations of the precarity of work in the Delta today, the 1986 market reforms in Vietnam – and the concomitants of such reforms for agriculture – figure prominently. In this regard, experts point in particular to the transformation of land tenure patterns since 1986; to increased competitive pressures; to the growing vulnerability of many agriculturalists to powerful "middlemen" and to the vagaries of world commodity prices; to increased rural inequality; and to the dismantling of the socialist welfare network, which led beginning in 1989 to steps toward privatization and to an attendant reduction in access for many in the countryside both to adequate health care and to tolerable disability/pension schemes. Some experts point as well to the declining status of Delta farmers, particularly rice growers, as the agricultural sector becomes more diversified, and Vietnam as a whole moves into higher value-added economic activities of one sort or another, and to the continued déclassé status of many minority Khmer and Cham farmers in the region. Although the market reforms have obviously brought numerous benefits to the Delta as well – Vietnam is now consistently number two behind Thailand as a rice exporter and, largely as a result, mean household income in rural parts of the Delta has risen significantly – such benefits have not been distributed evenly among the region's vast rural population (Kerkvliet et al. 1995; Ladinsky et al. 2000; Mekong Delta Poverty Analysis 2004; Taylor 2004; Adams 2005; Rama 2007; Varis 2008).

None of this is to suggest, however, that the increased precariousness of rice farming in the Delta is due solely to the market forces unleashed in 1986.[2] Delta rice farmers, like fellow agriculturalists in many parts of the developing world, also face an array of "legacy" problems resulting from history, as it were. The "path" taken by the Delta – a path marked by colonialism and exploitation, wars, and inefficient and often inept state economic policies – resulted inexorably in deficiencies in and/or malfunctioning of key infrastructural supports and institutional arrangements necessary for equable and sustained growth. In the rice sector – still the most important sector by far in the Delta – any short list of such problems would include the dearth of modern transportation and communications systems; milling, grading, and inspection capabilities of uneven quality; inadequate crop-storage facilities; insufficient (and often expensive) mechanisms of agricultural finance; the absence of "income-smoothing" tools such as crop insurance and futures contracts; inefficient, nontransparent land markets characterized by high transactions costs; rudimentary provisions for rural education; underfunded agricultural-research infrastructure, the capacity of which – despite some very creative

[2] Much of the material in this section is derived from the following interviews with experts on rice production in the Mekong Delta: Interview, Vo Tong Xuan, February 3, 2009, Ho Chi Minh City, Vietnam; Interview, Senior Staff, Mekong Delta Development Research Institute (MDI), February 4, 2009, Can Tho, Vietnam; Interview with Nguyen Xuan Lai and Luu Hong Man, Cuulong Delta Rice Research Institute, February 5, 2009, Thoi Thanh, Vietnam.

initiatives designed to reach practicing farmers – is uneven at best; and oligopolistic agricultural middlemen and intermediaries. Some critics also point to corruption in the private sector and public sector alike (Brocheux 1995; Minh and Kawaguchi 2002; Smith et al. 2007).

Exacerbating – and in some cases growing out of – the concerns mentioned above are another set of problems related to rice cultivation regimens and practices in the Delta. Holdings are generally tiny in the delta (generally about a hectare or two per farm household), which inhibits mechanization. When mechanized equipment is employed in rice production in the region – and more is being used each year – such equipment is all too often still rudimentary and of questionable quality, and yields a relatively high postharvest loss. For the most part, agriculture in the delta, as in the rest of Vietnam, is still labor intensive – almost 56% of the country's labor force is still engaged in agriculture – but labor productivity is low. Indeed, despite the fact that more than half of the country's labor force remains in the agricultural sector, that sector constitutes only about 21.4% of GDP. By way of contrast, in the US agriculture accounts for about 0.5% of the labor force and 1.2% of GDP. Moreover, though many agriculturalists are largely idle in the Delta for long periods during the year, during the busy harvest season labor shortages invariably arise in various parts of the region (Kerkvliet et al. 1995; Ladinsky et al. 2000; Mekong Delta Poverty Analysis 2004; Taylor 2004; Adams 2005; Rama 2007; Varis 2008; VietNamNet Bridge (2010); CIA World Factbook 2010).

Clearly, then, the list of problems plaguing the agricultural sector of the Mekong Delta is a long one. It should be noted – indeed, stressed – though, that the region's farmers have proved themselves at once resilient and in many ways utterly ingenious in the way that they have adapted over time to make ends meet in this fertile, but difficult and constantly changing agricultural environment. It should be noted, too, that in recent years they have been getting some important institutional help, which help is certainly welcome, even if it offers no panacea for the region's myriad agricultural ills. For example, several research centers in the South have been collaborating with Delta farmers in the creation of a range of new seed varieties. Expensive hybrids are not yet in wide use, but the new seed strains are nonetheless expensive relative to other costs, and, truth be told, decisions regarding variety and seed choice are often akin to risky games of chance where farmers with very incomplete information are compelled to bet on what market demand will be like during harvest time. The problems faced by farmers and traders in 2008 illustrate this point. In that year, Delta farmers bet big on a high-yielding but low-grade new variety, IR50404, and then found that there was little demand for it among exporters, who instead wanted so-called fragrant rice. Farmers had trouble selling IR 50404 in the domestic market as well – and ultimately often found themselves selling their crops at makeshift stalls in Can Tho and HCMC markets and, at very low prices, at roadside stands on the outskirts of HCMC.[3]

[3] See the interviews cited in note 2. Also see "Scramble to Get Rid of Rice at Low Prices," *Thanh Nien*, November 15, 2008, http://www.thanhniennews.com/features/?catid=10&newsid=43758.

Delta farmers are getting institutional help from outside the area as well. Agronomists and soil scientists in other parts of Vietnam – and from research centers such as the International Rice Research Institute (IRRI) in Los Baños, The Philippines – are also doing important work on rice-varietal improvement, more sustainable practices, and soil reclamation in the Delta. Here, too, the results have thus far been rather mixed. Important research findings often fail to reach Delta farmers, are viewed with skepticism by them, or in fact *are* irrelevant to the highly textured environmental conditions of different locales in the rice-growing areas of the Mekong Delta. Consequently, so-called soil-mining practices are unfortunately still all too common, accompanied by the widespread use of fertilizers – sometimes commercial, but generally bio-fertilizers of one sort or another – and the widespread and indiscriminant use of dangerous pesticides and herbicides. On the other hand, it should also be pointed out that a growing cadre of progressive farmers in the Delta have experimented with different crop mixes and rotations, with older varieties of rice which yield less but have greater resistance to mono-crop problems, with more responsible use of agricultural chemicals, and with Integrated Pest Management at least to some degree.

What is the upshot of such crosscutting developments on rice yields and quality? In general, rice yields in the Delta, though rising, remain far lower than in leading producing and exporting regions, including nearby Thailand, the world's leading exporter. And the quality of Vietnamese rice remains open to question, too, at least in the minds of most consumers in wealthier markets, forcing Vietnam to sell most of its rice cheaply, mainly in poorer developing countries in Asia, Africa, and other parts of the world.[4]

And what about the broader effects on risiculture in the Delta and on the region's rice farmers themselves? For decades now, development experts, agricultural policymakers, and research scientists in the Delta have been pushing hard to get the region's farmers to modernize, to consolidate plots into bigger holdings and to move, at the margin, out of rice and into higher value-added mixed farming/aquaculture systems, if not out of agriculture altogether. Regarding this last point: Many experts have assumed that the principal problems impeding the Delta's development grow out of the fact that far too high a proportion of the region's labor force is still lodged in the relatively inefficient agricultural sector of the economy. Many farmers have, explicitly or implicitly, acted on this assumption, quitting agriculture or scaling back to part-time agricultural status – often setting back agricultural reform efforts, in so doing – in order to join the swelling ranks of laborers in urban industrial zones within or outside the Delta. Whether or not the experts' assumptions about exiting agriculture are in fact valid is debatable, however, for without a sufficient quantity of viable and sustainable off-farm employment opportunities, life can become even more precarious for erstwhile farmers.

[4] See interviews cited in note 2.

Despite some small successes, it is fair to say that reformers have much more work to do before the Delta's agricultural sector "takes off" and begins to reach its formidable potential. Small farmers in the region, particularly poorer, less educated ones in more isolated parts of the Delta continue to struggle the most in terms of integrating their agricultural schemes and marketing arrangements into larger economic systems of one sort or another. Further complicating agricultural reform at all levels in the Delta are demographic concerns – not only absolute numbers, but density, and population growth rates – and far-reaching environmental issues. The legacy of environmental despoliation in the past is a significant problem, obviously, but one that arguably pales in comparison to the potentially dire threats posed by climate change and by dam-building initiatives (mainly Chinese) on the upper Mekong, which will complicate and likely render even more precarious the lives not just of farmers but of everyone who calls the Mekong Delta home (Kerkvliet et al. 1995; Ladinsky et al. 2000; Mekong Delta Poverty Analysis 2004; Taylor 2004; Adams 2005; Rama 2007; Varis 2008; Sajise et al. 2010).

Not surprisingly, over time attempts to stabilize and structure the fluid balance of water and soils in the Delta so that rice can be cultivated intensively – two and sometimes three crops a year – have in turn made this cultivation unsustainable in some locations. Soil fertility is declining in parts of the Delta, for example, especially in those areas where diking and dams have thwarted annual silt-laden inundations. In other places (such as the Plain of Reeds), the release of acids by the digging of new canals or by attempts to cultivate soils that are already acidic have compounded the problem that is a perpetual one for irrigated agriculture: Soils that become too acidic for crop cultivation, in this case, the cultivation of rice. In some locales adjacent to major branches of the Mekong, changes in water dynamics and in shorelines because of massive excavations of sand for sale to developers in places such as Singapore have also caused problems. Moreover, wetlands and forests that are important buffers to rice lands have shrunk as rice cultivation has expanded, and biodiversity in general has suffered the same fate in the Delta as it has everywhere that monocultures have expanded.[5]

Pollution of several kinds compromises the quality of Mekong environments as well. Facilities for water and waste treatment are minimal in the Delta, and problems arising from pesticide flush have become serious in some places in the Delta. Similarly, as manufacturing production increases in the Delta, the industrial pollution of air and water has become a recognizable problem and one that has been poorly regulated. Each of these problems is amplified by neglect; the government, lacking both capacity and capability, has not yet developed comprehensive land-use plans for the Delta, nor in spite of a couple of highly publicized cases demonstrated much vigor in attempting to eradicate the sources of environmental degradation (Kerkvliet et al. 1995; Ladinsky et al. 2000; Mekong Delta

[5] One study of the Mekong Delta concludes that about 42,000 hectares of forestland were lost to rice and shrimp culture expansion between 2001 and 2005. See Nguyen (2008).

Poverty Analysis 2004; Taylor 2004; Adams 2005; Rama 2007; Varis 2008; Sajise et al. 2010).

All of the above problems are increased by changes in the very foundation of Delta agriculture, that is, in the environmental conditions that make it possible. The threat to the current configuration of people and environments in the Delta by climate change is especially striking. Climate, as one scientist tagged it, is a "chaotic beast," and scientific efforts to make predictions about where climate change will take us are therefore not precise. There is still much uncertainty about the most likely climate-change scenario for the Delta. That said, given the ecological fragility of this low-lying region and its vulnerability to sea level rise, saline intrusions and changes in flooding patterns, it is no wonder that the effects of climate change, both the effects already experienced and those projected for the future, are sources of considerable anxiety in the Delta today.

With most of the region just a few meters above sea level, even small increases in sea level due to climate change – in this case global warming – can have profoundly dislocating and disruptive effects. A rise in sea level of just 1 m could displace multitudes and, in so doing, cause untold harm to millions of agriculturalists in the Delta, while perhaps providing smaller numbers with new economic opportunities. For example, with such an increase, salt water would intrude further upon the Delta, increasing soil salinity to levels that would make considerable swaths of the Delta unsuitable for rice production. The global warming trend we have seen over past decades is already endangering rice cultivation in the Delta, as increasing night temperatures cause increased plant stress and thence lower yields. If rising sea levels and increased salt-water intrusion in lower reaches of the Mekong might help some inhabitants in the region – people employed in aquaculture, for example – such benefits would hardly compensate for the millions of rice cultivators adversely affected by climate change (Kerkvliet et al. 1995; Ladinsky et al. 2000; Taylor 2004; Adams 2005; Rama 2007; Varis 2008; Sajise et al. 2010).[6]

Further complicating the environmental difficulties posed by rising sea levels and climate change are the myriad problems posed by the building of upriver dams. Although dam-building initiatives by the Chinese have quite rightfully received most of the attention, dams built by the Thais, Laotians, and Cambodians on tributaries of the Mekong have created problems as well. Since the first Chinese dam on the Mekong (at Manwan) began operating in 1993, the flow of water into the Delta from the north has been considerably reduced. Moreover, levels of sediment discharge and seasonal water flows into the Delta have been profoundly affected. As more Chinese dams are built on the river – two others are now in operation, another is scheduled for completion in 2012, and plans are being made for as many as eight or nine more – problems will only concatenate. According to a recent

[6] At least one study, of rice production and climate change in China, has explained that rising CO2 levels would extend the rice-growing season, reduce low-temperature injury, and allow some northern expansion of rice-growing, although the authors also acknowledge that any gains might be reduced through increased plant diseases and pest injury and a concomitant increase in the use of pesticides, as well as a decrease in water resources available. See Yao et al. (2009).

report by the Mekong River Commission (MRC), water levels in the Mekong River are today at record low levels, endangering the livelihoods, if not the lives of scores of millions in a number of different countries in Southeast Asia. While low levels of rainfall across the region in recent years have contributed to the problem, many experts place the burden of the blame on China's aggressive dam-building campaign over the past two decades, a campaign some have seen as an attempt by the Chinese at "hydro hegemony" over its downriver neighbors in Southeast Asia (Straits Times [Singapore] 2010; Osborne 2006, 223–278; Osborne 2007; Zeitoun 2007; Pomeranz 2009; The Economist 2010).

Delta farmers, dependent not only on water for irrigation but also on soil-enriching sediments from the river, obviously stand to lose, as do fishermen and women. The Mekong basin is the most important inland fishing zone in the world, and dam-induced changes in the flow – velocity, displacement, turbidity, and temperature – can have, indeed, are already having major impacts on the fisheries of the Mekong. Fish yields are falling; certain species (great freshwater catfish and freshwater dolphins in particular) are becoming increasingly endangered; and the complex seasonal rhythms of the Tonlé Sap, Cambodia's Great Lake – the fish from which are responsible for some 60% of the protein intake of the Cambodian population – is being disrupted. Thus, with problems arising from rising sea levels and salt-water intrusion, on the one hand, and, on the other, from upstream changes in the Mekong due to dam-building (as well as deforestation and the release into the Mekong of various and sundry toxins in rapidly industrializing southwest China), life for farmers, fishing populations, and everyone else in the Delta is in an environmental sense increasingly precarious (Straits Times [Singapore] 2010; Osborne 2006, 223–278; Osborne 2007, 2009; Zeitoun and Warner 2006; Zeitoun 2007; Pomeranz 2009; The Economist 2010).

In order to underscore both the complexities of rice production in the Mekong Delta *and* the precariousness of the venture even in the best of circumstances, let us look briefly at a few specific cases – based on farmer informants – from different parts of the region. In summarizing what might be called their business plans or, more properly perhaps, livelihood strategies, we hope also to demonstrate just how difficult it is to make sweeping generalizations regarding agriculture in such a variegated, fluid, and dynamic environment.[7]

Case I: The proprietors of a rice farm in Tan Thanh District in Long An Province (on the Plain of Reeds) that is less than two hectares in size and has been recently cleared from melaleuca forest and planted with seed from the district agricultural office, grow glutinous rice for market – which is sold to rice buyers by an oral contract while the crop is still in the field. The buyers coordinate the harvest for a percentage of the crop and pay in cash. A nearby road makes it easier for these growers to connect with buyers and for buyers to bring mechanical threshers and other equipment to

[7] All three field visits and interviews, along with supplementary interviews of seed-sellers, a rice miller, a tool and implements maker, and several rice traders, were conducted over a 4-day period (February 3–6, 2009), in Long An, Dong Thap, and An Giang provinces.

harvest the crop. These farmers also raise ducks and chickens and fish in the field for market and household consumption – one informant on the farm said he raised enough fish in the fields and irrigation canals to "feed his family."

Case II: Another farm family near Canal #12, also in Long An Province. This family, which carved its farm out of maleleuca forests, has been able to gain a competitive advantage through a variety of means: By working a larger (four hectare) rice plot; by storing rice in its house and then bargaining with the stream of buyers plying the adjacent canal until the patriarch procures a good price; by using hired labor and machines – a Japanese hand-held tractor for tilling, a hand-pulled seeder, a small thresher – and by operating without loans. This farm was also very much a family operation. The hard-working head of the family, who had no education, was clearly the backbone of the farm's success. But he had been able to support the education of both of his daughters, who then were able to contribute additional skills as well as labor to what appeared to be a very successful family strategy for operating a market-oriented family farm. An additional note: High soil acidity has been a problem in the past in modern rice-growing operations on the Plain of Reeds, but the canal for the time being provided fresh water that allowed farmers such as this one to adequately flush the fields. Another survival strategy, though not one that provided immediate returns, was to return rice fields to melaleuca forests – with a crop in 5–10 years. Melaleuca provides firewood and has traditionally been used in construction to frame homes in Vietnam, so it retains a market value at the same time that it can tolerate acid soils, flooding, and just about everything else.[8]

Case III: This farm, a 0.5 hectare parcel in the Lap Vo District of Dong Thap Province, was part of a larger complex of adjacent fields. The farmers of this and adjacent parcels lived elsewhere, and grew two and sometimes three crops of rice – or, alternatively, a third crop, during the dry season, of cucumbers, soybeans, peppers, sugarcane, or corn. The land was temporarily inundated during the wet season, and ditched to be irrigated during the dry season. These farmers sold to buyers who traveled a nearby road (close enough that rice could be effectively transported from the field along the dikes in bags by motorbikes). The farmer we interviewed was growing a high-yield (10 ton/ha) glutinous rice that was resilient and easy to grow but that did not fetch premium market prices (see above on IR50404). The rice was harvested and threshed by machine by crews who moved from parcel to parcel and charged a percentage of the harvest in return (1/2 bag per 25 bags, or 2% of the crop). The threshing crew told us that its thresher "belonged in a museum," but postharvest loss was greatly reduced by women gleaners who worked the piles of straw left by the threshing crew to recover rice for their own use.

[8] Although farmers who have adequate space in their homes to store rice usually gain a competitive market advantage by their ability to hold rice off the market until prices are best, prices are sometimes so volatile after a bumper crop or because the variety they have grown has low value that it erases this advantage. See, for example, Le An (2009).

These three cases, though a limited cross-section of Delta rice farmers, exemplify a range of possibilities for combinations of water and soil resources, access to expertise, transportation, and markets, and supplementary incomes and resources. The principals in at least one of the cases – the family that farmed 4 hectares near Canal #2 in Long An Province – appeared to be quite successful. The family was able to send the daughters to postsecondary schools, had a television, computer (though no Internet was available on the family's side of the canal), and a propane stove in its home. The family had just paid for a large wedding party for one of the daughters (and brought the son-in-law into the family enterprise). But all members of the household – like the informants in the other two cases – were dependent on the vicissitudes of the local rice field environments, on larger climate and water-flow dynamics, and on gambles made with seed choices each year. They also were at the mercy, no matter how precise and hardheaded the negotiations, of the rice buyers who gave them a price for their crops and who were also often responsible for arranging the conditions of harvest and transport. Even their best efforts were a gamble, in an environment and an economy where unpredictability and the lack of a safety net made even the best calculations also a matter of luck.

To return to a question we touched on earlier: Is rice farming becoming more precarious in the region? We think so. To be sure, rice cultivation in the Delta has always been a balancing act with the environment – and when population levels are relatively low and farmers are producing mainly for household consumption on lands that are regularly flooded and have good soils, it is a balancing act that generally works. But with increased commercialization, the growing intensity of production – intensity signaled in part by the diking of lands to add a third crop of rice or of vegetables or maize for cattle feed – and the return after many decades of an export orientation, the balancing act is becoming increasingly perilous for many. The perennial problems relating to the need for farmers to adjust on the fly to year-to-year environmental fluctuations in an extremely fluid and dynamic ecosystem have now been complicated and compounded by climate change and dam-building, and by fluctuations of considerable magnitude and scale in the economic and sociopolitical realms. New environmental and engineering concerns and the new political economy of rice production in the Delta, that is to say, pose additional challenges to farmers already burdened enough by matters relating to water and soil.

References

Adams SJ (2005) Vietnam's health care system: a macroeconomic perspective. Paper presented at international symposium on health care systems in Asia, Hitotsubashi University, Tokyo, January 21–22, 2005. https://www.imf.org/external/country/VNM/rr/sp/012105.pdf

Biggs D (2003) Problematic progress: reading environmental and social change in the Mekong Delta. J Southeast Asian Stud 34:77–96

Biggs D (2008) Breaking from the colonial mold: water engineering and the failure of nation-building in the plain of reeds, Vietnam. Technol Cult 49:599–623

Biggs D, Miller F, Hoanh CT, Molle F (2010) The Delta machine: water management in the Vietnamese Mekong Delta in historical and contemporary perspectives. In: Molle F, Foran T, Käkőnen M (eds) Contested waterscapes in the Mekong region: hydropower, livelihoods and governance. Institute of Southeast Asian Studies, Singapore, pp 203–225

Biggs D (2011) Quagmire: nation-building and nature in the Mekong Delta. University of Washington Press, Seattle

Brocheux P (1995) The Mekong Delta: ecology, economy, and revolution, 1860–1960. Center for Southeast Asian Studies, University of Wisconsin, Madison

Cooke N, Tana L (eds) (2004) Water frontier: commerce and the Chinese in the lower Mekong region, 1750–1880. National University of Singapore Publishing, Singapore

Economist (2010) Dammed if they do: China's hydropower plans are a test of its avowed good neighbourliness. July 10, 46

Interviews [with authors] Ho Chi Minh City and Mekong Delta. 3–6 Feb 2009

Kalleberg AL (2009) Precarious work, insecure workers: employee relations in transition. Am Sociol Rev 74:1–22

Kerkvliet BJT, Porter DJ (eds) (1995) Vietnam's rural transformation. Institute of Southeast Asian Studies, Singapore

Ladinsky JL, Nguyen HT, Volk N (2000) Changes in the health care system of Vietnam in response to the emerging market economy. J Public Health Policy 21:82–98

Le An (2009) Bumper crop brings misery to Mekong. Thanh Nien, 13 Sept 2009. http://www.thanhniennews.com/features/?catid=10&newsid=52475

Mekong Delta Poverty Analysis. Final Report. October 2004. Australian Government, AusAID

Minh HNT, Kawaguchi T (2002) Overview of rice production system in the Mekong Delta – Vietnam. J Fac Agri Kyushu U 47:221–231

Nguyen ST (2008) Determinants of land-use change: a case study from the lower Mekong of Vietnam. Electronic Green J 1. http://repositories.cdlib.org/uclalib/egj/vol1/iss27/art3

Osborne M (2006) The Mekong: turbulent past, uncertain future. Allen & Unwin, Brisbane

Osborne M (2007) The water politics of China and Southeast Asia II: rivers, dams, cargo boats and the environment. http://www.lowyinstitute.org/Publication.asp?pid=589

Osborne M (2009) The Mekong: river under threat. Lowy Institute Paper. 27. New South Wales: Lowy Institute for International Policy

Pomeranz K (2009) The Great Himalayan watershed: agrarian crisis, mega-dams and the environment. New Left Review 58. http://www.newleftreview.org/?view=2788

Rama M (2007) Social protection and labor markets in Vietnam. Paper, Commission on Growth and Development, Labor Markets and Growth, 15 Oct 2007. Washington, D.C. http://www.growthcommission.org/storage/cgdev/documents/LaborMarkets/Rama.pdf

Robequain C (1939) L'Évolution Économique de L'Indochine et du Vietnam. Centre D'Études de Politique Étrangère, Paris

Ross A (2009) Nice work if you can get it: life and labor in precarious times. New York University Press, New York

Sajise PE, Ticsay MV, Jr Saguiguit GC (eds) (2010) Moving forward: Southeast Asian perspectives on climate change and biodiversity. Institute of Southeast Asian Studies, Singapore

Sansom RL (1970) The economics of insurgency in the Mekong Delta of Vietnam. MIT Press, Cambridge

Smith W et al (2007) The impact of land market processes on the poor in Vietnam. Survey Review 39:3–20

Straits Times [Singapore] (2010) China Help on Mekong Sought, 7 Mar 2010. http://www.straitstimes.com/BreakingNews/SEAsia/Story/STIStory_499145.html

Taylor P (ed) (2004) Social inequality in Vietnam and the challenge to reform. Institute of Southeast Asian Studies, Singapore

Thanh Nien (2008) Scramble to get rid of rice at low prices, 15 Nov 2008. http://www.thanhniennews.com/features/?catid=10&newsid=43758

United States Government (2010) CIA World Factbook https://www.cia.gov/library/publications/the-world-factbook/geos/vm.html

Varis O (2008) Poverty, economic growth, deprivation, and water: the cases of Cambodia and Vietnam. AMBIO: A J Hum Environ 37:225–231

VietNamNet Bridge (2010) Agricultural mechanization – urgent need in the Mekong Delta, 16 Dec 2010. http://english.vietnamnet.vn/social/2008/12/818776/

Xuan VT, Matsui S (1998) Development of farming systems in the Mekong Delta of Vietnam. JIRCAS, CTU, and CLRRI, Ho Chi Minh City

Yao H, Zhang W, Yu Y, Sun W, Chen J (2009) A primary assessment of climate change impact on rice production in China. IOP conference series: Earth and environmental science, 2pp

Zeitoun M (2007) Hydro-hegemony: towards a radical view of transboundary hydropolitics. Paper presented at 3rd international workshop on hydro-hegemony, 12–13 May 2007. Centre for Environmental Policy and Governance, London School of Economics and Political Science. http://www.soas.ac.uk/waterissues/papers/file39700.pdf

Zeitoun M, Warner J (2006) Hydro-hegemony – a framework for analysis of trans-boundary water conflicts. Water Policy 8:435–460

Chapter 8
Analysis of Labour Migration Flows in the Mekong Delta of Vietnam

Huynh Truong Huy and Le Nguyen Doan Khoi

Abstract Migration occurs as a response to economic development as well as to social, cultural, and environmental causes. This chapter explores migration patterns in the Mekong Delta (MD) region of Vietnam by using information in the Vietnam Household Living Standard Surveys for 2004 and 2006. The main purposes of the chapter are: (1) to gain insight into migration flows in the Mekong Delta region and (2) to identify the determinants of the migration flows regarding commune-related variables. From the description and the multivariate analysis, there are several interesting findings. First, the number of migrants in the Mekong Delta region has been increasing in recent years, and their destinations are the leading industrial and commercial cities in Vietnam, including Ho Chi Minh and Binh Duong. Second, "push" factors such as poverty and challenges in farm production are the key causes of migration flows. In addition, policy makers should take advantage of economic development programs as an adjustment tool in migration matters.

Keywords Migration • Migrant • Origin • Destination

Abbreviations

FDI	Foreign Direct Investment
GSO	Vietnam General Statistical Office
MD	Mekong Delta
NGOs	Nongovernmental Organizations
IQR	Interquartile Range
UNDP	United Nations Development Program
VHLSS	Vietnam Household Living Standard Survey
VIF	Variance Inflation Factor

H.T. Huy (✉) and Le N.D. Khoi
School of Economics and Business Administration, Can Tho University,
No. 211, Street 30/4, Can Tho City, Vietnam
e-mail: huyseba@gmail.com; lndkhoi@ctu.edu.vn

M.A. Stewart and P.A. Coclanis (eds.), *Environmental Change and Agricultural Sustainability in the Mekong Delta*, Advances in Global Change Research 45,
DOI 10.1007/978-94-007-0934-8_8, © Springer Science+Business Media B.V. 2011

8.1 Introduction

8.1.1 Migration, Climate Changes, and Development

Migration, climate changes, and development have been major concerns of both researchers and policy makers in recent years in coastal countries such as Vietnam. Obviously, they have a causal relationship. Migration is usually viewed as one of the most frequently used strategies of rural households facing the negative effects of climate changes, and migration is also a useful tool for helping poor households to improve livelihoods and reduce poverty.

Migration's role in economic development has been widely recognized. Migration and particularly internal migration are seen as essentials for growth and a crucial strategy for escaping poverty for many poor and rural families in developing countries. As is widely known, the economic transition toward the expansion of urban centers and manufactures has strongly attracted a large number of labourers from less-developed surrounding regions; as a result, rural migrant labourers gradually engage in urban labour market, as shown by Lewis (1954). Simultaneously, an expansion of manufactures in urban areas has created employment opportunities for rural migrants to improve their livelihoods and well-being. Therefore, from an economic perspective, migration and development are complements. Development induces migration, and migration contributes to development.

In comparison with international migration, internal migration is likely more important in developing countries where a majority of the population lives in rural areas and where there is limited development of human capital. In the 2009 report on human development, the UNDP estimated that there were about 740 million internal migrants in the world, four times as many as the number that moved internationally (UNDP 2009). Additionally, data gathered from some countries in Asia by Deshingkar and Grimm (2005) show that in 2001 nearly 120 million people moved internally in China against about 450,000 international migrants. In Vietnam, during the period 1994–1999, 4.35 million people were estimated to have migrated internally, while the number of international migration was fewer than 300,000. In the literature, one finds that the relationship between migration (particularly internal migration) and development is widely recognized. This pattern of migration not only contributes to economic growth in the receiving area, but also has positive effects on livelihood strategies and poverty reduction in the originating area. More specifically, the contributions of migration to the development are usually summarized in the following way. For the destination area, migration can affect the average level of human capital, both in terms of quantity and quality, leading to an increase in the size and quality of the urban labour force (one side of the so-called brain drain). For the originating area, remittances from migrants as an additional source of income improve livelihoods of people remaining in the originating area, stimulate consumption, and improve the area's financial ability to cope with vulnerabilities.

For some countries in Asia, there is a gradual change in internal migration from rural – rural to rural – urban. Formerly, a majority of people often moved from one village to another village within a country to seek an availability of fertile soils for their farm work. More recently, an economic transition took place in most countries in Asia toward manufacturers and exports that has had strong effects on the internal migration trends. Most notably, this transition has led a large number of rural migrants to migrate to urban areas, as scholars and policy makers increasingly recognized. For example, the share of rural to urban migration in Thailand increased from 14.3% in the period of 1975–1980 to 18.4% after 10 years; the share of rural to rural migration decreased from 52% to 40.9% in the same period (Guest 2003). In Vietnam, rural to rural migration took place after the reunification of the country along with the planned migration programs. Such programs aimed to reclaim an abundance of fertile lands in the Highlands, areas with low population density. According to statistics of the GSO[1], the number of rural to urban migrants was 1.18 million people, accounting for 27.2% of all migrants in the period of 1994–1999. Although recent estimates of this migration are not available yet, the pattern of rural to urban migration has been widely recognized. According to estimates of the GSO in 2006, five cities – Chi Minh, Binh Duong (in the Southeast), Ha Noi, Quang Ninh (in the Red Delta), and Da Nang (in the Central coast) – are the most popular destinations for migrants. These cities are seen as the leading centers of industry and commerce in the whole country. In addition, Deshingkar and Grimm (2005) also documented that this same pattern of migration also has been increasing in Laos, China, and Cambodia in recent years, along with the development of industrialization and urbanization in these countries. Therefore, one can infer that rural to urban migration plays a crucial role in economic development and has been an interesting concern to many researchers in the domain of economics.

8.1.2 Commune[2]-Related Factors Affecting Internal Migration: A Review

As is widely known, the reasons of migration are usually attributed to a combination of "push" and "pull" factors. Especially for rural people with low qualification, the "push" factors of origin such as high population density, poverty, unemployment, and damages from natural disasters seem more important to induce them to migrate than the "pull" factors of the destination. In both literature and empirical evidences,

[1] General Statistical Office of Vietnam.

[2] A Commune is a third-level administrative unit in Vietnam. Normally, each commune consists of some villages, with its population of around 8,000 (Vietnam Household Living Standard Survey 2006).

commune-related factors that determine internal migration flows may be classified into four main categories, as following:

- Principal factors
- Economic factors
- Policy factors
- Environmental factors

8.1.2.1 Principal Factors

Principal factors refer to variables such as the density of population, geographic location (remote area), and ethnicity. More specifically, the relationship between dense population and migration has been long seen as a dual phenomenon. Rural to urban migration leads to higher density in urban areas where there are employment opportunities. Conversely, high population density is often associated with lack of employment opportunities and productive land in rural areas. It is apparent that in rural areas, high population density may induce people to migrate due to labour surplus in agriculture. Meanwhile, nonfarm work is usually so small that work of this type is unable to absorb the labour surplus. Moreover, based on a review of determinants of internal migration, Ivan (2008) also highlighted the fact that the density of population in the area of origin has a positive effect on internal migration, and this variable is usually seen as a "push" factor in migration.

Because of the variety of ethnic groups and geographic locations in most countries of Asia, researchers often study migration patterns across ethnic groups and geographic regions. For example, in a study of decision-making in rural to urban migration in China, Zhao (1999) found that minorities tended to move out of their villages less frequently. Other evidence on migration suggests that poor villages or remote villages also tend to have higher numbers of migrants (Deshingkar and Grimm 2005), because migration has long been seen as a tool to escape from poverty. However, people living in remote areas, far from the central province or district, can also be less migratory because of a lack of information, weak roads, and the high costs of migration. In their study of seasonal migration and improved living standards in Vietnam, Brauw and Harigaya (2007) also mentioned policies encouraging migration for people in remote areas and minorities in the Northern Uplands and the Highlands.

8.1.2.2 Economic Factors

Ever since the time of the earliest literature on migration, according to Ravenstein, the major motivations of migration were said to be economic and demographic. Over the years, some researchers such as Lewis (1954), Lee (1966), and Todaro (1969) closely examined economic variables (mainly the wage gap) as a key determinant for the migration decision.

Empirical evidence on internal migration in some Asian countries by IOM (2005) has shown that the income gap among sectors is usually regarded as the key cause of migration flows, including migration away from agriculture to nonagriculture or migration from rural to urban areas. More specifically, migration flows into nonagricultural sectors are seen as a result of the economic transition along with industrialization. As is already known, it is apparent that the economic transition often causes disparities in development and income across regions in countries. Certainly, higher wages in the nonagricultural sector in urban areas is a magnet that draws rural people to cities. Meanwhile, agricultural work is often seen as low-paid employment due largely to backward technology and the frequent negative impacts of climate changes that stimulate people to seek other employment opportunities outside of agriculture. In China, for example, scholars and policy makers predict that there are still 150 million surplus labourers in agriculture who will be moved into the nonagricultural sector in the next one or two decades.

Additionally, rural to urban migration has been viewed as a positive livelihood strategy for the poor in rural areas. Because of limited access to land, capital, and education, many poor, rural households often choose migration as the last survival option. Throughout his study of internal migration and poverty in Asia, Skeldon (2003) highlighted the fact that poverty is a root cause of migration. A large number of people moved from lesser developed areas (mainly agriculture) to more developed areas (mainly manufacturing).

As for agriculture production, there are numerous and diverse constraints to farmers, ranging from changes in agro-product prices, market access to capital needs and farming techniques, to diseases and climate changes. It is apparent that farmers' incomes from agriculture are usually more vulnerable to those constraints and risks; for example, a majority of farm households in Vietnam are very small in scale.[3] In order to overcome those constraints, farm households often tend to diversify sources of income by reallocating labourers and sending household members to urban areas for nonfarm work. According to the empirical evidence on migration in Bangladesh, Hossain (2001) reported that 25.6% of the migrants had engaged in the agricultural sector before migrating. After the economic crisis in 1997, Thailand, a farming-based country, had over 1.2 million people who went from agriculture to manufacturing, construction, commerce, and service work. Another study on internal migration in Vietnam (IOM 2005) indicated that rural people, who engaged in agriculture and fishery, often face sharp seasonal fluctuations and long slack periods. Therefore, returns from agriculture often only provide for basic food needs, while other costs of living are quite dependent on remittances from migrants.

8.1.2.3 Policy Factors

Migration occurs as a response to economic development in each country, province, and commune. A rapid increase of rural to urban migration has created many challenges to policy makers in maintaining the agricultural sector.

[3] According to the survey result of GSO in 2006, Vietnamese farm households had an average area of 0.77 ha.

Therefore, a set of policies regarding job creation for rural people is essential today, especially nonfarm rural employment ranging from traditional local industries (handicrafts) to services (repair of farm equipment). Such activities may also increase the attractiveness of employment in rural areas and consequently reduce the pattern of migration toward urban areas (Turnham 1993; Rhoda 1983).

8.1.2.4 Environmental Factors

Apart from socioeconomic factors, what else determines the trends and levels of migration? Recently, the external effects of environmental changes have been taken into account and seen as key determinants in explaining patterns of internal migration (mostly rural to urban migration). Citing 2005 reports from international organizations, Brown (2008) suggested that 50 million people in 2010 would be forced from their homes and off the land by a range of serious environmental problems, including pollution, land degradation, droughts, flooding, and natural disasters.

Regarding Vietnam, it is one of the most disaster-prone countries in the world. According to the study of flooding in the MD of Vietnam, Ninh (2007) warned that flood and inundation have their most profound effects on the agricultural products of the region (mainly the rice crop). Annually, the flood season usually lasts between September and mid-November, but inundation tends to be longer and deeper due to the rising level of seawater. Undoubtedly, land degradation and damages from natural disasters are currently the most salient causes of rural to urban migration in most agrarian economies such as Vietnam.

8.1.3 Empirical Evidence on Internal Migration in Vietnam

As for Vietnam and particularly the MD, internal migration has been increasingly recognized since the beginnings of the economic transition. As already described, the unbalanced state of development and inequality in living standards across regions of Vietnam and rural and urban areas have proven the key causes of rural to urban migration during the last decades. According to Vietnam's 1999 population census, nearly 4.35 million people changed their places of residence between 1994 and 1999, accounting for 6.5% of a population of 6.9 million. The proportion changing places of residence was only 2% in the period 1984–1989. It is apparent that internal migration is not a new phenomenon in Vietnam, but it has become an interesting concern of researchers and policy makers in the field of migration and development. Therefore, in this chapter, several studies of internal-migration patterns (macro-data analysis) in Vietnam are summarized.

Many previous studies used GSO census data to examine the pattern of internal migration across provinces and regions in Vietnam. For example, Anh et al. (1997),

Phan and Coxhead (2007), and Loi (2005) note that in the census, migrants are defined as the number of persons, over 5 years old in the survey period, who changed their place of residence during the last 5 years.

In their study of internal migration in Vietnam, Anh et al. also used the census data in 1999 to describe migration trends and to identify the determinants of the number of migrants across provinces. The authors indicated that the Highlands and the Southeast were the most promising destinations for migrants from other regions in Vietnam due to many advantages relating to fertile land and the development of manufacturing and commerce. In addition, they used a range of macro-data variables in studying origins such as density of population, industry sector, distance, and policy to explain the trend of interprovincial migration. Two of their many important findings were that increases in living standards and in the distance between origin and destination have negative effects on the number of interprovincial migrants. The authors explained that distance is seen as an intervening obstacle in the migration process because of lack of availability of information and transportation across provinces or regions in Vietnam. However, they also acknowledged the limitations of the macro-data included in their analysis. They had no further information relating to motivation of migrants, employment, and the outcomes of migration.

Like the earlier work of Anh et al. (1997), Phan and Coxhead (2007) also used census data from two periods – 1984–1989 and 1994–1999 – first, to investigate the patterns and determinants of interprovincial and interregional migration in Vietnam, and then to measure the impact of migration on income inequality across provinces. How did they undertake their measurements? They employed the gravity equation, which is a popular model to estimate the spatial movement of people between origin and destination. In so doing, they assumed, that migration between two regions is similar to the gravitational force between physical objects and thus proportional to the "potential" of the regions, but inversely related to the "distance" between regions. In the model, the authors also applied the Ordinary Least Square method to estimate the number of interprovincial migrants (with the logarithmic transformation); besides, they defined income inequality in terms of the ratio of per capita income between the richer province (destination) and the poorer province (origin). The results of their estimations indicated that migration flows tended to increase during the two periods of the census. The trend of those migration flows can be explained by the differences between the origin and the destination in variables such as per capita income, as well as by bus distance. Not surprisingly, the Highlands and the Southeast were still the most popular destinations for most migrants, while in the 1990s the MD and Central Coast were the most frequent places of origin. The authors demonstrated that provinces with a higher per capita income tend to attract more migrants. As estimated, each additional increase of 1% in per capita income in the destination province leads to an increase by 1.5% in the migration rate. Furthermore, they also revealed that migration acts as a tool to mitigate inequality in income across provinces. For example, a 1% increase in the out-migration rate leads to a decline in interprovincial income inequality by 0.01%.

In contrast with the previous existing studies on internal migration, Loi (2005) used the case-study method to explain the pattern of rural to urban migration across regions in Vietnam. The author highlighted three important factors causing rural people to migrate. First, population and employment pressure is seen as one of the key reasons for migration. More specifically, the share of population in the working age of 15–59 steadily increased from 47% in 1979 to 57.5% in 1999 and 65% in 2009, equivalent to 1.37 million persons per year between 2004 and 2009. Simultaneously, the increase of working-aged persons mostly took place in rural areas which were characterized by an agricultural sector with low returns. Therefore, migration flows from rural to urban areas have become a common phenomenon in the recent decade. Second, like to other countries in Asia, industrialization and urbanization process in Vietnam have also played important roles in causing a large number of rural people to migrate. Third, the author indicated that stagnation in traditional rural industries (mainly traditional handicrafts) in most villages of Vietnam caused a mass number of people to leave for urban areas to seek other jobs. It was found that of the decline of traditional industries was mainly caused by the severe competition of low-cost modern industrial products.

In short, by examining the existing literature on internal migration in some countries in Asia and Vietnam, we find that migration patterns are generally investigated via the underlying macro-data variables relating both to the originating area and destination. Until recently, there nevertheless has been an absence of specific studies on internal migration at the regional level – for example, in the MD. Therefore, the main objective of this chapter is to investigate the factors that are associated with rural to urban migration in the MD region. More specifically, this chapter attempts to exploit the VHLSS[4] series in 2004 and 2006, along with commune-related data, and attempts to ascertain whether or not findings in this area are similar to other recent findings on other parts of Vietnam.

The remaining part of this chapter is organized as follows. Sect. 8.2 presents an overview of internal migration flows in the MD region between 2002 and 2007. Section 8.3 provides an explanation of the data from the VHLSS series and the commune-data variables. Section 8.4 relates to model specification and the determinants of migration in the MD region. Finally, in Sect. 8.5 some conclusions are drawn.

8.2 General Migration Flows of the Mekong Delta Region

Internal migration in Vietnam, particularly in the MD region, is increasing over time for a variety of reasons such as uneven economic development, demographic imbalances, disparities in living standards, etc. According to the census of the

[4] Vietnam Household Living Standards Survey.

GSO in 1999, the MD was second in migration in Vietnam after the Red Delta in the North. Vietnam's economic transition, particularly the transition of the MD, plays an important role in migration trends in the region. One manifestation of this transition, regional disparities, is a key reason for migration. Because of such disparities, a massive number of people left rural areas for urban areas, hoping for better jobs and more income. Census data between 2003 and 2007 from the GSO showed that the rate of out-migration in the MD region increased steadily from 0.23% in 2003 to 0.79% in 2007; in comparison, the rate for the country as a whole rose from 0.25% in 2003 to 0.75% in 2007. Among those flows, migration away from rural areas and the agricultural sector emerged as a major pattern in the MD region. As a result, the area experienced declines in the rural share of the population and the agricultural proportion of the labour force in the same period (see Appendix 1) (Table 8.1).

The most popular questions in most scholarly studies on migration are: (1) why did people move, and (2) where did migrants originate from, and where are their destinations? In the following section, we use a variety of primary and secondary sources to interpret the region's migration patterns in recent years. Some socioeconomic indicators such as density, income per capita, share of farm income, and poverty rate are incorporated into the analysis.

Since nearly 80% population of the MD lived in rural areas and mainly depended on the agricultural sector until recently, it can be inferred that migration patterns in the region are mostly related to population and employment pressures in rural parts of the MD. With a population growth rate of 1.01% a year in recent decades, population density has increased steadily in the region over the last 15 years, from 383 persons per sq. km in 1994 to 436 persons per sq. km in 2008 (GSO 2008). Moreover, the region has about 110,000 additional people entering the labour force annually. Therefore, population and employment pressures have become central concerns of policy makers in the MD.

Furthermore, the economic transition away from agriculture and toward manufacturing and services has had considerable effects on the labour force in rural areas, as previously shown. More specifically, a proportion of the rural population *must* be forced away from the agricultural sector due to the expansion of industrialization and urbanization. In the meantime, traditional industries in rural areas are in stagnant situations and are gradually being eliminated from the local area

Table 8.1 Population and migration of the Mekong Delta (MD) region

	Population	In-migration	Out-migration	Rate of in-migration	Rate of out-migration	Net migration rate
Year	Person	Person	Person	%	%	%
2003	16,713,700	9,359	38,943	0.06	0.23	–0.18
2004	17,047,579	9,617	48,248	0.06	0.28	–0.23
2005	17,221,063	46,612	76,657	0.27	0.45	–0.17
2006	17,389,927	50,720	121,476	0.29	0.70	–0.41
2007	17,420,195	56,507	137,788	0.32	0.79	–0.47

Source: GSO 2007

because of low capacity, competitive pressures, their relatively small scales of production, and the thinness of local markets. Clearly, the pattern of interprovincial migration out of agriculture is a characteristic phenomenon during the initial period of economic transition. Recent findings relating to rural-to-urban migration in China corroborate this point (Zhang and Song 2003).

Basic information displayed in Table 8.2 provides us with comparisons of several interesting indicators across regions. This information plays an important role in helping us to interpret migration patterns in the MD region. It is apparent that the Southeast region, with a higher level of income per capita, is still the most popular destination for migrants from other regions of the country, as known in previous studies between 1994 and 1999. Regarding the MD region, 37.7% of the income per capita of MD residents was dependent on the agricultural sector, compared to 24.5% in the country as a whole. Simultaneously, the rate of underemployment in rural areas of this region was also higher than the average level in the country as a whole. Therefore, it is not surprising that rural people in the MD would face employment challenges in rural areas if they were not forced out of the agricultural sector altogether. Although the poverty rate in the MD has declined rapidly in recent years, average monthly income per capita in the MD region was still lower than the average level of the country as a whole, especially in comparison to the Southeast region. Therefore, one can conjecture with some confidence that migration patterns in the MD region are associated with socioeconomic indicators.

Where do the migrants come from and what are their destinations? Here, we use data of out-migration from the census of the GSO in 2006[5] to describe flows and magnitude of migration in the MD region.

Table 8.2 Socioeconomic indicators by regions in Vietnam

	Density	Monthly income per capita	Share of monthly income per capita from agriculture	Under-employment rate[a] in rural areas	Poverty rate
Year	Person/km^2	1,000 VND	%	%	%
Red Delta	933	653	19.0	8.2	10.1
Northeast	} 118	511	32.4	} 2.5	22.2
Northwest		372	45.9		39.4
North Coast	} 207	418	30.4	} 6.3	26.6
South Coast		550	21.0		17.2
Highlands	92	522	45.9	5.6	24.0
Southeast	543	1,064	11.1	3.7	4.6
Mekong Delta	436	627	37.7	7.1	13.0
Whole country	260	636	24.8	6.1	15.5

Source: GSO 2006, 2008

[a]According to the GSO, this rate is defined as the percentage of underemployed persons in the total employed population; underemployed persons refer to those who worked less than 35 hours a week when they were willing to work more time

[5]The ending of policies mandating that migrants notify local authorities when leaving the commune has led to a lack of statistical information regarding migration destination since 2007.

Table 8.3 Socioeconomic indicators of migrants' regions of origin

Origin	Net migration rate[a] (%)	Monthly income per capita[b] (1,000 VND)	Share of income from agriculture[b] (%)	Poverty rate[b] (%)	Economic feature[c] –
Hau Giang	–0.73	609	42.9	15.0	Pure agriculture
Soc Trang	–0.68	495	50.3	19.5	Agriculture
Vinh Long	–0.63	580	36.3	11.0	Agriculture and industry
Ben Tre	–0.62	614	38.8	16.2	Agriculture and seafood
Can Tho	–0.61	780	23.6	7.5	Industry and service
An Giang	–0.45	691	33.1	9.7	Service and agriculture
Dong Thap	–0.39	609	39.6	12.1	Pure agriculture
Bac Lieu	–0.32	610	45.7	15.7	Seafood and agriculture
Long An	–0.30	627	32.8	8.7	Agriculture and industry
Tra Vinh	–0.28	509	40.6	21.8	Pure agriculture
Tien Giang	–0.27	630	34.2	13.2	Agriculture and industry
Ca Mau	–0.25	666	47.8	14.0	Seafood and agriculture
Kien Giang	–0.03	675	39.0	10.8	Agriculture and seafood
Mekong Delta	–0.41	627	37.7	13.0	–

Source: GSO 2006

[a] Census

[b] VHLSS

[c] See VCCI-Can Tho 2006

8.2.1 *Migrant Origins*

According to the 2006 census of the GSO in, Hau Giang had the highest migration rate in the MD region, with a net migration rate of –0.73%. Hau Giang was followed by Soc Trang and Vinh Long. Not surprisingly, these places had lower monthly incomes per capita than the average level for the entire region. Moreover, agriculture, a low-return sector, played an important role in economy of these provinces. It is worth noting that a high rate of migration tends to be closely associated with poverty and the relative size of the agricultural sector. As shown earlier, most rural households engaged in agricultural production but had small landholdings per household. As a result, rural households are often more vulnerable and end up forced into poverty by economic shocks and natural disasters; additionally social security and unemployment compensation are virtually absent, and savings are too low for poor rural households to cope during periods without income coming in. Consequently, Vietnam and particularly the MD region are good locations to illustrate the relationship of agriculture, poverty, and migration (Table 8.3).

8.2.2 *Over 73% of Migrants in the MD Left the Region Altogether*

As expected, the Southeast region is still the most promising destination for most migrants out of the MD region. In this regard, Ho Chi Minh City and Binh Duong Province are the most attractive destinations in the Southeast region. Intra-provincial

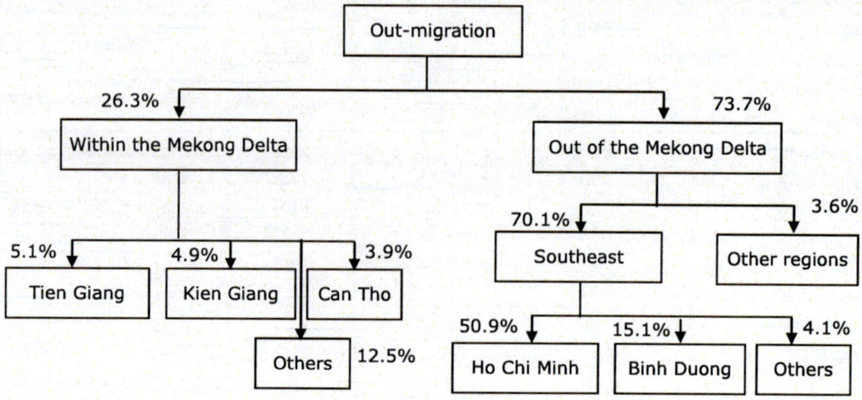

Fig. 8.1 Out-migration flows of the Mekong Delta (MD) region

migration within the MD region accounts for only around one-fourth of the total out-migrants. However, due to lack of availability of information concerning motivations for migration, it is reasonable to use basic information on socioeconomic status and to infer that opportunities in those destinations were the key "pull" factors driving migration (Fig. 8.1).

8.2.2.1 Migration Within the MD Region

It can be assumed that most migrants moving *within* the region are females and first-time migrants. In an earlier migration study, (Loi 2005), the author found that most migrants who had moved only a short distance were the first migrants in a household or were seasonal migrants. Female migrants in particular tend to move only short distances for several reasons. First, female migrants often find work more slowly than do male migrants (UNFPA 2007). Therefore, these first-time, female migrants usually move to destinations closer to work to save on costs. Second, for married women, migration has more constraints due to care of family and children, making short-distance moves less burdensome than long-distance moves.

8.2.2.2 Migration Out of the MD Region

The information regarding migration displayed in Table 8.4 shows that Ho Chi Minh City, as mentioned earlier, is the most attractive destination for most migrants from Vietnam, including the MD region. It absorbed over 50% of the total number of migrants from the MD region. Based on the 2004 census, migrants accounted for over 30% of the total population of this city. Among those migrants, migrants from rural areas made up around 80%. More specifically, 31.5% of those migrants were from the MD region, and 17.7% came from the Red Delta region. Therefore, it is

Table 8.4 Basic information on the destinations of migrants from the MD region

Destination	Destinations of MD migrants[a] (%)	Monthly income per capita[b] (1,000 VND)	Share of income from non-farm labour[b] (%)	Proportion of business units in the region*[c] (%)
Ho Chi Minh	50.9	1,480	30.6	79.0
Binh Duong	15.1	1,215	33.1	7.7
Tien Giang	5.1	630	23.5	11.3
Kien Giang	4.9	675	20.4	14.0
Can Tho	3.9	780	27.7	11.8
Mekong Delta	–	627	22.2	–

Note: *Ho Chi Minh and Binh Duong are compared in the Southeastern region. While Tien Giang, Kien Giang, and Can Tho are compared in the Mekong Delta

Source: GSO

[a] Census 2006

[b] VLHSS 2006

[c] Survey of enterprises between 2000 and 2007

apparent that rural migrants from the MD region have played an important role in the labour market and in the economic development of the most dynamic city in Vietnam.

Similarly, Binh Duong is also one of the most economically developed provinces in the Southeast region. This province has been industrializing at a high rate of speed and has a large number of industrial zones. According to statistical data relating FDI between 1988 and 2008, Binh Duong was the second leading province in this region as regards the absorption of FDI projects in terms of the number of projects and the amount of capital (after Ho Chi Minh City).

In short, the most important patterns relating to migration in the MD region are: the rapid increase of rural-to-urban migration (which has obvious effects on the distribution of population and on the labour force in rural areas); and the fact that a majority of migrants tend to move to industrial and commercial cities, where living standards and economic conditions more generally are higher.

8.3 Description of the VHLSS

In Vietnam, as in most developing countries, gaining access to data on migration is a challenge to researchers. As a result, many studies on this topic have primarily relied on multiple-purpose surveys, especially the VHLSS series.

Here we will use the commune-based data of the VHLSS series from 2004 and 2006 to examine whether the migration patterns for the MD region comport with the general picture of migration in Vietnam, as previously described. First, however, we shall explain the VHLSS dataset and its commune-related variables.

The VHLSS series was conducted every 2 years by the GSO in collaboration with the World Bank. The VHLSS is a comprehensive nationwide survey comprising two parts: commune and household. For the commune survey, information on a variety of topics is included, such as population, the number of people moving out of and moving within the commune, economic conditions, education, and public infrastructure. The household survey collected information on multiple dimensions of living conditions such as demography, education, health, employment, income, expenditure, physical capital, and so on.

The strength of this dataset grows out of the fact that it is a nationwide survey with a large number of observations: 3,063 communes and 9,189 households in the 2006 survey. The VHLSS is nevertheless not a specific survey relating to migration, which results in some limitations in this analysis. First, information relating to the destination of migrants who left the commune is not available; as a result, we cannot examine the "pull" factors of the destination. Second, average income indicators at the commune level – a key determinant of migration – are also absent in this dataset. Third, bus distance, as an intervening obstacle to migration, is not available. Therefore, an alternative measure is used in this analysis: the remote commune, defined as the least-economically-developed commune due to distance from cities and weak infrastructure.

Generally speaking, data on interprovincial migration in Vietnam and particularly in the MD region are quite scarce. In recent years, the VHLSS series have come to be seen as the most useful data source for most studies of migration in Vietnam. In this chapter, we only use the samples of the MD region to investigate the pattern of migration along with commune-related variables such as density, wages in the agricultural sector, poverty rates, economic conditions, constraints on agricultural production, geographic features, and policy questions.

8.4 Determinants of Migration: An Empirical Analysis

8.4.1 Model Specification

An interesting review on intra-provincial migration in developing countries presented by Lucas (1997) is still valid and empirically rich. The pattern of internal migration is generally interpreted via a range of macro-related variables using the gravity equation. The strength of the gravity model lies in its generality, which makes it possible to make tests on a number of issues relating to migration theory. Therefore, it has been widely used to study internal migration, especially across locations and regions in the country.

As previously shown, the VHLSS is not designed specifically for migration, which leads to a lack of availability of information on differences in income, the unemployment rate, and the distance between place of origin and destination. Therefore, the pattern of migration flows in this analysis can only be explained by a range of commune-related variables relating to origins, namely, socioeconomic factors and public policy. In various macro studies on migration in Vietnam, the number of migrants measured by the aggregate data is often treated as a dependent variable (Anh et al. 1997; Phan and Coxhead 2007).

However, to avoid differences in the population scale across origins, the pattern of migration is measured by a ratio of the number of migrants per 10,000 people in each commune.

Here, a general model is shown as follows:

$$M_i = \beta_0 + \beta_1 X_1 + \cdots + \beta_n X_n + e \tag{8.1}$$

Where M_i is the out-migration rate of the commune; β are unknown coefficients of independent variables; $X_{1..n}$ are independent variables, main categories such as principal factors, economic status, policy, and environmental factors; and e is the error term.

In this analysis, the Eq. 8.1 is transformed into the logarithmic model. In fact, the popularity of the logarithmic transformation (or semi-logarithm) is due to the following reasons. First, the logarithmic model often obtains a higher R square than is the case in the common linear model, especially when the relationship between variables of the model is not linear. Second, logarithmic transformation is one of the most widely used techniques, which not only has the effect of making the distribution of the transformed variable closer to the normal distribution, but also stabilizes the variance (a remedy for heteroskedasticity), which together leads to the robust standard error of the model (Wooldridge 2009; Chatterjee et al. 2000).

Consequently, we obtain the transformed model of migration (Eq. 8.2), as follows:

$$\mathrm{Ln}\, M_i = \beta_0 + \beta_1 X_1 + \cdots + \beta_n X_n + e \tag{8.2}$$

Where $\mathrm{Ln} M_i$ refers to the natural logarithm of rate of out-migration in the commune.

In this analysis, three estimated models are analyzed separately to examine the relative effects of each category of factors on migration patterns in the commune. In other words, the first model estimates the contributions of principal factors to the movement of population out of the commune, including density (1,000 person/km^2), farm wages (1,000 VND per day), the poverty rate (ratio of poor households to 100 households), remoteness of the area (classified as a remote commune by the Government's regulation), and ethnicity (1 for the Kinh, 0: otherwise). To the second model will be added economic factors to analyze the relative effects of these variables (mostly dummy) on the pattern of migration. These variables include: economic character of the commune (agriculture, industry, and service), qualitative assessment of increase in income of residents living in the commune in the last 5 years, the question of whether or not economic units in the commune attract local labour, and the presence of four of the biggest constraints for rural people in agricultural production – lack of capital; frequent price fluctuations for agro-products; the impossibility of accessing markets; lack of farm technique. And the last model

Table 8.5 Description and expected sign of variables in the models

Variables	Description	Expected effect
Principal factors		
Density	Population density of commune (1,000 person/km^2)	+
Farm wage	Farm wage a day (1,000 VND)	+, –
Poverty	Poverty rate (by 100 households)	+
Remote	Commune classified as remote area (see Appendix 2)	–
Kinh ethnic	Kinh is a dominant ethnic group in Vietnam, accounting for 80% of the total population.	+
Economic factors		
Agriculture[a]	Agriculture as the main sector of commune (coded by 1)	+
Industry[a]	Industry as the main sector of commune (coded by 1)	–
Fishery [a]	Fishery as the main sector of commune (coded by 1)	–
Living	Increase in income of resident in the last 5 years (coded by 1)	–
Labour attraction	Economic units in commune created employment to local labourers (coded by 1)	–
Lack of capital	Farmers face lack of capital in production (coded by 1)	+
Price fluctuation	Farmers frequently price fluctuations in agro-products (coded by 1)	+
Access to market	Farmers face difficulties in access to market (coded by 1)	+
Lack of technique	No access to advanced techniques (coded by 1)	+
Policy and disaster factor		
Job creation	There are programs for job creation in the commune (coded by 1)	–
Economic development	There are projects for local economic development in the commune (coded by 1)	–
Disaster	Rate of households granted economic support due to impacts of natural disaster	+

will incorporate additional factors relating to policy and the impacts of disaster, such as the presence of job-creation projects, rates of local economic development in the communes, and the proportion of households granted support and subsidies due to the impact of natural disasters in the commune.

In short, the main goal of the proposed models is to test the hypothesis regarding whether or not principal factors, economics, policy, and other considerations affect the migration flows of the commune. Table 8.5 provides a description of the variables and the direction of expected effects in the models.

8.4.2 *Descriptive Analysis*

Table 8.6 presents summary statistics relating to the main variables in the estimated models. At a statistical level of 5%, there were almost no differences in migration flows and density between 2004 and 2006 in the selected communes.

Table 8.6 Descriptive statistics of main indicators at commune level

	2004		2006		Mean difference between
Variable	Mean	Std.Dev.	Mean	Std.Dev.	2004 and 2006 at 5% level
Ln(migration rate)	3.82	1.21	3.74	1.27	n.s
Density	0.56	0.34	0.56	0.46	n.s
Farm wage	31.93	5.86	38.90	7.85	***
Poverty rate	9.71	8.37	17.58	10.70	***

Note: n.s, *** denotes not significant at 5% and significant at 1% level

However, a parallel increase of farm wages and the poverty rate took place during this period that resulted in a significant divergence in these variables. It is also surprising that the poverty rate in the analysis tended to increase in the two periods, while the poverty situation of the country and the MD region has been falling over time, as previously shown. However, this indicator was still appropriate with an average poverty rate of the country of 18.1% in 2004 and 15.5% in 2006 (GSO 2008). In addition, some other factors used in the models are shown in more detail in Appendix 3a, 3b.

Several hypotheses are proposed, as follows:

(1) Does an increase in the poverty rate induce people to migrate? In the literature on migration, migration is seen as a positive livelihood strategy for households to escape from poverty situations.
(2) How does an increase in farm wages affect migration patterns – positively or negatively? In theory, an increase in wages in a certain sector makes it possible to encourage people to work in that sector. In fact, whether or not that sector can attract labourers is dependent on the farm wages in other sectors.
(3) Do the production difficulties of farmers drive them to move out of their communes? In this regard, note that farmers' incomes are almost entirely dependent on the agricultural sector, which is seen as a low-return occupation and a more vulnerable vocation in times of economic shocks and natural disasters.

8.4.3 Estimated Results and Discussions

Table 8.7 displays the empirical results of the three models through the Ordinary Least Squares (OLS) estimation methodology, together with the robust standard error. In general, these models have demonstrated statistically significant correlations between migration flows and explanatory factors relating to place of origin. More specifically, a majority of the estimated coefficients in the three models for 2006 are likely higher than those for 2004. In other words, migration flows in 2006 would be explained to a relatively greater degree by the factors relating to the place of origin. In short, it is important that we examine each model to gain

Table 8.7 Estimations of migration flows of the MD region

Variables	Model 1 2004	Model 1 2006	Model 2 2004	Model 2 2006	Model 3 2004	Model 3 2006
Principal factors						
Density	0.129	0.206	0.098	0.118	0.093	0.126
	(0.07)	(1.48)	(0.51)	(1.14)	(0.48)	(1.14)
Farm wage	0.002	0.022**	0.007	0.021**	0.006	0.021**
	(0.16)	(2.47)	(0.62)	(2.20)	(0.56)	(2.18)
Poverty	0.005	0.018***	0.003	0.017***	0.004	0.017***
	(0.69)	(3.09)	(0.38)	(2.74)	(0.52)	(2.70)
Remote	−0.025	−0.325**	−0.014	−0.258*	−0.008	−0.272*
	(−0.19)	(−2.36)	(−0.11)	(−1.87)	(−0.06)	(−1.95)
Kinh ethnic	0.726***	0.537*	0.735**	0.659**	0.714**	0.626*
	(2.62)	(1.62)	(2.55)	(1.99)	(2.48)	(1.87)
Economic factors						
Agriculture			−0.814**	−1.008**	−0.851**	−0.918**
			(−2.34)	(−2.36)	(−2.47)	(−2.25)
Industry			−1.137***	−2.229***	−1.050**	−2.171***
			(−2.84)	(−3.02)	(−2.55)	(−2.98)
Fishery			−1.043***	−1.805***	−1.087***	−1.728***
			(−2.68)	(−3.73)	(−2.81)	(−3.73)
Living			−1.433***	−0.638***	−1.383***	−0.624***
			(−5.07)	(−3.49)	(−4.06)	(−3.83)
Labour attraction			−0.061	−0.109	−0.057	−0.101
			(−0.45)	(−0.82)	(−0.43)	(−0.75)
Lack of capital			0.183	0.252*	0.185	0.262**
			(1.11)	(1.90)	(1.11)	(1.98)
Price fluctuation			0.205	0.278	0.221	0.277
			(1.20)	(1.45)	(1.28)	(1.44)
Access to market			−0.001	0.468**	0.042	0.466**
			(−0.00)	(2.44)	(0.18)	(2.42)
Lack of technique			−0.343	0.292	−0.322	0.275
			(−1.10)	(1.32)	(−1.05)	(1.24)
Policy and disaster factor						
Job creation					−0.160	−0.170
					(−1.16)	(−1.24)
Economic development					−0.212	−0.085
					(−1.46)	(−0.51)
Disaster					1.342	0.997
					(0.97)	(0.43)
Constant	2.978***	2.041***	5.028***	3.640***	5.108***	3.609***
	(6.27)	(3.78)	(8.87)	(5.33)	(8.62)	(5.37)
Number of observations	451	480	451	480	451	480
Prob>F	0.000	0.000	0.000	0.000	0.000	0.000
R-squared	0.02	0.06	0.051	0.122	0.059	0.125

***, **,* significant level at 1%, 5%, and 10%
t-statistics are in the parentheses

clearer insights relating to the relative impact of each group of factors on the pattern of migration in this analysis.

The first model only regresses the principal factors in the communes between 2004 and 2006. As a result, a number of factors with significant effects on the migration flows in 2006 were higher than in 2004. Those factors are the farm wages, poverty rate, remote area variable, and Kinh ethnic group variable. While density of population is not presented, it has a significant effect on migration patterns, as expected from the empirical review. It is worth noting that an increase in farm wages also induces migration out of the commune, but with small effect. In fact, an increase of farm wages may relate positively to the pattern of migration, if there is still a gap in wages between the farm sector and other nonfarm activities or between rural and urban areas. Another reason is that farmers may migrate temporarily after the crop season in order to obtain more money needed to invest in their farm for coming crops (Rhoda 1983). Therefore, the relationship between migration and farm wages (or agricultural development) is relatively complex and perhaps dependent on a number of other factors.

Undoubtedly, this analysis demonstrates that the poverty situation is still an important factor driving people to migrate. More specifically, the poverty rate has a positive effect on the pattern of migration out of the commune, and this finding is significant at the 1% level. This relationship has been widely known and is found in both the theoretical literature and empirical studies in the field of migration.

However, communes classified as remote communes in 2006 are significantly and negatively correlated with the pattern of migration, which implies that people living in remote areas are likely less migratory than others living closer to the centers of the province. As mentioned in the literature review and the descriptive section of this chapter, spatial and geographic distance have been long seen as important obstacles to migration because of limited access to information and the inadequacy of road networks. Note that the effect of this factor in 2006 was much larger than in 2004. This may be seen as a consequence of the government's programs regarding rural development in remote areas and for ethnic minorities. The previous studies on intra-provincial migration between 1994 and 1999 in Vietnam by Anh et al. (1997) and Phan and Coxhead (2007) also support this conclusion.

Furthermore, the Kinh people are likely more migratory than other minorities in the communes of the MD region. There is for two main reasons. First, because the Kinh people are dominant in Vietnam, as already shown. Thus, they always play a crucial role in the migration flows. Second, because other ethnic groups in the MD region such as the Chinese, Khmer, and Cham have various cultural features and lifestyles that affect migration. For example, the Chinese usually work for themselves or for their relatives and other Chinese families in the village. The Khmer and Cham are often influenced by cultural and religious concerns. For example, there are many annual cultural and religious festivals for the Khmer and Cham that almost forces them to return to their home communes to join in those activities; as a result, those who have migrated sometimes face unemployment after each return. Additionally, ethnic minorities are usually less communicative than the Kinh people because most of them live in areas with disadvantaged economic conditions. Therefore, the fact that ethnic minorities comprise a small number of the migrant total is mainly due to cultural, religious, and social factors.

The second model, which adds economic factors concerning migration, shows that many of those additional variables are significant; and they have generally not changed in direction and statistical significance, as did the principal factors included in Model 1. It is apparent that different economic conditions of each commune have various effects on the magnitude of migration; more specifically, the pattern of migration out of a commune would be much decreased if the economy of that commune was dependent on industry and fishery development. On the other hand, people in the agriculture-based communes tended to be more migratory than others in the nonagriculture-based communes. That implies a gap in income or wages across sectors in these communes, which is also relevant to the general situation of the country.

Besides the level of socioeconomic development of the commune, both living standards and employment play important and significant roles in bringing about a decline in the pattern of migration out of the commune. Some labourers, mainly women, often find jobs in the commune due largely to reasons of family and marriage. As shown earlier, farmers seem to be faced with challenges, ranging from investment to production technology to marketing. There are two potential explanations for these concerns. First, agricultural production is dependent on natural factors like weather, water source, and diseases, etc., that usually render farmers very vulnerable. Second, improvements in the rural market such as crop insurance and market linkages are almost entirely absent. Such limitations result in frequent failures of farmers to gain access to markets and to prevent or recover from economic shocks. In model 2, the lack of capital and poor access to the market are two factors for farmers that have significant and positive effects on the pattern of migration out of the commune at the 5% and 10% levels.

In the last model, factors concerning policy and the impact of disasters are included in the estimation to investigate whether or not these factors are associated with the pattern of migration out of the commune. Although not all these factors are significant, their estimated signs are quite relevant to both the theoretical and empirical literatures. More specifically, the presence of rural development programs established by the government and NGOs has played an important role in generating employment and reducing migration from rural areas. Meanwhile, the impact of natural disasters is also seen as a cause of migration.

In sum, the determinants of migration from the communes in the MD region are fairly relevant to both the theoretical literature and to empirical findings in most recent studies of Vietnam and other countries in Asia. Undoubtedly, "push" factors at the region of origin are closely associated with the pattern of migration. Such factors include poverty, disasters, failures in production and the market, socioeconomic background, and even ethnicity in the commune.

8.4.4 *Diagnostics of the Model*

An econometric model usually has underlying assumptions, and regression diagnostics are common techniques for exploring problems in the analysis and for determining whether certain assumptions can be accepted. Although basic techniques were used in the estimating models to improve the "goodness of fit" of the model, including logarithm transformation and Robust standard error, checking other assumptions is also worthwhile. Therefore, several methods are used for checking the main assumptions of the estimating model, namely, normality of residuals, homogeneity of variance, and multi-collinearity.

Checking normality of residuals is necessary in regression analysis in order to ascertain whether the residuals are close to a normal distribution. We can use histogram graphs of residuals from the estimating model to make an inference of this matter. Another alternative is a test written by Lawrence, called interquartile range (IQR), to examine the symmetry of the distribution. In this analysis only the third model in 2006 is used for testing assumptions due to its full representativeness for all factors. The result of testing shows that value of IQR, is in a range of low inner fence and high inner fence, from which we can conclude that the distribution seems fairly symmetric, and the residuals are close to a normal distribution (see Appendix 4).

Checking for multi-collinearity[6] and homogeneity of variance[7] is worth noting in a regression analysis and especially for cross-sectional data. The variance inflation factor (VIF) measures the impact of collinearity among the independent variables in the estimating model. Estimations displayed in Table 8.8 show that mean VIF in all models is not so high as to drop any variables out of the model. However, these indicators tend to increase when we add more variables to the model (e.g., Model 2 and Model 3). Such increases are mainly caused by two dummy variables (agriculture and fishery) with their VIF of 28.4 and 25.9, respectively, in the Model 3 of 2006. Therefore, we finally have to drop two of

Table 8.8 Diagnostics of multi-collinearity and heteroskedasticity

	Model 1		Model 2		Model 3		Model 3[a]
	2004	2006	2004	2006	2004	2006	2006
Mean VIF	1.21	1.17	3.66	5.35	3.25	4.66	1.21
White's test Prob>Chi2	0.790	0.228	0.159	0.075	0.117	0.019	0.080

Note: [a] this is the model 3 but excluding two dummy variables of agriculture and fishery

[6] Multi-collinearity refers to a situation in which two or more explanatory variables in a multiple regression model are highly correlated. One of the assumptions of a regression model is that the explanatory variables are not mutually correlated.

[7] Another assumption of the regression model is that the standard deviations of the error terms are constant and not dependent on the explanatory variables. In other words, variation in the error term in the model should not be due to the explanatory variables.

these variables out of Model 3; and as a result, the mean VIF of the modified model (Model 3[a]) is strongly reduced and only remains 1.21.

Additionally, there are two widely used techniques to test homogeneity of variance: the Goldfeld–Quandt test and White's test. The Goldfeld–Quandt test assumes that the heterskedasticity has a linear relationship with one of the independent variables; while, White's test is more general without specifying the nature of the heteroskedasticity. This is the most important reason why White's test was used for this analysis. Generally, White's test has the null hypothesis of homoskedasticity (or no heteroskedasticity) and against the null hypothesis: heteroskedasticity. In short, most of the p-values of White's test are not significant at the 5% level (except for Model 3 in 2006). This means that there is not sufficient evidence to reject the null hypothesis at a 5% significance level. In other words, the errors of the model have a constant variance. It is apparent that using the semi-logarithm transformation seems a fairly appropriate way to measure the determinants of the migration pattern. Finally, the most relevant model for migration patterns in the MD region is Model 3* of 2006 without two dummy variables (agriculture and fishery) which is not only representative for all explanatory factors, but also meets the assumptions of a regression analysis.

8.5 Concluding Remarks

This chapter has explored two main questions. It sought to ascertain the determinants of the migration flows out of the commune and whether or not the migration flows of the MD region are relevant to the general situation regarding migration in Vietnam.

Regarding the first question, it is apparent that migration flows in this region have steadily increased in recent years. More importantly, migrants moved from the disadvantaged economic areas (i.e., the low income and agriculture-based economies) to advantaged economic areas (the high income and manufacture-based economies). Ho Chi Minh City and Binh Duong Province are the most attractive destinations for migrants in the country, including in the MD region. Additionally, the determinants of migration flows are revealing. First, "push" factors such as poverty and challenges in farm production are the main reasons associated with the migration flows of people out of the commune. As mentioned in the literature, the result of our estimation also shows that spatial distance is seen as an obstacle to migration; in addition, improved living standards may reduce migration. Second, policy makers can take advantage of tools relating to economic development to adjust migration patterns out of the commune.

In contrast with previous studies of interprovincial migration, the determinants of migration flows in the MD region also have many relevant points concerning the "push" factors at the place of origin. However, there have been different levels of effect among these studies due largely to differences in the number of migrants, socioeconomic factors, and temporal and geographic features. In sum, the findings presented in this chapter are expected to bring us a better and deeper insight into migration patterns at the commune level in the MD region in the present context.

Acknowledgments We have learned a great deal from Prof. Walter Nonneman of the Faculty of Economics, Antwerp University, who gave us technical advice. In addition, we gratefully acknowledge the many participants at the conference "Environmental Change, Agricultural Sustainability, and Economic Development in the Mekong Delta of Vietnam," held on 25–27 March 2010 at Can Tho University, Vietnam, who gave us valuable ideas for this chapter. Finally, many thanks are sent to the editors for their kind comments.

8.6 Appendices

8.6.1 Appendix 1: Migration, Population, and Labour in the MD (Unit: %) (Source: GSO 2008)

	2003	2004	2005	2006	2007
Rate of out-migration	0.23	0.28	0.45	0.7	0.79
Share of rural population	80.2	79.8	79.1	78.9	78.8
Share of agricultural labour	61.9	60.1	59.7	55.4	51.7

Source: GSO 2008

8.6.2 Appendix 2

Summary of Decision No. 587 on benchmarks of poor and far-away communes issued on April 22, 2002 by the Ministry of Labour, Invalids, and Social Affairs.

A commune is seen as a poor commune if such commune has some of the following characteristics:

1. A poverty household rate of 25% or above.
2. A public infrastructure that is absent in up to three of six items including roads, schools, health-care units, clean piped waters, electricity, market systems. More detailed:

   - Less than 30% of households provided with clean piped water.
   - Less than 50% of households using electricity.
   - No paved road to the center of the commune.
   - Less than 70% of total required classrooms for schooling.
   - No health-care unit in the commune.
   - No market to meet consumers' demands.

8.6.3 *Appendix 3a: Summary of Statistics in 2004*

Variables	Obs	Mean	Std. Dev.	Min	Max
Mig-Rate	451	86.735	154.878	0	2147.56
LnMig-Rate	451	3.826	1.217	0	7.67
Density	451	0.565	0.341	0.03	1.96
Farm wage	451	31.933	5.860	5	50.00
Poverty	451	9.707	8.375	0	62.99
Remote	451	0.372	0.484	0	1
Kinh ethnic	451	0.942	0.233	0	1
Agri	451	0.878	0.327	0	1
Indus	451	0.002	0.047	0	1
Fish	451	0.108	0.311	0	1
Living	451	0.993	0.081	0	1
Labour absorption	451	0.620	0.485	0	1
Lack of capital	451	0.487	0.500	0	1
Price fluctuation	451	0.188	0.391	0	1
Access to market	451	0.102	0.302	0	1
Farm technique	451	0.057	0.233	0	1
Job creation	451	0.215	0.411	0	1
Economic development	451	0.255	0.436	0	1
Disasters	451	0.013	0.044	0	0.65

8.6.4 *Appendix 3b: Summary of Statistics in 2006*

Variables	Obs	Mean	Std. Dev	Min	Max
Mig-Rate	480	78.775	109.646	0	1008.79
Ln Mig-Rate	480	3.738	1.273	0	6.92
Density	480	0.557	0.466	0.05	7.86
Farm wage	480	38.900	7.854	0	60.00
Poverty	480	17.585	10.706	0	74.27
Remote	480	0.329	0.470	0	1
Kinh ethnic	480	0.947	0.222	0	1
Agri	480	0.868	0.338	0	1
Indus	480	0.014	0.120	0	1
Fish	480	0.113	0.316	0	1
Living	480	0.993	0.078	0	1
Labour absorption	480	0.668	0.471	0	1
Lack of capital	480	0.475	0.499	0	1
Price fluctuation	480	0.152	0.359	0	1
Access to market	480	0.083	0.276	0	1
Farm technique	480	0.058	0.234	0	1
Job creation	480	0.235	0.424	0	1
Economic development	480	0.173	0.378	0	1
Disasters	480	0.007	0.024	0	0.26

8.6.5 *Appendix 4: Testing for Normality*

Mean = 7.2e-10	Std.dev. = 1.218	(n = 480)
Median = 0.1155	Pseudo std.dev. = 0.9994	(IQR = 1.348)
10 trim = 0.0753		
	Low	High
inner fences	–2.635	2.757
Number mild outliers	18	5
% mild outliers	3.76%	1.04%
outer fences	–4.657	4.78
Number severe outliers	0	0
% severe outliers	0.00%	0.00%

References

Anh DN, Goldstein S, McNally J (1997) Internal migration and development in Vietnam. International Migration Review 31:312–337

Brown O (2008) Migration and climate change. IOM migration research series, 31, Geneva: International Organization for Migration

Chatterjee S, Hadi A, Price B (2000) Regression analysis by example, 3rd edn. Wiley, New York

De Brauw A, Harigaya T (2007) Seasonal migration and improving living standards in Vietnam. Ameri J Agri Econ 89:430–447

Deshingkar P, Grimm S (2005) Internal migration and development: a global perspective, IOM Migration Research Series 19. International Organization for Migration, Geneva

General Statistical Office (2006) Vietnam's nationwide survey on agriculture and aquaculture in Vietnam. General Statistical Office, Hanoi

General Statistical Office (2007) The 2007 population change and family planning survey. General Statistical Office, Ha noi

General Statistical Office (2008) Statistical yearbook of Vietnam in 2008. General Statistical Office, Hanoi

Guest P (2003) Bridging the gap: internal migration in Asia. Paper prepared for conference on African migration in comparative perspective, Johannesburg, South Africa, June 4–7

Hossain MZ (2001) Rural-urban migration in Bangladesh: a micro-level study. Paper presented in a poster session on internal migration at the Brazil IUSSP Conference, August 20–24

International Organization for Migration (2005) Migration, development and poverty reduction in Asia. International Organization for Migration, Geneva

Ivan E (2008) Internal migration: a review of the literature. MPRA paper 8783. Munich: Personal RePEc Archive

Lee SE (1966) A theory of migration. Demography 1:47–57

Lewis WA (1954) Economic development with unlimited supplies of labor. Manchester Sch 22:139–191

Loi CC (2005) Rural to urban migration in Vietnam. Discussion Paper, Institute of Developing Economies, Japan External Trade Organization

Lucas REB (1997) Internal migration in developing countries. In: Rosenzweig MR, Stark O (eds) Handbook of population and family economics. Elsevier Science, Amsterdam/New York, pp 721–798

Ninh NH (2007) Flooding in Mekong River Delta, Vietnam. Human Development Report 2007/2008, Human Development Report Office, Occasional paper, 2007/53. New York: United Nations Development Programme

Phan P, Coxhead I (2007) Inter-provincial migration and inequality during Vietnam's transition. Staff Paper, 507. Madison: University of Wisconsin-Madison

Rhoda R (1983) Rural development and urban migration: can we keep them down on the farm? Int Migr Rev 17:34–64

Skeldon R (2003) Migration and poverty. Paper presented at the conference on African migration and urbanization in comparative perspective, Johannesburg, South Africa, June 4–7

Todaro PM (1969) A model of labor migration and urban unemployment in less developed countries. Am Econ Rev 59:138–148

Turnham D (1993) Rural development policies and their effects on migration. OECD Development Centre, Paris

UNDP (2009) Overcoming barriers: human mobility and development. Human Development Report 2009. Available at http://hdr.undp.org/en/media/HDR_2009_EN_Complete.pdf

UNFPA (2007) Internal migration in Vietnam: the current situation." Report of United Nations population fund in Vietnam

VCCI-Can Tho (2006) Vietnam Chamber of Commerce and Industry – Can Tho Branch, available at http://www.mekongdelta.com.vn/mekongdelta/Solieu/GDP.htm#B?ng_2.2

Wooldridge J (2009) Introductory econometrics: a modern approach, 4th edn. South-Western, Mason

Zhang HK, Song S (2003) Rural-urban migration and urbanization in China: evidence from time-series and cross-section analyses. China Econ Rev 14:386–400

Zhao Y (1999) Leaving the countryside: rural to urban migration decision in China. American Economic Review 89:281–286

Chapter 9
Dynamic Resilience of Peri-Urban Agriculturalists in the Mekong Delta Under Pressures of Socio-Economic Transformation and Climate Change

Matthias Garschagen, Fabrice G. Renaud, and Jörn Birkmann

Abstract Globally and in Vietnam, coupled social-ecological systems in the peri-urban fringes are amongst the most dynamic as well as strained systems as they are at the same time drivers and results of comprehensive transformation processes. Based on a literature and policy review, we argue that – globally and especially in Vietnam – there is a neglect of the specific needs and challenges faced by populations in the peri-urban interface. At the same time, the chapter suggests that, in the Mekong Delta in particular, agriculture-based population groups in the peri-urban areas are at risk of being negatively affected by overlapping trends of (a) socio-economic transformation (including, e.g. liberalisation or urbanisation), (b) biophysical degradation (including, e.g. pesticide residues in water bodies) and (c) climate change impacts. This hypothesis is tested based on empirical research in one of the most rapidly urbanising districts in the Mekong Delta, that is, Cai Rang District in the peri-urban fringe of Can Tho City. We find that formerly agriculture-based population groups in this district are facing substantial decreases in resilience due in particular to the effects of expropriation, relocation and inadequate compensation schemes. In addition, the population of the district will experience changes in water-related pollution patterns as sources of contaminants are likely to shift from mainly agrichemicals pollution to microbiological contaminations and pollution from other 'new' pollutants such as endocrine disruptors, all linked to changes in land use patterns and industrialisation. On top of this, climate change is in the future likely to imply increasing difficulties for those groups which – in view of the already deteriorated baseline resilience – pose substantial risks of tipping into a crisis situation due to a lack of resources and options for adaptation and coping. This can be true for single households, extended family networks or entire groups (e.g. populations in specific relocation clusters). Yet, the chapter argues that resilience effects may not be understood in a functionalistic or deterministic manner, meaning that entire systems *per se* will inevitably move along a given resilience

M. Garschagen (✉), F.G. Renaud, and J. Birkmann
WISDOM Project, United Nations University, Institute for Environment and Human Security (UNU-EHS), Hermann-Ehlers-Str. 10, 53113 Bonn, Germany
e-mail: garschagen@ehs.unu.edu

M.A. Stewart and P.A. Coclanis (eds.), *Environmental Change and Agricultural Sustainability in the Mekong Delta*, Advances in Global Change Research 45,
DOI 10.1007/978-94-007-0934-8_9,

trajectory. Rather, a review of overlaying trends in Vietnam as well as empirical case study analysis shows that resilience pathways are highly dynamic and depend on small-scale factors setting the directions. Resilience, therefore, can be differentiated between households or individuals. Acknowledging the importance of those small directive elements within resilience pathways opens up opportunities for resilience management and governance.

Keywords Resilience • Peri-urban interface • Agriculture • Transformation • Climate change • Vietnam

9.1 Introduction

Agriculture is amongst the land use types with the most direct and intensive interactions between humans and their biophysical environment. Over the last decades, various theoretical schools have developed which allow us to illustrate the dependencies of humans on ecosystem goods and services and which focus on sustainable environmental management (e.g. human ecology, political ecology, footprint research; Young 1974; Blaikie and Brookfield 1987; Robbins 2004; Wackernagel and Rees 1996; Millennium Ecosystem Assessment 2005). Land use management plays a key role in these discourses. However, in many studies the regions with most intensive land use transformations are left out from analysis: that is, the peri-urban interface with its dynamic hybrids of urban and rural land use patterns. Yet, these areas play an increasingly important role. First, they are of substantial quantitative relevance given strong urbanisation trends globally and, in particular, in developing countries and emerging economies. Second, peri-urban areas in developing countries often suffer from (a) growing biophysical and socio-economic stress resulting from agricultural intensification, (b) utilisation for urban waste disposal and (c) economic and social transformation – often coming along with land reforms or resettlement – which imply significant livelihood challenges for the original (often agriculture-based) population.

In addition to this existing stress, climate change will add to the perturbations in peri-urban areas as negative impacts on both rather rural and rather urban profiles are likely to be felt simultaneously. As with agriculture in other settings, peri-urban agricultural producers will feel the impacts of climate change in very direct terms due to the fact that changes in temperature-, precipitation- and flooding-patterns directly influence agricultural production. At the same time, peri-urban areas are at great risk with respect to environmental-technical-disasters related to man-made or natural hazards (of which the latter can be induced or intensified by climate change). For example, industrial production units or waste dumps – which bare a substantial risk of environmental pollution if flooded or destroyed by storms – are often located in the outskirts of cities. In addition, peri-urban areas are likely to be at the forefront with respect to receiving the increasing numbers of migrants who will no longer be able to sustain their livelihoods in rural areas due to climate change and who move into urban areas in search of new income sources. They in turn compete for land and increase both the pressures for the original population and the stress on bioyphysical systems.

Despite this present and future stress on peri-urban social-ecological systems, the resilience of peri-urban agriculture-based populations and their capacities to respond to existing and future challenges are in many parts of the world not well understood. Yet, a detailed understanding of these processes is necessary in order to allow for adequate policy and decision making. This is of particular importance since the reasonable management of those areas can make the difference between achieving a (rather) sustainable coupled social–ecological urban-hinterland system or an unsustainable and highly vulnerable system.

This chapter, therefore, generates lessons learned from an in-depth, interdisciplinary analysis on the resilience of households in Cai Rang District, Can Tho City. This district is one of the most rapidly transforming peri-urban areas in Vietnam, which in turn is one of the countries most intensively affected by overlaying dynamics of socio-economic transformation, ecosystem change, urbanisation and climate change risk.

9.2 The Peri-Urban Interface: Neglected Hybrid in an Ontologically Dichotomous Landscape

In much of the conventional scientific and political discourse, the terms 'urban' and 'rural' are used to represent mental conceptions of some archetype landscapes and related characteristics (Simon et al. 2006). These archetype conceptions are very often used in a dichotomous and mutually exclusive manner, thereby, constituting the pre-analytic vision for much of the past and current policy making, planning or research – often lacking critical validity- and reality-checks. Yet, in an increasingly urbanising world – in particular in developing countries – this ill-perceived dichotomy has to be challenged as it systematically contributes to a neglect and misconception of the peri-urban interface, that is, where the city fringes intermingle with their hinterland to build hybrid mosaics of land use, economic activity, demographic density, administration, socio-cultural dispersion and biophysical environments.

The peri-urban interface is thereby at the same time driver and result of cross-scale global change phenomena (comprising ecological, social, economic and cultural dimensions) (Johnston et al. 2002), as manifested in site-selections of diverse economic actors as well as in various flows of people and resources. This can, for example, include inbound migration of (rural) poor in search of cheap and accessible grounds for settlements (often informal or illegal). At the same time, there might be urban rich and upper middle-class people who are looking for uncongested sites for housing and leisure activities (often linked to the desire of being 'surrounded by nature'). In addition, industrial enterprises increasingly choose to build production sites in the peri-urban areas as the infrastructure inside cities is inadequate or land prices might be too high. Moreover, the urban fringes often host critical infrastructure elements such as airports, ring roads, power stations, transmitter stations, waste dumps or sewage treatment plants.

The above-mentioned aspects make clear that defining the peri-urban interface through static threshold indicators (e.g. population density or percentage-shares in land use patterns) would only be of limited use and would be prone to producing a false sense of clear-cut demarcations and uniform entities. Key elements for defining and grasping the complex peri-urban interface and the respective areas rather have to be seen in (a) a continuum understanding and (b) a process understanding (Simon et al. 2006: 10). This implies that, 'theoretically, a peri-urban zone may change in width and the steepness of what we might call its rural-urban gradient over quite short periods of time, depending on the nature of pressures within the growing metropolis and of migration towards it' (ibid.).

Despite the fact that this hybrid landscape constitutes the environment for a rising number of people, particularly in the global South, remarkably little attention has been given to the question as to how far it shapes livelihoods in the peri-urban areas. It has been noted that much of the published work on livelihoods focuses on an (allegedly) purely urban or purely rural setting, even though the livelihood framework in general recognises that livelihood strategies often embrace hybrid activities in rural, peri-urban and urban areas (Simon et al. 2006: 8). In this context it has also been argued that, in developing countries, poor people in the peri-urban interface are particularly vulnerable as they are affected by the 'worst of both worlds' (Birley and Lock 1998) since the peri-urban interface is subject to the negative externalities of both, the nearby urban and rural areas (Allen 2006: 30). Very often, peri-urban poor depend heavily on natural resources and ecosystem services as their livelihoods are in most cases (entirely or partly) based on agriculture, horticulture, animal husbandry, forestry or fishery. Hence, these groups are to a great extent affected by environmental degradation in the peri-urban fringes. They, thus, often face a double burden from this degradation, that is, the impacts on the livelihood base in addition to direct health risks (Allen and Davila 2002). At the same time, poor peri-urban dwellers are likely to suffer from steep increases in land prices that can in many parts of the world be observed in the course of urban sprawl and growth. Very often, those dwellers are, therefore, priced out by middle- or upper-income groups and the business sector or are even expropriated by the government, often with poor and insufficient compensation regimes, in particular in cases of unclear or informal land title. Resulting from this, the affected groups are in many cases forced into temporary shelters and lose their opportunity for subsistence farming or income generation (Allen and Davila 2002).

These problems are aggravated by the fact that peri-urban areas face substantial challenges with respect to policy formulation and implementation. First, the peri-urban interface as defined above usually comprises areas within different administrative units. They are frequently located in parts within the boundaries of urban administrative units and rural districts, respectively. Hence, responsibilities, e.g. for land use zoning, infrastructure development or environmental management, are distributed amongst different authorities resulting in a fragmented planning and management landscape that hampers integrative planning (Simon et al. 2006: 12). Second, Davila noted that environmental management for the peri-urban area is often 'falling between the stools' not only with respect to geographical dimensions but also in regard to the conceptions and configurations of planning and management (Davila 2006: 45). Policies with a primarily spatial focus (such as urban or regional master plans) in most cases do not comprise an explicit environmental dimension and are usually designed on the basis of territorial boundaries that do not follow the logic and organisation of ecosystems (ibid.). Many sectoral or macroeconomic policies, on the other hand, are lacking an explicit environmental focus, despite having a great impact on the environment in peri-urban areas. Poverty reduction strategies or socio-economic development plans often do not give explicit consideration to environmental issues but rather to the strengthening of market mechanisms, production and trade – which do depend on and affect environmental conditions, especially in the peri-urban fringe (Davila 2006: 50).

However, although policies directed towards the peri-urban interface *per se* are lacking, it has been argued that the response should not be to call for exactly these. To do so would risk adding another layer of actors, new bureaucracy and additional fiscal and regulatory burdens and would complicate integrated planning and management even further (Davila 2006: 53). Rather, policy-makers should call for a greater awareness of the specific needs and challenges faced by peri-urban ecosystems and populations and towards the effects that many spatial and sectoral policies imply for the peri-urban interface. In this context, increased consideration needs to be given to impact assessments and the integration of policies and plans across horizontal, vertical and sectoral divides.

9.3 Vietnam: Dimensions of Transformation, Climate Change and Resilience Effects

In the recent past, Vietnam has been undergoing multifaceted transformation processes in the economic, political, social, cultural and environmental domains which influence each other in multiple ways and which have strong impacts on the specific development pathways and vulnerabilities of different population groups. Those changes and resilience effects manifest themselves in particular in the country's peri-urban areas, due to the speed, magnitude and new qualities of overlaying processes in these regions.

Vietnam is also highly exposed to various future climate change impacts. These will likely have substantial (potentially devastating) impacts on the resilience of certain population groups – that means on the balance between, on the one hand, the level of stress and the magnitude and frequency of perturbations these groups are exposed to, and, on the other hand, their capacities to cope and adapt.

The following paragraphs explore the dimensions of transformation in detail. One section is devoted each to transformations within the agricultural sector and urbanisation, repsectively. Climate change risk will be illustrated in some more detail. Finally, the resulting resilience landscape for Vietnam is sketched out. The latter will be done in rather general terms here, in order to prepare the grounds for the more specific analysis on dynamic resilience pathways within the peri-urban case study presented afterwards.

9.3.1 *Political and Economic Transformation*

In response to chronic agricultural and industrial underproduction, thriving black markets, declining purchasing power, and the impending breakdown of the centrally controlled economy, the government of Vietnam initiated an economic and political reform process in the late 1970s towards a more market-oriented economy (Luong 2003: 8 et seq.; Dang 2007: 10 et seq.). Several plenary sessions of the Party Central Committee as well as Party Congresses during the late 1970s and early 1980s passed directives and decrees that implicitly initiated a shift towards a market-based economy. However, the breakthrough happened at the 6th Party Congress (December 1986) which embraced renewal (*doi moi*) of the political, economic, social and cultural system as an official policy line (Trong 2007: 25). The *doi moi*-process was continued in the late 1980s and 1990s with numerous additional legislative and regulative reforms, which strengthened the influence and freedom of private economic activity in the fields of, for example, investment, trade and property rights or land titles. The result has been strong economic growth with an average of 8% annually between the years 1990 and 1997 and with the number of private domestic enterprises rising from 318 to 5,714 between 1988 and 1998 (Luong 2003: 11).

9.3.2 *Urbanisation*

In combination with the economic transformation in the context of *doi moi*, Vietnam has over the last decades been experiencing a strong push towards urbanisation, which is expected to continue or even intensify in the future. While in 1985 (i.e. shortly before the official commencement of *doi moi*) less than 20% of the country's population was living in cities and towns (equalling some 11.5 million people), the figure has risen to nearly 30% in 2010 (accounting for over 26 million) (UN/DESA 2008). With the average annual urban growth rate only slowly falling from 2.9% (for the decade 2010–2020) to 1.7% (2040–2050), Vietnam's urban growth is expected to remain substantially above the Southeast-Asian average over the next decades. This means that an expected 42, 49 and 57% of the country's population will be urban by 2030, 2040 and 2050 respectively, then equalling some 46, 58 and 68 million people (ibid.). Cities and towns will both increase their density of population, buildings, and infrastructure as well as spread in their geographical extent. Regarding the latter, it has been estimated that currently around 100 km^2 of land are converted from agriculture to urban land use every year (Coulthart et al. 2006: x).

Because of the combination of densification and sprawl as well as the related effects of resource consumption and pollution, cities will have an increasing environmental impact on their own territory as well as on the surrounding peri-urban and rural hinterlands. At the same time, the concentration of populations and infrastructure can imply opportunities for a more sustainable land-use given the potential scale effects and synergies (e.g. with respect to transportation, cooling of buildings or sewerage infrastructure). Good planning and management in particular in peri-urban and urban areas is, therefore, of increasing importance with respect to overall (i.e. cross-sector and cross-scale) sustainable development of the country.

Urbanisation in Vietnam has to be seen in close correlation to the sensitive question of the (normative) role of cities within the social and economic fabric of the Socialist Republic. As with many other non-European socialist governments, the Vietnamese political elite for a long time had an ambivalent attitude towards urbanisation – which, given the specific history of Vietnam, contributed to different urbanisation pathways in the North and South of the country; the effects of these can be observed until today (c.f. Drakakis-Smith and Kilgour 2001: 220 et seq.). On the one hand, there has been recognition that urbanisation is an important prerequisite for the promotion of industrialisation which is in turn key to building up an independent economy and a strong national defence (Turley 1977: 624). On the other hand, the Vietnamese economy has long been heavily based on agriculture. Urban industries have played a comparatively minor role and agriculture has been a major defining element in the national identity. Linked to this, the socialist movement in Vietnam was very much rural-based (Drakakis-Smith and Kilgour 2001: 219) – which is certainly linked to the importance of agriculture in the country's social fabric and national history even before the arrival of socialism. In particular, Southern cities (and especially Ho Chi Minh City) were, after the reunification, seen as 'lairs of American imperialism and its puppets', which has in the early years contributed to an anti-urban bias amongst many cadres of the reunified socialist Vietnam (Turley 1977: 622).

This ambiguity translated for a long time in a somewhat vague and inexplicit handling of urban areas and their specific development challenges. The questions of whether urbanisation should be rather stimulated or prevented by official policy and how much attention should be given to funding and developing urban areas were therefore not answered coherently for many years. Resulting from this was a poor administrative and legal framework for the planning and management of urban areas which in combination with an overall lack of financial resources (internally as well as in regard to foreign investment) led to an 'urban neglect' within the first years after reunification, resulting in substantial shortcomings with respect to infrastructure and housing development (Yeung 2007: 272 et seq.; Coulthart et al. 2006). In addition, outside the large cities like Ho Chi Minh City, Hanoi or Da Nang, urban management and planning was often not recognised as a profession in its own right. Hence, medium-sized and small cities were despite substantial (informal) growth and specific urban challenges often administered by political cadres with a background in rural management.

In 1998, after more than 10 years of substantial economic growth following the commencement of *doi moi*, the national government passed an Orientation Master Plan for Urban Development to 2020, which collated an explicitly urban development strategy in response to the challenges for sustainable development of Vietnam's cities (SRV 1998). The plan sets out the goal of a more balanced urban development in which the growth of the large urban agglomerations shall be slowed down in order to prevent excessive regional disparities and the emergence of mega-urban areas that grow beyond control. At the same time, small and medium cities shall be fostered and specific cities promoted as industrial hubs, constituting locomotives of economic growth. Competitive elements in the search for national funds have been strengthened in tandem with substantial financial and administrative decentralisations. These decentralisations have in the following years been further supported through additional legislative reforms, most importantly the Amendment to the State Budget Law (2002), the new Land Law (2003), the new Law on Construction (2003) and the new Law on Urban Planning (2009) (SRV 2002, 2003a, b, 2009, respectively) which all entail shifts of tasks and responsibilities of urban planning and management from the national government to administrations on the provincial and district level. In spite of the increased attention towards urban areas and development challenges, peri-urban areas with their specific conditions and needs continue to be neglected and still fall through the cracks of rural and urban policy. The Orientation Master Plan, for example, does not give explicit consideration to the peri-urban interface, nor does the new Law on Urban Planning.

9.3.3 Agriculture

Given its high importance for the country's economy, its key role as a basis for livelihoods and its great normative significance within the ideological fabric of the Republic's political system, the agricultural sector is of particular importance for being able to understand Vietnam's development and renewal process as well as the resulting resilience effects.

Agriculture plays a key role in the economy of Vietnam and the Mekong Delta is the most intensively farmed area in the country. The Delta provides the majority of the production of various commodities (GSO 2009a, b). Yet, the extent of productive agricultural land is threatened in two ways: first by climate change impacts (see below); second by urbanisation, as many rural areas (particularly – but not only – Can Tho City) are transformed into peri-urban areas and peri-urban areas become more urbanised.

In rural and peri-urban areas of the Mekong Delta, intensive agricultural production has already had impacts on ecosystems. For example, pesticide monitoring programmes that were conducted in 2008 and 2009 at field outlets and in channels used for irrigation, in one strictly rural area in An Long District, Dong Thap Province, and one peri-urban area in Cai Rang District, Can Tho Province, indicated that a broad range of recently used pesticides are co-occurring at detectable levels in these systems (Toan et al. 2009). This can potentially have direct negative effects on aquatic ecosystems. But more importantly, this could have negative repercussions on human health as ongoing studies (research of Pham Van Toan, in UNU-EHS 2010) show that pesticides are detected in drinking water samples from water taken out of canals – which during the dry season are a major source for drinking water.

Going through an urbanisation process automatically implies land use changes. These are not restricted to agricultural land being set aside for urban land uses (e.g. buildings, infrastructures), but may as well imply shifts in production systems on the land that remains devoted to agriculture. An example could be a shift from rice production to fruit tree orchards: With improved access to markets, fruit can be sold more easily and provide higher returns to farmers. Shifts in land use patterns in rapidly urbanising areas such as Cai Rang imply pollution problems too. For example, with the loss of agricultural land, pesticides – as pollutants to the water system – may be progressively replaced, for instance, by bacteriological contaminations linked to inadequate sewerage systems which service new infrastructure developments. Aquatic ecosystems and people who rely on these may then progressively be exposed to new pollution problems. The link between ecosystems, the services they provide to communities, and the social component of this coupled system can be disrupted, requiring some adaptation from the communities.

Shifts in pollution problems can occur concomitantly with effects of climate change. If we focus on the agricultural sector only, Sebesvari et al. (Chap. 13, this volume) show the various ways in which climate change can affect pesticide pollution problems through the many interconnected processes of land use change, pest occurrences and farmers' behaviour. Concomitant changes can also take place if pollution problems shift from, for example, pesticide pollution to bacteriological pollution. At the same time, populations in maladapted peri-urban areas may face increased flooding problems linked either to the effects of the combination of sea level rise and changes in discharge patterns of the Mekong River or changes in rainfall patterns that may overwhelm drainage systems; flooding in general can intensify or multiply the problems of pollution. Here again, communities and ecosystems will have to adapt, if a relationship between them is to be preserved.

9.3.4 Climate Change Risk

A number of recent studies identify Vietnam to be amongst the countries most at risk from climate change impacts (e.g. Dasgupta et al. 2007; McGranahan et al. 2007; Carew-Reid 2008). With its long coastline and large deltas featuring high concentrations of population, agriculture and industry, Vietnam has high degrees of exposure to intensifying natural hazard such as floods or typhoons. In addition, underlying creeping changes in precipitation patterns and temperature but also sea level rise and resulting salinisation of water bodies and soils pose substantial risks, especially to agricultural production. The Mekong Delta in particular faces substantial risks given its low-lying topography and its key function in terms of agricultural production for Vietnam and the global food market.

According to the Climate Change and Sea Level Rise Scenarios developed by the Ministry of Natural Resources and Environment, a sea level rise of 75 cm – under a medium emission scenario expected until 2100 – would directly inundate 19% of the Delta, based on current protection measures and hydraulic infrastructure (MoNRE 2009). However, indirect impacts such as salinisation and changing hydrological regimes would affect a much larger share of the Delta. The same scenario would imply an increase in the annual mean temperature of 2.0°C by 2100 for the southern parts of Vietnam (ibid.). Overall annual rainfall would increase by 1.5% on average in Southern Vietnam with larger seasonal fluctuations, with larger precipitation decreases in the dry season and concomitant increases in the rainy season (ibid.: 27).

A combination of these and other changes on a regional-scale (changing monsoon and El Nino patterns and human activity such as hydraulic infrastructure for embankments or hydropower stations) and changes in flood patterns have already been observed. These changes are expected to be ongoing, which is likely to increase the frequency as well as magnitude of extreme flooding events (Tran et al. 2008; Wassmann et al. 2004; CFSC 2004). Moreover, typhoon activity – so far predominant in the northern and central parts of Vietnam (Kleinen 2007; Kelly et al. 2001) – is expected to intensify also in southern Vietnam (CFSC 2004).

9.3.5 Resilience Landscape

In all, the transformation processes described above result in a complex resilience landscape with overlaying, yet differentiated resilience effects for particular population groups. While Vietnam has experienced average economic growth rates of 7.5% over the last years and a general decrease in poverty rates (c.f. Carew-Reid 2008) the market liberalisation reforms under *doi moi* and accompanying land reforms have led to an increase in socio-economic disparities (Taylor 2004; Adger 2000a; Waibel 2005) and produced high numbers of landless farmers, particularly in the Mekong Delta (Marsh and MacAulay 2006). Statistical surveys have, for example, shown that the number of landless households has increased from 12,500 in 1994 to around 1,000,000 in 1998 (ibid.: 7). It has in addition been shown that strong correlations exist between poverty on the one hand and land title and size on the other (Tuan 2010). In this respect, Marsh and MacAulay further note that the general decrease of people living in poverty – which is observed across different measurement approaches – has to be taken with caution since 'a high percentage of the population is bunched just above the poverty line and a relatively small deterioration in living standards would be sufficient to push them below the poverty line again' (2006: 8).

Many studies emphasise that poverty is not to be mistaken as an equivalent measure to vulnerability or resilience, but that it certainly describes an important factor of vulnerability. Poverty regulates access to an increasing basket of services and assets within the socio-economic fabric as it changes in the direction of a market-based economy and the individualisation and privatisation of social security networks (Taylor 2004), including in particular reforms in health care coverage and social insurance schemes (c.f. Ekman et al. 2008; Wagstaff 2007).

A further factor contributing to the complex resilience landscape is to be found in intensifying export orientation and the integration into global commodity markets, which besides growing export revenues also imply soaring dependencies on global food markets, and, hence, increased vulnerability to price fluctuations. In this respect, agricultural producers in Vietnam are subject to what Adger et al. call nested and teleconnected vulnerabilities in the sense that vulnerabilities of people within one place can be linked and influenced through, for example, shifts in policies or market trends originating in other (distant) places (Adger et al. 2009). Lynn et al., therefore, call for a paradigm shift towards focusing on 'hot systems' instead of using the conventional and more static notion of 'hot spots' (Lynn et al. 2010). O'Brien and Leichenko have linked the pressures that climate change and globalisation can simultaneously have on agricultural communities and have introduced the concept of double exposure, the idea that each of those trends may create perturbations that no longer can be considered separately but which interact and overlay to create a new, complex nexus of increased exposure and vulnerability (O'Brien and Leichenko 2000). All these conceptual approaches give fruitful guidance for exploring and understanding the interactions of overlaying processes that create the dynamic and complex resilience landscape that is specific to Vietnam and especially the peri-urban areas in the Mekong Delta.

In addition, the intensification of agriculture can – as shown above – imply substantial risks of ecosystem degradation, putting at risk particular agricultural population groups which directly depend on the productivity of those services – a problem which has been observed to be of increasing relevance in the Mekong Delta (Ni et al. 2001). In combination with climate change impacts, those degradations may pose the risk of unexpected feedbacks in coupled social-ecological systems, including, for example, pest outbreaks or sudden drops in carrying capacity. Such events could in turn cause decreases in productivity and yield, putting the livelihood bases of agricultural populations at risk.

Investigators who have given attention to natural hazards and climate change have further argued that transformations in the political system and related shortcomings with respect to the layout and implementation of political decentralisation have caused a reduction in collective action for risk management, despite the emergence of civil institutions which have begun to shape a new landscape of institutional adaptation (Adger 1999; Tran et al. 2008). As a result, increased vulnerabilities can be observed amongst those groups who do not have sufficient resources to substitute the formerly socialised elements of risk management with action on an individual or household level (Adger 2000a). In addition, climate change requires increased efforts and resources for adaptation, which will likely exceed the capacities of certain population groups, increase their baseline vulnerability, and increase their risk of tipping into serious crises in cases of extraordinary natural hazards (Garschagen et al. 2009).

9.4 Case Study: Cai Rang District

Cai Rang District is one of nine districts within Can Tho City, which is the biggest city and economic centre of the Mekong Delta, located about 180 km southwest of Ho Chi Minh City (compare Fig. 9.1). In 2009, Can Tho City had a total population of 1.19 million, of which 86,000 were living in Cai Rang. This district belongs to the less populated districts of Can Tho, peripheral to the neighbouring Ninh Kieu District at the urban core of the Province and accounting for 244,000 inhabitants (compare Fig. 9.1). However, Cai Rang has recently commenced the significant build-up of residential and industrial developments, which will make it one of the most rapidly urbanising districts in the Mekong Delta over the next several years. The development master plan until 2025 envisages developing Cai Rang into an urban-port-industrial zone by 2025 (SRV 2006). This plan includes the development of new residential quarters for 120,000–150,000 people, covering an area of 700–800 ha, in particular in Hung Phu, Hung Thanh and Phu An wards (ibid.: 5.b). Furthermore, two industrial parks (called Hung Phu I and II) are planned, covering

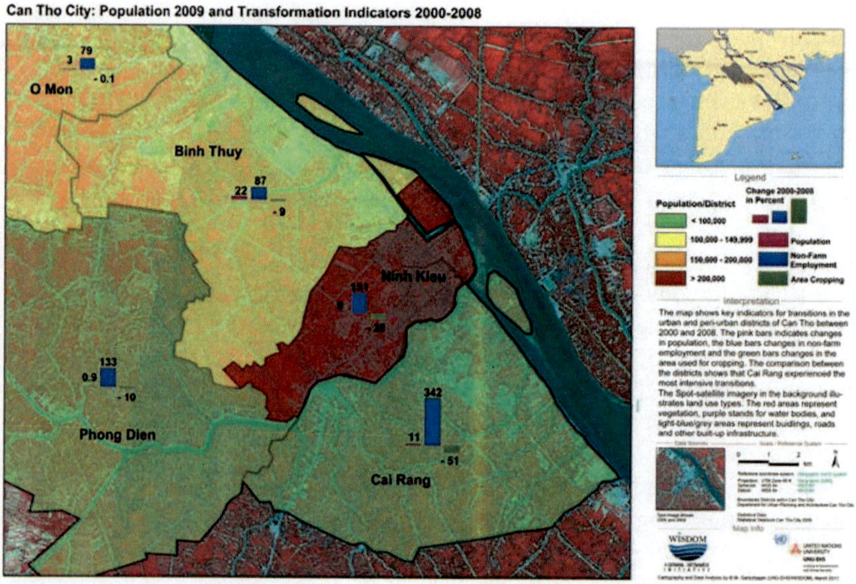

Fig. 9.1 Map of selected districts in Can Tho City and selected transformation indicators from Statistics Office Can Tho City 2010

an area of 600–700 ha. Attached to these, a new port (Cai Cui Port) is being built at the Hau River in the south of Cai Rang; this is planned to become a seaport, capable of handling large-sized ocean freight vessels. The port will become the commercial gateway that makes the shipment of goods produced in the Mekong Delta much easier and will improve the competitiveness of the agricultural and industrial producers in the region. In addition to development for residential and economic uses, the master plan sets aside 120 ha for a new cultural centre, as well as land for several other facilities such as commercial centres, hospitals, and golf courses or roads (SRV 2006). One particular characteristic of the district's development is the new bridge for Highway 1 (connecting the Mekong Delta to Ho Chi Minh City) across the Hau River lands in Cai Rang. This is causing many important distributor roads as well as new parts of Highway 1 to run through the district, and these infrastructure changes will likely have a catalytic effect on Cai Rang's formal as well as informal economic development.

Until the recent past, the majority of Cai Rang's population based their livelihoods on a combination of (a) agricultural activities (for subsistence purposes and as the main contributor to income generation), (b) other activities in the industry, service, or trade sector (often being located in the urban core of Can Tho, i.e. in Ninh Kieu District or adjacent wards of Cai Rang and Binh Thuy) or (c) on remittances or social welfare (compare Fig. 9.1). By and large, the livelihoods of Cai Rang's population is characterised by the aforementioned mixture of income-earning activities in the peri-urban interface and this does not fit into prevailing conceptions of rural versus urban livelihood patterns.

Because many of the new developments have already been under construction for a couple of years, transitions in the land use statistics of Cai Rang can already be observed. In particular, the area for cropping has been shrinking substantially, amounting to a reduction of more than 50% between 2000 and 2008 (Statistics Office Can Tho City 2010) (compare Fig. 9.1). Our field observations and discussions with households in 2008 revealed that some agricultural land was left uncultivated or orchards were left unattended because land was earmarked for development, despite the fact that construction (of roads, bridges or residential areas) had not yet started. Related to this, the number of people employed in the non-farm sector increased almost 350% between 2000 and 2008, with the population growing 11% during the same period of time (Statistics Office Can Tho City 2010) (compare Fig. 9.1).

Current and future land use change dynamics will directly and indirectly impact the environment and the provisioning of ecosystem goods and services. Large portions of Cai Rang are currently a major construction site where environ mental norms are likely not respected to the letter. In addition to construction-related pollution, land use changes will alter land and water pollution patterns in various ways. As already indicated above, our research related to pesticide pollution clearly showed that agrichemicals can be found in concentrations exceeding water quality guideline values for some of the monitored compounds. This is because pesticide management is absent in the intensive agricultural system in place in Cai Rang which includes rotations of vegetable crops with rice as well as fruit orchards (research by Pham Van Toan, in UNU-EHS 2010). With some land left fallow and fruit orchards left un-managed, the quantity of agrichemicals used in the district (or at least in some localities of the district) is most likely lower than under normal circumstances. This can currently have beneficial impacts for land and water ecosystems as well as on human health. But given the current development trends, new types of contaminants or higher concentrations of contaminants are likely to emerge, including microbiological contaminants should untreated sewerage be poured directly into the canal and river systems. Emerging contaminants such as endocrine disruptors, which can be found in urban and industrial settings in the Mekong Delta – but less so in Cai Rang as of now (Hoa N.T., PhD researcher in the WISDOM project, data not yet published) – could also become prevelant. This could make it increasingly difficult for the remaining farmers (who have managed to sustain their agricultural land and production) to rely on the available water for drinking and irrigation purposes. Finally, with urbanisation has come the sealing of land surfaces. Should urban development patterns in Cai Rang follow the same approaches of those in Ninh Kieu District, the infiltration capacity of soils will drastically decline. With the potential aggravating effects of climate change on slow-onset tidal flooding as well as on the intensity and frequency of heavy rain events in the region, flooding will most likely affect the district in a more pronounced way in the future. Padgham (2009) identified flooding as one of the highest risk factors for peri-urban agriculture together with access to water for irrigation and the potential magnification of urban heat island effects. This could lead to secondary impacts including additional pollution of land and water systems in the region through the inundation of infrastructures such as petrol stations, construction sites, and the like. The remaining farming communities in the region are therefore likely to have to deal with different types of water pollution in the long run (both for irrigation and consumption purposes) and a likely increase of seasonal flooding caused by the double constraint of climate change and urbanisation. They will also have to adapt their farming practices to deal with potentially increased pest problems linked to climate-change-related increases in extreme climate events, changes in soil water contents, and other factors (see, e.g. Padgham 2009).

In addition to the stress added to local livelihoods through those environmental issues, agriculturalists in Cai Rang face substantial economic pressures caused by peri-urban developments. The statistics on land use change dynamics given above already point towards what appears to be the most pressing challenge in the moment: Many former residents in the agricultural sector lose their agricultural land and have to resettle away from their original plot of land. Underlying this situation is a complex political economy including elements of being priced out, expropriated, relocated and poorly compensated. Long before the current construction master plan had been officially approved by the Prime Minister in September 2006, private property developers and real estate companies started to acquire land plots from farmers and other residents. Many of the households interviewed in 2009 reported that respective transactions with those companies go back 6–9 years (i.e. between the years 2000 and 2003).

As all land in Vietnam eventually belongs to the people (i.e. it is property of the state), the legislative framework for land transactions in the context of new developments establishes that the government gives the right to the developer (either a private company or the state itself in cases of roads, harbours and other public developments) to build the development as specified in the construction master plan. The developer then has to arrange the clearance of the land as well as the compensation for it in dialogue with the former land holders. The latter are under the Land Law entitled to have long-term individual land use rights, including the right to lease, transfer and mortgage land. The clearance and compensation procedure is supported by a site clearance committee with strong involvement of the local government which is supposed to mediate between the interests of the developer and the former residents or users. In addition, the government publishes a yearly updated price list as a yardstick for compensation rates in specific locations and for different land use categories. In theory, these should be a close representation of the actual market value of the respective land plots (for a more detailed description of the legal framework regarding compensation refer to the Land Law of 2003 (SRV 2003b) and the Circular 114/2004/TT-BTC on determining land prices or Han and Vu (2008)).

In Cai Rang, two schemes of compensation have been applied which in many cases are combined to build one hybrid mode. The first approach envisages that households receive monetary compensation representing the value equivalence of the land as well as of special assets such as the house or fruit trees. This scheme is locally referred to as 'money for land'. The second scheme in general aims at compensating in kind with a new plot of land and a house within a new residential cluster constructed by the developer ('land for land'). In addition, a very common hybrid form in Cai Rang envisages that households receive monetary compensation and can use this for purchasing a plot of land and/or a house in the new residential cluster.

Using semi-structured household interviews in Cai Rang in 2009, a number of shortcomings in the implementation of the land clearance and compensation schemes can be identified which imply substantial challenges and livelihood problems for the respective households. First, the compensation offered is considered inadequate by the majority of households interviewed. A commonly reported point of criticism is that the compensation rate is much below the actual prices for land use certificates in the respective area at the given time. Hence, the amount of money received (in the case of monetary compensation) was stated to be insufficient for purchasing an adequate plot of land elsewhere in return. Interestingly, this criticism was also registered with respect to the plots of land and/or houses in the new residential cluster, due to a stark difference in the compensation per square metre on the one hand and the price for the plot in the residential cluster on the other. The compensation scheme of one of the private developers, for example, sets out that land plots of 60 m^2 are reserved in a new residential cluster for those households that have to be relocated. In the hypothetical case that such a household had a plot of 60 m^2 residential land prior to the resettlement, the household can get the exact amount with the new plot in the residential cluster and, additionally, compensation for rebuilding the house, with the latter rate depending on the type of house in the old location as well as on the rebuilding and construction rates defined by the local authorities under the pricing scheme. However, if the original plot of residential land was smaller than 60 m², the household has to pay the difference for buying the new 60 m² parcel. The price for this was defined to be 2.5 million VND under the given scheme, of which the local People's Committee contributes 50%. However, the compensation rate for the old land is – under the same scheme – only 500,000 VND, resulting in a difference of 750,000 VND/m², a cost that has to be borne by the household itself. In addition, the compensation rates for the housing stock were reported to be insufficient for rebuilding an adequate house of the same size and quality. The compensation of additional agricultural land is handled separately under this scheme with a one-time compensation of 150,000 VND/m². On the one hand, this money can be used in order to support the purchase and construction of the land parcel and house for residential purposes. On the other hand, this compensation was stated to be much too low, considering that it served as a basis for long-term income generation in most households and that the purchase of adequate plots of agricultural land would be much more expensive, if available in the close vicinity at all.

Second, the vast majority of affected households stated that they experienced substantial delays between the transfer of the land title and the actual payment of compensation as well as the resettlement or the finalisation of resettlement areas – all potentially leading to significant microeconomic problems. As indicated above, most of the households transferred the official land use title between the years 2000 and 2003. However, much of the new residential areas have only been developed (clearance of land plots, construction of roads, drainage and sewerage systems, hauling of electricity connections etc.) years later, in most cases not before 2007. While most of the households affected by this delay were permitted to remain living on their old land until the actual date of resettlement, the transfer of agricultural land implied that agricultural production was no longer, or only in very limited terms, permitted or possible during those years. This caused a loss of income generation for the affected households, given the lack of alternative income sources. Many of the affected households were, therefore, forced to use up their savings, e.g. the first instalments of the compensation money, which were in fact meant for constructing houses or for purchasing land titles in the new place. Alternatively, they had to take on debts, most often with relatives due to the lack of access to bank loans.

Thirdly, in addition to those hard economic facts, most households interviewed expressed concerns in an additional dimension revolving around emotional bonds to their original land as well as social networks. The loss of the land threatens the continuation of the old income-earning activities and the foundation of personal and community identity – in particular for the case of farmers, but also within the localised service or trade sector. Resettlement also in most cases contributes to a dissolution or reshuffling of social networks and, therefore social capital, be it with respect to, for example, the schools of the children, long-grown neighbourhood and family bonds, or the circle of customers in business relations. Many households stated their strong opposition to being relocated and would prefer to stay in the old place if they were given the choice.

9.5 Discussion: Dynamic Pathways in Resilience

The review of large-scale trends resulting from transformation and climate change in Vietnam in combination with the case study analysis of developments in the peri-urban areas of Cai Rang reveals the complex interaction of trends and developments at various scales and their particular influence on peri-urban populations, often comprising rather negative effects and challenges for the livelihood regimes of these groups.

However, the case study analysis also shows that shifts in the resilience[1] of peri-urban population groups against the background of transformation trends and climate change can be influenced by many small circumstances. Specific resilience pathways of single households are therefore the result of the particular combination of decisions, events and access-portfolios with which the respective household is faced. Every one of these factors, thereby, works as a type of switch setting the direction for another section of the individual resilience pathway. For the analysed case study district, the chain or conglomerate of relevant elements influencing resilience directions would, for example, include the following questions: How far do environmental degradation processes affect agricultural production? Have water resources, even if polluted before the transition process, become unusable for agricultural and other uses – or if they are used, what health threats do they pose? Is the agricultural production of the household affected by surrounding production or construction sites? Can diversification of rural livelihoods be envisaged? Is the household at all located within the area earmarked for new projects? Is the developer the state or a private company? If it is a private developer, which company is it, meaning which specific compensation scheme will be applied? Did the household live on a parcel for which it had a land use certificate or not? Is the compensation scheme linked to this title and how big are the compensation differences for those plots with certificates and those without? Is the household free to choose between 'money for land' or 'land for land' or is only one option possible due to resource constraints? When the household prefers to move into the new residential cluster, was the old land patch big enough to be able to afford this? Does the compensation and relocation happen fast enough to withstand the effects of income loss and inflation? When is a point reached where all savings or compensation instalments are depleted due to inflation and rise in land prices? How do global food prices change over the same time and how does this affect the revenue ratio of agricultural-based households? If the household receives a new plot of land, how exposed is this new plot to flooding and other natural hazards, particularly against the background of climate change-related trends?

---

[1] Comprehensive overviews of different schools of resilience thinking and their specific epistemological as well as ontological perspectives have been provided, for example, by Folke (2006), Janssen et al. (2006) or Gallopin (2006) and shall not be repeated here. For the context of our argument, resilience can sufficiently be defined in rather broad terms as the ability of a system or actors in a system to cope with stress and disturbance resulting from social, economic, cultural, political or environmental change and to reorganise so as to retain essential functions of the system (adapted from Walker et al. (2004) and Adger (2000b)).

These elements or switches within resilience pathways can be conceptualised as tipping points. The differences which appear on first sight rather small or unimportant may have a comparatively large effect on the overall livelihood and resilience situation of the given household (e.g. for the illustrated compensation scheme, the question whether the plot of land is a little bit smaller or larger than the land size in the resettlement cluster may have a substantial financial effect and may be decisive to the question whether or not the household can afford to settle in the relocation cluster). The chain of critical elements exemplified above indicates that households are at risk of steering into significant crises situations if they experience an accumulation of circumstances in which switches are set towards negative developments – that is if they, for instance, do not have land certificates and, at the same time, happen to be located within the project area of a developer who strictly only compensates for those households who do have such land title.

Yet, at the same time, the notion of switches and dynamic pathways evolving from this analysis challenges much of the conventional thinking in resilience and tipping point discourses. Resilience is often thought of as a rather monolithic characteristic of one particular system or group in the sense that, for example, farmers in *the* given area share the same resilience characteristics and parts that are coalesced in one binding system. This pre-analytic vision often implies strong (implicit or explicit) functionalistic or monocausal notions in the sense that, for example, certain population groups that have similar socio-economic characteristics and share the same biophysical landscape (and hazards) will move along the same development pathway and eventually end in a similar (often deteriorated) situation. While this observation might be true for certain empirical cases, regions or systems, it is certainly not true for all of them, as the analysis of very differentiated and dynamic resilience pathways in Cai Rang has shown.

Further, the notion of overlaying and interacting processes that affect resilience has so far mainly focused on researching and conceptualising those cases where all the interacting processes work in the same direction. Particular attention has been directed towards deteriorations in resilience, mainly in the sense of cascading effects which eventually lead to the collapse of a given system (Abel et al. 2006; Kinzig et al. 2006). Intermediate options where only some elements are negative and others point in positive directions, are under-represented in the prevailing conceptual discourse. The above-mentioned case related to water pollution can serve as an example of how this might work; in some parts of Cai Rang the occurrence of agrochemicals will decrease due to land use change while endocrine disruptors and other new contaminants from urban waste, industrial waste and sewerage could increase. Similarly, relocation can – if sufficiently implemented – also imply substantial advancements for the affected people with respect to, for example, access to electricity, water supply or sanitation infrastructure.

An additional – and in the view of the authors highly important – point has to be seen in the relevance and applicability of resilience thinking for policy making. It has been noted that resilience thinking is widely considered to be conceptually and analytically appealing; however, by and large it so far gives little concrete and applicable guidance for decision making, for example, on how to avoid tipping into a crisis or collapse in a given coupled social-ecological system in which overlapping power relations and actions are at work (Garschagen 2010; Renaud et al. 2010). It is also of relevance to note that tipping points are so far predominantly discussed with respect to rather big systems – with the discussion originating from the analysis of global systems in the fields of climatology and oceanography. The threats of tipping points in those large-scale systems, however, seem to have been overwhelming in relation to the implementation of policy responses on the ground – let alone regarding the identification and distribution of responsibilities.

The notion of small-scale tipping points, in the sense of turning points in the trajectories of household-level resilience pathways as presented here, in contrast, allows for the clear identification of cause–effect relations and, therefore, of responsibilities for specific elements in this resilience landscape. This conceptual and analytic notion has strong and direct relevance for developing concrete response options and countermeasures. It divides over-generalised and unwieldy challenges to and possibilities for resilience into small pieces which are much easier to manage and for which concrete solutions can be found comparatively easily. Working on this scale also helps to uncover responsibilities of different actor groups as well as options and capacities for solutions amongst these groups (e.g. the time frame for the implementation of compensation and the issuance of new land titles are within the responsibilities of the developer; however, the local government can through the clearance committee – in theory – influence the timing and implementation of those decisions).

Identifying and explicitly analysing turning points and potential tipping points in resilience trajectories surely cannot be seen as a panacea for solving all the problems related to issues of power struggles in the complex political economy or political ecology at work. It is argued here, however, that explicitly uncovering these turning points and their potentially negative effects is a first step towards strengthening important attributes conventionally understood to be amongst the defining elements of good governance, that is, transparency and accountability in decision making (c.f. UNDP 1997). This way, the notion of dynamic resilience pathways can have a guiding role in fostering integrative planning and management approaches in the process of comprehensive governance.

9.6 Conclusions and Outlook

This chapter brought together elements from different disciplines for exploring resilience trajectories of peri-urban farmers in the Mekong Delta. An emphasis was placed on dynamic continua as well as grey-shades between static or dichotomous viewpoints in policy making and research. This holds true, for example, for the focus on the hybrid peri-urban interface; for exploring resilience effects of certain developments in both positive as well as negative directions; for acknowledging fragmented and often contested responsibilities as well as the agency of different actors in a transforming actor landscape; and, most importantly, for conceptualising tipping points and resilience pathways as the result of the interaction of many small factors with multiple possible pathways for different agents (rather than applying a deterministic understanding of a monolithic system moving as a whole into one pre-defined direction).

It is argued here that a strong emphasis on such continuum thinking is necessary for effectively approaching the complex and highly dynamic developments and challenges in Vietnam and in particular within the peri-urban agricultural sector, which is and will be heavily affected by the overlapping impacts of socio-economic transformations (including significant urbanisation) and climate change. Challenges to the peri-urban interface in Vietnam, which so far have not been sufficiently addressed in research and policy making and which continue to be falling through the cracks of horizontally and sectorally fragmented administration and planning, need to be addressed more thoroughly in the future. The comprehensive focus on socio-economic transformation (and related development questions) and climate change impacts, further, calls for increased efforts to link these two spheres. Finally, the explicit focus on what we have called turning points or switches in individual resilience pathways advances the conceptual resilience discourse. This is too often characterised by a rather deterministic system thinking, and pays little attention to differences in household-level resilience amongst groups that are conceived as sharing identical levels of resilience due to being part of the same larger system. Our viewpoint allows for concrete guidance with respect to resilience assessment and policy making as it splits up the domain of household-level resilience into smaller elements, which then become possible to address and manage. The explicit focus on these switches helps to identify responsibilities for respective action as well as shortcomings and challenges. Identifying these shortcomings is the first step for effectively addressing and improving the related problems (with often relatively small changes being necessary to induce large improvements, as the example of compensation schemes has shown). This process deserves great attention in countries like Vietnam, where the spectrum of actors is rapidly transforming due to socio-economic and political change which in turn is increasing the need for mediating the responsibilities and rights of different stakeholders. However, at the same time this last point is likely to be one of the biggest challenges in Vietnam, as the transparency of decision-making processes and mechanisms for public participation remain highly contested.

References

Abel N, Cumming D, Anderies J (2006) Collapse and reorganization in social-ecological systems: questions, some ideas, and policy implications. Ecol Soc 11(1):17–42

Adger N (1999) Social vulnerability in climate change and extremes in coastal Vietnam. World Dev 27(2):249–269

Adger N (2000a) Institutional adaptation to environmental risk under the transformation in Vietnam. Ann Assoc Am Geogr 90(4):738–758

Adger N (2000b) Social and ecological resilience: are they related? Prog Hum Geogr 24(3):347–364

Adger N, Eakin H, Winkels A (2009) Nested and teleconnected vulnerabilities to environmental change. Front Ecol Environ 7(3):150–157

Allen A (2006) Understanding environmental change in the context of rural-urban interactions. In: McGregor D, Simon D, Thompson D (eds) The peri-urban interface: approaches to sustainable natural and human resource use. Earthscan, London, pp 30–43

Allen A, Davila J (2002) Mind the gap! Bridging the rural-urban divide. In: Allen A, Davila J (eds) Mind the gap! Bridging the rural-urban divide.- id21 Insights, 41. Available from: http://www.id21.org/insights/insights41/index.html. Accessed 2009-05-04

Birley M, Lock K (1998) Health and peri-urban natural resource production. Environ Urbanization 10(1):89–106

Blaikie P, Brookfield H (1987) Land degradation and society. Methuen and Co, New York

Carew-Reid J (2008) Rapid assessment of the extent and impact of sea level rise in Vietnam. International Centre for Environment Management (ICEM), Brisbane

CFSC [Committee for Flood and Storm Control, Socialist Republic of Vietnam] (2004) National report on disaster reduction in Vietnam. For the World Conference on Disaster Reduction, Kobe-Hyogo, Japan, 18–22 Jan 2005. Hanoi

Coulthart A, Quang N, Sharpe H (2006) Urban development strategy: meeting the challenges of rapid urbanization and the transition to a market oriented economy. Worldbank, Hanoi

Dang H (2007) What is doimoi? From where did it start? In: Viet Nam, twenty years of renewal. The Gioi, Hanoi, pp 1–23

Dasgupta S, Laplante B, Meisner C, Wheeler D, Yan J (2007) The impact of sea level rise on developing countries: a comparative analysis. World Bank Policy Research Working Paper 4136, World Bank, Washington DC

Davila J (2006) Falling between stools? Policies, strategies and the peri-urban interface. In: McGregor D, Simon D, Thompson D (eds) The peri-urban interface: approaches to sustainable natural and human resource use. Earthscan, London, pp 44–56

Drakakis-Smith D, Kilgour A (2001) Sustainable urbanisation and environmental issues in Vietnam. In: Adger N, Kelly M, Ninh NH (eds) Living with environmental change: social vulnerability, adaptation and resilience in Vietnam. Routledge, New York, pp 213–233

Ekman B, Liem NT, Duc HA, Axelson H (2008) Health insurance reform in Vietnam: a review of recent developments and future challenges. Health Policy Plann 2008:1–12

Folke C (2006) Resilience: the emergence of a perspective for social-ecological systems analyses. Global Environ Change 16(3):253–267

Gallopin G (2006) Linkages between vulnerability, resilience, and adaptive capacity. Global Environ Change 16(3):293–303

Garschagen M (2010) Potential humanitarian crises and climate change adaptation in the coupled social-ecological systems of the Mekong Delta, Vietnam. In: Shen X, Downing TE, Hamza M (eds) Tipping points in humanitarian crises: from hot spots to hot systems, 13. SOURCE, pp 45–55

Garschagen M, Binh NT, Thach LN (2009) Vietnam: the challenge of integrating disaster risk reduction and climate change adaptation. In: Birkmann J, Tetzlaff G, Zentel K (eds) Addressing the challenge: recommendations and quality criteria for linking disaster risk reduction and adaptation to climate change. DKKV Publication Series, 38

GSO [General Statistics Office] (2009a) Production of paddy by province. <http://www.gso.gov.vn/default_en.aspx?tabid=469&idmid=3&ItemID=7885>. 03 Jun 2009

GSO [General Statistics Office] (2009b) Production of farmed aquatic product by province. <http://www.gso.gov.vn/default_en.aspx?tabid=469&idmid=3&ItemID=7818>. 03 Jun 2009

Han SS, Vu KT (2008) Land acquisition in transitional Hanoi, Vietnam. Urban Stud 45(5&6):1097–1117

Janssen M, Schoon M, Ke W, Börner K (2006) Scholarly networks on resilience, vulnerability and adaptation within the human dimensions of global environmental change. Global Environ Change 16(3):240–253

Johnston R, Taylor P, Watts M (2002) Geographies of global change: remapping the world, 2nd edn. Blackwell, Malden

Kelly M, Hien HM, Lien TV (2001) Responding to El Niño and La Niña: averting tropical cyclone impacts. In: Adger N, Kelly M, Ninh NH (eds) Living with environmental change: social vulnerability, adaptation and resilience in Vietnam. Routledge, London

Kinzig AP, Ryan P, Etienne M, Allison H, Elmqvist T, Walker BH (2006) Resilience and regime shifts: assessing cascading effects. Ecol Soc 11(1):23–46

Kleinen J (2007) Historical perspectives on typhoons and tropical storms in the natural and socio-economic system of Nam Dinh (Vietnam). J Asian Earth Sci 29(4):523–531

Luong HV (2003) Postwar Vietnamese society: an overview of transformational dynamics. In: Luong HV (ed) Postwar Vietnam: dynamics of a transforming society. Rowman & Littlefield Publishers, Inc, Singapore, pp 1–26

Lynn S, Garschagen M, Lehmann J, Khan S, Drew G, Prasad V, Nkem J, Mushongah J (2010) Introducing a 'Hot System'-approach to tipping points in humanitarian crises. In: Shen X, Downing T, Hamza M (eds) Tipping points in humanitarian crises: from hot spots to hot systems, 13. SOURCE, pp 14–21

Marsh S, MacAulay T (2006) Land reform and the development of commercial agricultural in Vietnam: policy and issues. http://www.agrifood.info/review/2002/Marsh.pdf

McGranahan G, Balk D, Anderson B (2007) The rising tide: assessing the risk of climate change and human settlements in low elevation coastal zones. Environ Urbanization 19:17–37

Millennium Ecosystem Assessment (2005) Ecosystems and human well-being: synthesis. Island, Washington DC

MONRE [Ministry of Natural Resources and Environment] (2009) Climate change, sea level rise scenarios for Vietnam. Hanoi

Ni DV, Safford R, Maltby E (2001) Environmental change, ecosystem degradation and the value of wetland rehabilitation in the Mekong Delta. In: Adger N, Kelly M, Ninh NH (eds) Living with environmental change: social vulnerability, adaptation and resilience in Vietnam. Routledge, London, pp 122–135

O'Brien K, Leichenko R (2000) Double exposure: assessing the impacts of climate change within the context of economic globalisation. Global Environ Change 10:221–232

Padgham J (2009) Agricultural development under changing climate: opportunities and challenges for adaptation. Joint Discussion Paper 1. The World Bank, Washington, DC

Renaud FG, Birkmann J, Damm M, Gallopin G (2010) Understanding multiple thresholds of coupled social-ecological systems exposed to natural hazards as external shocks. Nat Hazards 55(3):749–763

Robbins P (2004) Political ecology. Blackwell, Malden

Simon D, McGregor D, Thompson D (2006) The search for peri-urban resources sustainability. In: McGregor D, Simon D, Thompson D (eds) The peri-urban interface: approaches to sustainable natural and human resource use. Earthscan, London, pp 1–17

SRV [Socialist Republic of Vietnam] (1998) Decision to approve the master plan for the development of Vietnam's urban centres until 2020. No: 10/1998/QD-TTg. Hanoi

SRV [Socialist Republic of Vietnam] (2002) Promulgation of the state budget law. No: 21/2002/L-CTN. Hanoi

SRV [Socialist Republic of Vietnam] (2003a) Law on construction. No: 16/2003/QH11. Hanoi

SRV [Socialist Republic of Vietnam] (2003b) Law on land. No: 13/2003/QH11. Hanoi

SRV [Socialist Republic of Vietnam] (2006) Decision on approving the master plan on construction of Can Tho City up to 2025. No: 207/2006/QD-TTg

SRV [Socialist Republic of Vietnam] (2009) Law on urban planning. No.32/2009/QH12. Hanoi

Statistics Office Can Tho City (2010) Statistical yearbook Can Tho City 2009. Can Tho City

Taylor P (2004) Social inequality in Vietnam and the challenge of reform. Nias, Singapore

Toan V P, Sebesvari Z, Rosendahl I, Renaud FG (2009) Residue concentrations of recently used pesticides in surface water at two representative study sites of the lower Mekong Delta, Vietnam. Pesticide behaviour in soils, water and air 14–16 Sep 2009, York, UK

Tran P, Marincioni F, Shaw R, Sarti M, An L (2008) Flood risk management in central Viet Nam: challenges and potentials. Nat Hazards 46(1):119–138

Trong NP (2007) Doimoi (Renewal): in a retrospect view for the sake of future. Viet Nam: twenty years of renewal. The Gioi, Hanoi, pp 24–45

Tuan VV (2010) Vulnerability of different socio-economic groups to flooding and climate change in rural Dong Thap. WISDOM Mid-Term Report to the BMBF, UNU-EHS, Bonn

Turley W (1977) Urban transition in south Vietnam. Pac Aff 49(4):607–624

UN/DESA [United Nations / Department of Economic and Social Affairs] (2008) World urbanization prospect: the 2007 revision. http://esa.un.org/unup/

UNDP [United Nations Development Programme] (1997) Governance for sustainable human development. A UNDP policy document

UNU-EHS [United Nations University, Institute for Environment and Human Security] (2010) Final Report of the contribution of UNU-EHS to the first phase of the BMBF-funded WISDOM Project. Report to the German Federal Ministry of Education and Research

Wackernagel M, Rees W (1996) Our ecological footprint: reducing human impact on earth. New Society Publishers, Gabriola Island

Wagstaff A (2007) Health insurance and the poor: initial impacts of Vietnam's heath care fund for the poor. World Bank Policy Research Working Paper, 4134. World Bank, Washington DC

Waibel M (2005) Vietnam: a new Asian economic tiger? Pazifik Forum 10:31–51

Walker B, Holling CS, Carpenter SR, Kinzig A (2004) Resilience, adaptability and transformability in social-ecological systems. Ecology and Society, 9(2): article 5

Wassmann R, Hien NX, Hoanh CT, Tuong TP (2004) Sea level rise affecting the Vietnamese Mekong Delta: water elevation in the flood season and implications for rice production. Clim Change 66(1–2):89–107

Yeung YM (2007) Vietnam: two decades of urban development. Eurasian Geogr Econ 48(3): 269–288

Young G (1974) Human ecology as an interdisciplinary concept. Adv Ecol Res 8:1–105

Chapter 10
Efficiency Analysis of Selected Farming Patterns: The Case of Irrigated Systems in the Mekong Delta of Vietnam

Quan Minh Nhut

Abstract The objective of this study is to measure the technical, allocative, and cost efficiency for farmers who grow crops following either rice-monocultural patterns or crop-rotation patterns in the non-flooded and flooded areas of the Mekong River Delta in Vietnam. The non-flooded areas are located within irrigated boundary systems, and the flooded areas are located outside these systems. In addition, the determinants of household income and productive efficiencies are identified in the study.

Related to productive efficiency, the measured results show that the crop-rotation farmers are more efficient in terms of technical and cost efficiency than the continuous-rice farmers and vice versa for allocative efficiency, for the case of non-flooded areas. Similarly, the mean efficiency score is greater with respect to technical, allocative, and cost efficiency for farmers employing the crop-rotation pattern in comparison with farmers following the continuous-rice pattern.

Regarding the factors influencing efficiencies, in the case of non-flooded areas, the estimated results show that although there are some differences in determinants of each component of total productivity, sex, age, education, share of female labor, and farming pattern are found to be the main factors driving changes in most of the components such as technical and allocative efficiency. In the case of flooded areas, all components of total productivity in terms of the technical, allocative, and cost efficiency are impacted by the variation in sex, age, and education.

Keywords Efficiency of continuous-rice and crop-rotation patterns • Data envelopment analysis

---

Q.M. Nhut (✉)
School of Economics and Business Administration, Can Tho University,
Campus 1 – 30/4 Street, Hung Loi Ward, Ninh Kieu District, Can Tho City, Vietnam
e-mail: qmnhut@ctu.edu.vn

M.A. Stewart and P.A. Coclanis (eds.), *Environmental Change and Agricultural Sustainability in the Mekong Delta*, Advances in Global Change Research 45,
DOI 10.1007/978-94-007-0934-8_10,

10.1 Introduction

Vietnam is an agricultural country. About 80% of the population living in rural areas and more than 74% of the workforce are employed in agricultural sectors. Agricultural products dominate exports and home consumption. They contribute a significant percentage to Vietnam's economic growth rate. The Mekong River Delta (MRD) region, which is located in the south of Vietnam, comprises about four million hectares or about 12% of the country's total area.[1] It has substantial potential for agricultural development. More specifically, the MRD contributes about 55–60% of nationwide agricultural production and about 65% of exported agricultural products. Looking at the rice sector, about 60% of Vietnam's total rice production and 65% of rice exports are provided by this region (Statistical Yearbook 2006).

In recent years, farmers in the MRD have wanted to shift from a *continuous-rice pattern* to a *crop-rotation pattern* in response to governmental encouragement and in order to increase profit, for almost all rice monoculturalists are plagued by low productivity and low incomes. However, for a variety of reasons, they also have had some difficulties in this transition. The first reason is that the agricultural sector tends to be small scale and uses traditional technology and extension services which are inadequately funded, leading to a shortage and poor distribution of agricultural inputs. The second reason is the low level of technical efficiency and the lack of appropriate knowledge regarding shifting cultivation from traditional rice crops (called continuous-rice patterns) to crop-rotation patterns, which farmers have wanted to apply in recent years.

In addition, some provinces such as Long An, Tien Giang, Vinh Long, Dong Thap, Can Tho, and An Giang are located along two large rivers – the Tien and Hau – which are derived from Tay Tang highland (China) flowing through Lao, Thailand, Myanmar, Cambodia, and Vietnam before running into the Pacific Ocean. Every year from September to November, the water levels of these rivers rise about 1–2 m higher than the normal levels in a year. This leads some areas of these provinces to be under water (these areas are called flooded areas). Such characteristics make it very difficult for farm households in those areas to practice agriculture during the flooded time period.

The central and local governments have established some policies to prevent deleterious impacts of flooding on agricultural production by building boundary-irrigated systems (called flood prevention dikes) in the regularly flooded areas. The effects of these policies, however, need to be evaluated by comparing profits and the productive efficiency of farm households that live within boundary-irrigated systems to profits and the productive efficiency of farm households that live outside the systems.

---

[1]The Mekong River Delta region is located in the south and consists of 13 provinces: Long An, Ben Tre, Tien Giang, Vinh Long, Dong Thap, Tra Vinh, Can Tho, Soc Trang, An Giang, Kien Giang, Bac Lieu, Hau Giang, and Ca Mau provinces.

Farmers in the MRD are facing the choice between traditional rice and crop-rotation patterns, and the government is considering whether or not to continue building boundary-irrigated systems. In light of these considerations, comparing the productive efficiency of the continuous-rice and crop-rotation patterns outside and outside boundary-irrigated systems seems appropriate and useful.

10.2 Data

The heads of farm households were personally interviewed by trained interviewers who are junior staff members of the School of Economics and Business Administration of Can Tho University under the supervision of the author.

Primary data was collected via the use of interview schedules designed to collect information on quantities and prices of inputs and outputs as well as socioeconomic/environmental factors from the selected farm households using multistage random-sampling techniques. Prior to actual data collection, the permissions to conduct the interviews in the sample villages were secured by the municipal and local executives. Such permissions were necessary to establish a good relationship and good cooperation with the respondents.

The surveys covered four provinces in the MRD, of which An Giang and Long An were in the flooded areas, and Tien Giang and Tra Vinh in the non-flooded areas of the MRD. These provinces were chosen because they all had large areas of agricultural land in production, together with a large number of mono-rice and crop-rotation farmers.

For each of the selected provinces, one district was then randomly chosen, except for An Giang Province wherein two districts were selected, giving a total of five selected districts, namely, Moc Hoa (Long An Province), Cai Lay (Tien Giang Province), Cau Ke (Tra Vinh Province), Cho Moi, and Tri Ton districts (An Giang Province).

In the last stage, because it was impossible to survey all farmers, samples were drawn so as to select farmer representatives from populations of farmers who followed continuous-rice patterns and who employed crop-rotation patterns in the 2004/2005 agricultural years. Through this sampling scheme, two groups of farmers were randomly selected for each of the five districts. Of these two groups, one consisted of 60 farmers who followed the continuous-rice pattern, and the other consisted of 60 farmers who employed the crop-rotation pattern. The surveys gave a total of 600 selected farmers: 300 farmers who employed continuous-rice patterns and 300 who employed crop-rotation patterns. However, the analyses were carried out using the responses from 262 and 242 farmers, respectively, for the continuous-rice and crop-rotation patterns. The remaining farmers were rejected due to inconsistencies in their responses.

Although the surveys were randomly conducted, in order to avoid any bias in the analyses, households were only selected if they used the same agricultural land areas in their production of crops throughout the year. This was consistent for both continuous-rice and crop-rotation patterns.

10.3 Methodology

10.3.1 Calculation of Technical, Allocative, and Cost Efficiency Using the DEA Model

Data envelopment analysis (DEA) is another approach to frontier estimation, which is based on the mathematical programming method instead of the econometric method to measure the production efficiency of a firm. DEA was first proposed by Farrell (1957), then by Boles (1966), Shephard (1970), and Afriat (1972). However, the idea was quite novel, and originally not much attention was paid to it. This situation lasted until the publication of the comprehensive papers of Charnes et al. in 1978, wherein the DEA model strongly focused on the input-oriented approach (i.e., the firm has more controls on inputs than on outputs) and relied on the constant returns to scale (CRS) circumstance. This model, called the CCR model, is also known as the CRS-DEA model.

In agricultural cultivation, the productivity of a farm household consists not only of the technical component such as TE or the component of operation scale such as SE, but also of the other components such as effects of resource allocation (allocative efficiency – AE) and the use of productive costs (cost/economic efficiency – CE). The AE measure is used to evaluate the farm household's ability in allocating and utilizing a mix of inputs in the optimal way with given relative prices and production technology. The CE is calculated by a combination of TE and AE, and hence, may be used to estimate the possibility of cost savings by the household when moving to the technically and allocatively efficient point with given input prices and technology.

In the case of available information on input prices, the TE, AE, and CE can be measured by using the CRS Input-oriented DEA Model (Coelli et al. 2005; Krasachat 2007). Consider the situation with N observations of farm households, each of the households produces S outputs using M different inputs. The LP problem must run N times, one for each household. For the particular household $p \ (p \in N)$, the DEA model is specified.[2]

$$\min_{\{\lambda, x^*_{jp}\}} \left\{ w'_{jp} \, x^*_{jp} \right\}$$

subject to:

$$\begin{aligned} -q_{rp} + \textstyle\sum_{i=1}^{n} \lambda q_{ri} \geq 0, &\quad r = 1,\ldots,s \, x^*_{jp} - \textstyle\sum_{i=1}^{n} \lambda x_{ji} \\ \geq 0, &\quad j = 1,\ldots,m \, \lambda_i \\ \geq 0, &\quad i = 1,2,\ldots,p,\ldots,n \end{aligned} \tag{10.1}$$

[2]Coelli et al. (2005).

where:

i is the parameter presenting the number of DMUs ($i=1, 2,\ldots, p,\ldots, n$);
j is the parameter presenting the number of inputs ($j=1, 2,\ldots, m$);
r is the parameter presenting the number of outputs ($r=1, 2,\ldots, s$); x_{jp}^* is the input vector j utilized by the household p with respect to production cost minimization, x_{jp}^* are calculated from the LP, w_{jp} is the input price vector j paid by the household p;
q_{rp} is amount of output r produced by the DMUp;
q_{ri} is $(N \times S)$ matrix of S outputs of each of N observing DMUs;
x_{ji} is $(N \times M)$ matrix of M inputs of each of N observing DMUs;
λ_i is an $N \times 1$ vector of weights which defines the linear combination of the DMUi and subsequently creating a *projected/virtual point* of the DMUp lying on the frontier. This projected point is the fully efficient potential point produced by the radial contraction of the input vector with unchanged output level.

10.3.2 Metafrontier and Relative Positions of Group Frontiers

In the preceding sections, the DEA models used to calculate the efficiency of farm households are provided with two separate specifications, that is, the CRS-DEA model and the VRS-DEA model. Actually, the production-efficiency score of a given household is measured by comparing its real position to the frontier of the observed data of the particular group (sample) to which it belongs. Different groups, however, may differ from each other in various respects such as size of landholding, labor, or capital, together with other characteristics relating to soil quality and socioeconomic environment. Thus, every group has a specific production frontier and differs from each other. This specific frontier is called the *group frontier* (O'Donnell et al. 2008). The idea behind the group frontier is that we can evaluate and compare the efficiency score of a farm household with others within a group of farmers.

In this study, apart from a comparison of the efficiency of each farmer within a group of farmers in a particular farming pattern, the efficiency of farm households should be compared with farmers having adopted the other farming pattern (i.e., between the two selected farming patterns – the continuous-rice and crop-rotation pattern). To do this, all farm households in both the continuous-rice and crop-rotation patterns, besides facing their specific group frontier, are assumed to share a common frontier. This assumption is plausible because, although each farming pattern has specific characteristics, both may be involved in the same activities with respect to agricultural infrastructure and supported policies. The common frontier is called the *metafrontier*. Generally, the metafrontier is defined to be the boundary of a set of all specific group frontiers (O'Donnell et al. 2008).

Given the DEA models along with the metafrontier technique, the TE scores of farmers can be compared not only with those of farmers within the group to which they belong but also with those of farmers from other groups (i.e., between two groups of farming patterns such as the continuous-rice and crop-rotation patterns) by calculating the *metatechnology ratio* (MTR) for each farming pattern. This ratio is defined by the following equation:

$$\mathrm{MTR}_{\mathrm{i}}(x,y) = \frac{E_i(x,y)}{E_i^g(x,y)} \tag{10.2}$$

where:

$\mathrm{MTR}_i(x,y)$ = metatechnology ratio for the i-th household ($i = 1, 2, \ldots, n$);
$E_i(x,y)$ = technical efficiency of the i-th farm household using the input vector, x, to produce the output vector, y, in comparison with the metafrontier;
E_i^g = technical efficiency of the i-th farm household with respect to the frontier of group g.

The relative position of each group frontier (i.e., farming pattern frontier) in comparison to the metafrontier is specified based on its mean MTR. Therefore, the relative positions of farming pattern frontiers are determined by comparing the mean MTRs of different groups of farming patterns. A specific group with a higher mean MTR has a closer position to the metafrontier.

10.4 Findings and Discussions

10.4.1 Household Efficiency Measurement Using CRS-DEA Model

In this section, efficiency analyses of farm households are presented in terms of the following aspects:

- Measuring and analyzing the production-efficiency scores of the farm households based on the estimated results of the CRS-DEA models
- Comparing the technical-efficiency scores of the farm households between the two selected farming patterns relying upon the calculated results of the MTRs

The average efficiency scores, the number of fully efficient households, and efficiency distributions of the farming patterns are depicted in Table 10.1.

10.4.1.1 Household Technical Efficiency Distributions of the Two Selected Farming Patterns

For farmers in the MRD, the cultivation of crops is considered to be the main job in earning a living. However, the productivity of a crop varies due to differences in

Table 10.1 Production efficiency of households and their distributions by farming patterns (*Source*: Measured from the surveyed data using DEAP software)

	Continuous-rice pattern			Crop-rotation pattern		
Parameters	TE	AE	CE	TE	AE	CE
Non-flooded areas						
Efficiency scores						
1.00	41	2	2	57	5	5
0.90–0.99	19	19	11	34	9	7
0.80–0.89	26	55	21	39	38	17
0.70–0.79	41	42	20	14	31	31
0.60–0.69	25	33	33	2	37	32
0.50–0.59	5	6	50	0	24	41
0.40–0.49	0	0	17	0	2	12
<0.40	0	0	3	0	0	1
Mean	0.83	0.78	0.65	0.92	0.73	0.67
Min	0.52	0.53	0.36	0.60	0.47	0.34
Max	1.00	1.00	1.00	1.00	1.00	1.00
SD	0.14	0.11	0.16	0.09	0.13	0.15
No. of households	157			146		
Flooded areas						
Efficiency scores						
1.00	30	3	3	44	4	4
0.90–0.99	20	16	7	21	11	7
0.80–0.89	26	48	16	24	30	15
0.70–0.79	20	31	31	6	36	30
0.60–0.69	7	4	28	1	14	33
0.50–0.59	1	3	17	0	1	7
0.40–0.49	1	0	2	0	0	0
<0.40	0	0	1	0	0	0
Mean	0.87	0.82	0.71	0.93	0.79	0.74
Min	0.46	0.55	0.31	0.68	0.53	0.50
Max	1.00	1.00	1.00	1.00	1.00	1.00
SD	0.13	0.09	0.13	0.08	0.10	0.12
No. of households	105			96		

Note: *SD* – standard deviation

production technology, differences in efficiency of the production process, and differences in the environment in which production occurs. In this section, I will present results that reflect the ability of the farm households to use the best practices and available technology in the most effective way. This effect is referred to as the technical contribution (TE) to its productivity.

In Table 10.1, the average TE score is less than one for both continuous-rice and crop-rotation patterns. The estimated TE scores of farmers following the continuous-rice pattern range between 0.52 and 1.00, with a mean and standard deviation of 0.83 and 0.14, respectively. Similarly, the TE scores of farmers employing the crop-rotation pattern vary from 0.60 to 1.00, together with a mean of 0.92 and standard deviation of 0.09.

Table 10.2 Mean metatechnology ratio of households by farming patterns (*Source*: Calculated from the surveyed data)

Farming patterns	Non-flooded areas	Flooded areas
Continuous-rice pattern	0.96	0.91
Crop-rotation pattern	0.98	0.98

With regard to the TE distributions, for the continuous-rice pattern, about 41 households are found to be fully technically efficient. The remaining households mostly get high scores of TE that range between 0.70 and 0.90, and a few have low TE scores (from 0.50 to 0.70). Analogously, most farmers applying the crop-rotation pattern attain very good achievements corresponding to a TE score of 1.00 for 57 households and values between 0.70 and 0.90 for the remainder.

Given such results together with the average MTR for the TE listed in Table 10.2, we may conclude that farmers employing the crop-rotation pattern are a little more technically efficient than farmers following the continuous-rice pattern. This result is consistent with the studies of Latruffe et al. (2000), who found that livestock farmers get higher TE scores than crop farmers and Linh (2006), who discovered that diversified farmers are more technically efficient than farmers who mainly cultivate rice.

According to the analyses of the responses from the surveys and local authorities, there are some main causes for the difference such as (1) the crop-rotation farmers have more ready access to production technology than do the continuous-rice farmers (e.g., about 52% of interviewed farmers employing crop rotation and 42% of interviewed farmers growing rice continuously had been trained on appropriate production technology); (2) the formal-education levels of the crop-rotation farmers are somewhat better than that of the continuous-rice farmers; and (3) the number of male household heads is relatively greater for the crop-rotation farmers. These factors have positive effects on the TE of the farm households (refer to the estimated results of the Tobit model that are presented in Table 10.5 in the later section).

10.4.1.2 Household Allocative Efficiency Distributions of the Two Selected Farming Patterns

This section focuses on discussing the calculation of allocative efficiency (AE) that is used to measure how optimal the input resources are that are allocated in the production process, with given relative prices and production technology.

As the results in Table 10.1 indicate, the average AE score of the farm households is 0.78 and 0.73 corresponding to the continuous-rice and crop-rotation patterns. For the continuous-rice pattern, although only two households are fully AE, most of the remaining farmers are fairly efficient in making optimal decisions in allocation of their input mixes. This is illustrated by the AE scores which concentrate on values between 0.60 and 0.90. The results for the crop-rotation farmers are some different. Five households among the crop-rotation farmers are found to be allocatively efficient, and the variation of AE scores of other farmers is relatively large with the ranges between 0.50 and 0.90.

Allocative inefficiency is likely derived from low levels of education and lack of market price information on agricultural products (most farmers just obtain primary school levels and get price information from relatives and neighbors). Education is also found to be positively statistically significant for the explanation in change of AE from the Tobit's estimation (refer to Table 10.4 in the later section).

10.4.1.3 Household Cost Efficiency Distributions of the Two Selected Farming Patterns

Finally, an overall effect of technical and price factors on the production of farm households can be determined by calculating a combination of TE and AE. This measure indicates the amount of money that farmers can save if they employ technically and allocatively optimal production technology (i.e., cost minimization).

The CE scores and their distributions for both farming patterns are presented in Table 10.1. The variations of CE are quite large, ranging between 0.36 and 1.00 for the continuous-rice pattern, and from 0.34 to 1.00 for the crop-rotation pattern. The results imply that there are *cost inefficiencies* in their operations. Particularly, the scores of cost inefficiency range from 0.00 to 0.64 and from 0.00 to 0.66 for the continuous-rice and crop-rotation farmers, respectively. There is also substantial opportunity for farmers in both farming patterns to increase their profits by reducing the costs related to technical and allocative inefficiency.

With regard to Table 10.1, we see that the average CE scores are 0.65 and 0.67, respectively, for the continuous-rice and crop-rotation patterns. This means that if the average efficiency household in the sample of continuous-rice farmers is to achieve a cost-efficiency level at the highest efficiency of its counterpart, then that average household can receive a cost savings of 35% (i.e., 1 – [0.65/1.00]). The same calculation for the lowest-efficiency household suggests a gain in cost efficiency of 64% (i.e., 1 – [0.36/1.00]). Similarly, the average household and the most inefficient household in the sample of crop-rotation farmers can realize a cost savings of 33% and 66%, respectively (i.e., 1 – [0.67/1.00] and 1 – [0.34/1.00]).

10.4.2 Identifying the Sources of Production Efficiency Using the Tobit Model

In the sections above, the TE, AE, CE, and SE of farm households for both farming patterns have been measured and analyzed. These results provide us with significant information relating to the ability of farm households to employ the best practices and technology, make optimal decisions on the input mix, and minimize the cost of production, as well as to increase productivity by moving toward the point of technically optimal productive scale. However, they do not tell us anything about the causes of these efficiencies. Therefore, this section focuses on the determinants of productive efficiency.

10.4.2.1 A Practical Model for Calculation of the Sources of Productive Efficiency

In our estimations, ten socioeconomic/environmental variables are included in the Tobit regression function. They are age, education, training, female-labor share, land area, credit, land ownership, sex, and two district dummy variables (local 1 and local 2). The model is defined as follows:

$$\begin{aligned}\text{Efficiency} = {} & \beta_1(\text{Age}) + \beta_2(\text{Education}) + \beta_3(\text{Training}) \\ & + \beta_4(\text{Share of fe\_labour}) + \beta_5(\text{Area}) \\ & + \beta_6(\text{Credit}) + \beta_7(\text{Landownership}) \\ & + \beta_8(\text{Sex}) + \beta_9(\text{Local 1}) + \beta_{10}(\text{Local 2})\end{aligned} \tag{10.3}$$

The meaning and the expected sign of the variables are described as follows.

Efficiency denotes the components of productivity such as TE, AE, CE, and SE that have been obtained from the DEA models in the previous sections. The efficiency scores have values between zero and one.

Age reflects the age of the household head. Unlike old farmers, young farmers are active and have reasonable levels of formal education, which may help them in understanding and applying new technology and skills of production. They, however, likely lack experience and are impatient in their decisions about cultivation, which the older farmers are not. For this reason, farmers may get higher efficiency in production if they are not too old. Therefore, the expected sign of coefficient for this variable is negative.

Education expresses the number of formal years of schooling for the household head. The expected sign of this variable is positive because the real effects of low and high levels of education on farmers are clear.

Training represents the hours of agricultural advice received by the household head. This variable is often used to calculate the effects of training courses provided by the agricultural-extension centers. We expect that farmers obtain greater efficiency if they have taken part in training courses. The more training hours they have, the higher the levels of efficiency. Hence, the expected sign for this variable is positive. It is clear that when farmers regularly attend training courses they get more knowledge of production technology and apply these in their cultivation activities.

Female-labor share is measured by the ratio of the number of female laborers to the total number of laborers in production. This variable is used to capture the different effects of male and female labors. In rural areas of the MDR, men often tend to undertake fieldwork while women do the housework. Therefore, male farmers probably have more cultivation experience than do female farmers. This leads to productive efficiency which may be increased with the increasing use of male farmers. Thus, the expected sign of this variable should be negative.

Land area is the size of land under production. This variable is used to determine the influences of differences in operation scales on efficiency. Normally, the larger land areas may result in higher productivity.

Credit is a dummy variable whereby a value of one is given if the household has a bank loan, and a value of zero otherwise. Basically, efficiency will increase with the use of credit. The availability of credit will reduce the constraints on production due to limited finance, and hence, it may help farmers increase their cultivation efficiency. In addition, when farmers get loans they have to pay interest to the banks; this may induce them to work harder. Therefore, the expected sign of this variable should be positive.

Ownership of land is designed as a dummy variable with a value of one if at least part of the land farmed is based on a tenancy arrangement, and a value of zero otherwise. In fact, there are two main reasons for farmers in the MRD to rent land: they have no land to culture and/or because they want to expand their scale of operation. In both cases, efficiency is likely to be increased because farmers often tend to work hard when their cultivation land is rented and they have more opportunities to use the best practices and available technology in the most effective way with larger scales of operation. Thus, the expected sign is positive.

Sex is a dummy variable for gender, which has a value of one if the household head is male, and a value of zero otherwise.

The last two factors are related to the *district dummy variables*, *Local 1* and *Local 2*, which are used to explain the impacts of different locations (Cho Moi, Cai Lay, and Cau Ke districts) on productive efficiency. For Local 1, a farmer gets a value of one if located in Cho Moi, and a value of zero otherwise. For Local 2, a farmer gets a value of one if located in Cai Lay, and a value of zero otherwise. The productive efficiency can vary due to differences in the natural factors which are not included in the model, such as soil fertility and weather.

The model (10.3) runs four times separately in order to identify the determinants of TE, AE, CE, and SE of the farm households.

10.4.2.2 Empirical Results

First, the accurate representation of the data for the Tobit regression models must be verified. Using $\alpha = 0.05$ together with 303 observations ($N = 303$) and 11 variables (including the dependent variable, $M = 11$), the absolute value of the test statistic (z) is greater than the critical value $t_{1-0.025}(303-11)$ for all of the four models. Therefore, the Tobit regression functions are all statistically accepted for identifying the determinants of productive efficiency.

For the sake of simplicity, the analyses will be conducted separately in the following sections.

Determinants of Technical Efficiency

The results of the test show that four variables are found to be statistically significant for explanation of changes in TE. They are sex, age, education, and Local 1.

In fact, the positive relationship between TE and sex is quite appropriate. This indicates that TE can increase if the production process is implemented and managed by male farmers. Normally, male farmers likely reach a higher level of efficiency than female farmers for several reasons: (1) male farmers are probably more robust and have more farming experience than female farmers, (2) male farmers often obtain higher educational levels than female farmers, and (3) in the traditional culture characteristic of the MRD, most major decisions are made by male farmers. Such characteristics help male farmers improve their ability to use experience, skill, knowledge, and available technology in the production process. The estimated result is similar to the results reported by Illukpitiya (2005), Linh (2006), Oladeebo et al. (2007), and Shehu et al. (2007).

The sign of age parameter is somewhat surprising. The positive sign of its coefficient means that the TE probably tends to increase with an increase in age of the farmers. However, the survey data shows that the average age of household heads in the non-flooded areas is about 45 years. Farmers of that age can still enjoy relatively good health, and they have experience which they continue to accumulate. This data can be used to interpret the positive relationship between TE and age, in this case. Some empirical studies have reported similar results. For example, Oladeebo et al. (2007) found that age has a positive impact on TE for both men and women rice farmers in Nigeria. However, it should also be noted that other studies, such as Linh (2006), have come up with different findings.

With regard to the positive impact of education on TE, our result tells us that the higher the levels of education achieved by the farmers, the greater the efficiency they reached. Obviously, when they obtain higher levels of formal education, they will master more basic knowledge, and they can learn new production technologies through self-study and/or attending training courses. This estimate is quite consistent with the empirical studies of Illukpitiya (2005), Linh (2006), Oladeebo et al. (2007), and Shehu et al. (2007).

Related to the variation of TE due to differences in the natural properties of the cultivation areas such as climate and soil quality, the positive sign of the coefficient of local 1 implies that farmers in Cho Moi district are more technically efficient than farmers in Cau Ke and Cai Lay districts (Table 10.3).

Determinants of Allocative Efficiency

The estimates of the explanatory variables with respect to coefficients, statistic values (z), and p values are presented in Table 10.4.

Making comparisons between the statistic values (values z in Table 10.4) and the critical value obtained from the statistical table relating to the Subjects t-distribution at 5% significance for the data on 11 variables of 303 observations indicates that

Table 10.3 The truncated estimates for the sources of technical efficiency of farm households in the non-flooded areas

		Technical efficiency					
		Non-flooded areas			Flooded areas		
Variables		Coef.	Z	P[|Z|>z]	Coef.	Z	P[|Z|>z]
Sex	β_1	0.334*	10.413	0.000	0.237*	6.168	0.000
Age	β_2	0.008*	13.567	0.000	0.008*	10.436	0.000
Credit	β_3	0.034	1.624	0.104	0.053**	1.909	0.056
Land ownership	β_4	0.054	1.435	0.151	0.019	0.298	0.766
Education	β_5	0.020*	7.147	0.000	0.024*	6.425	0.000
Training	β_6	−0.001	−1.062	0.288	0.000	0.356	0.722
Share of female labor	β_7	−0.002	−0.564	0.573	0.079	1.104	0.270
Area	β_8	−0.012	−0.763	0.445	0.005	0.756	0.450
Local 1	β_9	0.083*	2.848	0.004	0.091*	3.001	0.003
Local 2	β_{10}	−0.025	−1.036	0.300			
Sigma	σ	0.171	24.617	0.000	0.181	20.050	0.000
Log likelihood function	LLF	104.774			58.696		

Note: * and ** is significant at 1% and 5%, respectively

Table 10.4 The truncated estimates for the sources of allocative efficiency of farm households in the non-flooded areas

		Allocative efficiency					
		Non-flooded areas			Flooded areas		
Variables		Coef.	Z	P[|Z|>z]	Coef.	Z	P[|Z|>z]
Sex	β_1	0.211*	7.622	0.000	0.196*	6.040	0.000
Age	β_2	0.008*	15.133	0.000	0.008*	11.651	0.000
Credit	β_3	0.019	1.069	0.285	0.065*	2.768	0.006
Land ownership	β_4	0.009	0.289	0.773	0.046	0.881	0.378
Education	β_5	0.019*	7.796	0.000	0.017*	5.499	0.000
Training	β_6	0.000	−0.476	0.634	0.001	0.944	0.345
Share of female labor	β_7	−0.003	−1.098	0.272	0.089	1.472	0.141
Area	β_8	0.006	0.428	0.669	0.009***	1.712	0.087
Local 1	β_9	0.118*	4.686	0.000	0.063**	2.469	0.014
Local 2	β_{10}	0.004	0.207	0.836			
Sigma	σ	0.147	24.617	0.000	0.153	20.050	0.000
Log likelihood function	LLF	150.101			92.414		

Note: *, ** and *** is significant at 1%, 5%, and 10%, respectively

four variables appear to have positive effects on the AE of farm households: sex, age, education, and the district dummy variable Local 1.

The estimated results show that the AE tends to increase when the production process is implemented and managed by males and by more highly educated farmers.

Regarding the effects of education, farmers with higher levels of education likely have more opportunities to make better decisions in allocating input resources, given relative prices and production technology. This may lead to higher AE scores. The other reasons are as previously interpreted.

Two other factors that also positively impact the AE are age of household head and cultivation location (Local 1). Obviously, it is difficult for farmers living in different locations to hire productive resources at the same prices; hence, for farmers living in different areas, there is a difference in making proportional combinations of resources in the production process. The explanation of the effect of age is similar to that in the previous discussion.

Determinants of Cost Efficiency

Table 10.5 shows that the CE of farmers is influenced by the effects of five different factors: sex, age, education, and two district dummy variables (Local 1 and Local 2).

Sex, age, and education are found to have positive impacts on the changes in CE of farm households. We are not surprised with these estimates because CE is a combination of TE and AE, so the factors that affect TE and AE probably also affect CE. The reason for this can be explained in the same way as in the cases of TE and AE.

Regarding the district dummy variables, the positive effect on CE belongs to Local 1 and vice versa for Local 2. The estimated coefficients indicate that farmers in Cho Moi are more effective with respect to cost minimization than are farmers in Cau Ke. The lowest CE is found with farmers in the Cai Lay district.

Table 10.5 The truncated estimates for the sources of cost efficiency of farm households in the non-flooded areas

		Cost efficiency					
		Non-flooded areas			Flooded areas		
Variables		Coef.	Z	P[|Z|>z]	Coef.	Z	P[|Z|>z]
Sex	β_1	0.234*	8.375	0.000	0.199*	5.652	0.000
Age	β_2	0.006*	11.471	0.000	0.006*	8.894	0.000
Credit	β_3	0.024	1.309	0.191	0.049**	1.921	0.055
Land ownership	β_4	0.011	0.348	0.728	0.042	0.728	0.467
Education	β_5	0.016*	6.553	0.000	0.018*	5.305	0.000
Training	β_6	0.000	−0.564	0.573	0.001	0.879	0.379
Share of female labor	β_7	−0.001	−0.243	0.808	0.016	0.242	0.809
Area	β_8	−0.008	−0.593	0.553	0.007	1.231	0.218
Local 1	β_9	0.153*	6 002	0.000	0.079*	2.869	0.004
Local 2	β_{10}	−0.044**	−2.082	0.037			
Sigma	σ	0.149	24.617	0.000	0.165	20.050	0.000
Log likelihood function	LLF	146.681			76.750		

Note: * and ** is significant at 1% and 5%, respectively

10.5 Conclusions

This study focuses on estimating productive efficiency with respect to TE, AE, and CE, as well as on the causes for these efficiencies for farm households that were following either the continuous-rice pattern or employing the crop-rotation pattern in the non-flooded and flooded areas of the MRD in Vietnam.

Turning first to the efficiency of farm households, in both non-flooded and flooded areas, crop-rotation farmers are likely more efficient than continuous-rice farmers in terms of TE, according to the mean MTRs. This is consistent with the computed results of similar studies conducted by Latruffe et al. (2000), who found that livestock farmers get higher TE scores than crop farmers, and Linh (2006), who found that diversified farmers are more technically efficient than farmers who grow mainly rice. Regarding other components of productivity such as AE and CE, farmers employing both patterns mostly attain high efficiency in comparison with the best farmers of one farming pattern.

With regard to the factors influencing efficiencies, in the case of non-flooded areas, the estimated results show that although there are some differences in determinants of each component of total productivity; sex, age, and education are found to be the main factors that can bring about changes in all components (TE, AE, and CE). Similarly, in the case of flooded areas, all components of total productivity are impacted by sex, age, credit, and education.

References

Afriat SN (1972) Efficiency estimation of production functions. Int Econ Rev 13:568–598

Boles JN (1966) Effiency squared – efficiency computation of efficiency indexes. Proceedings of the 39th annual meeting of the westen farm economics association, pp. 137–142

Charnes A et al (1978) Measuring the efficiency of decision making units. Eur J Oper Res 2:429–444

Coelli TJ et al (2005) An introduction to efficiency and productivity analysis, 2nd edn. Springer, New York

Farrell, MJ (1957) The measurement of productive efficiency. J Roy Stac Soc. Series A, 120(3):253–290

Illukpitiya P (2005) Technical efficiency in agriculture and dependency on forest resources: An economic analysis of rural households and the conservation of natural forests in Sri Lanka. University of Hawaii

Krasachat W (2007) Economic efficiency of feedlot cattle farms in Thailand. King Mongkut's Institute of Technology Ladkrabang, Bangkok

Latruffe L et al (2000) Determinants of technical efficiency of crop and livestock farms in Poland. University of Paris, Paris

Linh HV (2006) Effiency of rice farming households in Vietnam: A DEA with bootstrap and stochastic frontier application. Department of Applied Economics, University of Minnesota

O'Donnell C et al (2008) Metafrontier frameworks for the study of firm-level efficiencies and technology ratios. Empirical Econ 34:231–255

Oladeebo JO and AA Fajuyigbe (2007) Technical efficiency of men and women upland rice farmers in Osun, Nigeria. J Hum Ecol 22(2):93–100

Shehu JF and SI Mshelia (2007) Productivity and technical efficiency of small-scale rice farmers in Adamawa State, Nigeria'. J Agric Soc Sci, 1813–2235

Shephard RW (1970) Theory of Cost and Production Functions, Princeton University Press, Princeton

Statistical year book (2006) General statistics office of Vietnam

Chapter 11
Living with Floods: An Evaluation of the Resettlement Program of the Mekong Delta of Vietnam*

Vo Thanh Danh and Shahbaz Mushtaq

Abstract The Vietnamese Mekong Delta (VMD) is a region increasingly affected by flooding. In 1996, the Vietnamese government launched an ambitious Living With the Flood (LWF) program. The objective of the program was to build dwelling houses for residents relocated from the VMD's flood areas. The program has built more than 1,000 resettlement clusters (RCs) for the 200,000 households and 1 million people previously living in the now permanently flooded areas. Total investment capital has been about US $200 million. While the LWF policy is accepted and popular within the VMD, there are many reasons that make the resettlement program both successful and unsuccessful. Identifying the factors influencing the effectiveness of the program will help local authorities to develop appropriate measures to improve the resettlement program. The rationale of this study was to review and conduct a qualitative analysis of the effectiveness of the resettlement program. The policy recommendation drawn from the study is that to effectively cope with yearly floods in the long run, people need to adjust their habits and their social and economic activities, towards living with floods and gaining benefit from them, rather than preventing them. At the national level, economic development strategies and planning in the VMD need to be consistent with the LWF policy. At the local level, authorities need to persuade and assist people to better adapt to flooding.

Keywords Living with flood policy • Resettlement cluster • Resettlement path • Resettlement program

---

*This paper is a part of the project funded by The Economy and Environment Program for South East Asia (EEPSEA)-IDRC.

Vo Thanh Danh (✉)
School of Economics and Business Administration, Cantho University of Vietnam, Vietnam
e-mail: vtdanh@st.ctu.edu.vn

S. Mushtaq
Australian Centre for Sustainable Catchments, University of Southern Queensland, Toowoomba 4350, Australia

M.A. Stewart and P.A. Coclanis (eds.), *Environmental Change and Agricultural Sustainability in the Mekong Delta*, Advances in Global Change Research 45,
DOI 10.1007/978-94-007-0934-8_11, © Springer Science+Business Media B.V. 2011

Abbreviations

VMD	Vietnamese Mekong Delta
VND	Vietnam Dong
LWF	Living With the Flood
RC	Resettlement Cluster
RP	Resettlement Path
SLR	Sea Level Rise
IDI	In-Depth Interview
FGD	Focus Group Discussion
MRC	Mekong River Commission
VRSAP	Vietnam River Systems and Plains

11.1 Introduction

The Vietnamese Mekong Delta (VMD) has a natural area of 4 million hectares, and the population of 18.2 million people accounts for 21% of the total population of Vietnam. The VMD is a low-lying delta, and sea level rise (SLR), one of the calamities of global climate change, creates a serious threat to the population of the area. Reiner et al. (2004) used a hydraulic model under two SLR scenarios of 20 and 45 cm to compute water levels during the flooding season. The results showed that the average increase in water level in the VMD under these scenarios would be 14.1 to 32.2 cm, respectively. These results suggest that between 0.6 million and 2.3 million of the 4 million hectares of the Delta are vulnerable to flooding due to SLR. Ninh and Hong (1993) summarized the impacts of SLR in Vietnam. First, SLR would result in the loss of land, especially fertile agricultural lands in low-lying areas such as the VMD. SLR would have significant consequences for the livelihoods and socio-economic development of people living in the VMD. Regions previously not permanently inundated by sea water would become unsuitable for agricultural production. Second, salinization would increase. In particular, irrigated rice paddies would be affected by the increased intrusion of saline water. Third, vulnerability to flooding would increase, and people living in the VMD's increased-flooding areas would have to be relocated.

Half of the area of the VMD is inundated during the flood season. More than 2 million households with 11 million people are affected by flooding in the flooded area, for 3–4 months each year. Of the 800,000 or so houses flooded annually, more than 200,000 are inundated by 2 m or more. The flooded area in the VMD is approximately 1,828,000 ha and is home to more than 50% of the population in the region. The flooded area is the low-lying region, with more than 61% of the flooded areas less than 1 m above sea-water level. The flood period starts in July and ends in December. The inundation level is from 0.5 to 4.0 m. In economic terms, the flooded area today constitutes 67% of the VMD's GDP, 75% of the total value of agricultural-fishery-forestry production, 79% of the total value of industrial production, 80% of the total value of services, 80% of the total value of rice exports, and more than 50% of the total value of Vietnam's agricultural exports. Although flooding causes deaths and disruption, as well as serious damage to houses, schools, hospitals, and infrastructure, flooding also has benefits such as alluvial deposition, fishery and water-environment improvement, and flushing of acid soils in rice fields.

Over the past 100 years, people in the VMD have learned to live with flooding and to adapt in order to avoid damage and gain benefits from floods. Currently, the Living With the Flood (LWF) concept is institutionalized as the LWF policy. According to the LWF policy, the VMD's socio-economic development strategy will be interdependent with the flood-control masterplan. There are two flood-related programs implemented in the flooded areas: a dyke program and a resettlement program. Program selection depends on the level of inundation of the flooded area in question. Flooded areas in the VMD are divided into two parts: a deep inundated area and a shallow inundated area with flooding below the 1-m inundation level. The LWF measure recommended depends on the nature of each individual flooded area but also takes into account the time frame for either short-term or long-term solutions. In the long term, the main solution for the shallowly flooded areas is to maintain normal agricultural production and socio-economic activities via a network of small dykes which can be opened to allow some flooding, but also provides protection from floodwaters throughout the year. Shallow flooding enables three rice crops to be grown each year. In the shallow flooding area, the dyke system specifications are 2.5 m top × 4.5 m bottom × 2.0 m height. In the deeply flooded area, the main policy measure is a large dyke system built to prevent damage from floodwaters coming from the direction of the Cambodia border into the Dong Thap Muoi region and the Long Xuyen Quadrangle, and to speed the flow of flood waters into the West Sea. A strategy of periodic flood control is recommended, the intent of which is to slow the flood and maintain the shallow inundation level at the beginning of flooding to enable harvesting of the Summer–Autumn rice crop, while speeding up flood flows at the end of the flood period to enable on-time harvesting of the Winter–Spring rice crop. It should be noted, however, that in the deeply flooded area, the Autumn–Winter rice crop is lost.

In addition, the program involves the building of resettlement clusters (RCs) to allow permanent resettlement of people living in the deeply flooded areas. Since 1996 the government has launched an ambitious livelihood program. The objective of the program is to relocate households from the flooded areas. The elevation of the resettlement housing is a minimum of 0.5-m higher than the record flood level in year 2000. In the resettlement clusters, infrastructure such as dykes and roads and public buildings such as schools, markets, clinics, etc. are built. In the deeply flooded areas, a combination of canal construction and resettlement with the large dyke system or houses on stilts is recommended. The average size of each resettlement is 300 ha. Each commune in the flooded areas will have at least two resettlements. In the period 2001–2005, there were 1,043 resettlements for 200,000 households with 1 million people.

The rationale of this study is to review and qualitatively analyze the effectiveness of the resettlement program. Based on the results of the study, recommendations for the policy will be made and research gaps identified. The main objective is to evaluate the effectiveness of the resettlement program under the LWF policy in the VMD. The specific objectives are:

1. To review the LWF policy and the resettlement program,
2. To evaluate the impacts and constraints of the resettlement program,
3. To recommend actions.

11.2 Methodology

11.2.1 Research Design

The study used both qualitative and quantitative methods. Qualitative data were collected through in-depth interviews (IDI) and focus group discussions (FGD). The IDI survey was designed to collect subjective evaluations on the resettlement program's performance from local people living in the VMD. The IDI variables were: periodic progress of the program, management and monitoring, pros and cons, and self-evaluation. To facilitate the collection of relevant information, the IDI questionnaire was first sent to the commune authority and then discussed in a meeting with representatives of the commune authority. For FGD surveys, respondents were grouped into (1) those who currently live in the RCs and resettlement paths (RPs) and (2) those who live outside the RCs and the RPs. The FGD respondents were selected through nomination by the commune authority. For those who were living in the RCs, the selection criterion was that they had lived in the flooded and eroded areas before being resettled by the resettlement program. For those who were living in the RPs, the selection criterion was that they had been resettled in the RPs for at least 1 year. For those who were living outside the RCs and RPs, the selection criterion was that their houses were located near the RP and next to an eroded river bank. The FGD variables included ex-post income generation, reasons for living inside or outside of the RCs, pros and cons, opinions about the program, satisfaction, and recommendations. There were three FGD surveys: one for 15 people living in the RC of Long Hiep village, one for 12 people living in the RP of Long Hoa village, and another one for 18 people living in Tan Hau B2 village outside the RCs.

For the quantitative data, there were three household surveys: one for 20 people living in the eroded area, one for 20 people living in the RC, and another for 20 people living in the RPs. Variables collected in the first household survey for people living outside the RCs included their opinion about the impact of flooding on their lives, their adaptation to flooding, and their perception about the floods. Variables collected in the household surveys for people living in the RCs and RPs included changes in income, living conditions, and job opportunities.

11.2.2 Description of Study Site

The study was conducted in the Tan Chau district of An Giang province. An Giang province is located at the upstream limit of the VMD flooded area. The area of the province is 3,424 km^2, and its population is more than 2.1 million. Every year in June or early July, flooding from the Tien and Hau rivers inundates the An Giang region, with the water level at Tan Chau station reaching 3.0 m. Peak flooding generally occurs from the end of September to mid-October. According to observations dating from 1911, on average, there is a serious flood every 7 years. A flood is considered serious when the water level at Tan Chau station exceeds 4.6 m. In the 1990s serious floods in An Giang province occurred in 1991, 1994, 1995, and 1996. The dreadful flood in 2000, especially, had a record water level of 5.06 m (compared to the 1961 flood at 5.12 m) and is considered a "century" fatal flood with uncountable damages.

Tan Chau district is 15,994 ha in area, with a population of more than 150,000 people. Figure 11.1 represents the position of Tan Chau district in the administrative map of An Giang province. The eastern border of the district is the Tien River

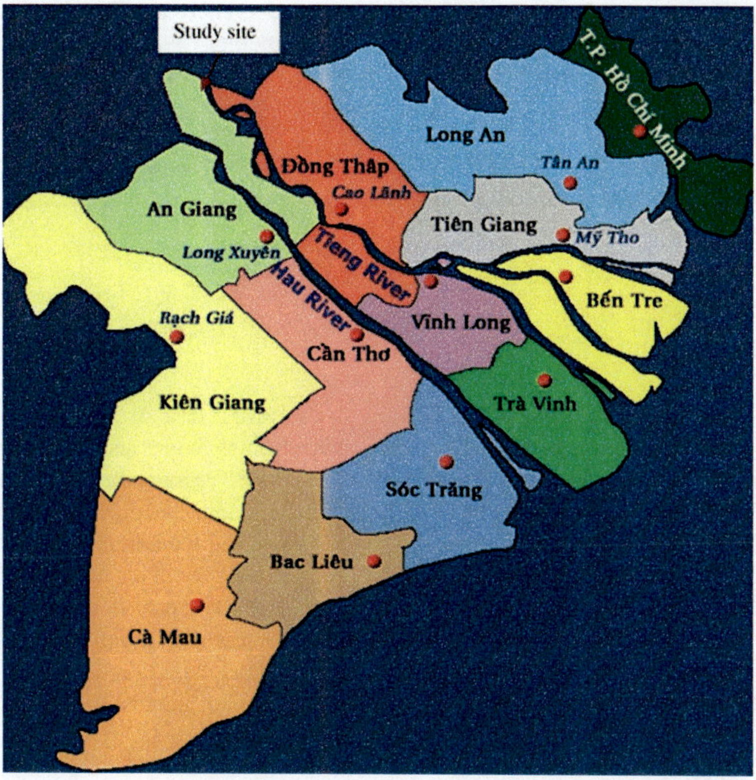

Fig. 11.1 Administrative map of An Giang province, with the study location indicated

of Dong Thap province. The western side shares a border with the An Phu district of An Giang. The southern side is next to the Phu Tan district of An Giang province. The northern border is with Cambodia. Tan Chau district is one of the most deeply inundated areas of An Giang province with an average inundation level of 2 m in the flood season. It is one of the two most seriously inundated areas in the VMD[1]. The selection of a study site located in the most flooded area helped us to better capture flood-related problems. As of July 2007, Tan Chau district accounted for 5,022 households out of the total of 6,250 "targeted" households moving to the RCs. Eleven communes of Tan Chau district had RCs or the RPs.

Whether people were moved to an RC or a RP depended on where they lived before relocation and the vulnerability of their households to flooding. Those who lived on the isolated island at the river mouth were seriously affected during the flood season. These were prioritized in the resettlement program and were asked to move to the RCs for reasons of safety. While many households faced the dangers of erosion and flooding, only the households at greatest risk were selected initially. However, many households were still at risk. When the government launched the LWF policy and resettlement program, an inter-commune road was raised as a river dyke to prevent flooding. Following this, RPs were built along the road, and people built their houses on land inside the RPs. Only a small proportion of resettlers were newcomers who lived in the eroded areas near the RPs. These people were selected by the local authority based on the vote of the community.

11.3 An Overview of the Living with Flood Policy and Resettlement Program

11.3.1 Flood Situation in the Mekong Delta

11.3.1.1 Nature of Flooding

Flooding is an annual natural phenomenon in the VMD. It starts upstream in the Mekong River, with waters rising due to heavy rains from typhoons and the Southwest monsoon. Flooding occurs when water in the Mekong River inundates Savannakhet and Pakse at southern Laos and Kratie at eastern Cambodia. The flood season lasts from June to October, and accounts for some 90% of the total annual flow in the river. Water levels in the Tien and Hau rivers (major tributaries of the Mekong River) start to rise at the end of June or early July. The peak of the annual flood is recorded at the end of September or early October. The flood season is usually over by the end of December or January. The flood in the VMD is called "lu lut." The "lu" implies that water is rising; the "lut" refers to the period of inundation.

[1] Another inundated area is Hong Ngu district of Dong Thap province.

According to Quang (2000), besides typhoons and the Southwest monsoon, there are four causes of flooding in the VMD, namely, hydroelectric dams upstream in China, population migration into the flooded areas, deforestation, and the network of irrigation dams. It is said that the dams in China affect only 2% of total water flow of the Mekong River; however, environmental and ecological impacts for the downstream region are still possible. Migration into the "Plain of Reeds" in the past and a change in farming practices have made the flooding more serious and longer. The reason for this is that the new residents created high-density canal systems in order to desalt the land for rice planting. Deforestation also contributes to increased flood frequency and severity.

To assist with flood management, a flood warning system in the VMD was built by the Mekong River Commission (MRC). According to the MRC standard, the "floating" season or non-flood situation occurs when the peak water level at Chau Doc station of Dong Thap province is between 3.8 and 4.2 m. Flooding occurs when the peak water level at Chau Doc station is higher than 4.2 m; when the water level at Chau Doc station is lower than 3.8 m, the region is said to be in drought. At Tan Chau station of An Giang province, the non-flood situation occurs when the peak water level is between 3.5 and 4.5 m. When the water level is above 4.5 m, it is considered a serious flood. Above 5 m, it is considered a dreadful flood. The floods in 1961 and 2000 both exceeded this level and caused serious damage in the region. Table 11.1 represents the flood warning system in the VMD. When flooding starts upstream in the Mekong River, the water in the main rivers rises. Once the water level at Tan Chau is between 2.8 and 3.0 m, water begins to inundate the Plain of Reeds and the Long Xuyen quadrangle (flood warning at the alarm level No. 1). When the water level at Tan Chau exceeds 3.0 m (flood warning at the alarm level No. 2), flood waters inundate the Plain of Reeds, the Long Xuyen triangle, and areas between the Tien and Hau rivers. If the water level exceeds 4.2 m, serious flooding is experienced that year (flood warning at alarm level No. 3).

An important consequence of flooding in the VMD is the presence and seasonal variation of the Great Lake in Cambodia. The Great Lake plays a role as a reservoir in regulating water volume in the flood season. At the beginning of the flood season, 80–90% of the water volume of the Mekong River discharges into the Mekong Delta, with the remaining 10–20% of the water discharging into the Great Lake. In October as the water level in the Mekong River falls, water from the Great Lake continues to discharge into the Mekong River's network – often until May of the following year. As a result, water levels rise slowly, on average, 3–5 cm/day. Therefore, the flood in the VMD is referred to as a "peaceful" flood.

Table 11.1 System of alarm levels for flooding in the Vietnamese Mekong Delta

Class	Tan Chau station (m)	Chau Doc station (m)
1	3.0	2.5
2	3.6	3.0
3	4.2 (alarm level)	3.5 (alarm level)

Source: Mekong River Commission 2003

Table 11.2 Estimated economic losses for the different damages (Unit: million $US) (*Source*: Ngoc 2003)

Probability of design flood (%)	Residential	Commercial	Agriculture	Infrastructure	Total damages
50	24	0	272	388	684
20	218	48,202	299	538	49,257
10	232	52,830	322	613	53,995
5	243	54,262	341	678	55,522
2	260	61,450	353	749	62,810
1	280	75,235	403	795	76,712
0.2	295	76,072	413	939	77,719
0.1	305	80,610	432	975	82,321

11.3.1.2 Damages of the Flood

In the last 50 years, on average, there is a flood every 2 years (i.e., the water level at Tan Chau station is above the third-class alarm level of 4.20 m). However, at times, flooding may occur 3 or 4 years in succession as in the periods 1937–1940, 1946–1949, 1994–1996, and 2000–2002. Observations since 1941 indicate that the 1961 flood was the biggest flood ever recorded in the VMD, with the water level at Tan Chau station at 5.12 m. However, the "century" flood in 2000, when the water level at Tan Chau station was 5.06 m, had a record-flooded area, inundation time, and water volume. Its economic losses were estimated to be US $500 million.

Flooding has significant consequences in the VMD. The inundation extends over several months, changing the economic activities and livelihoods of people in the flooded area. Infrastructure such as roads, small dykes, and canals are under water for several weeks. Flooding also causes damage due to erosion and landslides. In particular, casualties may affect people living in flooded areas. Ngoc (2003) applied the VRSAP model[2] to estimate the economic losses due to flooding in the VMD. The model used 1996 and 2000 data as the base cases. It predicts the flooding situation associated with different return periods. Flood records from 1935 to 2000 were used to analyze flood frequency and to estimate the peak discharges for different return periods. This information was used to evaluate flood damage. Table 11.2 presents estimated economic losses for the different kinds of damage.

11.3.2 Description of the Living with the Flood Policy

11.3.2.1 History of Coping with the Flood

The modern history of coping with flooding in the VMD can be divided into four periods. Prior to the nineteenth century, the pioneers of the new land chose to adapt naturally to the flood situation. At that time population density was 10–20

[2] VRSAP stands for The Vietnam River Systems and Plains.

persons/square kilometer. The majority of residents lived along the rivers or chose places that were not inundated. In inundated areas, people built houses on stilts. During this time, there was no irrigation infrastructure to manage flood waters.

In the period from the nineteenth century to the early twentieth century, as the population density reached 100 persons/square kilometer, resettlers created a new network of canals to exploit new land and partially adapt to the flood. Thoai Ha canal in the Long Xuyen quadrangle, 51-m wide and 31-km long, was built in 1818. Concurrently, in 1819 the Vinh Te canal (38.4 m by 98.3 km) was built to connect An Giang with Kien Giang.

In the period from 1950 to 1999, the VMD was exploited for the purpose of agricultural development. A network of irrigation and dykes to prevent flooding was established for the area, which then supported a population density of 415 persons/km^2. The total length of the dyke system was about 12,000 km. This system was designed to protect 1,200 residential clusters across 1.3 million hectares. In addition, a huge project was inaugurated to speed up the discharge of flood waters into the West Sea.

Since 2000, the concept of the LWF has become accepted with the realization that, besides damage, the flood also provides benefits to the people in the flooded area. By actively coping with the flood, people can mitigate the damages and gain more benefits. According to this concept, the socio-economic development of the VMD needs to be linked to the strategy of "controlling" the flood. The main objectives of controlling flooding in the VMD are to mitigate damages and to live with the flood. Preparing for adaptation to the inundation situation and minimizing changes to current flood flows are other objectives of this strategy.

11.3.2.2 Introduction of the Living with Flood Policy

Following the fatal flooding in 1978, a strategy of coping with flooding was set up. Through an ambitious program called "bring the flood to the west sea," a new canal system was created in the hope of controlling flooding. However, the flood situation in subsequent years indicated that serious floods in 1996, 2001, and 2002 could be partly due to the negative effects of this program. It is argued that this program is probably a main cause of other fatal floods in later years. As a result, there was a policy change toward mitigating the negative impacts of flooding rather than controlling the flood. This is popularly now called the LWF policy. Under the LWF policy, an adaptation strategy and response methods to cope with flooding were implemented. The Prime Minister's No. 99 Decision, dated September 2, 1996, is considered a keystone of the LWF. Furthermore, the Prime Minister's No. 173 Decision, dated November 6, 2001, stated that in order for growth in the VMD to take place, the populace needs to take advantage of the benefits and potential of the water resources of the Mekong River, as well as strengthen the capacity to cope with flooding. In addition, a permanent resettlement policy was designed to ensure that residents in the flooded area would not migrate during the flood time.

11.3.3 Introduction of Resettlement Program

Among the measures of the LWF policy, a resettlement program is considered one suitable solution. A resettlement, an RC or an RP, is usually built within a permanently flooded area. Socially, the main objective of the program is to normalize the living conditions for the people in the flooded areas. As a requirement, the base of a RC must be higher than 0.5–1.0 m above the 2000 flood level. The construction of the resettlement area typically includes public infrastructure such as a marketplace, school, clinic, communication system, and road, along with facilities for electricity, supplying water, etc. The RC may be designed along the river, canal, or road on the condition that it is not affected by flooding. However, in reality, building the RC is usually linked with dyke-system development. In the permanently-flooded areas it is required that there be at least two residential clusters built in a commune. Households dispersed across the flooded areas are able to get preferential treatment in accessing loans to raise the base of an existing house or to build a house on stilts.

Since 1996 the government has promulgated various policies related to building and developing the RCs in the VMD. The 105 Decision of the Prime Minister, dated September 2, 2002, stated that the program's beneficiaries are to be households in the flooded areas. Such households are able to get preferential loans for buying houses in the RCs. The preferential arrangements include a maximum zero-interest loan of 10,000,000 VND for buying a piece of land in an RC and a maximum annual 3%-interest loan of 7,000,000 VND for buying a house on that piece of land. The duration of preferential loans is 10–15 years. Those who benefit from the program are committed not to sell or mortgage the assets of land and house for a period of 10 years. Regarding the standard of house building in the RCs, the 2004 78 Decision of the Prime Minister required that houses be not less than 32 m^2 in area. For infrastructure construction within the RCs, the 04 Decree of the Ministry of Construction required a system of roads, communication facilities, electricity supply, water supply, and waste discharge to be completed. Internal roads enable connection with the road systems of the district and province. These roads are limited to 7 m in width, while main roads in the RCs are not to be larger than 5.5 m in width.

11.3.4 Results of the Resettlement Program's Implementation

Although establishing the settlement program is the right way to adapt to the permanent flood situation in the VMD, the program is not always viewed positively by people living in the affected areas. The success of the resettlement program depends on the development of infrastructure. Many RCs have no clean water supply or waste discharge system. In some cases, there is not even electricity or roads.[3]

[3]It is called the "four nos" situation, i.e., no electricity, no road, no clean water supply, and no waste discharge system.

The development of public facilities such as schools, kindergartens, clinics, post offices, and marketplaces has not progressed as expected. Another concern is the quality of infrastructure built. To meet the time frame of the program, a huge number of works had to be built very quickly. As a result, it is not surprising that the quality of many structures built is inadequate. In addition, living conditions in the RCs often act as a deterrent. The majority of resettlers are poor[4], with very little or no land. Before coming to the RCs, they had employment. Living in the resettlement cluster causes difficulties in getting on-farm employment. Tien (2004) learnt that 45% of respondents indicated that they received less income and had greater expenditures compared to a prior year when they lived outside the RCs. For the poor, a traditional farming model of garden-pond-cage – that is, a farming system combining fruit, aquaculture, and livestock – is commonly considered a way to generate extra income. Living inside the RCs prevents people from implementing this farming model.

Regarding living conditions in the RCs, Xe and Dang (2006) conducted a survey in selected households in the VMD. The purpose of the survey was to evaluate the socio-economic impacts of the resettlement program in An Giang province and Can Tho city. A field survey was implemented for 281 "inside" households (insider) and 81 "outside" households (outsider). Insiders were selected from 28 RCs in two provinces. The number of RCs accounted for 30% of the total number of RCs in the study site. The results are summarized below:

Change in land use. After living in the RCs, the number of households that owned land decreased, and the agricultural land area per household also declined because the resettlers generally sold a piece of land when they relocated. The reasons for selling land were financial difficulties (68%), job changes (33%), production capital deficiencies (11%), labor deficiencies (9%), and other reasons (13%).

Decrease in animal-husbandry activity. Animal-husbandry activity is commonly a main income source, besides rice production, of VMD farmers. Before moving to the RCs, more than 50% of households were involved in animal-husbandry activities. Because of regulations of the RCs, these activities are banned. However, a few households were observed to continue illegally.

Change in structure of employment. The resettlement program caused changes in the structure of employment, with employment decreasing in the agricultural sector relative to the non-agricultural sector. After coming to the RCs, the number of people getting jobs in non-agricultural activities increased by 5.8% but decreased by 12% in non-agricultural activities. Moreover, the unemployment rate increased to 5.6%.

Change in sources of income. The income per household was mostly unchanged in the before-and-after analysis framework. However, the structure of income sources changed. Off-farm incomes increased significantly while on-farm incomes, including employment activities, decreased. More specifically, income derived from the agricultural sector decreased from 5.5 VND million to 3.7 VND million per household,

[4]Tien et al. (2004) indicated that 64.4% of resettlers are poor.

and income from agricultural-employment activities dropped from 7.4 VND million to 5.6 million per household. Meanwhile, income derived from non-agricultural activities increased from 3.0 VND million to 7.1 VND million per household, and income derived from wage-earning activities increased from 4.7 VND million to 6.2 million per household.

Improvement in social conditions. Resettling in the RCs afforded people greater access to public services. In particular, the proportion of the population going to school increased quickly at all educational levels. Rates of children going to kindergarten and primary school increased from 46% to 96% and 67% to 94%, respectively. In addition, after moving to the RCs, the percentages of households using metered electricity and having access to a clean water supply increased from 26% to 84.7% and 12% to 68%, respectively. Other public services such as medical treatment and market activities were also improved.

Income comparison of Insiders and Outsiders. According to this survey, there were differences in incomes between the insider and the outsider groups. Generally, the average income per household was 0.8 VND million lower inside the RCs than outside. Income from cropping, husbandry, and on-farm employment decreased by 1.7 VND million, 0.9 VND million, and 2 VND million. Meanwhile, incomes from off-farm employment and services increased by 1.3 VND million and 3.9 VND million, respectively.

11.4 Findings and Discussion

11.4.1 The Resettlement Program at the Study Site

There are three types of stakeholders involved in the resettlement program: the provincial authority, the district authority, and the commune's people's committee. The role of the provincial authority is to direct the district authorities to implement policies such as planning the RCs and selecting the target groups for the RCs. The province is also the authority that directly receives the funding from the central government. The district authority is responsible for designing the RCs, selecting the place for the RCs, and determining the size of the RCs. The district authority is also responsible for managing the investment capital allocated for building the RCs. The commune's people's committee takes part in the monitoring and the allocation of the residents' plots in the RCs. Once the RC is completed, a new administrative unit is established and managed under the commune authority. Table 11.3 presents the result of the resettlement program in Tan Chau district. There were a total of 34 RCs built in inundated areas in the Tan Chau district. Most of the communes had 3–4 RCs. The majority of resettlers belonged to the four categories above. At the end of July 2007, 80% of the 6,250 targeted households had moved to live in the RCs. Eighty-five percent of these (5,236 households) paid the subsidized price of 7–10 VND million. The rest, which did not meet the criteria for the "4 Nos" groups, had to pay the higher

Table 11.3 Status-quo of resettlement clusters at Tan Chau district of An Giang province

Commune/ Town	Number of resettlement clusters	Number of houses	In which		Number of households	Proportion (%)
			Subsidized price	Market price		
(1)	(2)	(3) = (4) + (5)	(4)	(5)	(6)	(7) = (6)/(3)
Tan Chau	1	329	280	49	269	81.8
Long Phu	4	530	402	128	398	75.1
Phu Vinh	2	262	178	84	169	64.5
Le Chanh	3	611	456	155	432	70.7
Chau Phong	5	847	793	54	684	80.8
Long An	4	596	480	116	456	76.5
Tan An	3	865	749	116	731	84.5
Tan Thanh	3	622	482	140	480	77.2
Vinh Hoa	4	883	816	67	805	91.2
Vinh Xuong	3	512	412	100	410	80.1
Phu Loc	2	188	188	0	188	100.0
Total	34	6,250	5,236	1,014	5,022	80.4

Source: Tan Chau district's report on the resettlement program, 31 July 2007

market price of 30–50 VND million. The reason they had to pay the higher market price was to enable the government to get more funds to cover the cost of infrastructure, which was usually higher than the initial investment capital requirements.

Table 11.4 presents the result of the resettlement program at the commune level. There were two RCs and two RPs at Long An commune. There were 1,404 households, among the total of 3,202 households affected by flooding, that needed to be relocated. Up to the end of November 2007, there were 480 households already relocated in the RCs. About 70% of these had paid the subsidized price. It was predicted that, in the period 2008–2015, Long An would need to build 6 new RCs to relocate an additional 924 households. Among these, there were 224 households living in areas vulnerable to erosion and landslides along the Kinh Sang River, and 196 households living in areas vulnerable to flooding.

11.4.2 Program Manager's Evaluation

The management board of the resettlement program was an agent of the Tan Chau district's people's committee. It was a proxy of the provincial authority set up to manage and monitor the program and was responsible for technical and financial issues related to the resettlement program. Commune's people's committees helped the management board to select the place for constructing the RCs and persuaded and selected the target households to relocate to the RCs. In this survey, Long An commune was selected to represent the situation of the resettlement program in the Tan Chau district. According to the IDI survey, when the program started there were few people selected to live in the new place as, for some reason, people were unwilling to move. At that time, the construction of houses and infrastructure

Table 11.4 Results of the resettlement program at Long An commune

Resettlement cluster/path	Number of houses (designing for….)						Infrastructure		
	Total	Public	Market price	Subsidized price	Resettled	% Resettled	Electricity	Clean water	Road
(1)	(2)	(3)	(4)	(5)	(6)	(7) = (6)/(5)	(8)	(9)	(10)
Long Thanh	193	0	0	193	181	94.0	Completed	Completed	Completed
Tau Hau B2	129	1	37	91	88	97.0	Completed	Completed	Completed
Long Hiep	137	3	40	94	88	94.0	Completed	Completed	Completed
Long Hoa	152	6	44	102	99	97.0	Completed	Completed	Completed
Total	607	11	116	480	456	95.0			

Source: Tan Chau district's People Committee report (2007)

inside the RCs was incomplete, and there was little advantage in moving. Second, living conditions and income-earning ability were not improved by relocating to the new place. Living in one place and going to another place to get work caused difficulties for the resettlers. As a result, many of the poor did not want to join the RCs in the first phase of the program. Finally, living in a new place would mean a significant change in living habits, and this was not easily accepted by many people.

However, after the construction of houses and infrastructure in the RCs was mostly completed, the situation changed, and many people agreed to participate in the program. Currently, the Long An commune has relocated 480 households affected by flooding. In the period from 2008 to 2015, it will establish 6 new RCs and paths to relocate more than 900 target households of people still living in areas vulnerable to flooding and landslides. Seventy percent of the houses built will be set aside for those target households that are comprised mostly of the poor. Thirty percent of the houses left will be sold at market prices to the general public. Currently, most of the resettlers do not have agricultural land. Before living in the RCs, they were very poor. After relocating, their incomes and employments have not improved. However, the local authority is not responsible for generating off-farm employment in the RCs.

The infrastructure of the RCs, including the intra-commune road, electricity system, clean water-supply network, and waste-discharge system, was not completely designed. Generally, infrastructure in the RPs was better than that of the RCs. In the RCs of Long Hiep and Long Thạnh villages, the waste-discharge system was poor and needed to be improved; the in-house toilet systems did not work, and this was considered to be the most serious current problem in the RCs. At the moment, households can get a loan to make their own in-house toilet instead of using a badly designed in-house toilet system built by contractors. Table 11.5 summarizes the managers' evaluation of the RP.

11.4.3 Resettlers' Evaluation

To enable evaluation of the insiders' views about the resettlement program, two FGD surveys were done, one for those living in the RCs and the other for those living in the RPs. Variables included income-generating activities, reasons causing people to move into the resettlement cluster, evaluation of the advantages and disadvantages of living in the resettlement cluster, evaluation of the state of the resettlement cluster, people's satisfaction, and recommendations for the resettlement program.

The FGD result showed that before resettling in the RCs, people earned income from different sources such as on-farm and off-farm employment activities and fish-catching activities. There was no change in income sources and employment once they relocated in the RCs. However, because of the long distance between the old place and the new place (i.e., the RCs), some had not retained their agricultural land. This made their current livelihoods more difficult than before. The reasons for deciding to relocate to the RCs were the danger of flooding and landslides. The resettlement program gave them a chance to live better lives. They were supported

Table 11.5 Managers' evaluation of the resettlement program

Evaluation	Description
General evaluation	• Management board was a representative of the Tan Chau District's People's Committee in managing and monitoring the RP • Commune's People's Committee had responsibility in selecting the target households based on the provincial authority's regulations • Most of the target people were poor • Majority of the resettlers had no stable job. Main income sources were from on-farm tenant activities • Since relocating to the RCs and paths, livelihood of the resettlers was not considerably improved • There still were many households (affected seriously by flooding) that need to be moved to the RCs
Reasons that people did not prefer to take part in the program prior to 2005	• The poor's financial ability was limited House construction and infrastructure were not completed and had poor quality • Job seeking was not easy • People did not want to change customs and living conditions in the new place
Reasons that people preferred to take part in the program	• The poor were allowed to borrow money to buy the allotment and house • Infrastructure was now completed • People had access to clean water • People's perception of the impact of flooding and the benefits of the RP was considerably improved

Source: The survey (2007)

by the provision of a loan to buy a piece of land inside the RCs. Households relocating can access a zero-interest loan of 7,000,000–10,000,000 VND with the preferential 6-year time frame.

There were some advantages to living in the resettlement cluster. First, living conditions were better. People no longer had to move away when the flood arrived. Those that had to find jobs in other places felt that their children and old people were safer. Second, the quality of life was improved. Water and electricity were supplied, healthcare services were accessed easily, a road network was available, going to school was possible, and things in general were more convenient. All of these considerations were examples of improvements in the quality of life associated with living in the RCs. All respondents were very satisfied with living conditions in the resettlement cluster. However, there were still some disadvantages that made them disappointed. First, the distance to jobs was greater than before. It took more time getting to work outside the RCs. Second, for the poor, incomes had not increased and, in some cases, had even decreased due to extra expenditures for water and electricity bills which they did not have to pay before. These changes placed burdens on family budgets.

Besides the positive feedback, there were a lot of problems that needed to be resolved. First, the quality of house construction in the RCs was very poor. In-house toilet systems did not work, and although people participating in the FGD had made many complaints to the commune authority and the program's management board, this situation had not been addressed. Second, infrastructure in the RCs was not completely built. The water-discharge system did not function well during the flood season. Many areas in the RCs were inundated when either a heavy rain or the flood came. The poor quality of house construction and infrastructure within the RCs was due to the poor project implementation and the lack of participation of the beneficiary households. In reality, the contractor built the house, including an in-house toilet, at a cost equal to the price that the resettler paid. The contractor was assigned by the board of management, and the resettler had no involvement in the process. After the house was built, the resettler was informed when to receive the house. Consequently, with the low price of 10 VND million, a poor house was sold to the resettler. Third, there were no job opportunities for those who were living in the RCs. Table 11.6 summarize findings from the FGD surveys. Table 11.5 summarizes the FGD result of the insiders.

Table 11.6 FGD results of the households living in the resettlement cluster/path

Item	Before living in the resettlement cluster/path	After living in the resettlement cluster/path	Change[a]
Income sources	Tenant in agriculture and non-agriculture activities, fish catching	Tenant in agriculture and non-agriculture activities, fish catching	~
Agricultural land	Some of them had their own agricultural land	They sold their pieces of land because they could not travel the long distance to work	–
Subsistence activities	Besides main income sources, they raised vegetables, chicken, and pigs	They were not allowed to raise animals around their houses or inside the RCs	–
Living conditions	• Some of them had no houses • There was no electricity, clean water system, or health services in the old place • Children had difficulties getting to school during the flood season	• Basic living conditions were ensured • Electricity, clean water supplies, and health services were easily accessed • Children went to school easily	+
Living with the flood	• Most of them lived in the dangerous areas affected by erosion and the flood • During the flood season, their livelihoods were badly affected because they had to stay at home to take care of their children and the elderly	• They now could go out to find a job throughout the year without worrying about flooding affecting their children and the elderly	+

[a] + increase/– decrease/~ unchanged
Source: the survey (2007)

Table 11.7 Changes in income sources of the resettlers (Unit:%)

	Resettlement cluster			Resettlement path		
Income source	Before	After	Change[a]	Before	After	Change[a]
Crop planting	14.2	8.3	–	27.8	27.0	–
Animal raising	16.1	0	–	14.6	12.4	–
On-farm tenant	47.6	52.5	+	24.9	23.6	–
Off-farm tenant	22.1	39.2	+	33.7	37.0	+
Total	100.0	100.0		100	100	

[a]+ increase/– decrease
Source: The survey (2007)

Household surveys were implemented to evaluate the impacts of the resettlement program on the livelihood of the resettlers. Changes in income sources, use of facilities, and job opportunities were compared in the before-and-after analysis framework. Results in Table 11.7 show that there were four main income sources, namely, crop planting, animal raising, on-farm tenancy, and off-farm tenancy. Tenant jobs were the main sources of income even when people lived in the RCs. Tenant incomes tended to increase and were an especially important income source for those living in RCs. About 92% of income now came from tenant activities, while the figure was only 70% when FGD participants lived outside the RCs. Meanwhile, for those who lived in the RPs, income from tenant activities did not change so much and was about 60%. Income derived from animal raising decreased because regulations did not allow people to raise animals inside the RCs. These results were consistent with the Xe and Dang (2006) survey. That survey showed that household income was relatively unchanged after moving to RCs but that the structure of income sources changed. Off-farm incomes increased significantly while on-farm incomes, including employment activities, decreased. Per household incomes from the agricultural sector decreased from 5.5 VND million to 3.7 VND million, and per household incomes from agricultural employment activities decreased from 7.4 VND million to 5.6 million. On the other hand, per household income from non-agricultural activities and from wage-earner activities increased from 3.0 VND million to 7.1 VND million and 4.7 VND million to 6.2 million, respectively. It can be said that the structural change in income generation based mainly on tenant activities was a big challenge to the resettlement program.

With respect to job creation, most respondents said that job opportunities were not improved. In the RCs, on-farm tenant activities became more difficult. Sixty percent thought that the situation was worse than before, while 35% thought that the situation had not changed. One possible reason that made it more difficult to find on-farm work was the distance from the RCs to places where jobs are located. Many people lived on the island before moving to the RCs. Living in the new place made it difficult to get information on jobs in the old place, while job opportunities in the new place were not easy to find. Opportunities for finding off-farm work were a little better. Thirty-five percent said they had better chances of finding a job, while 25% said that they were experiencing difficulties. Those who got more off-farm jobs previously received income mainly from on-farm activities.

Table 11.8 Resettlers' evaluation of job opportunities in the resettlement cluster (Unit:%)

Job opportunity	Better	Worse	Unchanged	Total
In resettlement cluster				
• On-farm job	5	60	35	100
• Off-farm job	35	25	40	100
• Job opportunity in the flood season	0	85	15	100
In resettlement path				
• On-farm job	10	10	80	100
• Off-farm job	15	30	55	100
– Job opportunity in the flood season	35	35	30	100

Source: The survey (2007)

Table 11.9 Percentage of households accessing the facilities in the resettlement cluster (Unit:%)

	Resettlement cluster			Resettlement path		
Facility	Before	After	Change[a]	Before	After	Change[a]
Electricity	30	100	+	85	100	+
Clean water	10	100	+	35	95	+
In-house toilet	5	100	+	20	100	+

[a]+ increase/– decrease
Source: The survey (2007)

With the decrease in on-farm jobs, they were forced to get other off-farm jobs. In addition, 85% were having difficulty finding jobs during the flood season. For those who lived in the RPs, most did not see any improvement in finding jobs. Some thought that the job market was better, and others thought that it was worse than before. In the flood season, it was not clear whether job finding was difficult or not. Table 11.8 presents the results of evaluation on job opportunity.

Resettlers listed public services, such as better health care and communication, as one of the benefits of living in the RCs. Use of electricity and clean water increased. Most households now living in the RCs had access to clean water, while previously this was available to few. In addition, all had in-house toilets, although the quality of the toilet system was not ensured. Table 11.9 presents the before–after analyses of accessing public services and in-house toilet installation.

11.4.4 Evaluation from Outsiders

To get the opinion of outsiders about the resettlement program, a third FGD survey was conducted for people living in Tan Hau B2 village of Long An commune. This area is located near the RP of Long Hoa. It is annually affected by flooding. In the interview, questions about the status of the landowner, on-farm and off-farm income-generating activities, the reasons people did not move to the RCs, evaluation of the advantages and disadvantages of living outside the RCs, and subjective evaluation of the resettlement program came up in the discussion.

The results showed that most of the people interviewed had their own agricultural land. Besides income generated from agricultural production, all of them earned incomes from on-farm and off-farm tenant activities. Income earned from these tenant activities exceeded that earned from agricultural activities such as rice planting and cash crops, indicating that these people were mainly comprised of the poor. All of them expected to be targeted and selected by the commune authority to join the resettlement program. The main reason they had not joined the program was that they were not on the "4 Nos" list.

According to the respondents, there were advantages and disadvantages to living outside the RCs. One of the advantages was that they got jobs easily. Living near the workplace helped them to earn more income. Another benefit was stability. Custom and habit in their lives did not change unless they moved into the RPs. In addition, people living in the RCs were not allowed to raise animals while they could do so, providing a source of savings for the poor. Although there were advantages in living outside the RCs, the outsiders also had to cope with a number of disadvantages. First and most significantly, many of them were living near dangerously eroded areas, so it was not safe for them and their families during the flood season. Some of them said that because of safety problems they often had to quit their jobs to take care of their families and, as a result, lost their incomes during the flood season. Second, some of the outsiders said that their children were not able go to school during flood periods. Going to school in the flood season was a dangerous thing for their children. In the study site, traveling during the flood season was very difficult. They expected that the government would establish more new RPs so that they would be able to take part in the program. Table 11.10 summarizes the FGD results.

The FGD results showed that the outsiders were seriously affected by flooding and landslides. It was extremely dangerous for those who were living in the eroded areas along Kinh Sang River. To evaluate public awareness of the impact of flooding, a household survey was implemented in this area. Most people agreed that the landslides along the river bank were caused by the annual flood. During the flood season, almost all economic activities were interrupted, and living conditions were changed. One hundred percent of them thought that flooding was evil. They worried about their lives, assets, and jobs. All of the respondents were willing to move to another place for safety. The majority made similar recommendations for building the RPs to enable them to live with flooding. Table 11.11 presents some recommendations regarding how to cope with flooding.

Table 11.10 Advantages and disadvantages of living outside the resettlement path

Advantage	Disadvantage
• Able to raise animals around the house	• Vulnerable to flooding or landslide
• Near the workplace/easy to find a job	• Bad transportation conditions during the flood season
• No change in living conditions and customs	• Unable to go to school during the flood season
	• No clean water to use
	• No health-care services

Source: The survey (2007)

Table 11.11 Opinions about the way to live with the flood

Recommendation	Percentage
Establishing the resettlement path	70
Building a dyke system	5
Raising the houses' stilts	20
Moving to a higher place temporarily	5
Total	100

Source: The survey (2007)

11.5 Conclusion and Recommendation

11.5.1 Conclusion

People in the VMD recognized that flooding has both negative and positive effects. Besides huge damage, flooding brings a lot of benefits to the people, especially to the poor. People knew how to adapt and respond to flooding through many years of experience. Since the 1990s, the government has invested more effort and capital to cope with disastrous flooding in the VMD. Not until the end of the last decade was the new perception on the flood situation in the VMD – the so-called "living with the flood" policy – officially accepted. Under this policy, the VMD's socio-economic development strategy was linked with adaptation to the flood situation.

A resettlement program designed for the permanently inundated areas of the VMD was implemented in 2001. This was expected to be completed by the end of 2005. Total investment for the program was approximately US $200 million. However, the program is not yet complete. There were still 184 RCs not finished. The Vietnamese government decided to accelerate the program so that the program would be completed by the end of 2007. There are two kinds of resettlements in the program, namely RCs and RPs. RCs are built at one place, while the RPs are built along the inter-commune roads. In many RCs, public infrastructure such as an internal road system, marketplace, clinic, school, and so on remain incomplete. There are also problems with waste discharge systems in the RCs. In addition, there are financial problems faced by the program. Many resettlers are poor. Therefore, the capacity of resettlers to repay loans to the government remains an open policy question. Another issue is the employment problem. Review analyses and the FGD survey showed that the insiders' income decreased. Lack of improvement in livelihood has become a big concern for the insiders. To enable the program to be sustainable, financial support and job creation need to be priorities for policymakers.

Although there were still complaints about its implementation, most of the insiders felt satisfied with the program. Thanks to the program, they had better lives. For those living in the permanently inundated areas, the program was the best choice for the poor. The FGD survey for the outsiders also revealed that they expected to move to live in the RCs. There were still many people affected by flooding. Enlarging the program may need to be considered. However, there is a policy question regarding the appropriateness of the resettlement program in the context of the LWF policy. Flooding affects a large area of the VMD, while the RCs have been built only in the most severely-affected areas, as specified by the rules of the program. As a result of the financial limitations of the program, it is unable to provide a solution for the whole region. With global climate change, it was predicted that the flood situation in the VMD is likely to worsen. To protect the whole region of the VMD via this program, the government would need to invest billions of US dollars, while the results of the program are still unknown. At the moment, this investment decision is impossible since it is still too early to assess the success of the program. The resettlement program is currently coping with many problems such as financial constraints, living conditions, and the quality of infrastructure.

An alternative approach with regard to the resettlement program in the context of the LWF policy might be to establish temporary residential clusters instead of permanent RCs. During the flood season, people would move to residential clusters for safety and return home after flooding subsides. This approach acknowledges that flooding brings more benefits than evils. People could avoid flooding at high-risk times but harvest the benefits of the floods at all times. On the other hand, the government will be able to expand the program to cover a larger area and at greater scale with a reasonable investment. This will be very important to the VMD's flood situation if heavily affected by global climate change in the future.

11.5.2 Recommendation

In the long run, to effectively cope with flooding, people living in the inundated areas need to adjust their habits and social and economic activities in order to adapt. By acknowledging the fact that flooding will likely become more serious due to global climate change, the LWF policy is probably the best strategy. Exploiting the flood is better than preventing it. This approach allows the VMD to maintain the natural characteristics of the flood phenomenon that people have actually adapted to since they arrived in the area. The government plays a very important role in directing the residents of the VMD to react adaptively to the flood situation. At the national level, planning of economic development strategies in the VMD needs to be consistent with the LWF policy. At the local level, authorities need to persuade people to actively adapt to flooding and choose an appropriate way to live with flooding.

With regard to the resettlement program, the most important thing to do right now is to complete the construction of infrastructure in the RCs. Although the local authorities at the program site announced that the RCs are mostly completed, a number of real problems remain. The poor quality of public infrastructure limits the program's success. In particular, the state of water-discharge systems in the RCs is poor, and these frequently do not work properly during the flood season. Furthermore, the poor quality of house construction and poor in-house toilet systems are the biggest concerns of the insiders. It is also recommended that additional measures need to be addressed so that the program can be completed.

In order to ensure the quality of the RCs, some processes in designing, implementing, and monitoring the resettlement project need to be improved. Regarding the design of the RCs: the participation of the commune authority and the community is essential so that the construction of infrastructure and houses within the RCs is adequate to cope with the real conditions relevant to each location. RCs should not have the same design everywhere. In particular, the design of the discharge-treatment system for the RCs should be re-examined to ensure that these work during the flood season. In implementing the construction of the RCs, contractors are currently assigned by the management division of the district authority and not by the commune authority and the board of management of the RCs who know the situation and local conditions better. The problem of the poor quality of the RCs could be resolved by involving the commune authority and the board of management of the RCs in the process of project design and implementation. In monitoring the construction of infrastructure and houses in the RCs, the participation of households is an important factor to ensure the quality of construction. For the construction of in-house toilet systems, it is recommended that households be responsible for implementation instead of the contractors.

Enlargement of the program may be required in the future. While the demand for living in the RCs is potentially huge, examples in the study site showed that public awareness of the need for adaptation to the permanent inundation situation is increasing. Observed benefits accrued by insiders influence outsiders' willingness to join the program. Therefore, in the long run the government needs to continue enlarging the program to cover all kinds of people affected by flooding.

Although the program satisfies most of the poor, the fact that living conditions have not improved is a big concern of the insiders. The review analyses revealed that the insiders' income was lower than the outsiders' income (Xe and Dang 2006). Moreover, the FGD survey showed that the insiders experienced greater difficulties in getting jobs than they did before living in the RCs. So, measures to create new jobs – which are mainly off-farm activities – are needed. It is recommended that social programs such as rural-poverty alleviation, traditional "trade village" development, work for rural youth, etc., need to be linked to the resettlement program. This is very important from the policy maker's point of view. A stand-alone resettlement program is not enough; it also requires synchronous measures to enable the program to be successful.

Moreover, environmental consequences of the program need to be considered, as there are environmental impacts associated with the resettlement program. A poor water-discharge system and a badly designed in-house toilet system will cause negative impacts for the insiders. One issue relates to the health hazard, which becomes a very serious problem during the flood season. The costs of this health hazard must be evaluated in the context of the program. Broadly, a comparative analysis of insider–outsider health problems should be undertaken to assess the impacts of poor living conditions in the RCs.

To enable the program to be sustainable, the financial-support policy needs to be revised. In the context of the program, most of the target households are poor. In the survey, most of the poor are very concerned about their capacity to repay loans. At present, some of the borrowers are not meeting the time schedule for repayments to the bank and are currently incapable of repaying their loans. A revised institutional financial policy is necessary to help the poor to be able to repay loans.

This study was implemented in areas most affected by flooding in the VMD. The fact that flooding now comes more frequently and more seriously is potentially due to the impact of climate change. Living with flooding via the resettlement program in the VMD would be a lesson for other territories and places facing the threat of sea-level rise.

Finally, this pilot study simply describes what is happening with the LWF policy implementation and resettlement program in the VMD. Difficulties in designing, implementing, and monitoring the resettlement program need to be examined so that appropriate policy advisories can be evaluated. Factors affecting the successes and failures of the program should be analyzed comprehensively. It is recommended that in the future, more case studies should be conducted within the VMD in order to enable comparison between RCs in different places and to better evaluate the resettlement program. In this proposed research, factors including the year of resettlement construction, ethnic elements, poverty level, etc, are important criteria to consider when selecting RCs for study.

References

Ngoc P (2003) Flood damage assessment in the Mekong Delta, Vietnam. Unpublished MSc thesis, Asian Institution of Technology, Pathumthani, Thailand

Ninh NH, Hong PN (1993) Global warming and impact on Vietnam: sea level rise impact. Tiempo Resource Service. Norwich, United Kingdom. ihttp://www.tiempocyberclimate.org/portal/archive/vietnam/index.htm

Prime Minister's No. 78 Decision (2004) Adjusting and supplementing a number of mechanisms and policies in order to speed up the construction progress of population clusters and lines as well as dwelling houses in frequently flooded provinces in the Mekong Delta

Prime Minister's No. 99 Decision (1996) Long term orientation and the five year plan of 1996–2000 for development of irrigation, transport and construction in rural areas of the Mekong Delta

Prime Minister's No. 105 Decision (2002) Policies towards households in flooded areas that purchase on deferred payment stereobates and dwelling houses in residential clusters and lines in Mekong River Delta provinces

Prime Minister's No. 173 Decision (2001) Socio-economic development in the Mekong River Delta region in the 2001–2005 period

Quang NM (2000) *Floods in the Vietnamese Mekong Delta-A History*. Mekong Forum. http://vastvietnam.org/quang/qglutxn.html Date accessed: 26/7/2008

Reiner W, Hien NX, Hoanh CT, Tuong TP (2004) Sea level rise affecting Vietnamese Mekong Delta: water elevation in the flood season and implications for rice production. Climate Change 66:89–107

Tien DC (2004) Status-quo of the resettlement cluster in the Mekong Delta, Vietnam. Unpublished Report, Ho Chi Minh City University of Economics, Ho Chi Minh City.

Xe DV, Dang NH (2006) Analyses of socio-economic impacts of the resettlement clusters and recommendations for economic development strategy of the Vietnamese Mekong Delta. Cantho University, Cantho

Part III
Consequences of Environmental Change

Chapter 12
Climate Change in the Mekong River Delta and Key Concerns on Future Climate Threats

Le Anh Tuan and Suppakorn Chinvanno

Abstract The Mekong River Delta in Vietnam is the largest agriculture and aquaculture production region of the nation. As the most downstream part of the Mekong River to both the East Sea and the Gulf of Thailand, the majority of the Delta is slightly under 2 m above sea level. Historically and practically, the people of the Delta have settled in the highest densities along the river and banks of the connected canals. Human life, agriculture and aquaculture production, and domestic water supplies in the Delta depend highly on the meteorological and hydrological regimes of the region. However, Delta livelihoods are sensitive and could be threatened by climate change and hydrological cycles. Future climate projection from the regional climate model indicates that the Mekong River Delta region will likely be warmer in the future with longer and drier summers. Seasonal patterns could be altered under the influence of global warming. Moreover, changes in climate patterns in the upstream region of the Mekong River may affect the flood regime of the Mekong Delta, which may lead to an extension of the current boundaries of flooding patterns. These changes raise many concerns, especially in terms of those who make their living from agriculture and aquaculture, because of their significant potential for creating new environmental challenges in the Mekong River Delta.

Keywords Climate change • Scenarios • The Mekong River Delta • Flood • Threats

Le A. Tuan (✉)
DRAGON Institute – Mekong – CanTho University, Campus II, Street 3/2, CanTho City, Vietnam
e-mail: latuan@ctu.edu.vn

S. Chinvanno
Southeast Asia START Regional Center, Chulalongkorn University, Rm. 508, 5th Floor, Old SWU Pathumwan Bld. N.5, Bangkok, Thailand
e-mail: suppakorn@start.or.th

M.A. Stewart and P.A. Coclanis (eds.), *Environmental Change and Agricultural Sustainability in the Mekong Delta*, Advances in Global Change Research 45,
DOI 10.1007/978-94-007-0934-8_12,

12.1 Introduction: Future Climate Projections of the Mekong River Delta Through a Modeling Approach

Climate change, an inevitable effect of global warming, has become a global concern due to its potential consequences on various systems and sectors and because of its threat to human well-being (IPCC 2001). Understanding climate change is crucial to the proper planning of adaptation measures to cope with future risks. However, global warming is a slow process and requires rather long-term, future climate projections in order to clearly detect patterns of future climate change (IPCC 2007) and the impact of these on certain sectors within a specific area. Global circulation models (GCMs) have been developed (The IPCC Data distribution Centre 2003, available access on http://cera-www.dkrz.de/IPCC_DDC/IS92a/) and are commonly used to simulate future climate projections. However, the majority of simulation results available today from most GCMs are of a coarse scale due to technological limitations, and these are not an effective tool for climate change impact assessment at the local level. Therefore, regional climate projections in high resolution have been developed based on various techniques to address the scale requirements in climate change impact assessment. Typically, there are three types of techniques for obtaining high resolution regional climate change projections: statistical, dynamical, and hybrid (statistical–dynamical). The use of Regional Climate Models (RCMs) falls into the dynamical category (Jones et al. 2004). This chapter discusses the approach in dynamic downscaling of GCM data using the regional climate model to develop future climate projections for the Mekong River Delta (MD).

An RCM is a downscaling tool that adds fine scale (high resolution) information to the large-scale projections of a global GCM. While GCMs typically run according to horizontal scales of a few hundred kilometers, regional models can resolve features down to a much smaller scale of 50 km or less. This results in more accurate representation of many surface features, such as complex mountain topographies and coastlines. It also allows small islands and peninsulas to be represented realistically, whereas in a global model, the sizes of these would associate their climate with that of the surrounding ocean. RCMs are full climate models, and as such, are based on physical data. They represent most, if not all, of the processes, interactions, and feedbacks among climate system components represented in GCMs. They produce a comprehensive set of output data over the model domain. This study uses a regional climate model, namely PRECIS, for downscaling coarse scale GCM to derive climate change scenarios for the Mekong River Delta (Jones et al. 2004).

PRECIS is a regional climate model that was developed by the Hadley Centre for Climate Prediction and Research, and is based on the Hadley Centre's regional climate modeling system. It can be used as a downscaling tool that adds fine scale (high resolution) information to the large-scale projections of a global GCM. It has been ported to run on a PC (under Linux) with a simple user interface, so that experiments can easily be set-up over any region. PRECIS was developed in order to help generate high-resolution climate change information for as many regions of the world as possible. These scenarios can be used in impact, vulnerability, and adaptation studies (Simson et al. 2006).

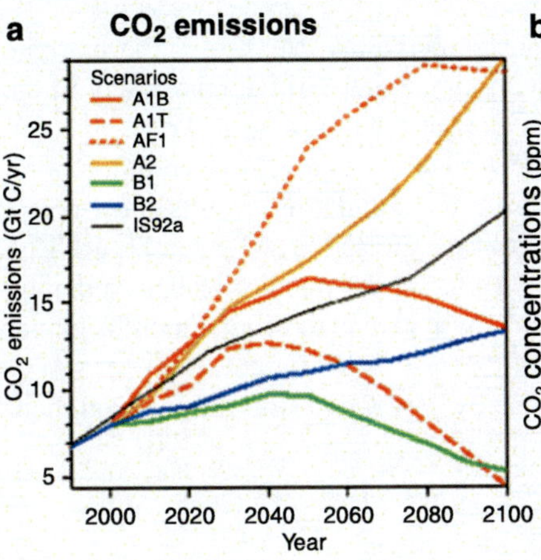

Fig. 12.1 Emissions and concentrations of CO_2 for the various reference scenarios by IPCC (**a**) CO_2 emissions, (**b**) CO_2 concentrations

Water vapor and carbon dioxide (CO_2) are the main gases responsible for the greenhouse gas (GHG) effect. The Intergovernmental Panel on Climate Change (IPCC) has considered the modulation of carbon dioxide emissions and concentrations as a fundamental factor in controlling climate variations over this time scale. Increasing atmospheric greenhouse gases is the key factor in global warming, and it is expected that GHGs will continue to increase in the future. Changes in the level of atmospheric GHGs depend on the level of human activity, which will affect GHGs emission in the future. The IPCC has developed CO_2 emission and concentration scenarios which project changing socio-economic conditions and associated emission levels to measure different plausible GHG emission and concentration outcomes (SRES 2000). See Fig. 12.1.

This study conducted dynamic downscaling based on an initial dataset from ECHAM4[1] GCM from the Max Planck Institute for Meteorology, Germany (http://cera-www.dkrz.de/IPCC_DDC/IS92a/Max-Planck-Institut/echam4opyc3.html) and used PRECIS RCM to simulate future climate scenarios for the Southeast Asia region at a resolution of .22° grid (approximately 25×25 km²) with daily time step.

12.2 Climate Change in the Mekong River Delta

Simulation results from the PRECIS regional climate model show that the Mekong River Delta will likely be a few degrees Celsius warmer in the 2030s than in the 1980s, the baseline period for comparison. Warmer temperatures can be seen in

[1]209ECMWF Atmospheric General Circulation Model coupled with University of Hamburg Ocean Circulation Model (http://www.ipcc-data.org/is92/echam4_info.html).

both the average maximum and minimum temperatures. Moreover, the extreme maximum temperature, that is, the maximum temperature of the hottest day in the year, will also be warmer by a few degrees Celsius (see Figs. 12.2–12.4).

Changes in the climate in the Mekong River Delta can also be seen in temporal terms, as well as in terms of magnitude. From the simulation, results show that it will not only be warmer, but the hot period is also expected to be longer. Figure 12.5 shows that the hot period, defined in this chapter as the number of days annually where the maximum temperature is over 35°C, will extend to about 2 months longer in the 2030s compared to that of the 1980s.

Annual precipitation is likely to decrease by 10–20% in the future throughout the Delta area (see Figs. 12.6 and 12.7).

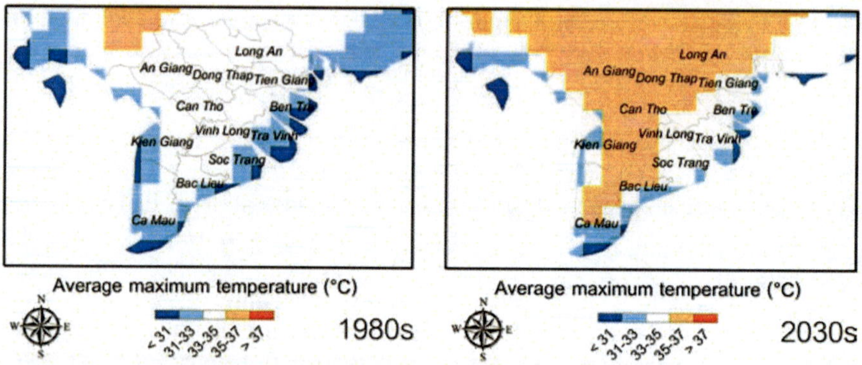

Fig. 12.2 Average maximum temperature in the Mekong River Delta in the 1980s and 2030s (simulated)

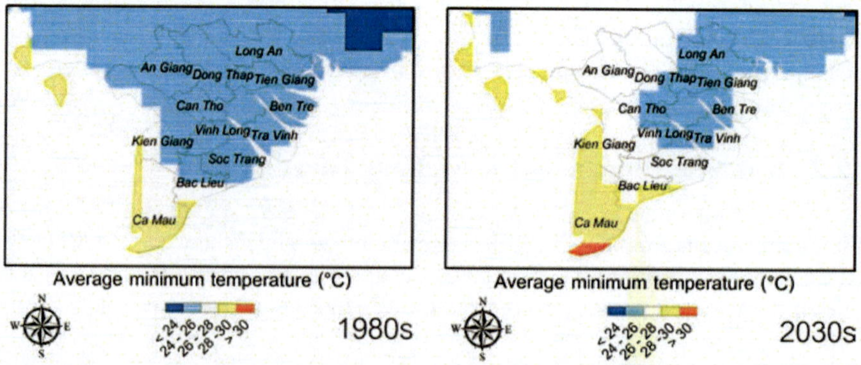

Fig. 12.3 Average minimum temperature in the Mekong River Delta in the 1980s and 2030s (simulated)

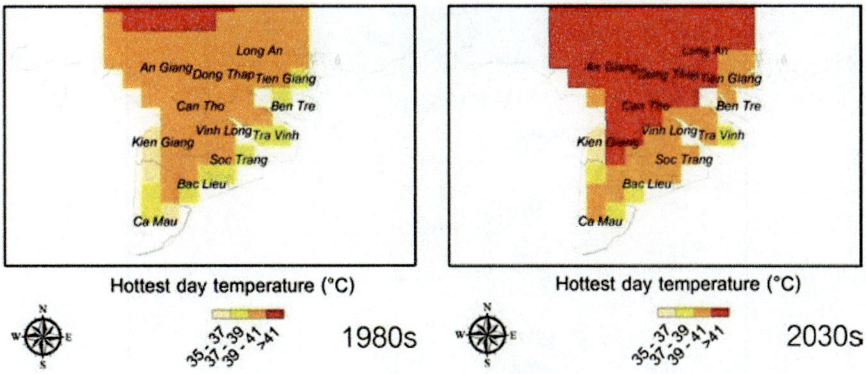

Fig. 12.4 Average maximum temperature of the hottest day of the year in the Mekong River Delta in the 1980s and 2030s (simulated)

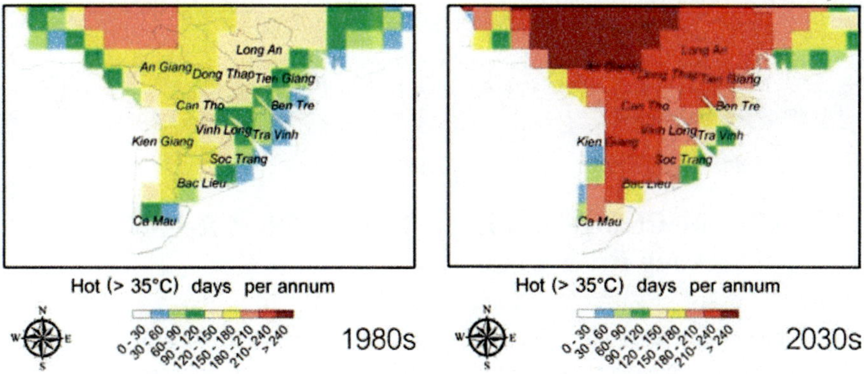

Fig. 12.5 Hot period (number of hot days in a year) in the Mekong River Delta in the 1980s and 2030s (simulated)

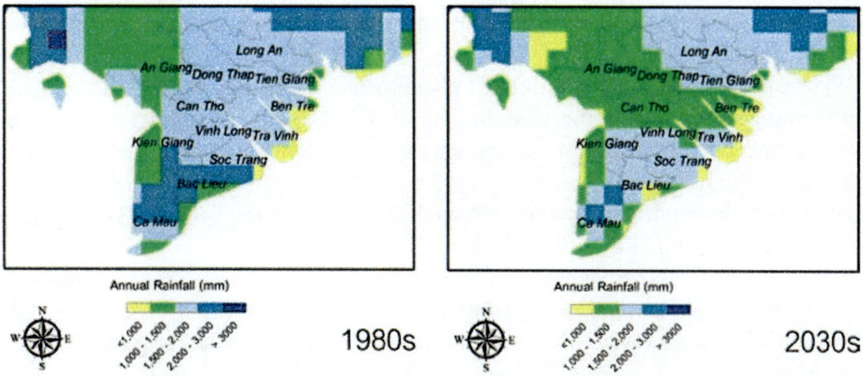

Fig. 12.6 Annual precipitation in the Mekong River Delta in the 1980s and 2030s (simulated)

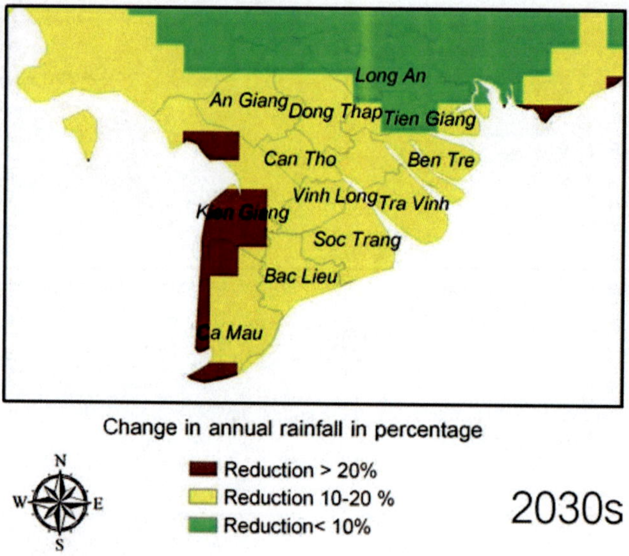

Fig. 12.7 Comparison of change in annual precipitation in the Mekong River Delta between the 1980s and 2030s (simulated)

12.3 Climate Change and Impact on the Flood Regime in Mekong River Delta

One of the key concerns with climate change in the Mekong River Delta is its impact on the flood regime, which may influence many economic sectors in the Delta area. The flood regime in the Mekong River Delta is determined by the regional flow variation that results from the influence of climate change, especially the change in annual precipitation, on the upper parts of the basin. Moreover, sea level rise induced by global warming would also affect the flood regime in the Delta, as well as produce more significant salinity intrusions and coastal erosion; these latter have also become major concerns in the Delta.

The impact of these climate-related phenomena on the hydrology of the studied areas was analyzed using the modeling approach. The schematic overview of the interactions among climatic, hydrological, and hydrodynamic models is presented in Fig. 12.8 (Water and Development Research Group, Helsinki University of Technology, Finland and Southeast Asia START Regional Center [SEA START RC], Chulalongkorn University, Thailand 2009).

As explained above, the future climate projection data was simulated by ECHAM4 Global Circulation Model under the IPCC SRES A2 GHG scenario and downscaled to high resolution using the PRECIS regional climate model, which was used as an input to the hydrological and ocean circulation model to determine changes in the future flood regime in the Mekong River Delta.

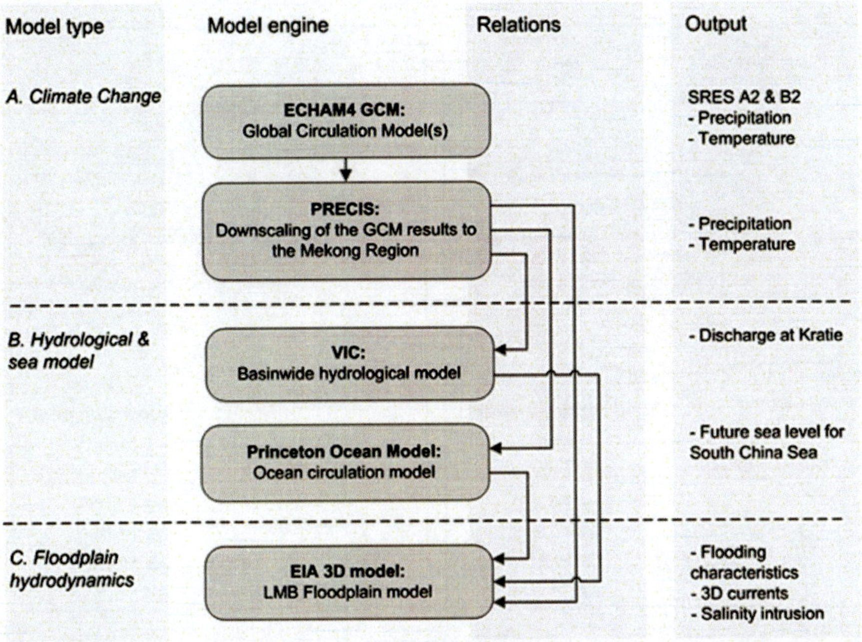

Fig. 12.8 Schematic overview on the interactions among climatic, hydrological, and hydrodynamic models

Regional climate projection data was used in a simulation of a basin-wide hydrological regime under the influence of future climate projections. The simulation was based on the Variable Infiltration Capacity (VIC) hydrological model, which is a macro-scale hydrologic model that solves full water and energy balances, originally developed by Xu Liang at the University of Washington (Liang et al. 1994). In addition, regional climate projection data was also used as an input to the Princeton Ocean Model in the simulation of future sea levels at the mouth of the Mekong River; these may vary under the influence of changing wind speed and wind direction (Blumberg and Mellor 1987). Future projections of changing sea levels due to the direct effect of global warming, for example, ocean water expansion, as indicated in the IPCC Fourth Assessment Report, was also taken into account in the analysis (IPCC 2007). Results from the regional climate model, the regional hydrological simulation, and the ocean circulation model were then fed into the hydrodynamic model, the EIA 3D model, for more detailed hydrological analysis of the Mekong River floodplain system. The model is able to describe the three-dimensional characteristics of the flooding, flow, water quality, and erosion and sedimentation in the lakes, reservoirs, river channels, and floodplains of the study area (Koponen et al. 2004).

Results from the 3-D hydrodynamic model provide guideline data to determine future changes in the flood regime in the Mekong River Delta. The model simulations

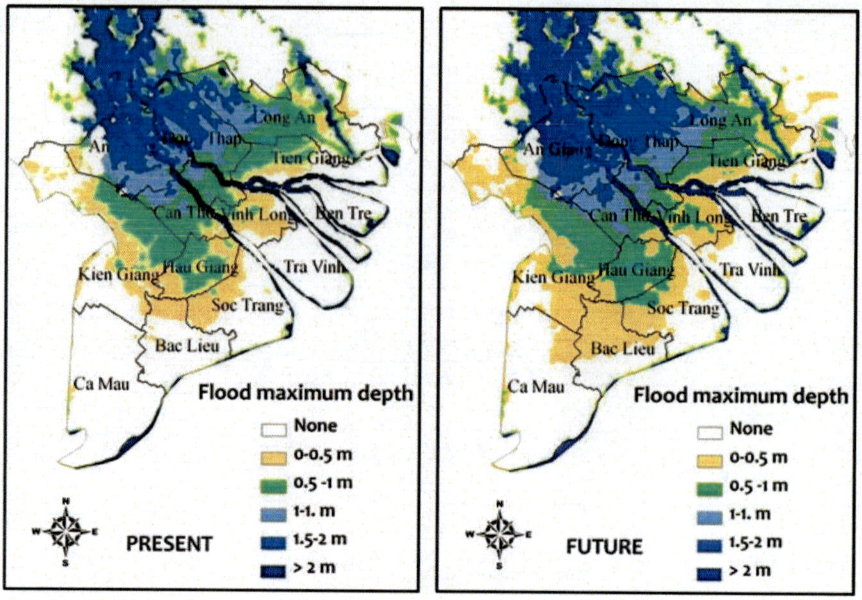

Fig. 12.9 Flood boundary in the Mekong River Delta in the 1980s and 2030s (simulated)

show an increasing trend in the annual maximum water depth and flooded area during the average and driest years. On the other hand, a clear trend is not visible in the wettest years. This change may have a significant impact on both agriculture and aquaculture (see Fig. 12.9).

12.4 Concerns About Future Changes in Climate and Flood Regimes in the Mekong River Delta

Many scientists and international organizations have classified the Mekong River Delta as a vulnerable area of climate change and sea level rise (Peter and Ruysschaert 2008; Dasgupta et al. 2007; IPCC 2007; UNDP 2007; ADB 1994). Climate change will have complex and significant impacts on rural production and the quality of life in general in the region. The increase in average maximum temperatures during the dry season in the future will lead to high evapo-transpiration. Salinity intrusions from the East Sea to land will become more serious. Secondary crop cultivation and freshwater aquaculture will face challenges due to the declines in water supply and irrigation sources. People will be more exposed to weather-related diseases. The reduction in precipitation, especially at the beginning and in the middle of the rainy season, will strongly affect the summer–autumn rice crop, leading to higher production costs. Yet, the increase of precipitation in the latter periods of the rainy season combined with early flooding will threaten the rainy season crop harvest. The boundary of significant floods is projected to expand to the southern part of the Mekong River Delta, toward areas in Bac Lieu and Ca Mau peninsula, which will then threaten current aquaculture operations. Shrimp farmers will face higher costs to install protection dykes around their shrimp ponds. The inundation periods in the upstream provinces will decrease. This flood regime will shorten the fishing period for the poor in the An Giang and Dong Thap flooding areas.

The future of the Mekong River Delta in terms of these consequences can be described generally as the following: areas devoted to agricultural production, such as rice fields, secondary crop fields, fruit orchards, and aquaculture will shrink and production and capacity will decrease, which in turn may threaten the food security of the nation. Rice farmers, shrimp farmers, salt farmers, and small agricultural businessmen will be significantly impacted due to the lack of essential nutrient sources, changes in the meaning of land ownership, and possibly access to information necessary for adaptation to changes in the climate and flood regime. The consequence will be that regional forest, land, water, wildlife, and mineral resources will be over-exploited. Many wetland protected areas such as Tram Chim, Upper U Minh, Lang Sen, Tra Su, Ha Tien, Vo Doi, Bai Boi, Dat Mui, and Lung Ngoc Hoang may also be exploited negatively. Certain organisms may face extinction while the populations and range of some insects, such as mosquitoes, may increase. Farmers in coastal regions seriously impacted by climate change and sea level rise may migrate to urban areas in the north and west of the Mekong River Delta (such as Chau Doc, Long Xuyen, Can Tho, Vinh Long, My Tho, Tan An). Migrations could lead to significant population increases, affect urban planning, and create new social problems, which in turn will have an environmental impact on urban areas.

12.5 Discussion and Conclusion

The Mekong River Delta is considered to be the most seriously impacted area in South East Asia in terms of intense environmental, social, and economic change. The changes in this area need to be studied and explained to strategy planners, policy makers, scientists, businessmen, local officers, and local people: all of those who will be affected by problems created by future changes as well as those who are trying to solve them. The development of policy for information sharing and of measures to mitigate and adapt to environmental changes are crucial. Hesitation, suspicion, and irresponsibility will yield ill consequences for future generations. While scientists have recognized the phenomenon of climate change and sea level rise, further analysis is still being carried out. It is essential to organize collaborative research on simulations of climate change in various time frames with different scenarios and equally essential to identify subjects who will be affected by climate change as well as the kinds of impacts they are likely to suffer. Some of the most important variables that must be considered at the same time are:

- The uncertainty of future change scenario-based study: this study, for example, is based on a single scenario which represents only a single plausible future.
- The risk and vulnerability of economic sectors in the Mekong River Delta: climate change impacts make up a chain of consequences which will affect sectors ranging from the bio-physical to the socio-economic. However, each sector may be at risk differently and may respond to future changes differently.
- Other long-term change under influences from other forces, especially those related to development and globalization, will affect the social and economic context of environmental change in the future and will affect the interaction among various social and economic sectors and sub-sectors and alter future vulnerability.
- The holistic approach in the development of adaptation strategies in the future will have to take changes in the social and economic contexts into consideration by looking into the impact of climate change on various systems and sectors and understanding how each sector's response to climate change may affect other sectors in order to come up with appropriate strategies for the region.

Acknowledgments We would like to thank the Asia-Pacific Network for Global Change Research (APN) for financial support of a research project, "Climate Change in Southeast Asia and Assessment on Impact, Vulnerability and Adaptation on Rice Production and Water Resource," which provided the foundation for the development of this chapter. Further thanks to the Met Office Hadley Centre, UK, for their technical support, especially for technical transfer assistance with regional climate modeling. Additional thanks to the Water and Development Research Group, Helsinki University of Technology, Finland, for joint study activities in the project "Water and Climate Change in the Lower Mekong Basin: Diagnosis and Recommendations for Adaptation," which has provided essential information for this chapter.

References

Asian Development Bank (1994) Climate change in Asia: Vietnam country report. Asian Development Bank, Manila

Blumberg AF, Mellor GL (1987) A description of a three-dimensional coastal ocean circulation model. In: Heaps NS (ed) Three-dimensional coastal ocean models. American Geophysical Union, Washington, DC, pp 1–16

Dasgupta S, Laplante B, Meisner C, Wheeler D, Yan J (2007) The impact of sea level rise on developing countries: a comparative analysis. World Bank Policy Research, Working Paper 4136, Washington, DC, Feb 2007

Intergovernmental Panel on Climate Change (2000) Special Report on Emission Scenarios (SRES). Cambridge University Press, Cambridge

Intergovernmental Panel on Climate Change (2001) Climate change 2001: impacts, adaptation and vulnerability. Cambridge University Press, Cambridge, Contribution of Working Group II to the Third Assessment Report of the Intergovernmental Panel on Climate Change (IPCC)

Intergovernmental Panel on Climate Change (2007) Climate change 2007: the physical science basis. IPCC Secretariat, Geneva

Jones RG, Noguer M, Hassell DC, Hudson D, Wilson S, Jenkins G, Mitchell JFB (2004) Generating high resolution climate change scenarios using PRECIS. Met Office Hadley Centre, Exeter, p 40, Apr 2004

Koponen J, Kummu M, Sarkkula J (2004) Modelling tonle sap lake environmental change. Paper presented at the SIL XXIX International Congress of Limnology, Lahti

Liang X, Lettenmaier DP, Wood EF, Burges SJ (1994) A simple hydrologically based model of land surface water and energy fluxes for general circulation model. J Geophys Res 99:14415–14428

Peter C, Ruysschaert G (2008) Climate change & human development in Vietnam: a case study for the human development report 2007/2008. Oxfam and UNDP

Simson W, Hassell D, Hein D, Jones R, Taylor R (2006) Installing using the Hadley Centre regional climate modeling system, PRECIS: version 1.4.6. Met Office Hadley Centre, Exeter

The IPCC Data Distribution Centre (2003) ECHAM4/OPYC3 Description [on-line]. Retrieved from http://cera-www.dkrz.de/IPCC_DDC/IS92a/Max-Planck-Institut/echa4opyc3.html

United Nations Development Programme (2007) Human Development Report 2007/8, Fighting climate change: human solidarity in a divided world. Palgrave MacMillan, New York

Water and Development Research Group, Helsinki University of Technology (TKK), Finland and Southeast Asia START Regional Center (SEA START RC), Chulalongkorn University, Thailand (2009) Water and climate change in the lower Mekong basin: diagnosis & recommendations for adaptation. Project Report for the Finnish Ministry for Foreign Affairs, 28 Apr 2009

Chapter 13
Climate Change Adaptation and Agrichemicals in the Mekong Delta, Vietnam

Zita Sebesvari, Thi Thu Huong Le, and Fabrice G. Renaud

Abstract Since the implementation of economic reforms in 1986, the Mekong Delta has experienced an extensive transformation process in its agricultural sector. This transformation has been characterized by agricultural intensification, the enhanced use of agrichemicals (fertilizer, pesticides), and emerging concerns for human health and the environment. The predicted impacts of climate change such as sea level rise, greater seasonal variability in precipitation and river flows, and elevated temperature and CO_2 concentration will all likely also influence the agricultural landscape and thus agrichemical use. Against the background of the anticipated climate change impacts in the Mekong Delta, this chapter aims to draw a scenario for future agrichemical use and attendant environmental problems. This scenario is achieved through a review of the main climate change-mediated drivers for agrichemical use, with a focus on land-use changes and changes in pest and disease patterns. In addition, the chapter identifies possible adaptation measures that may be implemented by the agricultural sector in the Mekong Delta and explores the potential environmental effects of these adaptation strategies.

Keywords Adaptation • Agrichemicals • Agriculture • Climate change • Mekong Delta

Z. Sebesvari (✉), T.T. Huong Le, and F.G. Renaud
Institute for Environment and Human Security, United Nations University,
Hermann-Ehlers-Str.10, 53113 Bonn, Germany
e-mail: sebesvari@ehs.unu.edu

M.A. Stewart and P.A. Coclanis (eds.), *Environmental Change and Agricultural Sustainability in the Mekong Delta*, Advances in Global Change Research 45,
DOI 10.1007/978-94-007-0934-8_13, © Springer Science+Business Media B.V. 2011

13.1 Introduction

13.1.1 Physical Conditions and Agriculture in the Mekong Delta

The Vietnamese portion of the Mekong Delta covers an area of 4.06 million ha and is home to 17.7 million people, of which 79% live in rural areas (GSO 2008). It is a flat and low-lying area (<4.0-m above mean sea level) with a complex network of rivers and channels. The hydrology of the Delta is influenced by tides from the South China Sea and to a lesser extent from the Gulf of Thailand. The climate is tropical monsoonal with a rainy season from May to October and a dry season from November to April. Average annual temperature is about 27°C. Mean annual rainfall is about 1,600 mm, of which more than 80% falls during the rainy season (Statistical Office of Can Tho City 2000, 2005, 2008). During this period, a large part of the Delta is inundated. In the dry season, the low discharge of the Mekong, tides from the South China Sea and from the Gulf of Thailand, and high water extraction rates cause salinity intrusion in the Delta (White 2002).

To date, the hydrology of the Delta is largely managed by sea dykes, embankments, sluice gates, and pumping stations used for irrigation purposes. These large-scale hydraulic structures have been established mainly to control floods in the upstream part of the Delta and saline intrusion in the coastal areas. The consequence has been a largely human-regulated water regime. Existing hydraulic control structures, soil fertility and productivity, and, since the implementation of economic reforms in 1986, agricultural modernization have successfully stimulated rice production in Vietnam. Once a rice importer, it is now the world's second largest rice exporter. Similarly, rapid growth of the aquaculture sector has enabled Vietnam to become one of the largest fishery export countries in the world (FAO 2008). To date, 63% of the Delta is used for agricultural production, a very high rate compared with 28% for the entire country and 38% for the Red River Delta (GSO 2008). The region supplies 55% of national rice production (2007, all three rice seasons) and 82% of the farmed shrimp production (GSO 2009a, b). This rapid growth was a direct result of intensification processes such as the introduction of three annual rice seasons instead of two in some areas of the Delta (Dang and Danh 2008) and the adaptation to the production of exportable goods (e.g., *Pangasius*). This success is based to a large extent on the enhanced use of agricultural chemicals such as fertilizers and pesticides in rice production (Meisner 2005; Dang and Danh 2008) and the use of processed feed, pesticides, and veterinary drugs in the aquaculture sector (Trong et al. 2002). The drawbacks of this development are widely recognized as environmental degradation and pollution (IRIN 2009), concerns about drinking water and food safety (Holland 2007; Neubacher 2007), failures in meeting international standards of exported goods (Yen 2006), and health concerns (Margni et al. 2002; Dasgupta et al. 2007).

13.1.2 Climate Change in the Mekong Delta

Climate change is an ongoing process in the Mekong Delta. Possible impacts can already be projected by reviewing the recent climate history of the Delta. In the past 10 years, the Delta experienced high floods in 2000, 2001, and 2002, including the historic high flood in 2000 and drought for four successive years (especially the drought combined with low river flow in 2004 and early 2005). Most recently, in early 2010, the Delta was hit by a severe drought with saline intrusion into areas which were never before impacted.

The Vietnamese Ministry of Natural Resources and Environment (MONRE) published recent climate change scenarios for Vietnam in 2009. According to the medium emission scenario [IPCC SRES B2, reference period 1980–1999 (IPCC 2007)] the annual mean temperature will increase by 0.6°C in South Vietnam by 2030. Temperature increase will be 0.5–0.6°C during the dry season (Dec–May) and 0.6–0.7°C during the wet season (Jun–Nov). For the same area and period the annual rainfall is predicted to increase by 0.4%. However, in the dry season (Dec–May) annual rainfall is predicted to decrease by 4.3% in average, while rainfall will increase during September–November by 1.7% on average, which will lead to enhanced seasonal variability. Sea level is predicted to rise by 17 cm by 2030 and 75 cm by the end of the century. Sea water intrusion in the southwest coastal provinces already influences 1.77 million hectares of land (45% of the region). Although sea water intrusion into low lying deltas is a natural phenomenon, sea level rise, climate influenced changes in river flow, and human activities such as increasing water extraction from aquifers of the Mekong Delta are predicted to expand greatly the area affected by sea water intrusion. A sea level rise of 75 cm by the end of the century would lead to an inundation of about 7,600 km² – that is, 19% of the Delta (MONRE 2009).

Using a regional climate model (PRECIS) under IPCC SRES A2 scenario (high emission scenario, reference period 1980s) Tuan and Suppakorn predict a higher average maximum and minimum temperature and a longer and drier dry season (2011, this publication, Chap. 11). Annual precipitation is predicted to decrease by 10–20%. The altered precipitation pattern would lead to an altered flood regime with an increasing annual maximum water depth and a wider area of the Delta impacted by floods. At the same time, the flooding period will be shorter in the upstream provinces of the Delta as compared to the 1980s.

13.1.3 Possible Climate Change Impacts on Agriculture

Croplands occupy about 18% of the earth's surface and are increasingly exposed to threats from climate change induced climatic variability. Changes in air temperature and rainfall pattern and resulting increases in the frequency and intensity of drought and flood events, as well as altered hydrological cycles, will all have implications for agricultural productivity (FAO 2007). A considerable number of publications discuss important expected impacts of climate change on agriculture such as changes in the production area, the migration of agro-ecosystems, physiological effects on crops influencing yield quantity and quality (Peng et al. 1995; IRRI 2007; Wassmann and Dobermann 2007; Battisti and Naylor 2009; Gregory et al. 2009; Padgham 2009), changes in soil and water resources, and the changing occurrence and severity of pest and disease outbreaks (Coakley et al. 1999; Chakraborty et al. 2000; Chakraborty and Pangga 2004; IRRI 2007). Despite the large number of studies of particular aspects of climate change impacts (e.g., on CO_2 fertilization), our understanding of climate change impact on agriculture is still limited. One of the reasons is that the processes described above all interact with one another through feedback loops. For example, water scarcity causes plants to be water-stressed and thus more susceptible to pests and diseases. In addition, climate change will likely trigger adaptation, i.e., changes in agricultural practices (e.g., planting more flood resistant or salt resistant varieties of rice, implementing better irrigation practices, increased use of agrichemicals). Some of these adaptation measures will likely have unwanted effects such as increasing water pollution and soil degradation. Holistic studies, which consider a broader range of changing parameters and possible feedback loops, are thus far largely missing. Additionally, there is a lack of information on climate predictions at the smaller scales (field scale), where most of the yield-relevant processes take place. We therefore face large uncertainties when attempting to determine the effects of climate change on aquiculture. Being aware of these limitations, this chapter aims to sketch a scenario for future agrichemical use and attendant environmental problems by reviewing the main climate change-mediated drivers for agrichemical use with a focus on land-use changes and changes in pest and disease patterns. In addition, the chapter identifies possible adaptation measures that may be implemented by the agricultural sector in the Mekong Delta and explores the potential environmental effects of these adaptation strategies.

13.2 Possible Impacts of Climate Change on Agrichemical Use

Climate change may influence agrichemical use in many ways. Temperature and rainfall patterns have been shown to directly influence the amount of pesticides used in the US (Chen and McCarl 2000). Climatic conditions also influence the efficiency of agrichemicals directly by, for example, influencing their retention time on the foliage. Climate change also affects the net area of crop production by inundation and saline intrusion in the Delta.

13.2.1 Possible Impacts on the Production Area in the Mekong Delta

Production area may be influenced by climate change in direct and indirect ways. Sea level rise (SLR) and changing flood patterns will result in a direct net loss of arable land in the Delta. A sea level rise of about 30 cm is expected by 2050 and of about 75 cm by the end of the twenty-first century based on a medium emission scenario (MONRE 2009). Inundations in the Mekong Delta caused by 20–40 cm of SLR would significantly affect all three rice cropping seasons by limiting the number of rice crops per year (Wassmann et al. 2004; White 2002). Beside inundation and permanent salinization, temporary saline intrusion will increasingly affect arable land. As a result of saline intrusion into arable land, salt will need to be washed-out with fresh water at the beginning of the rainy season. With an increasing extent (duration and area) of saline intrusion, it might prove difficult to reduce salinity which could result in salinity accumulation in soils. Saline intrusion was found to be the major factor leading to regional differences in rice cropping systems and land use patterns in the Delta (Kotera et al. 2008). Should there be no implementation of further structural adaptation measures, predicted inundation of about 19% of the Delta and increasing salinization of water and soil resources would lead to a significant limitation of land resources suitable for agriculture in the region. These limitations would likely lead to an enhanced pressure on the remaining arable land in terms of yield per hectare in order to maintain food and income (export) security which in turn would likely require further agricultural intensification with corresponding high pesticide and fertilizer use.

The Vietnamese government plans to undertake significant investments into the construction and upgrading of sea-dykes and sluice gates to respond to sea level rise (SRV 2008). Thus, predictions for the impacts of sea level rise on land use and agricultural production need to take into account a considerable number of technical adaptation measures. Such technical solutions are already applied in large areas of the Delta as a response to existing saline water intrusion. A series of dykes and sluice gates have been constructed in the Ca Mau Peninsula (southern part of the Mekong Delta) to enhance the production area of rice since 1993. The establishment of saline-intrusion control measures was a response to the Mekong Delta Master Plan (NEDECO 1993) and the Mekong Delta Water Resources Project's six provincial cross-boundary subprojects in the late 1990s (Evers and Benedikter 2009). The impacts of these measures on land use, management practices, and resulting impacts on the environment are already assessable. Reduced saline intrusion and the prolongation of the cropping season behind the sluice gates lead to progressive expansion of the area suitable for rice production and to intensification of the production by having two or three rice crops per year instead of one (Kam et al. 2001). This level of intensification increases the pressure on soil resources and favors pest outbreaks which then lead to drastically increased pesticide and fertilizer use. Since freshwater behind the sluices is limited and mainly stagnant in the dry season, water pollution by agrochemicals becomes a threat for human health, for aquatic ecosystems, and for agricultural production other than rice paddies. Aquaculture especially is affected by pollution with agrichemicals; this adds to the existing conflicts on water allocation between freshwater users for crop farming and brackish water users for shrimp aquaculture (Nhan et al. 2007). Additionally, leaching from acid sulfate soils tends to acidify the water behind closed sluice gates. In saline water-protected areas, acidic water and water scarcity in the dry season were found to be the limiting factors for rice production (Aizawa et al. 2007). Overall, intensification in agriculture behind water-control structures challenges the environmental sustainability of the Mekong Delta.

13.2.2 Possible Climate Change Impacts on Rice Plants

Influence of climate change, especially global warming and elevated CO_2 (ECO_2) on the growth and development of rice plants has been well documented (Baker et al. 1992, 1996; Peng et al. 1995; De Costa et al. 2006). Enhanced CO_2 concentrations alter physiological processes in rice plants such as photosynthetic rate or stomatal conductance. In an investigation, doubled CO_2 concentration from 330 to 660 μmol mol^{-1} increased total aboveground and root biomass and final yield due to an increase in net photosynthesis and tillering (Baker et al. 1996). These kinds of changes will have positive effects on rice production such as shortening the growth period by 10–12 days due to shorter vegetative phase and increasing the grain yield by 10–70% (Imai 1995; Allen et al. 1995; Ainsworth 2008) through carbon fertilization. However, global warming – without considering parameters such as carbon fertilization – is likely to influence rice production negatively (Baker et al. 1996; Peng et al. 2004; IRRI 2007) e.g., through more frequent occurrences of acute and chronic heat stress events for the plants (Ingram et al. 1995) or because of changes in evapotranspiration and the availability of water used for irrigation (Tao et al. 2008). Since global warming and ECO_2 are predicted to co-occur, investigations on the interactions between these two climatic variables are extremely important. Ingram et al. (1995) demonstrated that the interactions between CO_2 and temperature would be specific for rice varieties and production locations. For example, adverse effects of a warmer temperature coupled with enhanced CO_2 environments are likely to be greater for the tropics (in the Mekong Delta, for example) than for temperate regions. Benefits from the carbon fertilization process on rice can be overshadowed by a negative impact of higher temperature, which in turn leads to a greater sink demand due to increased growth and respiration rate (Gesch et al. 2001). In the sensitive development stages of rice such as the reproductive phase, higher night temperatures can significantly reduce fertilized spikelet percentage and consequently grain yield, even when the atmospheric concentration is doubled (Cheng et al. 2009). In comparison, ECO_2 coupled with higher temperature might lead to an increase in rice yield in sub-humid tropical climates (De Costa et al. 2006). The degree of response to climate change factors vary with rice cultivars (Shimono et al. 2009) and thus selection and breeding programs should consider adaptation qualities.

The impact of climate change on rice production has been extensively studied using simulation models (Horie et al. 2005), which has made it possible to predict changes at large scales. In an attempt to assess the impact of climate change in rice production in Asia, Masutomi et al. (2009) input future climate change scenarios based on the projections of many general circulation models (GCMs) for three SRES scenarios (18 GCMs for A1B, 14 GCMs for A2, and 17 GCMs for B1, reference period 1990s) into a crop model M-GAEZ and calculated the average change in rice production, taking into account the effect of CO_2 fertilization. The results showed that rice production would be affected in all atmospheric CO_2 scenarios and estimates of the impact of climate change significantly depend on the GCMs used. According to these simulations, rice production in Vietnam would decrease largely in the 2020s and 2080s but slightly increase in the 2050s for all scenarios used, although the degree of change in each period of these three SRES scenarios would be different.

In addition, climate change also affects plant pests and pathogens and agrichemical-use efficiency, and the inter-relationship between these factors would need to be taken into account for a more holistic assessment of climate change impacts on rice production.

13.2.3 *Possible Impacts on Rice Pests and Pathogens*

Damage caused by pests and diseases in general is one of the most important limiting factors in agriculture production. Yield losses due to pathogens and pests are estimated at around 16–18% of total worldwide production (Oerke 2006) of which losses to pathogens were estimated to around 220 billion US dollars (Agrios 2005). In rice production alone, global potential losses due to pathogens and pests account for 13–25% of attainable yield, and the actual losses vary considerably according to agro-ecological regions (Oerke 2006).

Numerous pests and pathogens cause different levels of damage to rice crops. The most common and important pests for rice are stem borers, brown planthoppers (BPH), rice leaf folders, and rice thrips (Pathak and Khan 1994). These pests can occur in one, several, or all stages of rice growth and cause great losses. Diseases with economic importance in rice are blast disease, sheath blight, sheath rot, brown spot, bakanae, bacterial leaf blight, and viral diseases, such as rice grassy stunt virus (RGSV). In the Mekong Delta, the most important pests and diseases in terms of economic loss are BPH, rice thrips, stem borers, leaf folders, rice blast and sheath disease, sheath rot, brown spot, and RGSV, as reported by the Plant Protection Department – Ministry of Agriculture and Rural Development (PPD – MARD).

Among the various factors affecting pest and disease incidence and development, weather conditions play a significant role and influence all stages of growth and development of host plants as well as the occurrence and severity of the disease (Chakraborty et al. 2000). The predicted changes in climate will likely alter the geographical and temporal distribution of plants, which in turn affect disease infection and development processes. Changes in temperature have significant effects on pest and disease distribution and development in most locations. Higher temperature favors population development of some pests and pathogens by increasing the number of life cycles per year. Higher temperature promotes plant growth in cool regions and thus provides more food and nutrient for pests and pathogens, whereas the impact is opposite in warmer locations. Thus, the effects will vary among different agro-ecological zones (Coakley et al. 1999). ECO_2 in the atmosphere will alter physiology and morphology of the host, resulting in changes in light interception, canopy structure, and microclimate, which will in turn affect disease epidemiology (Chakraborty et al. 2000; Ghini et al. 2008). Interactions between climate, plant host, and pathogens will determine the degree of impact on agriculture (Chakraborty and Pangga 2004), but will also be influenced by the effectiveness of applied management strategies (Chakraborty et al. 2000).

The effects of climate change on pests and diseases have been investigated in numerous regions of the world and these are reviewed in the following sections for some of the most important pests and diseases. However, studies on the effect of climate change on rice pests and diseases in Vietnam do not exist to the best of our knowledge.

13.2.4 Global Warming and Pests, Diseases – Examples

13.2.4.1 Rice Blast Disease

The impacts of global temperature change on rice blast disease *Pyricularia grisea* (sexual stage *Magnaporthe grisea*) in three agro-ecological zones, including the cool subtropics (Japan and northern China), subhumid and warm humid subtropics (southern China), and humid tropics (Philippines and Thailand) was studied by Luo et al. (1998a), using a combined simulation model (Coupling of CERES-Rice with BLASTSIM). The simulations suggested that temperature changes had significant impacts on disease development in most regions. In the cool subtropics such as Japan and northern China, warmer temperature resulted in more severe blast epidemics. In warm/cool humid subtropics, rising temperature reduced blast epidemics significantly. In contrast, lower temperature caused small difference in disease epidemics compared with currently prevailing temperatures. The effect of temperature change in the humid tropics was opposite to that in cool regions where daily temperature changes by –1°C and –3°C resulted in significantly more severe blast epidemics, and temperature increases of +1°C to +3°C reduced blast severity.

Using a similar simulation method, the effect of global temperature changes on risk of yield loss caused by rice blast disease was analyzed for five Asian countries (Japan, Korea, China, Thailand, and the Philippines). It was demonstrated that changes in temperature have a significant impact on the disease compared to that caused by changes in rainfall. The influence varied with agro-ecological zones but the disease would become more severe in cool, subtropical regions like Japan while the model predicted disease inhibition in humid tropics and subtropics such as in the Philippines (Luo et al. 1998b). As the Mekong Delta has a similar monsoon climate to the Philippines, a reduction of blast epidemics due to global warming would be expected. However, the interaction of global warming with other production factors such as ECO_2, land use, and pesticide-use patterns makes it difficult to predict the effects under real field conditions.

13.2.4.2 Rice Pests

Global warming would likely result in a redistribution and altered abundance of rice arthropod communities with a degree of influence depending on pest species and populations. A change in temperature can affect any stage of the life cycle of a pest, which in turn can affect pest survival, reproduction, and development. For example, adult survival of BPH remained unchanged between 25°C and 35°C but was significantly reduced at 40°C. It was demonstrated that warmer temperatures would increase BPH abundance in areas with temperatures below 30°C (Heong et al. 1995).

As temperature increases, insects that are directly limited by temperature will be able to expand to temperate regions rapidly. Insects of many tropical and subtropical species might move pole-ward from their current locations as long as their cold hardiness allows, because they usually lack diapause in their life cycles (Kiritani 2006). For example, it was predicted that the rice stem borer *Chilo suppressalis* would shift northward in Japan by about 300 km if the temperature rises by 2°C (Morimoto et al. 1998). Analysis of long-term insect population in Japan showed that an increase in winter temperature increased population of the rice stem borer and the green rice leafhopper (*Nephotettix cincticeps*). The degree of change was much larger for the green rice leafhopper than for the rice stem borer, indicating a difference in the number of generations per year (Yamamura and Yokozawa 2002). The rice stem borer is an important rice pest in the Mekong Delta (DARD Dong Thap 2006), but the effect of global warming on this pest under Mekong Delta conditions has not yet been studied. A study carried out by Kiritani (2006) in Japan predicted that rice stem borer may spread northward and will be more prevalent in the cooler time of the year, such as during the winter–spring crop (December to March) in the Mekong Delta.

Many rice pests are vectors for destructive viral diseases. Hence, changes in pest population due to global warming also influence the prevalence of some viral diseases associated with those pests. For example, the geographical occurrence of the rice strip virus disease (RSV) which is transmitted by the small brown planthopper (*Laodelphax striatellus*) could be shifted due to the synchronization of planthoppers with the cultivation of rice plants. The occurrence of this disease, therefore, is determined by the interactions among three organisms: the rice plant, RSV, and the vector. Global warming will change the time when planthoppers occur by accelerating their development. Thus, it is expected that the geographical area that is potentially vulnerable to disease prevalence will shift (Yamamura and Yokozawa 2002).

13.2.4.3 Natural Enemies of Pests

Global warming may also work in favor of natural enemies by increasing the number of generations more than in their host species irrespective of the type of agroecosystem. For example, the attack rate of *Cyrtorhinus lividipennis*, a natural egg predator of the BPH, increased in temperature range between 20 and 32°C. Thus, it was predicted that in areas with such temperatures there will be an increase in predation on BPH eggs. However, in areas with temperatures higher than 35°C – such as in the Mekong Delta in most times of the year – the ability of *C. lividipennis* to influence BPH population could be reduced (Song and Heong 1997). Biological control utilizing native natural enemies is expected to become a more important control tactic in the future (Kiritani 2006).

13.2.4.4 Brown Planthopper (BPH) and Rice Grassy Stunt Virus (RGSV)

In the last 5 years, the Mekong Delta has witnessed severe outbreaks of BPH and the virulent disease RGSV. BPH outbreaks occurred in almost all provinces in the Mekong Delta and caused infections of hundred thousand of hectares. The percentage of infected area of rice caused by some major rice pests and diseases in the Mekong Delta in the period 2004–2009 is illustrated in Fig. 13.1. There was a

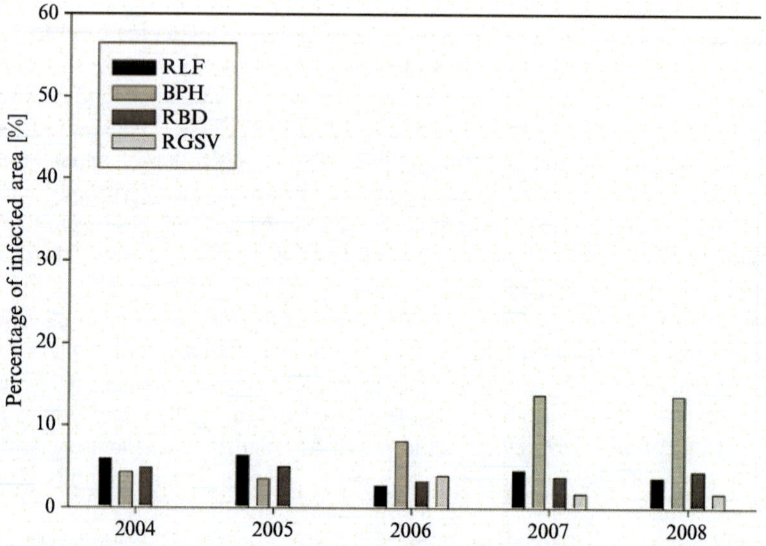

Fig. 13.1 Percentage of cultivated area affected by some important rice pests and diseases in Southern Vietnam in the period of 2004–2008. (Data collected by the Southern Plant Protection Centre). *BPH* brown planthopper, *RBD* rice blast disease, *RGSV* rice grassy stunt virus, *RLF* rice leaf folder

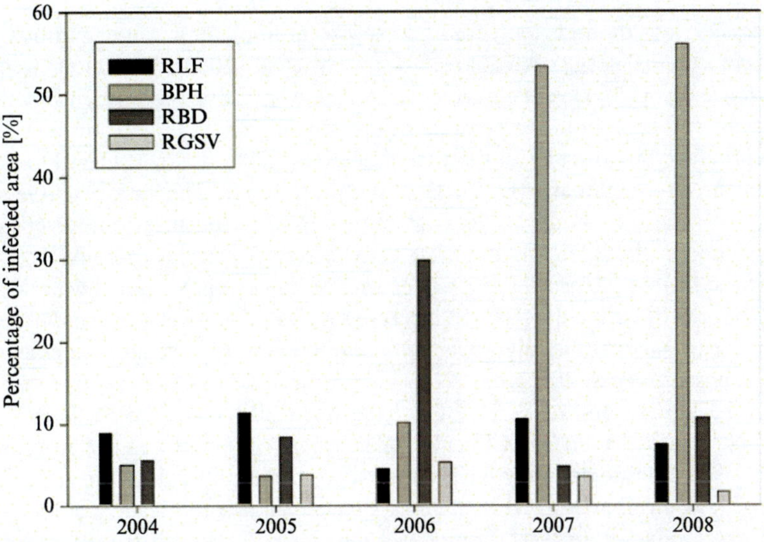

Fig. 13.2 Percentage of cultivated area affected by some important rice pests and diseases in Dong Thap province in the period of 2004–2008. (Data collected by the Department of Agriculture and Rural Development, Dong Thap province). *BPH* brown planthopper, *RBD* rice blast disease, *RGSV*, rice grassy stunt virus, *RLF* rice leaf folder

significant increase in the areas affected with BPH from 2006 to 2008. The viral disease RGSV also became prevalent in this period due to an increase in the area infected with BPH, the vector of the viral disease. Outbreaks of BPH and RGSV were even more severe in some provinces with intensive rice production such as Dong Thap, An Giang, and Tien Giang. Figure 13.2 refers to the percentage of infected areas caused by major pests and diseases of rice in Dong Thap province. The area infected by BPH increased drastically and reached 56% in the year 2008. Similarly, the area infected with RGSV also increased significantly.

The prevailing hot and humid weather in the Mekong Delta favors the development of BPH in general. However, under low population densities BPH is normally not considered as a pest, and the outbreak of this pest is generally rare (Heong 2008). Although climate change has often been quoted to play an important role in the recent severe outbreaks (Đào Xuân Học 2009), the reasons are probably manifold, and it is challenging to work out the share of climate change on these developments. First of all, intensive rice production with up to seven crops in 2 years creates a constant food supply for BPH. Second, management practices such as over-fertilization with urea, high seeding density, and the use of susceptible varieties favor pest development and distribution (Lam 2007). Third, intensive use of insecticides in the early stage of rice development harms natural enemies of BHP while developing a resistance to pesticides, and thus decreases the effectiveness of pesticides used (Heong et al. 2008). Fourth, extreme weather events such as abnormally low or high temperatures, severe rainfall, or drought periods may contribute to the pest outbreaks (Chakraborty et al. 2000). The more frequent occurrence of El Niño events, warmer temperature with a larger difference between minimum and maximum temperature in the Mekong Delta in general or in Dong Thap province in particular in the period of 2001–2008, was considered to play a role in the pest outbreaks (DONRE Dong Thap 2008). The strong northeast or southwest monsoon winds in 2006 also spread BPH to other uninfected areas (Agriviet.com 2008). In addition, interactions between two or more of the above mentioned factors can accelerate pest occurrence. For example, global warming can accelerate the life cycle of BHP and increase survival of nymphs' and adults' fecundity and egg hatch ability significantly, along with the higher nitrogen content of the rice host (Lu et al. 2005).

Similar severe outbreaks of BPH have been also observed in several rice growing regions of India and China in 2008. Recent abnormal weather patterns and the misuse of pesticides may have contributed to the unusual outbreaks of BPH (IRRI 2009). In Thailand, BPH outbreaks have caused significant yield losses in some rice-producing regions. Losses were estimated around 7–8 million tons of rice, which is equivalent to about 400 million US dollars in 2010 (excluding indirect costs like environmental pollution due to the enhanced use of pesticide) (Heong 2010). Results obtained from experimental and field research showed that BPH outbreaks are caused by the deterioration of ecosystem services which can be influenced by climatic events, drought or rainfall, pesticide applications, and agricultural practices (Heong 2008).

13.2.5 Elevated CO_2 (ECO_2)

The impacts of ECO_2 on pathogens are not easily determined due to the dynamic interactions between the host, the environment, and the pathogens (Ghini et al. 2008). In a recent review by Chakraborty and Pangga (2004), it was reported that of the 26 diseases studied, severity of 13 diseases would increase under ECO_2 whereas 4 remained unchanged and 9 would decrease. Lake and Wade (2009) demonstrated that doubled CO_2 concentration facilitated infection of the fungus *Erysiphe cichoracearum* (a causal agent of powdery mildew) on marrow whilst increased the susceptibility of the host to the pathogen infection through changes in stomatal density under controlled conditions. Barley powdery mildew was also in an earlier study demonstrated to cause more severe reduction in yield under higher CO_2 levels than the normal CO_2 concentration under controlled experiments (Hibberd et al. 1996). Some pathogens expressed higher fecundity or aggressiveness under elevated CO_2 environments (Chakraborty et al. 2000). In addition, ECO_2 or changes in temperature would also impact the resistance/susceptibility of the host.

The impact of ECO_2 on some important diseases of rice has been demonstrated in many studies. Rice plants grown under higher CO_2 concentrations were more susceptible to leaf blast than those in ambient CO_2, probably because of their lower leaf silicon content which may contribute to the increased susceptibility to leaf blast. The percentage of sheath blight diseased plants was also higher under ECO_2 condition compared to that under ambient CO_2 concentration. This was probably due to the higher number of tillers observed under ECO_2 concentration which may have increased the chance for the fungal sclerotia to adhere to the leaf sheath at the water surface (Kobayashi et al. 2006). Therefore, the positive effect of climate change on rice yield caused by carbon fertilization could be overshadowed by the negative influence of global warming on rice growth and more abundant and severe infestations of pests and pathogens. Most of the available studies have been conducted under controlled conditions; the effects might be different under field conditions (Ghini et al. 2008). Furthermore, most of the studies have been carried out on the effect of a single meteorological variable on the host or pathogen rather on the interactions between two or more factors (Coakley et al. 1999). Studies considering a combination of more than two climatic factors are not available. Based on the available information on the effect of climate change on rice pests and diseases elsewhere, it is difficult to draw any clear conclusions at the moment for the Mekong Delta. Therefore, comprehensive research needs to be carried out on this topic in the region.

13.3 Possible Impacts on Agrichemical Use Efficiency

Paddy rice production in the Mekong Delta increased steadily at a rate of 5.1% in the period of 1986–1995; pesticide use increased at a rate of 5% per annum. National demand for pesticides is around 50,000 tons, equivalent to about 500 million US dollars (BVSC 2010). Pesticide use in rice accounted for 65.5% of the total amount of pesticide use in agriculture, and insecticide use alone accounted for 85% of the total volume (Dung and Dung 1999).

Climate change will likely affect agrichemical use efficiency in several ways. First of all, warmer temperatures and high rainfall may change the dynamics of pesticide residues on crop foliage. High-rainfall intensity could reduce the retention of pesticide on the foliage by increasing wash-off rates (Schepers 1996). Higher temperature can reduce the effectiveness of certain classes of pesticides such as pyrethroids and spinosad (US GCRP 2009). The complex interaction between rainfall intensity, temperature, and host and disease will determine pesticide-use efficiency (Coakley et al. 1999). Statistical analyses of pesticide use in relation to higher temperature and rainfall in the US showed that the effect varied greatly among different crops. For some crops such as corn, cotton, soybeans, and potatoes, higher temperatures resulted in the increased use of pesticide. In contrast, the use of pesticides was reduced for wheat. More rainfall increased pesticide-usage costs for corn, wheat, soybeans, and potato (Chen and McCarl 2000).

Morphological and physiological changes in the host plants under changing climate would also have an impact on pesticide-use efficiency. Increases in canopy thickness and biomass under ECO_2 conditions would reduce the effectiveness of pesticides on the foliage and consequently would increase the amount of pesticide needed. Furthermore, changes in leaf morphology such as a thicker, epicuticular-wax layer on leaves could slow or reduce the uptake of pesticide and thus reduce pesticide-use efficiency (Coakley et al. 1999).

In general, the main climate drivers for changing pesticide efficiency would be most likely the predicted changes in rainfall intensity by causing an increased rate of pesticide applications to replace those removed from the plant surfaces and increased temperatures leading to faster degradation of the chemicals.

13.4 The Agrichemical Dimension of Climate Change Adaptation in the Mekong Delta

As reviewed in this chapter, climate change will likely decrease the area available for rice production in the Delta. As a consequence, food and income (export) security could be met only by increasing yields per unit of land (IPCC 2007). This will likely put the remaining production areas under pressure, which will lead to an intensification of agricultural practices – i.e., also via enhanced use of pesticides and fertilizers. This development could be amplified by direct negative impacts on crops and climate-driven changes in some host and pest/disease relationships leading to an increase in the occurrence and severity of some diseases and pests by a potentially decreasing efficiency of pesticides at the same time. Similarly, weeds that are adapted to hotter climates could potentially increasingly compete with crops (see e.g., Tungate et al. 2007 for an example from the US) leading to an increased use of herbicides. Our surveys conducted in 2008 and 2009 in the Mekong Delta revealed that farmers rely almost exclusively on chemical responses to any threats to their crops. This results in large amounts of pesticides applications as opposed to using alternative strategies such as integrated pest management (Toan et al. unpublished data). Thus again, the enhanced use of pesticides would be a likely response of the farmers to the expected yield losses induced by pests. These projections deserve attention since recent pesticide use and management practices in the Delta are already not sustainable from environmental and human health perspectives. Ongoing studies at two study sites in the Delta revealed that a broad range of recently-used pesticides is co-occurring at detectable levels in field discharges and channels used for irrigation, aquaculture, and personal hygiene throughout the year (Toan et al. 2009). The frequent detection of pesticides in the water systems of rural areas of the Delta indicates that surface water quality deserves special attention in these land-use settings, particularly because surface water often serves in rural areas as a drinking water source (GSO 2007).

However, surveys conducted in 2008 and 2009 showed that the improvement of water quality requires changes in farmers' practices in the use of pesticides and the management of pesticide waste (Toan et al. 2009). Farmers – not only in the Mekong Delta – tend to increase pesticide use in order to compensate for crop losses due to pests and diseases. This practice is economically not justified due to higher pesticide-use costs and the resulting pollution in the environment (Oerke 2006). Thus, investments in the training and education of farmers will be necessary to promote alternative management strategies and train farmers on effective and safe pesticide use. However, farmer schools and training in integrated pest management are already in place in the Delta and are strongly promoted by plant protection departments and extension offices. Although farmers usually claim to be aware of and apply these methods, their everyday practices show otherwise. The reasons for the gap between theory and practice are probably deep-rooted and should be overcome to enhance the impact of future training and education programs.

Beside the promotion of integrated pest management via training and education, further improvement in information generation and sharing will be necessary. This includes better seasonal and small-scale forecasts of weather together with crop models that can predict yield a few months ahead of the harvest. The information needs to reach the farm level possibly also via innovative communication ways such as text messages on mobile phones.

In terms of production areas, there is a need for a balance between protections of arable land on the one hand and the costs for the establishment of these protection measures (dykes, sluice gates) on the other hand. Protected areas need to have a long-term strategy to deal with the most probable enhanced pressure in terms of expected yield per land area. Soil exploitation and pest outbreaks are likely to occur if no adaptation strategies are put in place.

Concerning crops, the development of new rice varieties which are better adapted to a changing climate (e.g., more drought and/or salt tolerant varieties) and resist better to major pests and diseases such as BPH and RGSV would be advantageous with respect to decreasing or stabilizing pesticide demand. Also, an improvement in agricultural practices such as crop rotation instead of three rice crops per year would decrease the number of pest outbreaks by creating times where food and hosts are not available for pests.

For some regions a promotion of organic farming could be a viable adaption strategy and would reduce pesticide use at the same time. Organic farming tends to increase soil fertility and water retention capacity in the long run (Niggli et al. 2007). However, products from organic farming are generally more expensive to produce and thus the market capacity for these products will be most likely the limiting factor for their planting area.

While some of the above mentioned adaptation strategies will be likely taken up by farmers as a first-order response, other strategies need to be supported, facilitated, or even regulated. Table 13.1 summarizes the most likely climate change impacts, adaptation, and possible facilitation options in the agricultural sector.

Table 13.1 Selected adaptation options with a link to agrichemical use

Changing climate parameter	Possible impact	Possible adaptation at farm level	Possible impact of the adaptation on the environment	Facilitation, support, regulation
Increased frequency of extreme events (drought, heavy rains)	Crops more susceptible to pests and diseases	Increased use of pesticides, introduction of new pesticides	Negative impacts on water quality, biodiversity	Training, education on integrated pest-management, pest and disease-resistant crops
Increased mean temperatures, changes in rainfall patterns, humidity	Increased occurrence of diseases, pests, weeds, and vectors carrying diseases	Use of crop varieties resistant to pests and diseases	Positive impacts on water quality	Provide seasonal and small scale weather forecasts at farm level
	Increased occurrence of secondary infections			Improved pest and disease monitoring and surveillance
	Increased risk of severe outbreaks			Support pest predators and competitors e.g. through the use of field margins
Increased rainfall intensity	Reduced top soil quality due to increased runoff and soil erosion. Nutrient and pesticide loss	Increased use of fertilizers and pesticides	Negative impacts on water quality, biodiversity	Use of buffer strips to reduce runoff. Consider organic farming
	Reduced efficiency of pesticides (wash up from foliage)			Training on selection of pesticides used, integrated pest management, organic farming
Sea level rise	Inundation or salinization of soils and water recourses, loss of arable land	More saline-tolerant rice varieties and shrimp farming in the area of saline intrusion	Negative impacts on water quality, biodiversity	Balance the needs for shrimp farming and rice production, flexible water-management systems
		Agricultural intensification in the remaining area with enhanced use of pesticides and fertilizers		Crop rotation or rice-aquaculture instead of three crops of rice

13.5 Conclusions

The consequences of climate change will likely lead to increased pesticide use in the Mekong Delta. The significant limitation of arable land driven by inundation and soil salinization in the future would likely lead to an enhanced pressure on the remaining arable land in terms of yield per hectare. Rice production, it needs to be remembered, is a matter of national policy in Vietnam (Ryan and Garrett 2003). Yield optimization will in turn likely require further agricultural intensification with corresponding high pesticide and fertilizer use. On the other hand, the impacts of climate change will increase the risk for some pest and disease outbreaks, while decreasing it for others. The risk will be probably most pronounced when several influencing factors co-occur and lead to unanticipated outbreaks accompanied by growing uncertainty in the function of the production systems. One consequence, particularly in view of current practices and trends, will be an increase in agrichemical use and the resulting increased pollution of the environment in general and water systems in particular. It is thus timely to improve pest-management strategies in the Delta as one of many different approaches (development of new varieties, crop rotations, changes in farming systems, etc.) to adapt to the potential effects of climate change. The benefits would be immediate, resulting in more targeted pest management, reduced costs at the farm level, and reduced pollution of land and water systems. The latter would then directly reduce the exposure of humans and freshwater ecosystems, and therefore afford increased protection of the most vulnerable communities in the Delta who rely on the freshwater system for their everyday livelihoods. On the other hand, agricultural systems and prevailing pests and diseases may change more dynamically in the future. More research and policy attention will be required to be able to keep pace with the challenges the future will bring to agriculture in the Mekong Delta.

References

Agrios GN (2005) Plant pathology. Elsevier Academic Press, London

Agriviet.com (2008) Rầy nâu trong canh tác lúa ở ĐBSCL. Retrieved from http://agriviet.com/nd/1183-ray-nau-trong-canh-tac-lua-o-dbscl

Ainsworth EA (2008) Rice production in a changing climate: a meta-analysis of responses to elevated carbon dioxide and elevated ozone concentration. Glob Change Biol 14:1642–1650

Aizawa M, Chi MTT, Thanh NK, Hao NV, Can ND, Kurokura H, Kobayashi K (2007) Adaptation of rice farming to the changes in water regime due to dykes and sluice gates built against salinity intrusion in Mekong Delta. In: Furumai H, Katayama H, Kurisu F, Satoh H, Ohgaki S, Thanh NC (eds) Southeast Asia water environment 2. IWA Publishing, UK, pp 83–91

Allen LH Jr, Baker JT, Alerecht SL, Boote KJ, Pan D, Vu JCV (1995) Carbon dioxide and temperature effect on rice. In: Peng S, Ingram KT, Neue HU, Ziska LH (eds) Climate change and rice. Springer/International Rice Research Institute, Berlin/Manila, pp 278–290

Baker JT, Allen LH Jr, Boote KJ (1992) Temperature effects on rice at elevated CO_2 concentration. J Exp Bot 43(252):959–964

Baker JT, Allen LH Jr, Boote KJ, Pickering NB (1996) Assessment of rice response to global climate change: CO_2 and temperature. In: Koch GW, Mooney HA (eds) Carbon dioxide and terrestrial ecosystems. Academic, San Diego, pp 265–282

Battisti DS, Naylor RL (2009) Historical warnings of future food insecurity with unprecedented seasonal heat. Science 323:240–244

BVSC (2010) Phân tích ngành thuốc bảo vệ thực vật. Bao moi. Retrieved from http://www.baomoi.com/Info/Phan-tich-nganh-thuoc-bao-ve-thuc-vat/127/3853154.epi

Chakraborty S, Pangga IB (2004) Plant disease and climate change. In: Gillings M, Holmes A (eds) Plant microbiology. BIOS Scientific, London, pp 163–180

Chakraborty S, Tiedemann AV, Teng PS (2000) Climate change: potential impact on plant diseases. Environ Pollut 108(3):317–326

Chen C, McCarl B (2000) Pesticide usage as influenced by climate: a statistical investigation across the U.S. Retrieved from http://www.usgcrp.gov/usgcrp/nacc/agriculture/chen2.pdf. Accessed 26 Feb 2010

Cheng W, Sakai H, Yagi K, Hasegawa T (2009) Interactions of elevated [CO_2] and night temperature on rice growth and yield. Agric For Meteorol 149:51–58

Coakley S, Scherm H, Chakraborty S (1999) Climate change and plant disease management. Annu Rev Phytopathol 37:399–426

Dang NH, Danh VT (2008) An overview of the development of the agricultural economy in the Mekong Delta. In: Lensink R, Mai VN (eds) Economic development of the Mekong Delta in Vietnam, CDS Research Paper No. 27

Đào Xuân Học (2009) Kế hoạch thích ứng với biến đổi khí hậu trong lĩnh vực nông nghiệp và phát triển nông thôn. Hội thảo Việt Nam thích ứng với Biến đổi khí hậu, ngày 31/07/2009 tại Hội An – Quảng Nam. Retrieved from http://www.vncold.vn/modules/cms/upload/10/PhatTrienNuoc/090729_Hoithao_DN09/Hoithao_DN09.pdf

DARD Dong Thap (Department of Agriculture and Rural Development) (2006) Dong Thap province plant protection department report. Cao Lanh, Dong Thap

Dasgupta S, Meisner C, Wheeler D, Xuyen K, Lam NT (2007) Pesticide poisoning of farm workers: implications of blood test results from Vietnam. Int J Hyg Environ Health 210:121–132

De Costa WAJM, Weerakoon WMW, Herath HMLK (2006) Physiology of yield determination of rice under elevated carbon dioxide at high temperatures in a subhumid tropical climate. Field Crop Res 96:336–347

DONRE Dong Thap (Department of Natural Resources and Environment) (2008) Report 2008. Cao Lanh, Dong Thap, Vietnam

Dung NH, Dung TTT (1999) Economic and health consequences of pesticide use in paddy production in the Mekong delta, Vietnam. EEPSEA Research Report rr1999021. Retrieved from http://203.116.43.77/publications/research1/ACF124.html

Evers HD, Benedikter S (2009) Strategic group formation in the Mekong Delta – The development of a modern hydraulic society. ZEF Working Papers Series, 1–27

FAO (Food and Agriculture Organization of the United Nations) (2007) Adaptation to climate change in agriculture, forestry and fisheries: perspective, framework and priorities. Retrieved from ftp://ftp.fao.org/docrep/fao/009/j9271e/j9271e.pdf

FAO (Food and Agriculture Organization of the United Nations) (2008) The state of world fisheries and aquaculture. Retrieved from ftp://ftp.fao.org/docrep/fao/011/i0250e/i0250e.pdf

Gesch RW, Vu JCV, Allen LH, Boote KJ (2001) Photosynthetic responses of rice and soybean to elevated CO_2 and temperature. Recent Res Develop Plant Physiol 2:125–137

Ghini R, Hamada E, Bettiol W (2008) Climate change and plant disease. Sci Agric 65:98–107

Gregory PJ, Johnson SN, Newton AC, Ingram JSI (2009) Integrating pests and pathogens into the climate change/food security debate. J Exp Bot 60(10):2827–2838

GSO (Vietnam General Statistics Office) (2007) The 2006 rural, agricultural and fishery census, vol 2, Agriculture situation. Statistical Publishing House, Hanoi

GSO (Vietnam General Statistical Office) (2008) Statistical handbook 2008. Statistical Publishing House, Hanoi, Vietnam. Retrieved from http://www.gso.gov.vn/default_en.aspx?tabid=466&idmid=3&ItemID=8597

GSO (Vietnam General Statistics Office) (2009a) Production of paddy by province. Retrieved from http://www.gso.gov.vn/default_en.aspx?tabid=469&idmid=3

GSO (Vietnam General Statistics Office) (2009b) Production of farmed aquatic product by province. Retrieved from http://www.gso.gov.vn/default_en.aspx?tabid=469&idmid=3

Heong KL (2008) Are brown planthopper outbreaks due to deteriorated ecosystem services in rice fields? In: Heong KL, Huan NH (eds) The proceedings of the final consultation workshop: scoping study to identify research and implementation issues related to management of the Brown Planthopper/virus problem in rice in Vietnam, Ho Chi Minh City, 8 Jan 2008

Heong KL (2010) Thai government to provide B2bn to compensate farmers hit by BPH. Blog of the Rice Planthopper Project. Retrieved from http://ricehoppers.net/2010/02/10/thai-government-to-provide-b2bn-to-compensate-farmers-hit-by-bph

Heong KL, Song YH, Pimsamarn S, Zhang R, Bae SD (1995) Global warming and rice arthropod communities. In: Peng S, Ingram KT, Neue HU, Ziska LH (eds) Climate change and rice. Springer/International Rice Research Institute, Berlin/Manila, pp 326–336

Heong KL, Chien HV, Huan NH (2008) Contributing factors to the brown planthopper – virus outbreaks in Vietnam – Scoping for opportunities in ecological management strategies. In: Heong KL, Huan NH (eds) The proceedings of the final consultation workshop – scoping study to identify research and implementation issues related to management of the Brown Planthopper/virus problem in rice in Vietnam, Ho Chi Minh City, 8 Jan 2008

Hibberd JM, Whitbread R, Farrar JF (1996) Effect of elevated concentrations of CO2 on infection of barley by *Erysiphe graminis*. Physiol Mol Plant Pathol 48:37–53

Holland J (2007) Walking a tightrope. Seafood International 14 July 2007

Horie T, Yoshida H, Kawatsu S, Katsura K, Homma K, Shiraiwa T (2005) Effect of elevated atmospheric CO2 concentration and increased temperature on rice: implications for Asian rice production. In: Toriyama K, Heong KL, Hardy B (eds) Proceedings of the world rice research conference: rice is life: scientific Perspective for the 21st Century, Tsukuba

Imai K (1995) Physiological response of rice to carbon dioxide, temperature, and nutrients. In: Peng S, Ingram KT, Neue HU, Ziska LH (eds) Climate change and rice. Springer/International Rice Research Institute, Berlin/Manila, pp 252–257

Ingram KT, Manalo PA, Namuco OS, Pamplona RR, Weerakoon WM (1995) In: Peng S, Ingram KT, Neue HU, Ziska LH (eds) Climate change and rice. Springer/International Rice Research Institute, Berlin/Manila, pp 258–277

IPCC (2007) Climate change 2007: impacts, adaptation, and vulnerability. Contribution of Working Group II to the Third Assessment Report of the Intergovernmental Panel on Climate Change. In: Parry ML, Canziani OF, Palutikof JP, van der Linden PJ, Hanson CE (eds) Cambridge University Press, Cambridge

IRIN (Integrated Regional Information Networks) (2009) Vietnam. Even bottled water unsafe. Retrieved from http://www.irinnews.org/Report.aspx?ReportId=83965

IRRI (2007) Coping with climate change. Rice today 6(3):10–15

IRRI (2009) Vietnam to host international rice congress. Rice today 8(1):6

Kam SP, Hoanh CT, Tuong TP, Khiem NT, Dung LC, Phong ND, Barr J, Ben DC (2001) Managing water and land resources under conflicting demands of shrimp and rice production for sustainable livelihoods in the Mekong River Delta, Vietnam. In: INRM2001 workshop on integrated management for sustainable agriculture, forestry and fisheries, Cali

Kiritani K (2006) Predicting impacts of global warming on population dynamics and distribution of arthropods in Japan. Popul Ecol 48:5–12

Kobayashi T, Ishiguro K, Nakajima T, Kim HY, Okada M, Kobayashi K (2006) Effects of elevated atmospheric CO_2 concentration on the interaction of rice blast and sheath blight. Phytopathology 96(4):425–431

Kotera A, Sakamoto T, Khang ND, Yokozawa M (2008) Regional consequences of seawater intrusion on rice productivity and land use in coastal area of the Mekong River Delta. Jpn Agric Res Q 42(4):267–274

Lake JA, Wade RN (2009) Plant-pathogen interactions and elevated CO_2: morphological changes in favour of pathogens. J Exp Bot 60(11):3123–3131

Lam PV (2007) Cơ sở khoa học của các giải pháp phòng trừ chống dịch rầy nâu, bệnh virus lúa cỏ và lúa lùn xoăn lá. Retrieved from http://www.lrc.ctu.edu.vn/pdoc/34/149-draynau.pdf

Lu ZX, Heong KL, Yu XP, Hu C (2005) Effects of nitrogen on the tolerance of brown planthopper, *Nilaparvata lugens*, to adverse environmental factors. Insect Sci 12:121–128

Luo Y, Teng PS, Fabellar NG, TeBeest DO (1998a) The effects of global temperature change on rice leaf blast epidemics: a simulation study in three agroecological zones. Agric Ecosyst Environ 68:187–196

Luo Y, Teng PS, Fabellar NG, TeBeest DO (1998b) Risk analysis of yield losses caused by rice leaf blast associated with temperature changes above and below for five Asian countries. Agric Ecosyst Environ 68:197–205

Margni M, Rossier D, Crettaz P, Jolliet O (2002) Life cycle impact assessment of pesticides on human health and ecosystems. Agric Ecosyst Environ 93:379–392

Masutomi Y, Takahashi K, Harasawa H, Matsuoka Y (2009) Impact assessment of climate change on rice production in Asia in comprehensive consideration of process/parameter uncertainty in general circulation models. Agric Ecosyst Environ 131:281–291

Meisner C (2005) Pesticide use and poverty in Vietnam, PEN Report, ADB. Retrieved from http://econ.worldbank.org

Ministry of Natural Resources and Environment (MONRE) (2009) Climate change, sea level rise scenarios for Vietnam. Ministry of Natural Resources and Environment, Hanoi

Morimoto N, Imura O, Kiura T (1998) Potential effects of global warming on the occurrence of Japanese pest insects. Appl Entomol Zool 33:147–155

NEDECO (1993) Master plan for the Mekong Delta in Vietnam. A perspective for sustainable development of land and water resources. A Report to Government of Vietnam, State Planning Committee, World Bank, Mekong Secretariat, UNDP, Netherlands Engineering Consultants, Arnheim

Neubacher H (2007) Pangasius (2007): growth on shaky grounds? Seafood International 14–15 Aug 2007:27–30

Nhan DK, Be NV, Trung NH (2007) Water use and competition in the Mekong Delta, Vietnam. In: Be TT, SinhBT, Miller F (eds) Challenges to sustainable development in the Mekong Delta: regional and national policy issues and research needs. The Sustainable Mekong Research Network, pp 143–188

Niggli U, Schmid H, Fliessbach A (2007) Organic farming and climate change. Retrieved from http://www.blauen-institut.ch/tx_blu/tp/tpg/g1695_itc-climate.pdf

Oerke EC (2006) Crop losses to pests. J Agric Sci 144:31–43

Padgham J (2009) Agricultural development under a changing climate: opportunities and challenges for adaptation. The World Bank Joint Department Discussion Paper, Washington, DC

Pathak MD, Khan RZ (1994) Insect pests of rice. International Rice Research Institute, Manila

Peng S, Ingram KT, Neue HU, Ziska LH (eds) (1995) Climate change and rice. Springer/International Rice Research Institute, Berlin/Manila

Peng S, Huang J, Sheehy JE, Laza RC, Visperas RM, Zhong X, Centeno GS, Khush GS, Cassman KG (2004) Rice yields decline with higher night temperature from global warming. Proc Natl Acad Sci USA 101:9971–9975. doi:10.1073/pnas.0403720101

PPD – MARD (Plant Protection Department – Ministry of Agriculture and Rural Development). Warning pests (in Vietnamese). Retrieved from http://www.ppd.gov.vn/?module=warning_pests

Ryan J G, Garrett JL (2003) The impact of policy research: lessons on attribution and evaluation from IPFRI. Impact Assessment Discussion Paper Number 20. Retrieved from http://portals.wdi.wur.nl/files/docs/ppme/iadp20.pdf

Schepers HTAM (1996) Effect of rain on efficacy of fungicide deposits on potato against Phytophthora infestans. Potato Res 39:541–550

Shimono H, Okada M, Yamakawa Y, Nakamura H, Kobayashi K, Hasegawa T (2009) Genotypic variation in rice yield enhancement by elevated CO_2 relates to growth before heading, and not to maturity group. J Exp Bot 60(2):523–532

Song YH, Heong KL (1997) Changes in searching responses with temperature of *Cyrtorhinus lividipennis* Reuter (Hemiptera: Miridae) on the eggs of the brown plant hopper, *Nilaparvata lugens* (Stal) (Homoptera: Delphacidae). Res Popul Ecol 39:201–206

SRV [Socialist Republic of Vietnam] (2008) Issuance of the action plan framework for adaptation to climate change in the agriculture and rural development sector 2008–2020. Decision No: 2730/2008/QD-BNNKHCN, Hanoi

Statistical Office of Can Tho City (2000, 2005, 2008) Statistical yearbooks. Statistical Office of Can Tho, Can Tho City

Tao F, Hayashi Y, Zhang Z, Sakamoto T, Yokozawa M (2008) Global warming, rice production, and water use in China: developing a probabilistic assessment. Agric For Meteorol 148:94–110

Toan PV, Sebesvari Z, Rosendahl I, Renaud FG (2009) Residue concentrations of recently used pesticides in surface water at two representative study sites of the Lower Mekong Delta, Vietnam. In: Pesticide behaviour in soils, water and air, 14–16 Sept 2009, York

Trong TQ, Hao NV, Griffiths D (2002) Status of Pangasiid aquaculture in Vietnam. MRCS Technical Paper No.2, Mekong River Commission, Phnom Penh

Tuan LA, Suppakorn C (2011) Climate change in the Mekong River Delta and key concerns on future climate threats. In: Mart A Stewart, Peter A Coclanis(eds) Environmental change and agricultural sustainability in the Mekong Delta. Springer, Dordrecht, pp 207–217

Tungate KD, Isarel DW, Watson DM, Rufty TW (2007) Potential changes in weed competitiveness in an agroecological system with elevated temperatures. Environ Exp Bot 60:42–47

US GCRP (US Global Change Research Program) (2009) Global climate change impacts in the United States. Report on Agriculture (pp 71–78). Retrieved from http://downloads.globalchange.gov/usimpacts/pdfs/climate-impacts-report.pdf

Wassmann R, Dobermann A (2007) Climate change adaptation through rice production in regions with high poverty level. In: ICRISAT and CGIAR 35th anniversary symposium on Climate-proofing innovation for poverty reduction and food security, Hyderabad, 22–24 Nov 2007

Wassmann R, Nguyen XH, Chu TH, To PT (2004) Sea-level rise affecting the Vietnamese Mekong Delta: water elevation in the flood season and implications for rice production. Clim Change 66:89–107

White I (2002) Water management in the Mekong Delta: changes, conflicts and opportunities. IHP-VI Technical Documents in Hydrology 61, UNESCO, Paris

Yamamura K, Yokozawa M (2002) Prediction of a geographical shift in the prevalence of rice stripe virus disease transmitted by the small brown planthopper, *Laodelphax striatellus* (Fallèn) (Hemiptera: Delphacidae), under global warming. Appl Entomol Zool 37:181–190

Yen H (2006) Japan exasperated by substandard VN seafood shipments. Vietnam Net. Retrieved from http://english.vietnamnet.vn/biz/2006/08/605629. 25 Aug 2006

Chapter 14
Role of Villagers in Building Community Resilience Through Disaster Risk Management: A Case Study of a Flood-Prone Village on the Banks of the Mekong River in Cambodia

Serey Sok, Louis Lebel, Ram C. Bastakoti, Sokkalyan Thau, and Sela Samath

Abstract Floods are a normal feature of the monsoon season around the Tonle Sap Great Lake and areas neighboring the Mekong River in Cambodia. In some years, floods are much more severe than others; in particular, when there is a combination of high discharge from upstream and a high run-off from rainfall within the area. Actions taken by villagers could help build resilience at household and community levels against these more serious floods and thus reduce the risks of disasters, loss of life, and costly damage. In a detailed study of impacts and responses to floods in Angkor Ang village in Prey Veng Province, we found evidence of major barriers to effective actions by villagers and other actors working at community and other levels. Building and maintaining resilience in flood-prone communities requires that attention be given to local capacity and knowledge, differences in wealth or poverty, gender relations, and local participation. Moreover, these factors need to be examined in various social processes before, during, and after major floods.

Keywords Community resilience • Gender • Poverty • Floods • Cambodia • Mekong

---

S. Sok (✉)
Research Office, Royal University of Phnom Penh, Phnom Penh, Cambodia
e-mail: sereyhawaii@yahoo.com

L. Lebel
Unit for Social and Environmental Research, Chiang Mai University, Chiang Mai, Thailand
e-mail: llebel@loxinfo.co.th

R.C. Bastakoti
Water Engineering and Management, Asian Institute of Technology,
Klong Luang, Pathumthani, Thailand
email: rcbastakoti@gmail.com

S. Thau and S. Samath
Research Team for ProVention Consortium Research and Action Grants,
Hosted by Asian Disaster Preparedness Center (ADPC), Bangkok, Thailand
e-mail: tintin-thau@yahoo.com; samathsela@yahoo.com

M.A. Stewart and P.A. Coclanis (eds.), *Environmental Change and Agricultural Sustainability in the Mekong Delta*, Advances in Global Change Research 45,
DOI 10.1007/978-94-007-0934-8_14,

14.1 Introduction

Several decades ago – when most households pursued livelihoods in tune with the regular and seasonal pulse of flood waters and relatively little physical infrastructure was at risk – annual flooding in Cambodia usually brought more benefits than costs (Kummu and Sarkkula 2008; Baran and Myschowoda 2009; Friend 2009). Floods, when higher than usual, affected only a small fraction of the population, caused only modest damage, and were years apart (ADPC 2002). Since 1999, however, devastating floods have become a common occurrence and have caused an increase in deaths, losses, and damage. Several recent floods have had severe impacts on agricultural production and public infrastructure. A common perception is that floods are becoming more extreme, frequent, and less predictable (NCDM 2008). These changes in the flood regime may be due to climate change as well as the combined impacts of human activities – for instance, changing patterns of agricultural land-use and water-use, efforts to regulate and divert flows of water, and deforestation (ADPC 2007; NCDM 2008).

Between 2000 and 2002, a series of severe flood events struck the Mekong Delta (Few 2003). The floods in 2000 were the worst in at least 70 years. More than 3.4 million people in 22 provinces were affected. About 347 people died; 80% of these were children. Physical damage was estimated at US$157 million (Council of Ministers 2000). Floods in 2001 killed another 62 people and caused estimated losses of US$ 36 million (Helmers et al. 2003). The 2002 floods killed 29 people, and the total extent of damage was over US$12 million (NCDM 2006). In 2003 and 2004, in contrast, the seasonal flooding was less than normal and led to concerns about drought (MRC 2005). In those 2 years, Cambodia also had a strong dry season, and lakes in the deepest part of the flood plain completely dried out. While some areas of the country were affected by drought, rice crop losses did not result in a national disaster (Hak 2005).

The huge losses in terms of human death, physical infrastructures, agricultural yields, and livestock may be an indicator that households and local communities are no longer be resilient to floods. Frequent disasters contribute to the high levels (30%) of current poverty (NIS 2008) and represent an on-going challenge for human development. Poverty is also an important factor in determining vulnerability to floods. Building community resilience is a big challenge in poor villages where most residents have no or limited access to natural, human, financial, material, and social resources. Community-based disaster management is recognized by the Cambodian government (NCDM 2004) as important to reducing vulnerabilities and improving capacities to cope with adverse consequences of floods. At the same time, sustainability of interventions by external governments and non-state agencies is an important concern as there is always the danger of losing impetus when outside support ends (Weerasinghe 2007).

Floods, it should be emphasized, are a normal feature of the monsoon season around the Tonle Sap Great Lake and banks of the Mekong River. The seasonal flood pulse is particularly important to fisheries but also for the replenishment of agricultural soils (Kummu and Sarkkula 2008; Baran et al. 2007). Floods become disasters for rice farmers when they come too early in the crop growing cycle and destroy rice seedlings before transplanting, or when water levels are too high for too long and destroy established wet season rice crops (Helmers and Jegillos 2002). Two distinct types of floods can be recognized. First, “river floods,” caused by high rainfall upstream in the Mekong River basin, usually affect provinces bordering the Mekong River in the south-east. Second, “combined floods” caused by the combination of high discharge in the Mekong River and high run-off from heavy rains within the Tonle Sap Lake region and southern provinces have major impacts on provinces around the lake and the southern provinces (ADPC 2002). This chapter explores the impacts on households living along the Mekong River in Cambodia and the responses by various actors to reduce risks of flood-related disasters. Special emphasis is placed on understanding local participation in disaster-risk management.

14.2 Study Area and Methods

This chapter is based on work carried out at several administrative levels in Prey Veng province in mid-2008. Prey Veng is in the plain region and has a border with Vietnam (Fig. 14.1a). It has a total population of just less than one million, of which more than 80% are farmers and 14% are fishermen according to provincial statistics.

a

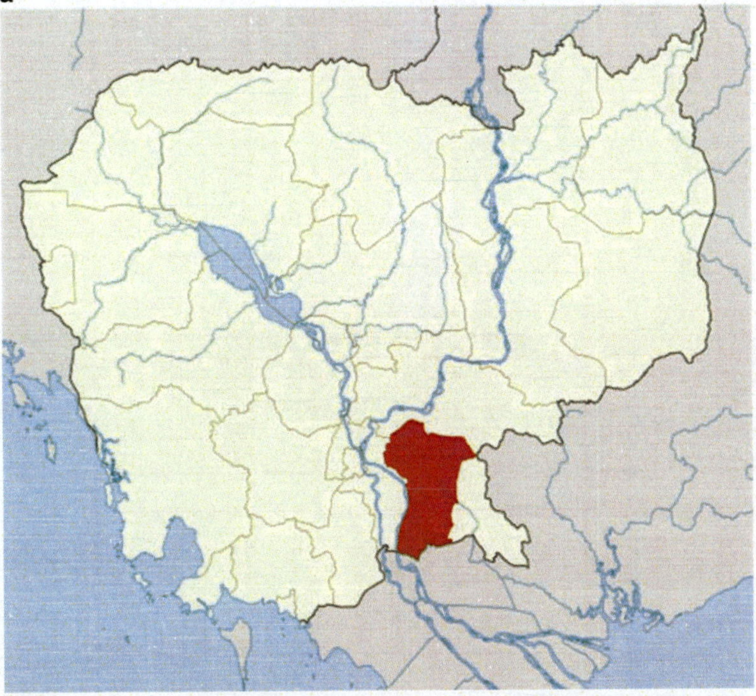

Fig. 14.1 (**a**) Map of Cambodia showing location of Prey Veng Province, Mekong River channels, other major tributaries, and Tonle Sap Lake

b

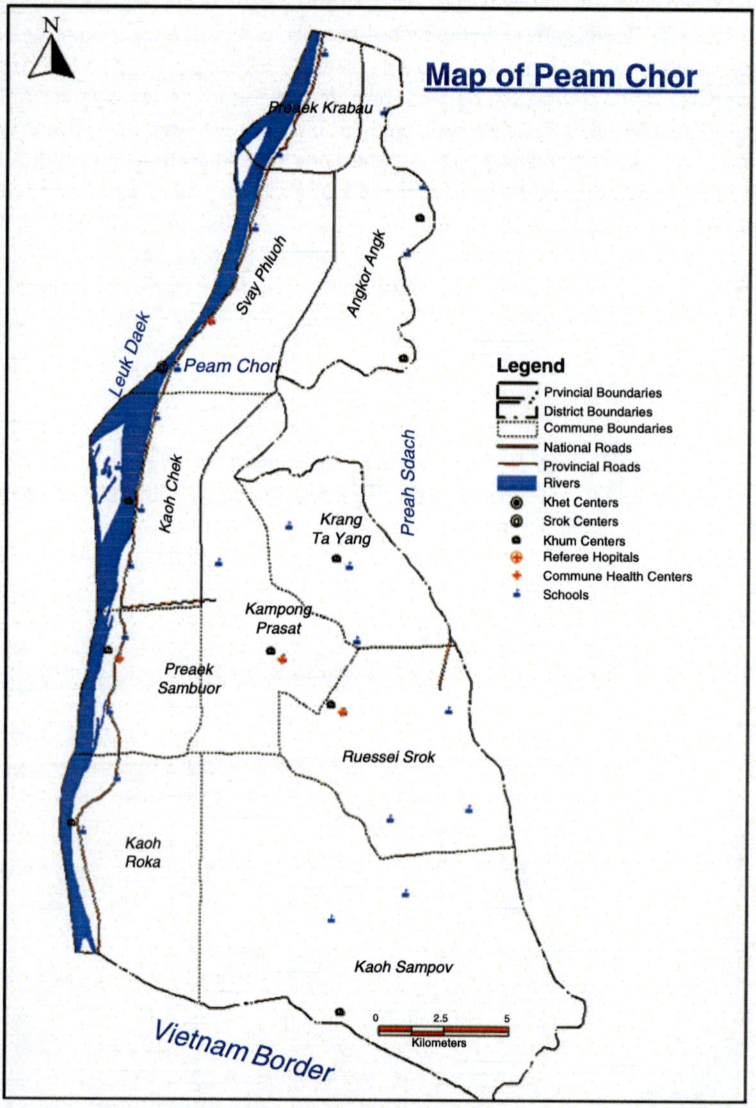

Fig. 14.1 (continued) (**b**) Map of Peam Chour District in Prey Veng Province (Source: ADPC, Phnom Penh Office)

Peam Chour district is bound by the Mekong River and the tributary Tonle Touch (Fig. 14.1b). The district has a total population of just over 67,000 made up of four main ethnicities: Khmer, Vietnamese, Cham, and Malay (MoAFF 2003–2007). Most people consider themselves farmers (83%) or fishermen (6%).

At the commune level, Angkor Ang commune includes five villages. We studied one of these – Angkor Ang village – more closely. It had 397 households. A sample of 140 households was selected for interviewing using stratified random sampling across two wealth classes – poorer households and better-off households – and gender. The wealthier class strata was identified by local leaders and villagers in a community meeting using a wealth ranking approach based on assets; in particular, by the type of home. These were relative groupings – in this community, a "better-off" household may still be close to the poverty lines at the national level. Information was also gathered through field observations, by attending community meetings, and through mapping exercises with residents.

In the data analysis, both qualitative and quantitative analyses were used. The statistical package for social science (SPSS) software was used for data processing and analysis for both descriptive analysis and statistical analysis. A key method used in this study was logistic regression for exploring binary "outcome" variables describing disaster-risk management measures, services and participation in community activities, and variables describing household and informant characteristics. As a multivariate technique this approach has the added advantage of being able to consider patterns of association and taking into account associations with other confounding variables.

14.3 Results

For the decade 1999–2008, the most serious flood in Prey Veng was clearly in 2000. The floods in 2001 and 2002 were also more serious than other years (Table 14.1). In 2000, water levels at the Mekong-Neak Loueng station peaked at just over 8.0 m

Table 14.1 Severity of floods and droughts during 1999–2008 in Prey Veng Province and flood damage at national level in Cambodia (Source: NCDM 2009, 2010)

Year	Drought severity	Flood severity	Flood damage (US$ million)
1999	Low	Low	n.d.
2000	Low	High	161
2001	Low	Medium	36
2002	Low	Medium	12
2003	High	Low	n.d.
2004	High	Low	55
2005	Medium	Low	4
2006	Low	Low	12
2007	Low	Low	9
2008	Low	Low	6
2009	Low	Low	n.d.

in mid-September, whereas the mean peak for 1990–2008 was around 7.0 m at the end of September (MWH 2010). The annual reports of the National Committee for Disaster Management (NCDM) for 2006–2009 indicate there was little or no flood damage in the country during those years that could be linked to hydrological conditions along the Mekong mainstream, the Bassac, or in the Tonle Sap Basin (NCDM 2010).

14.3.1 Flood Disaster Management in Peam Chour District

Due to its location, Peam Chour district is regularly exposed to floods from the Mekong River and its tributaries. As part of the National Strategy on Disaster Management flood-prone districts establish a committee to oversee disaster risk reduction programmes. The Peam Chour committee has as its chair the District Governor, and as its vice chairs, Deputy Governors. Military and other district officers make up the other committee members. The committee works closely with other national and international organizations.

Information on the annual effects of flooding has been recorded by the Peam Chour Committee. According to their reports, virtually the entire (94%) of the district is "vulnerable" (NIS 2008). Angor Ang commune is considered the most affected, and the entire population is classified as "vulnerable." Floods in 2000, 2001, and 2002, for example, had major impacts (Table 14.2).

Table 14.2 Impacts of flood disasters in Peam Chour district (Source: ADPC 2007)

Damages/victims	2000	2001	2002
Affected communes	10 (50 villages)	10 (40 villages)	10 (35 villages)
Human deaths	12	08	0
Human injuries	N/A	N/A	360
Affected individuals	59,283	1,528	4,626
Animal deaths	246	139	N/A
Affected houses	123	N/A	3,245
Damaged houses	123	162	09
Damaged wet season rice	3,624 ha	1,785 ha	1,060 ha
Damaged crops	1,905 ha	2,840 ha	115 ha
Damaged roads/trails	10,482 m	7,258 m	16,000 m
Affected schools	13	32	N/A
Damaged schools	N/A	1	N/A
Damaged dams	3,271 m	N/A	11,487 m
Damaged canals	4,317 m	N/A	N/A
Damaged water gates	13	N/A	N/A
Damaged bridges	8	5	11

N/A Not Available

Floods in 2001 killed eight people and damaged rice and food crops. It was remarkable that there were no human deaths in the 2001 flood, as the destruction of roads and trails was high (DCDM-Peam Chour 2007). The sequence of devastating events increased awareness about, and led to several efforts to improve, flood disaster-risk management at the district level. Governmental agencies, the Red Cross, and development partners have paid attention to capacity building, training, and early warning systems. Several national and international organizations work in the Peam Chour district, including UNICEF, Community Economic Development Assistance Corporation (CEDAC), CARE, Socioeconomic and Data Applications Center (SEDAC), and Asian Pacific Development Center (ADPC), as well as some micro-finance institutions such as PRASAC, AmRath (EMT), and ACLEDA Bank Plc. Many have contributed to rehabilitation and recovery work stemming from major flood events or to more general efforts at alleviating poverty' (DCDM-Peam Chour 2009). But none have permanent offices in the study village.

14.3.2 Flood Disaster in Angkor Ang Village

The pattern of flood severity during the period 1999–2007 in Angkor Ang Village was similar to that at the district level (Table 14.2). The flood in 2000 was the most severe and killed six people. In the period prior to this; in 1997, three people died in another severe flood year; and in 1998, one person died in what was a low-flood year. In the 2000 floods, water levels rose over 1.5 m in just one day. After the 2000 flood event, people became more aware of the risks and more interested in prevention and preparedness. There have been no flood deaths since 2000, although agricultural activities and infrastructure have still been impacted. "Combined" floods with high peak flood heights occurred in 1997, 2000, 2001, and 2002. In other years, floods were a simple river-flood type. These findings suggest that local people have become more resilient and have learnt to prepare or cope with unusual annual floods.

At the community level, early warning systems include a local water level meter with a red mark and an information board near the commune head's house (Fig. 14.2).

In focus group discussions held in mid-2008, villagers were clearly aware – often from recent experiences, but also from traditional knowledge – about the potential impacts of floods. At the same time, efforts to reduce risks at the community level were limited as a result of the lack of both local initiative and on-going external support. Villagers were particularly at a loss about what to do in the more unusual and severe "combined floods" as opposed to the more routine "river floods." The main flood-disaster response was focused on the event alone and on meeting urgent, emergency needs, like piling up sandbags to hold back water and transporting people to safety. Neighbors also helped fix damaged homes. One indicator of underlying limitations at the community level was that villagers were afraid property would be stolen if they left their houses during a flood.

Fig. 14.2 Information board with water-level reports in front of the commune head's home

Table 14.3 Impacts of 2007 floods on different types of households in Angkor Ang village ($n=70$ in each group)

Impacts	Poorer households (%)	Better-off households (%)
Fear that missing family members had died	83	89
Illness or disease	73	61
Property/facilities lost	69	51
Crop/rice destroyed	24	17
Home damaged	37	10

14.3.3 Household Impacts

Interviews with individual households provided more detailed information on the impacts of the 2007 floods (Table 14.3). The most common concern was fear that temporarily missing family members had died; although in the end, nobody from the village died from the floods. Health problems following floods were also common for all types of households. Problems with sanitation and in securing clean drinking water resulted in diarrhea and other water-related illnesses. Most villagers drink and use water for cooking without boiling it. Contamination from agricultural chemicals may also have been a factor in the emergence of health problems after the flood. Access to health care is also a factor; the nearest primary health care facility was in a town more than 30 km^2 away, with relatively high transport costs for poor families. Although women did not suffer more than men from health problems, their burdens as care-takers were higher. Analysis by logistic regression suggests that households with illness or disease impacts were less likely (Odds Ratio or O.R. = 0.44, $P < 0.05$) to contain five or more family members.

Households with significant property or facility losses were more likely to be poor (Table 14.3, O.R. = 2.4, $P<0.05$) and much less likely to have large families (O.R. = 0.20, $P<0.001$). There were no significant associations between having crops or rice fields destroyed and income level, gender, family size, or levels of male or female education.

The largest difference between poor and wealthier households was that the former were more likely to suffer from damages to their homes due to floods and associated storms (Table 14.3, O.R. = 5.3, $P<0.001$). Those with stronger houses also did not need to retreat to a place of higher elevation during high flood waters. While most houses in this area are built on stilts, stronger houses often use concrete pillars with strong foundations and have well attached walls and roofs that are much less affected by flood waters and winds in associated storms (Fig. 14.3).

Apart from their roles in specific tasks, we also asked if there were certain activities for which men or women had to spend additional time in order to recover livelihoods and regain household security after floods (Table 14.4). This dimension of

Fig. 14.3 (**a**) Flimsy houses typically built by poorer households. (**b**) Stronger houses built by better-off households

Table 14.4 Comparison of activities for men and women which took additional time for them to recover their livelihoods

Livelihood recovery activities	Women (% of households)	Men (% of households)	Adjusted odds ratio
Agriculture	54	64	0.63
Livestock	57	60	0.91
Firewood collection	37	79	0.07
Fishing gear	9	6	1.5
Home utensils	76	46	**4.2**
Rice stock	20	50	**0.22**
Off-site income work	3	9	0.32

Odds ratios from logistic regression analysis were adjusted for income level, family size, and education variables. Significant odds ratios are shown in bold font "impact" might also be thought of as mitigation responses to reduce the impacts of floods (see next section). Men were more likely to spend extra time in the recovery of rice stock than women (O.R. = 0.22, $P<0.001$); whereas women were more likely to spend extra time on preparing utilities (O.R. = 4.2, $P<0.01$). For all other activities – agriculture, livestock, fishing, and off-site work, there were no gender differences (Table 14.4). Households requiring additional time to recover agriculture (O.R. = 2.8, $P<0.01$) or prepare utilities (O.R. = 2.1, $P<0.05$) were more likely to be better-off. Households requiring more time for fuel collection were likely to be poorer (O.R. = 0.06, $P<0.001$).

14.3.4 Household Flood Disaster Risk Management

Four types of measures for reducing flood-disaster risks were considered (Table 14.5). Associations between these measures and several household and individual characteristics were explored using logistic regression. Preventive measures were significantly more likely in better-off than poorer households (O.R. = 7.3, $P<0.001$) but not associated with differences in gender, family size, or achievement of formal education by men or women. We looked in more detail at the roles of men and women in a set of specific, prevention-related activities prior to floods. For most tasks such as preparedness planning, mitigation planning, impact assessment, social mapping, community meetings, public campaigns, and voluntary work there were no gender differences. With respect to capacity building, women, however, were less likely to be involved than men (O.R. = 0.28, $P<0.01$). Information sharing was less likely for women (O.R. = 0.42, $P<0.05$) and more likely in better-off households (O.R. = 2.4, $P<0.05$).

The facilities and infrastructure available in the community for rescue and emergency responses were limited for both types of households in the village (Table 14.5). They included a small rescue boat provided by CARE International and two areas of higher ground where people retreated in the high floods of 2000–2002. The local response strategy mainly involved the use of sand bags, but these were only effective in low and moderate floods. There was no difference in rescue and emergency response measures by income level (Table 14.5), gender, family size, or achievement of formal education by men or women.

Table 14.5 Indicators of performance in flood-disaster risk management in Angkor Ang village-measures taken before, during, and after floods by individuals and households

Performance indicators	Poorer households (%)	Better-off households (%)
Flood disaster risk reduction		
Preventive measures	20	63
Rescue and emergency responses	9	13
Rehabilitation and reconstruction of public infrastructure	11	20
Flood impact mitigation	23	37

We looked in more detail at the roles of men and women in a set of specific coping and emergency-related activities during floods. Women had larger roles in collecting valuable property and taking care of family members (Table 14.6). Men had larger roles in helping flood victims and getting new information. A few other associations were significant. Women with formal education were more likely to be engaged in collecting valuable property (O.R. = 3.5, $P<0.01$). Women taking on the roles of tending to family members and securing daily food was less likely in better-off households (O.R. = 0.01 and O.R. = 0.13, $P<0.001$, respectively) presumably because these households tended to be much less impacted by flood events.

Since the government is unable to afford the costs for the rehabilitation and reconstruction of public infrastructures damaged in floods, local authorities have solicited contributions from the local residents and NGOs. Better-off households have more resources to contribute financially or in the form of materials. Poor households pay more with their labor and time to disaster management. Overall, the contributions of poorer and better-off households were not significantly different (Table 14.5). Local support for rehabilitation work prioritizes repairs to schools and roads and important public infrastructure in the period before external support arrives. There was no difference in rescue and emergency response measures by income class (Table 14.5), gender, or achievement of formal education by men or women.

There was no association between likelihood of undertaking flood mitigation measures and income level (Table 14.5), gender, family size, or achievement of formal education by men or women. We also looked in more detail at the roles of men and women for a set of specific mitigation-related activities after floods (Table 14.7). Health care and prevention was more likely to involve women than men, as was the acquiring of loans for flood relief measures. Health care services were less needed by better-off households (O.R. = 0.33, $P<0.01$). Helping others, working in rehabilitation and construction, and collecting property was done more by men (Table 14.7).

Table 14.6 Comparison of the roles of men and women during floods ($n=70$ in each group)

Roles	Women (% of households)	Men (% of households)	Adjusted odds ratio
Collecting valuable property	81	54	**4.6**
Taking care of family members	49	29	**13.3**
Helping other victims	34	80	**0.13**
Taking care of livestock	17	24	0.61
Getting new information	34	64	**0.30**
Sharing information	33	47	0.57
Seeking daily food	47	57	0.58

Odds ratio adjusted for income class, family size, and education variables. Significant odds ratios are shown in bold font

Table 14.7 Comparison of the roles of men and women after the floods

Roles	Women (% of households)	Men (% of households)	Adjusted odds ratio
Health care service	66	24	**7.4**
Support and relief	16	29	0.47
Rehabilitation/construction	36	79	**0.15**
Soliciting loans/funding	59	30	**3.4**
Collecting of property	29	53	**0.35**
Helping others	17	69	**0.09**
Health prevention	66	29	**4.8**

Odds ratio adjusted for income class, family size, and education variables. Significant odds ratios are shown in bold font

Table 14.8 Indicators of performance in flood-disaster risk management in Angkor Ang village: quality of work and services provided by agencies and the levels of local participation

Performance indicators	Poorer households (%)	Better-off households (%)
Work and services		
Government agencies	18	10
Humanitarian aid agencies	11	7
Commune council	34	33
Non-government organizations	20	17
Local public participation		
Awareness of activities	81	76
Involvement in decision making	51	60
Participation in activities	67	61

14.3.5 Agency Services and Public Participation

Poorer and better-off households in Angkor Ang village had similar perceptions of the quality of work and services performed by different types of agencies (Table 14.8). No evidence of association was detected between the organizations/agencies and the candidate predicators of gender, family size, and the formal education of men and women (determined through logistic regression). In discussions, villagers said that external assistance has only been prominent in major floods and when there had been deaths.

Three indicators of local participation were considered in this study: awareness, decision-making, and involvement in activities (Table 14.8). Levels of participation in disaster risk management activities were similar for poorer and better-off households (Table 14.8). No significant associations were detected between the three participation variables and the other candidate predicators (determined used logistic regression); including, gender, family size, and whether or not men and women heads of household had a formal education. It should be noted that almost half the women (47%) and a third of the men (29%) in the study group had no formal education. Awareness and engagement in decision-making were not closely related to education levels, as might be expected.

Awareness was higher than engagement with regard to decision-making. Awareness functioned largely in terms of local experience; knowledge about modern technologies and other measures to reduce risk was more limited. Participation in decision-making in the community, overall, was modest, and rarely extended beyond attending meetings run by external agencies. According to information we gathered at a community meeting with the villagers and the village head, the participation of local villagers in disaster-risk management was dependent on external initiatives, such as activities run by NGOs. For many poor households, participation in meetings run by local NGOs is often driven by incentives, such as allowances for "transport." When there is no incentive, they do not participate.

14.4 Discussion

Seasonal floods are part of normal life for communities living along the Mekong River in Cambodia. In some years, floods are much more severe than others, especially when there is the combination of a high discharge from upstream and a high run-off from rainfall. Actions taken by villagers could help build resilience at the household and community levels to these more serious floods and thus reduce the risks of disasters which cause loss of life and costly property damage.

Based primarily on our findings in Angkor Ang village on the banks of the Mekong River in Cambodia, but with some additional insights from impacts and responses to floods at larger jurisdictional levels, we conclude that the effectiveness of actions by villagers and others to build and maintain flood resilience in the community has been modest. Five groups of factors appear to influence effectiveness and thus should be targeted in future interventions. First, local resources and knowledge may simply be insufficient to deal with extreme and unusual events. There is evidence that the severe "combined" flood in 2000 was beyond the response capacities of the community. Facilities for emergency relief and rescue operations were very limited. Early warning systems at the time were inadequate, and the community was not well prepared for an event of such magnitude. In subsequent years, awareness was much higher, but local initiatives remained hampered by lack of leadership and resources. In these unusually difficult flood situations, cross-level links to the district, provincial, and national levels of support are very important, but they appear to have been under-developed. Committee structures to improve the coordination of disaster management have since been established, but their effectiveness has not yet been tested by another more extreme flood event.

Second, poverty at the household level has important multi-dimensional consequences for vulnerabilities and response capacities. Information sharing before floods, for example, was more likely in better-off households. Poorer households were less likely to take preventative measures because they did not have adequate resources to do so. Poorer household also had homes that were not as well constructed and thus vulnerable to flood water and wind damage. Damage to homes was a major problem for poorer households and reflects the importance of poverty reduction and rural development initiatives, more generally, in reducing burdens and risks from flood-related disasters.

Third, interventions need to take into account existing gender relations and norms. For example, in pre-flood, preventative activities, women were less likely to be involved in capacity building and information-sharing activities. Women are also much more likely to have roles related to health management. Given the prevalence of illness and diseases after floods, this implies that interventions should pay special attention to women's roles and experience as care-givers, while not ignoring the need to draw the attention of men to health issues. Acquiring loans for recovery was another domain where the role of women was prominent and thus important to take into account in building community resilience to flood disasters. Insufficient consideration of gender relations may lead to flood-disaster risk interventions that leave women even more dependent on men than before.

Fourth, meaningful participation of women and men in disaster-risk management needs to be expanded. The most important barrier when initiatives are externally driven is that the benefits of participating in planning and training meetings are often unclear. Villagers who attend meetings are passive, as they feel intimidated by conveners from government and non-government organizations with much higher education levels. They are also skeptical that their inputs and efforts would influence decisions. Some participants also worry about costs, are concerned about the loss of their time, and are thus only interested in attending if there is sufficient compensation for "transport" or other clear incentives. Important barriers when initiatives are more internally initiated and driven include the lack of trust – the latter illustrated by people's fear of leaving homes and property unattended and thus placing themselves at greater risk from rising flood waters.

Fifth, the capacity and quality of engagement in community-based organizations needs to be improved. If villagers could meet in advance to discuss problems and issues faced in the community, collective action would be easier, including when they must deal with external agencies. Currently, community-based organizations struggle to prioritize problems, needs, and issues; and as a consequence are not as effective as they might be in influencing the annual commune-investment-plan development and leveraging support from NGOs and humanitarian agencies. Most issues and solutions remain based in the efforts of individuals and families with little opportunity for collective community action. Many of the problems faced by women, the elderly, and children are left out as well, as men dominate interactions.

In summary, building resilient communities in locations like Angkor Ang requires attention to local capacity and knowledge, differences in income level, gender relations, and local participation. These issues need to be examined in the social processes that take place before, during, and after major floods. Monitoring and evaluation systems are also critical, as they provide opportunities for learning to take place within and among communities at different jurisdictional levels, without having to wait for disaster to strike at home.

References

ADPC (2002) Mitigating flood risk in Cambodian communities empowering communities to manage disaster risk. The Asian Urban Disaster Mitigation Program, The Asian Disaster Preparedness Center, Pathumthani

ADPC (2007) Disaster management policy, legal and institutional arrangement and plans in Cambodia. The Asian Disaster Preparedness Center, Pathumthani

Baran E, Myschowoda C (2009) Dams and fisheries in the Mekong basin. Aquat Ecosyst Health Manage 12(3):227–234

Baran E, Jantunen T, Kieok CC (2007) Values of inland fisheries in the Mekong River basin. World Fish Center, Phnom Penh

Council of Ministers (2000) Flood damage estimates. Royal Government of Cambodia, Nov 2000

DCDM-Peam Chour (2007) Flood disaster preparedness programme. Peam Chour District, Prey Veng Province 2006–2007, Phnom Penh

DCDM-Peam Chour (2009) Flood disaster preparedness programme, Prey Veng Province 2009–2011, Phnom Penh

Few R (2003) Flooding, vulnerability and coping strategies: local responses to a global threat. Prog Dev Stud 3:43–58

Friend RM (2009) Fishing for influence: fisheries science and evidence in water resources development in the Mekong basin. Water Altern 2(2):167–182

Hak M (2005) Cambodia country report for third annual Mekong Flood Forum, Lao PDR, 7–8 Apr 2005

Helmers K, Jegillos S (2002) Linkages between flood and drought disasters; flood impact on women and girls, Prey Veng Province Cambodia. CARE-International, Phnom Penh

Helmers K, Gibson J, Wallgren P (2003). Rural Sources of Income and Livelihood Strategies Study, Final report, Washington DC

Kummu M, Sarkkula J (2008) Impact of the Mekong River flow alteration on the Tonle Sap flood pulse. Ambio 37(3):185–192

MoAFF (2003–2007) Profile of Prey Veng. The Ministry of Agriculture, Forestry and Fisheries, Phnom Penh

MRC (2005) Impact of the 2005 flood in Cambodia. Impact of the 2005 flood – country reports. The Mekong River Commission Secretariat, Phnom Penh

MWH (2010) Department of hydrology and river work. The Ministry of Water and Hydrology. On-line: http://www.dhrw-cam.org/hydrograph.php?action=image&hy_cat_id=&hydroid=72. Accessed Aug 2010

NCDM (2004) Policy document. The National Committee for Disaster Management, Phnom Penh

NCDM (2006) Disaster information management and database. The National Committee For Disaster Management, Phnom Penh

NCDM (2008) The disaster situation and disaster risk management in Cambodia. The National Committee for Disaster Management, Phnom Penh

NCDM (2009). Cambodia country report 2008. The National Committee for Disaster Management, Phnom Penh.

NCDM (2010) Cambodia country report 2009. The National Committee for Disaster Management, Phnom Penh

NIS (2008) General population census of Cambodia 2008. The National Institute of Statistics, The Ministry of Planning, Phnom Penh

Weerasinghe B (2007) 'Critical mass' for community disaster resilience. In: Gunawardene N, Noronha F (eds) Communicating disasters – an Asia Pacific resource book. TVE Asia Pacific/UNDP, New York, pp 112–114

Part IV
Human Responses to Environmental Change

Chapter 15
Community-Based Fish Culture: A Viable Coping Strategy for Farmers in the Mekong Delta?

Olivier M. Joffre, N. Sheriff, H.H. Ngai, and N.V. Hao

Abstract Floodplains are characterized by a period of several months when the land is not available for agriculture and large and open areas are used for fisheries. Enclosures in the flooded areas can be utilized to produce a crop of stocked fish, in addition to naturally occurring self-recruited species. The WorldFish Center and the Research Institute for Aquaculture no2 (RIA 2) tested options for community-based fish culture in floodplain enclosures in the Mekong Delta. The trials yielded fish production in the range of 61–179 kg ha^{-1}. Results indicate that the models tested are sensitive and dependent on flood patterns and limitations imposed by the rice culture calendar. Other technical challenges included a short grow-out period and fingerling size. These initial trials have shown that community-based fish culture is an innovative approach for the Mekong Delta and has the potential to provide an alternative livelihood option in the face of environmental change and development. To increase uptake, the technical design of the approach could be further optimized, and mechanisms for community participation could be enhanced to increase economic incentives for adoption of the technology by farmers.

Keywords Environmental change • Floodplain aquaculture • Mekong Delta

O.M. Joffre (✉)
WorldFish Center, Greater Mekong Regional Office, #35 Street 71, Phnom Penh, Cambodia
e-mail: o.joffre@cgiar.org

N. Sheriff
Policy, Economics and Social Sciences Discipline, WorldFish Center, Jalan Batu Maung, Batu Maung, Bayan Lepas, Penang 11960, Malaysia

H.H. Ngai
National Breeding Center for Southern Freshwater Aquaculture, Research Institute for Aquaculture No. 2, An Thai Trung Commune, Cai Be District, Tien Giang Province, Vietnam

N.V. Hao
Research Institute for Aquaculture No. 2, Ministry of Fisheries, 116 Nguyen Dinh Chieu St., District 1, Ho-Chi-Minh City, Vietnam

M.A. Stewart and P.A. Coclanis (eds.), *Environmental Change and Agricultural Sustainability in the Mekong Delta*, Advances in Global Change Research 45,
DOI 10.1007/978-94-007-0934-8_15, © Springer Science+Business Media B.V. 2011

15.1 The Delta Environment

In the Vietnamese Mekong Delta, an area of 1.2–1.9 million hectares is annually flooded; one million hectares are inundated by floods of more than 1 m. Yet this flooded environment has been exposed to developments that have transformed the floodplains, bringing them under increased control. The intensification of rice culture, including the spread of high-yielding rice varieties (HYV) and the development of flood control, drainage, and irrigation systems in floodplains and Deltas has led to the replacement of deep water rice with two or even three HYV irrigated rice crops in the dry season, followed by a fallow period during the flood (Kakonen 2008). In recent years, the floodplain has provided an important resource for the development of fresh water aquaculture, particularly for the intensive production of pangasids. Between 2000 and 2006, the aquaculture area in the Mekong Delta increased by 39%, and in 2007 aquaculture in the Mekong Delta represented 72% of the national production. However, intensive aquaculture in cages or ponds is not necessarily an option accessible to poor farmers in the floodplain, and the environmental cost of these production systems is significant (GTZ 2005).

The Mekong Delta is characterized by constant change and very high land and water productivity (Pech and Kengo 2008). Climate change is likely to bring more change to this dynamic environment, as recent studies on climate change indicate that the runoff throughout the Mekong Basin is expected to increase by 21% (Eastham et al. 2008), and annual flood volumes are likely to increase with greater peak flows and longer duration of flooding compared to historical conditions. It is estimated that the average area of flooding in the Mekong Delta is likely to increase by an annual average of 3,800 km^2. At the same time, projection scenarios for agricultural productivity and population growth relate a high probability of food scarcity (Eastham et al. 2008). Floodplain aquaculture and more generally agrarian systems will have to evolve in order to cope with these environmental changes and to sustain not only economic growth but also the livelihoods of farmers in the Mekong Delta.

15.2 Community-Based Aquaculture

15.2.1 Design and Approach

Flooded areas are considered to be relatively "unproductive," as valuable agricultural land is submerged for part of the year, creating open access water bodies. Yet these seasonal water bodies provide an opportunity to increase water and land productivity by integrating fish culture in seasonally inundated rice fields. Enhanced water productivity is the basis for the community-based fish culture concept, which has been tested by the WorldFish Center and national research partners in five countries[1] since 2005.

---

[1]Bangladesh, Cambodia, Vietnam, China, Mali.

Community-based fish culture aims to provide an option for households in seasonally-flooded areas to benefit from fish culture, where the costs of individual aquaculture systems are prohibitive. Governed by the same technical design principles as individual aquaculture systems, such as size and stocking density of cultured fish, community-based aquaculture also introduces institutions for managing fish culture on a collective basis. Pilot testing of the approach therefore involved both a technical and an institutional component, with the outcome of fish culture equally dependent on successful community management as it was on the profitable production of fish.

In Bangladesh, decades of experience in community-based fisheries have supported the development of fish culture systems, implemented on a community basis, which are now generating important benefits for both landowners and landless fishers. The success of the approach suggested that other floodplain communities in Asia and Africa may also benefit from increased productivity during the flood season. This has led to the dissemination and testing of community-based fish culture in Cambodia, Vietnam, China, and Mali.

In the Mekong Delta, fingerlings were stocked in flooded rice fields after the rice harvest in August, during the early stage of the flood. Flood control infrastructure in the form of dikes surrounding rice fields provided a boundary for fish culture sites. Participation in the fish culture trials was based on ownership of land within the boundary of the selected project site, with members organized in a group to oversee the fish culture activities and to accomplish basic administration and record keeping. Fish are harvested and sold when the rice fields are drained, at the end of the flood in December. Between 2006 and 2009, nine sites in Can Tho, Dong Thap, An Giang, and Vinh Long provinces were selected to test the fish culture model.[2] Fish culture groups were provided with funds in the first year of fish culture trials to purchase fingerlings and materials for maintaining the site boundary. If fish culture was successful, it was expected that some of the income would be returned to a central fund to support the purchase of fingerlings the following year, encouraging self-sufficiency of the farmer group.

Assistance and technical support were provided by the local Department of Fisheries (DoF) and RIA 2. At some sites, additional financial support for dike maintenance and improvement was provided by the commune.

15.2.2 Site Selection

We present the results and analysis of some of the project sites in Can Tho, Vinh Long, and Dong Thap provinces. The technical design, including the area or fish species stocked and the benefit sharing agreements, is presented in Tables 15.1 and 15.2. Fish culture was implemented in areas where two rice crops per year or

[2] In 2006 in four sites in Dong Thap (one site – Phu Cuong – Tam Nong), An Giang (one site – Vinh Hanh – Chau Thanh) and Can Tho (two sites – D1 Thanh Thang – Vinh Thanh and Thoi Dong – Co Do). In 2007 in four sites (three in Can Tho D1, C2 Thanh Thang – Vinh Thanh, Truong Phu B Co Do) and one in Vinh Long (Tan Hung, Binh Tan) In 2008 in one site in Dong Thap (Truong Xuan Thap Muoi). In 2009 two sites in Dong Thap province (Truong Xuan, Hung Thanh – Thap Muoi) and one site in Can Tho (D1 Thanh Thang – Vinh Thanh).

Table 15.1 Characteristics of the project sites in Vietnam

	D1	C2	Trung Phu B	Hung Binh	Truong Xuan
Province	Can Tho c	Can Tho	Can Tho city	Vinh Long	Dong Thap
District	Vinh Thanh	Vinh Thanh	Co Do	Binh minh	Thap Mouy
Rice cropping	Double	Double	Triple	Triple	Double
Flood period	Aug. to Nov.	Aug. to Nov.	Sept. to Nov.	Sept. to Nov.	Aug. to Nov.
Max water level in 2007 or 2008	102 cm, 1st week of November	106 cm, 1st week of November	60 cm, 1st week of November	63 cm, 1st week of November	>100 cm in November (2008)
Project site area (ha)	65 (2005–2007) 19 (2009)	48	39	26	90 (2008) 120 (2009)
Sharing benefit	Area owned (2006, 2007) Membership (2009)	Area owned	Area owned	Membership	Membership
Households involved in the project	34 (2005, 2006) 30 (2007) 11 (2009)	28	28	5	13 (2008) 7 (2009)

Table 15.2 Production (kg), yield (kg ha^{-1}) and economic results of the community based fish culture (1 USD = 17,429 vnd in 2008; 1 USD = 16,000 vnd in 2005, 2006, and 2007)

	Year of culture	Cultured fish production (kg)	Cultured fish yield (kg ha^{-1})	Wild fish yield (kg ha^{-1})	Operational cost (USD ha^{-1})	Net return (USD ha^{-1})
Vietnam						
D1	2006	11,271	173	m.d	74	–4
D1	2007	8,052	124	38	39	41
D1	2009	3,403	179	31	21	–9
C2	2007	5,511	114	10	65	13
Trung Phu B	2007	4,935	126	10	21	16
Hung Binh	2007	2,191	84	8	50	–23
Truong Xuan	2008	5,900	61	6 + 31[a]	31	–6
Truong Xuan	2009	10,822	90	9 + 12[b]	41	–8

m.d missing data

[a] 6 kg ha^{-1} harvested by the group and 31 kg ha^{-1} estimated harvest by landowners when the water level was lower than rice field's dike

[b] 9 kg ha^{-1} harvested by the group and 12 kg ha^{-1} estimated harvest by landowners when the water level was lower than rice field's dike

three rice crops per year were produced. In the latter case, the water level and flood period are shorter, with two sites with a flood depth of less than 100 cm and rice fields flooded later in the year – in September rather than August.

Regulations governing access to the water body during the fish-culture period, duties of each member of the group, and benefit sharing arrangements were discussed during a general meeting facilitated by RIA no 2 at the start of the trial. In all sites, fishing in the fish-culture area during fish grow-out was prohibited, for members and for non-members.

The model is based on extensive fish culture of species suited for floodplain aquaculture, such as common carp (*Cyprinus carpio*), bighead carp (*Hypophthalmichthys nobilis*), and silver carp (*Hypophthalmichthys molitrix*). In a few cases grass carp (*Ctenopharyngodon idella*) was also stocked. Only in one site, in Can Tho province, high value species such as snakehead (*Channa striata*) and red tilapia (*Oreochromis sp)* were nursed and stocked. Production was based on the availability of natural food in the water body.

In addition, a series of surveys was implemented in some of the project sites to understand the constraining and enabling factors for collective aquaculture in floodplains. The survey was based on semi-structured interviews with project beneficiaries and non-beneficiaries to understand the process of project implementation, including technical and economic aspects, but also incentives and constraints faced by project members. In each site more than 50% of the beneficiaries and at least ten households not involved in the project were interviewed. In total, 130 households were interviewed (67 households involved in the project and 63 households not involved).

15.3 Results and Discussion

15.3.1 Benefit from Community-Based Fish Culture

Fish culture on a community basis in the Mekong Delta produced varying results in terms of production, income, and subsequent adoption of the technology by farmers. Compared with non-fish culture areas, the fish production of flooded rice fields can be increased. Additional benefits for rice field ecology and the impact on rice culture can be observed. The culture of fish in rice fields could reduce the cost of soil preparation that uses inputs, as fewer pesticides, insecticides, and less fertilizer and rice seeds may be needed.

A straightforward comparison of production across sites is complicated by the variations in size and characteristics of project sites and the constraints that affected production at each location. Fish production from fish stocking ranged from 61 to 179 kg ha^{-1}, with an associated net benefit of 50 USD ha^{-1} (Table 15.2).

However, while these figures are comparable to outputs in some locations in Bangladesh, the perception of success differed amongst project participants in the two countries, with farmers in the Mekong Delta showing a preference for alternative livelihood options available during the flood season that yield a higher income.

15.3.2 *Supporting and Constraining Factors*

As the technical and institutional design components of the project have been introduced to each of the project countries, mechanisms have been needed to permit flexibility in the research approach and adapt the technology to local conditions. The project concept was based on the principles of adaptive management, an iterative process that encourages the review and adaptation of activities after each harvest cycle. However, at a number of sites, participants chose to discontinue fish culture after one culture cycle.

Several technical, economic, and social factors explain these results, including the technical design of the fish culture system, flood characteristics, the dike system, the integration of fish culture within rice-based production systems, as well as marketing aspects and social characteristics of the fish culture group.

15.3.3 *Technical Factors*

15.3.3.1 Intensive vs. Extensive Culture

On this scale, it was found that extensive culture, based on naturally available feed, was more appropriate than semi-intensive culture using commercial, pelleted feed. The provision of feed increased the operational cost of production to 74 USD ha^{-1} compared with extensive culture of lower value species where the operational cost was 21 USD ha^{-1}. Moreover, production results were not high enough to cover the cost of nursing, with a low survival rate of 10% and 1% for red tilapia and snakehead, respectively. The survival rate was between 13% and 33% on average in the different sites for other species, with an average fingerling size of around 6 g per fish.

15.3.3.2 Stocking Size

Fish stocking is particularly important in these extensive, relatively open systems, where self-recruiting wild fish are also present. Selecting the correct size of fish for stocking is dependent upon a critical balance between cost and survival. Small fingerlings may reduce costs, but predation by wild fish increases mortality. Larger fingerlings may shower greater survival rates, but at higher overall costs. The research showed that fingerlings larger than 6 g should be selected for stocking.

15.3.3.3 Environmental Factors

Flood duration, height, and flood delay were important factors in determining production. With a production system based only on natural water productivity, water levels need to be sufficiently high to create suitable conditions for fish culture.

In Can Tho and Vinh Long provinces, the water level reached only 60 and 63 cm in 2007, limiting fish growth. Generally, fish culture requires a water temperature between 25° and 35°, which can be excessive if the water level is too shallow (Halwart and Gupta 2004). In addition, pH needs to range from 6.5 to 9.0, which can be a constraint in the Mekong Delta, where acid sulfate soil can generate acidity. Therefore, water level is an important factor for site selection as well as for potential water pH changes.

Floods arrived later than expected, delaying fish stocking and thus reducing the grow-out period. This development highlights the dependency of the model on fluctuating environmental conditions and its integration in a broader production system in both space and time.

Areas supporting double rice culture are more suitable for the development of community-based fish culture when compared to triple rice crop areas. The size of fingerlings should also be adapted to a short grow-out period to reach marketable size in less than 12 weeks of growth. Fish culture in seasonally inundated rice fields depends on the rice culture calendar. Fish culture can begin in nurseries, while rice is still cultivated, but in most cases fish were stocked after the rice harvest as water levels began to rise. Fish harvesting is timed according to flood recession, which indicates the start of rice field drainage and of rice culture. Those project members and non-members who had land inside the fish culture areas prioritized their rice culture activities and were willing to harvest the fish as early as possible in order to start the next rice crop. Rice culture after the flood period was never delayed to allow for a longer fish grow-out period, indicating the preference and value of rice culture over fish culture. The coordination of the rice cropping calendar necessitates the drainage of the entire cultivated area surrounded by dikes. As a result, fish culture cannot be extended and fish must be harvested. In cases where not all the landowners are involved in the fish culture group, water management for rice culture can generate conflicts. Rice culture is of high economic importance for farmers and of national interest, particularly in the light of the 2007–2008 food crises and the high price of rice. Rice production is therefore a priority in these areas, and fish culture is considered a secondary activity. On average, the net return from dry season irrigated rice (December to April) can generate 860–1,612 USD ha^{-1} ($n=28$), according to soil type, rice variety, and market price.

15.3.4 Socio-Economic Factors

An optimum market environment is required to maximize the benefits from fish-culture production. The potential benefits available from fish production in the flooded area were undermined by the effects of market supply and demand and by fluctuations in fish market prices. When asked, marketing of the harvested fish was ranked as one the main challenges of the model by 26%, 33%, and 18% of the respondents participating in the project in three hamlets in Can Tho province. The period of harvesting in November and early December coincided with the bulk of the wild fish harvest and was when the fish supply is most abundant. This lowered fish prices on the regional market; for example, common carp (*Cyprinus carpio*) brought 1.14 USD ha^{-1} in April but only 0.34 USD ha^{-1} in November. Harvesting fish when the supply is lower could significantly improve the value of fish production. However, this would require that the fish are held until the optimum time for harvest; and with land at a premium, the conversion of valuable agricultural land to a holding pond was not a feasible option. Introducing a wider range of stakeholders in the group, including landholders or commercial fish producers with holding facilities for large volumes of fish, may be one option if issues of power and elite capture could be controlled.

Participants considered that fish culture demanded too much of their time in the form of guarding duties, meetings, and harvesting. Comparatively, the different types of off-farm and non-farm labor can provide higher incomes during the flood period, with farmers engaging in a range of activities according to opportunities in fishing and rice post-harvest processing or construction work (Table 15.3). For villagers, the project was perceived as a new technique, without a guarantee of results and with potential benefits earned only at the end of the flood season, while other activities provided daily income for household needs. This is a common issue in aquaculture and agriculture systems. Households rely upon a range of activities that meet different household needs. Activities that provide a daily source of income are critical to most households who lack savings. These activities provide the greatest contribution to the households when they complement one another within the household portfolio.

The operational costs of fish culture also varied according to the rules established by each fish culture group. At two sites in Can Tho province, labor to guard and harvest were paid, with costs accounting for 10–20% of the total cost of production, while in other groups these activities were undertaken by group members at no cost.

Table 15.3 Salaries and incomes from different non-farm and off-farm activities during the fish-culture period (1 USD=17,429 Vietnamese Dong) (Results presented are based on a semi-structured interview of 130 in Vietnam (project and non-project households))

Activities	Salary or Net return (USD)	Average days hired/fishing	Seasonal income (USD)
Hired labor in rice field	2.3–4.6 USD day^{-1}	17.5	40–80
Hired labor in forest sector	2 USD day^{-1}	15–60	30–120
Digging	4 USD day^{-1}	10–30	40–120
Rice post harvest	2.3–2.8 USD day^{-1}	12.5–17.5	36–50
Construction work	3.4 USD day^{-1}	Variable	Not estimated
Fishing	0.3–5.7 USD day^{-1}	Variable	Not estimated
Renting out land for duck raising	5.7–28 USD $month^{-1}$	30	5.7–28
Duck raising (100 heads)	86–143 USD	–	86–143
Lotus culture	0–1,147 USD	–	0–1,147

The indirect benefit of fish culture on the operational cost of rice culture in this project needs to be investigated further. Other studies have found no negative effects of fish on rice production when cultured concurrently (Vromant et al. 2002; Sinh 1995), with a generally positive impact on farm income (Berg 2002; Duong et al. 1998; Gupta et al. 1998; Mai et al. 1992). Such benefits can be considered a valuable contribution of this model to a long-term environmental protection strategy.

15.4 Conclusions

15.4.1 Is Community-Based Aquaculture a Viable Option for Coping with Change in the Mekong Delta?

Community-based fish culture is a new and innovative technology for the Mekong Delta. Unlike Bangladesh, where community-based fisheries management has a long history, the community-based fish culture project in Vietnam represents the first attempt to encourage farmers to pool their resources and work collectively for fish production in flooded areas. As such, the results of the trials should be considered within this context, as an approach with a potential for optimization under the appropriate environmental, social, and economic conditions.

As the Mekong Delta undergoes rapid change, farmers will be required to adapt to changing circumstances that are both man-made and natural in origin. A policy of expanding commercial aquaculture and encouraging farmers and fishers to adopt alternative livelihoods, including employment on aquaculture farms, could make community-based aquaculture a more attractive option for farmers if it means they are able to retain access to land and livelihood. Following improvements in the technical design, this low-cost, community-based fish culture model and the associated benefits to rural farmers in terms of food security and increased income may gain the support of national and local authorities seeking to improve the well-being of the rural poor, as well as increase solidarity in communities. Furthermore, in our case extensive culture does not require inputs other than fingerlings. The environmental costs of extensive fish culture are lower than intensive, export-oriented aquaculture based on high inputs. This later type of aquaculture contributes to water pollution with the use of commercial feed and chemical inputs. Financial investment and the operational costs of community-based fish culture are minimal compared to more intensive fish culture, which in the case of catfish culture in the Mekong Delta, for example, has an average cost estimated to be higher than 1,000 USD ha^{-1} (Phuong et al. 2007).

Considering the potential environmental changes in the region, with a longer flood period, higher water levels, and declining fisheries (Eastham et al. 2008), fish culture in seasonally inundated rice field is an interesting option for development. The increased height and duration of flooding in areas that currently produce three rice crops per year may reduce production to two crops per year. Under these circumstances, community-based fish culture could provide a suitable, alternative use of inundated lands and a coping strategy for farmers as they adjust to a new environment. Unlike individual aquaculture systems, where decision-making powers regarding technical design and preferences for participation rest with one individual or household, community-based fish culture requires that both the technical design and the institutional approach for managing the fish culture system and sharing its benefits are equally successful. Collective action requires a substantial investment of time and sufficient social capital amongst the participants. The benefits of working together must outweigh those of the next best alternative occupation. Testing both components simultaneously requires sufficient motivation in the community group to overcome production constraints and that participants are able to accept a degree of risk and the possibility of failure in the early stages of testing.

The reluctance of groups to continue the trials for more than 1 year suggests that: (1) sufficient livelihood alternatives which generated greater benefits than fish culture were available to the participants; (2) alternative options delivered livelihood benefits that fish culture could not provide; (3) and that demands on labor were sufficiently high that participants had to choose between fish culture and other options.

Community-based fish culture is a complex and fragile system. Its success depends both on technical and socio-economic factors. Even if several groups in the project discontinue testing of this model, community-based fish culture can provide a viable option for poor farmers in terms of food security and income, particularly within the volatile environmental and economic conditions of the Mekong Delta. The following section present some possible options to adapt collective fish culture to the specific context of the Mekong Delta for a higher adoption rate of this technology.

15.4.2 Recommendations and Lessons Learned

The results show that in the Mekong Delta where flood protection infrastructure to delimit fish culture areas is well developed, extensive fish culture over a large area is possible. However, the model is still dependent on flood patterns; low water levels or delayed floods limit fish growth.

Double rice crop areas are preferable for fish culture development. These allow a longer fish-culture period as they tend to be characterized by floods of more than 1 m and are therefore more suitable to support fish culture. Fish culture between two rice crops requires the involvement and the agreement of all the rice farmers of the area to coordinate water management and extend the fish growth period as long as possible. The growth period, which cannot be extended due to rice culture, will influence the choice of fingerling size. The fingerling size should be determined in order to reduce operational costs by using small sized fingerling, but the size needs to be sufficient to allow a high survival rate and to reach marketable size at harvest time.

Marketing channels and low fish prices during harvest were found to be one of the main challenges limiting the financial benefits of the model. Improvement of marketing channels can be achieved over successive harvests and the delay of the harvest until fish prices are higher in January or even February.

Smaller production units involving fewer farmers could be more effective when compared to large production units where farmers have to market large amounts of fish in a short period. We can hypothesize that small units of less than 10 ha managed by smaller groups and delimited by fences (within a larger unit delimited by dikes and embankments) can be easier to harvest in several stages, accompanied by selective marketing of market-sized fishes. Delayed marketing can be envisaged in cases where farmers own homestead ponds, with the stocking of non-marketable sized fish into a pond pending further growth and higher market prices.

In addition, smaller units of production will reduce group size and the likelihood of conflict. With smaller groups, governance mechanisms to promote transparency, equity, and sharing of labor might be easier to implement. For example, trials in one community in Can Tho province in 2009 support this hypothesis. The group was comprised of 11 members. Fish production of 179 kg ha^{-1} was recorded; and, according to members, a smaller group composed of relatives is easier to manage, with less potential for conflict over divergent ideas on management strategies and technical options.

These initial trials have shown that community-based fish culture is an innovative approach for the Mekong Delta that has the potential to provide an alternative livelihood option in the face of environmental change and development. To increase uptake, the technical design of the approach could be further optimized, and mechanisms for community participation could be enhanced to increase economic incentives, which are an important factor for adoption of the technology by farmers.

References

Berg H (2002) Rice monoculture and integrated rice-fish farming in the Mekong Delta. Vietnam – economic and ecological considerations. Ecol Econ 41:95–107

Duong LT, Nahn DK, Rothius A, Quang PM, Giau TQ, Chi HH, Thuy LT, Hoa NV, Sinh LX (1998) Integrated rice-fish culture in the Mekong Delta of Vietnam: problems, constraints and opportunities for sustainable agriculture. In: Xuan V-T, Matsui S (eds) Development of farming systems in the Mekong Delta of Vietnam. Japan International Research Center for Agricultural Sciences, Can Tho University & Cuu Long Delta Rice Research Institute. Ho Chi Minh Publishing House, Ho Chi Minh City, pp 235–271

Eastham J, Mpelasoka F, Mainuddin M, Ticehurst C, Dyce P, Hodgson G, Ali R, Kirby M (2008) Mekong River Basin water resources assessment: impacts of climate change. Water for a Healthy Country National Research Flagship report CSIRO

GTZ (2005) Pangasius farming practices in Vietnam – a study in view of sustainability issues. GTZ, Vietnam

Gupta MV, Sollows JD, Abdul Mazid M, Rahman A, Hussain MG, Dey MM (1998) Integrating aquaculture with rice farming in Bangladesh: feasibility and economic viability, its adoption and impact. ICLARM Technical Report 55, p 90

Halwart M, Gupta MV (eds) (2004) Culture of fish in rice fields. FAO, WorldFish Center, Rome, 83 p

Kakonen M (2008) Mekong Delta at the crossroads: more control or adaptation? Ambio 37(3):205–212

Mai TD, Le TD, Dang KS, Pham NM Nguyen ND (1992) Rice field aquaculture systems in the Mekong Delta, Vietnam: potential and reality. In: dela Cruz CR, Lightfoot C, Costa-Pierce BA, Carangal VR, Bimbao MP (eds) Rice-fish research and development in Asia, Conference Proceedings 24, ICLARM, Manila, 457 p

Pech S, Kengo S (2008) Population growth and natural resources pressures in the Mekong River basin. Ambio 37(3):219–224

Phuong NT, Sinh LX, Thinh NQ, Chau HH, Anh CT, Hau NM (2007) Economics of small-scale aquaculture feeding systems: the economics of different Catfish feeding practices in Vietnam. Rome, FAO Fisheries Technical Paper No. 505

Sinh LX (1995) The effects of aquaculture on farm household economy: a case study of O Mon district, Can Tho Province, Vietnam. Master thesis, Asian Institute of Technology, Bangkok, p 162

Vromant N, Duong LT, Ollevier F (2002) Effect of fish on the yield and yield components of rice in integrated concurrent rice-fish systems. J Agric Sci 138:63–71

Chapter 16
From Rice to Shrimp: Ecological Change and Human Adaptation in the Mekong Delta of Vietnam

Ngo Thi Phuong Lan

Abstract In the last few decades, structural transformation in agriculture has been considered a key to rural economic development in Vietnam. In the Mekong Delta, the most important rice basket of Vietnam, farmers have diversified their economic activities and engaged in production which involves high risks, partly due to global market price fluctuations. This change in livelihood patterns has resulted in a significant change in the Delta's ecology.

This chapter analyzes socio-economic changes of human adaptation to a new living environment in two shrimp-farming communities in the Mekong Delta, one in the lower part (Ca Mau Province) and the other in the upper part of the Delta (Long An Province). It examines the changes in livelihood and local ecology when farmers shift from conventional rice cultivation to high-value shrimp-farming and thus, from fresh water to saline water ecology. The chapter concludes that the human environment is a social process in which people constantly shape their landscape and have to adapt to the "created" environment by changing their socioeconomic lives. In the shift from rice to shrimp, when people cannot adapt to ecological changes locally, labor migration is the best solution for their livelihood. This phenomenon can be seen as an indicator of agricultural unsustainability. My comparative study seeks to contribute to the understanding of socio-economic changes from an environmental perspective.

Keywords Ecological change • Structural transformation in agriculture • Gender division of labor • Human adaptation • Shrimp farming in Vietnam

N.T.P. Lan (✉)
Department of Anthropology, University of Social Sciences and Humanities, Vietnam National University – Ho Chi Minh City, 10–12 Dinh Tien Hoang Street, District 1, Ho Chi Minh City, Vietnam
e-mail: ngophuonglan@yahoo.com

M.A. Stewart and P.A. Coclanis (eds.), *Environmental Change and Agricultural Sustainability in the Mekong Delta*, Advances in Global Change Research 45,
DOI 10.1007/978-94-007-0934-8_16,

16.1 Introduction

As a fertile Delta that is constantly raised by the silt from the Mekong River, the Mekong Delta has the greatest potential for agricultural production in Vietnam. As part of the process of increasing integration into the world market, this Delta has undergone significant changes in agricultural production. With various support policies, trade liberalization, and technological advancement, the Delta's economic activities have shifted to a market system and to commercially high-value agricultural production. Consequently, although the Delta only makes up 12% of the country's area, annually it accounts for 50% of the rice production, 90% of the exported rice, 80% of aquaculture production, and 60% of the exported aquacultural products of the country (Vietnam Economic News 2008). Along with economic growth, the Delta and its inhabitants' livelihood patterns have undergone significant changes in various aspects. In the last few decades, the structural transformation in agriculture has become a prominent phenomenon in the Mekong Delta.

By analyzing the shift from rice cultivation to shrimp farming in the Mekong Delta, this chapter suggests that along with the increasing integration into the world market, the delta ecology has experienced dramatic changes. Delta inhabitants play an important role in shaping their living environment. This ecological change, in turn, shapes human lives. People have to use their social and material resources to adapt to the newly "created" environment.

16.2 Research Methods and Case Study Sites

This chapter presents the initial findings of my research based on qualitative methods, including interviews with shrimp farmers and local authorities at different levels, as well as participant observation during my fieldwork in 2009 and early 2010 at the study sites.

My study examines two shrimp-farming communities, one in Long An Province and the other in Ca Mau Province in Vietnam. These two communities engage in extensive and advanced extensive shrimp culture, two major types of shrimp culture in the Mekong Delta. Extensive shrimp culture (colloquially called traditional extensive culture) is characterized by low capital investment in pond building, pond preparation, and shrimp-raising techniques. Shrimps are allowed to grow naturally, without being fed processed food. Shrimp fry (post larvae) are released at low density, from one to two shrimps/m^2. Shrimps rely completely on natural food. In the whole raising process, farmers only apply such simple pond-preparation measures as *derris elliptica* to kill the various fish competing with shrimps for food or killing shrimp fry, lime powder ($CaCO_3$) to reduce alum, and occasionally chemical fertilizers such as NPK to create algae as a natural food for shrimps in the pond. Having a large amount of land but limited capital, farmers apply these measures only to some extent, not as intensively as suggested by the government's aquaculture experts. In contrast to other types of shrimp culture in which shrimp fry are released and harvested once per crop, in the extensive type, people usually release fry every 1 or 2 months and harvest shrimps every day or every 1 or 2 weeks.

With a much-improved technology over extensive shrimp culture, the advanced extensive type requires considerable investment in both capital and raising methods. Besides applying more intensively the measures characteristic of extensive culture, farmers adopting the advanced extensive type also use processed pellets or self-processed food and vitamins to feed shrimps, as well as various chemicals such as sterilizer and bio-enzymes in pond preparation and shrimp raising. Pond dikes and ditches are prepared annually. In principle, the density of shrimp post-larvae is from three to five shrimps/m^2. However, in Long An, in order to achieve high productivity despite limited land, people release fry with a stock density much higher than the encouraged level, from 10 to 20 shrimps/m^2. As water in this type of shrimp culture is polluted by processed food, it must be changed frequently by using pumps. Besides the two aforementioned types of shrimp culture, a minority of famers also practice industrial shrimp farming which requires a much higher capital investment, normally 40 times higher than the extensive type and 5 times higher than the advanced extensive type per hectare per year.

The two study sites are both saline-affected areas, with 6 months of saline water and 6 months of fresh water every year. Their inhabitants previously practiced one-crop rice cultivation. The shift from rice cultivation, which relies on a fresh-water ecology, to shrimp farming, which relies on a saline-water ecology, leads to a change in farmers' living patterns as a human ecological adaptation. These two cases will present some social aspects of the structural transformation in agriculture in the Mekong Delta.

16.3 Shrimp Production in the Mekong Delta

Vietnamese Government Resolution 09/2000/NQ-CP, on June 15, 2000, provided policy guidelines for economic structure transformation and agricultural product marketing. Economic structure transformation (*chuyển dịch cơ cấu kinh tế*) is the process of choosing the structure, scale, and product types in agricultural production on the basis of local natural resources and market demand in order to achieve high economic and ecological efficiency.

Farmers in the Mekong Delta have a long tradition of market-oriented rice production. When the Sai Gon port was opened on February 22, 1860, rice and pepper were the two major exports from the south of Vietnam. During the period from 1860 to 1945, Vietnam was a well-known rice-exporting country in the region and in the world. Foreign markets for Vietnamese rice during this period included France, Europe, the United States, Indonesia, Singapore, the Philippines, China, and Japan (Nguyen Phan Quang 2004, 21).

Today, the Mekong Delta has shifted from rice mono-agriculture to a diversified economy, and it is gradually moving toward an industry- and service-based economy. Technological advances which provide high-productivity breeds, new fertilizers, and pesticide have played a crucial role in the transformation in the Delta. The structural transformation process in the Mekong Delta has taken place intensively through the shift from rice cultivation to other agricultural products like fruits and to aquaculture. The shift from rice cultivation to shrimp farming has proven to be an effective one. Thanks to the high value of shrimps, this shift has been seen as an appropriate way for farmers to overcome poverty or even to become rich, especially for farmers living in saline-affected land where they could normally cultivate only one low-yield rice crop a year.

Trade liberalization along with various policies in support of aquaculture has accelerated the development of shrimp production in the Delta. Trade liberalization opens a way for Vietnam shrimps to enter the world market. The high demand in the world market is a key element that encourages the delta's farmers to shift to high-value shrimp farming. Policies in favor of aquaculture in general and shrimp farming in particular provide the legal foundation for the development of this sector. The policy adopted in the Decision 773-TTg of December 21, 1994 aimed at building infrastructure for aquaculture through irrigation systems, shrimp nurseries, and large-scale farms, triggering the expansion of aquaculture in the 1994–1999 period. Since 1999, many other policies and programmes have been adopted to boost the country's shrimp exports, particularly the Aquaculture Development Programme for 1999–2010 (Decision 224/1999/TTg of 08/12/1999), Government Resolution No. 09 in 2000 encouraging crop diversification and allowing a conversion of unproductive rice land into aquaculture, and Decision 173/2001/QĐ-TTg of November 6, 2001 promoting further development of aquaculture as the direction for the economic development in the Mekong Delta (Mai et al. 2006, 6). In response to favorable policies and market conditions, shrimp production has become more developed in the Delta. Ca Mau Province, in particular, has converted 150,000 ha of rice land to shrimp ponds. The acreage of shrimp land in this province had reached 257,000 ha by 2008 and accounted for half of the Mekong Delta's shrimp farming area (Nhandan online 2008).

In order to meet increasing market demand for shrimps, farmers make the best of their natural competitive advantage by participating in commercial shrimp production. Unlike rice production in which farmers and their families can consume a part of their own rice crop, shrimps are raised entirely for the market. Farmers who live in mangrove areas mix mangrove logging with shrimp farming (for example, in Nam Can and Dam Doi districts of Ca Mau Province and in An Minh District of Kien Giang Province). Those who previously either cultivated one rice crop per year (e.g., in Long An, Soc Trang, Tra Vinh provinces) or produced salt in saline-affected areas (esp. in Bac Lieu Province) have switched to shrimp mono-farming or rice–shrimp rotation farming. Although encouraged to adopt a mixed mangrove-shrimp model and a rice–shrimp rotation system in which rice is cultivated during the rainy season, and shrimps are raised in the dry season, farmers tended to clear mangrove forests and to adopt shrimp mono-farming due to the high value of shrimps. In Cai Nuoc District of Ca Mau Province, in the 1997–2002 period, the shrimp-raising acreage increased by 8.5 times (Mai et al. 2006, 8); Dam Doi District in 1995–1999 was the first place to "destroy saline-prevention dams," and "to bring saline water illegally to rice fields in order to raise shrimps" (Nhandan online 2008).

At present, profits from shrimp farming are decreasing due to shrimp diseases and declining prices. However, up to November 2008, shrimp export revenues still accounted for the largest share of Vietnam's revenues from aquatic product exports (Aquatic Information Center 2008). This suggests that despite its high risks, shrimp farming is still attractive to Mekong Delta farmers. In the Mekong Delta, shrimps are raised in the coastal provinces, including Long An, Tien Giang, Ben Tre, Kien Giang, Tra Vinh, Soc Trang, Bac Lieu, and Ca Mau. Among them, Ca Mau has the largest shrimp-farming acreage while Tien Giang and Long An have the smallest. The 2000–2002 period was a prosperous one for shrimp production in Vietnam in general and in the Mekong Delta in particular, as shrimp production and farming acreage increased significantly. In Ca Mau, the biggest shrimp-farming province in the Mekong Delta, shrimp land increased 2.5 times, from 90,551 ha in 1999 to 239, 398 ha in 2003 (Mai et al. 2006, 7). The total reached 257,000 ha in 2008 (website of Ca Mau Province 2008). With the significant increase in production, Vietnam has become one of the top five shrimp-exporting countries in the world.

Major export markets for Vietnam shrimps are the United States and Japan, which consume 70–80% of Vietnam's shrimp exports. Other important markets are European countries and other Asian countries such as China, Hong Kong, Taiwan, and Korea (Aquatic Information Center 2008).

The two main types of shrimp culture in the Mekong Delta are "extensive" and "advanced extensive." In the 2003–2004 period, extensive shrimp farming was adopted on 68% of the shrimp-farming land in the Delta, advanced shrimp culture on 27%, and intensive and semi-intensive on 5% of the land (Nguyen et al. 2004, 5). This structure has remained essentially the same since then; intensive and semi-intensive shrimp culture, adopted on 46,257 ha, makes up only 8% of the shrimp farming acreage. Extensive and advanced extensive shrimp cultures are still dominant. In Ca Mau and Bac Lieu provinces, these two types of shrimp cultures have been adopted on over 90% of the shrimp-farming acreage (Ca Mau People's Committee 2008; General Statistics Office 2006, 450). In my 2009 and early 2010 study in Hoa My, Cai Nuoc District, and Ca Mau Province, 91% of the studied shrimp-farming households had adopted extensive shrimp culture; while in Tan Chanh, the biggest shrimp-producing commune in Can Duoc District of Long An Province, advanced extensive culture was adopted by 98% of the households studied.

Spontaneousness is a common feature in the initial period of shrimp production. Gradually, shrimp culture practice has officially been introduced to saline-affected areas in the Mekong Delta in order to help local people to improve their livelihoods. The government, through its support policies, has played an important role in the shift to shrimp production. These policies sanction the adoption of shrimp farming in traditional rice cultivation areas. The 2000–2002 period witnessed a strong development of shrimp production and remarkable ecological and social changes in the Mekong Delta.

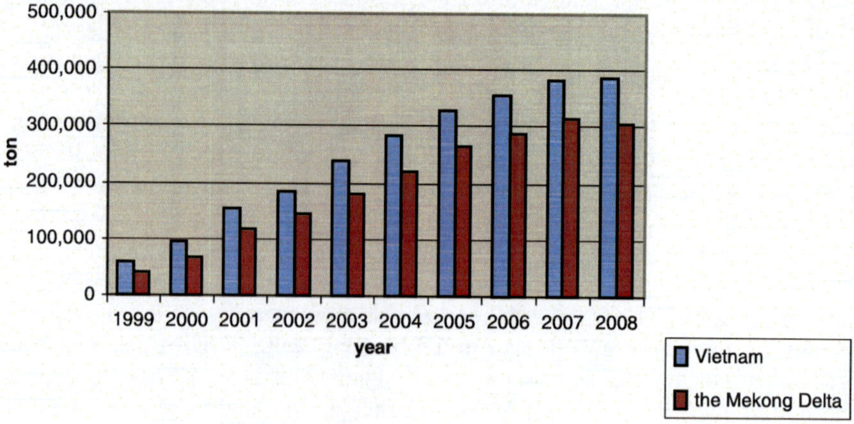

Fig. 16.1 The productivity of shrimps raised in the Mekong delta and the whole country (Source: General Statistics Office 2006, 2007, 2008)

In Tan Chanh Commune of Can Duoc District, Long An Province, shrimp farming was officially introduced to farmers as an alternative livelihood pattern in 1994. However, farmers were initially reluctant to adopt this new practice. They were doubtful of its success because they had no knowledge regarding how to raise shrimps. However, remarkable profits from the first shrimp crops of the few brave shrimp farmers led to an increasing conversion of rice fields to shrimp farms. At present, shrimp farming is the main agricultural activity in Tan Chanh Commune. Local shrimp farmers practice advanced extensive shrimp culture, which requires feeding shrimps with processed food. In Hoa My of Cai Nuoc, Ca Mau, there was no shrimp farming until 2001. Hoa My used to be a fresh-water area thanks to saline prevention dams. Subsequently, with the approval of the government, all rice fields and gardens were permitted to be converted to shrimp ponds. Influenced by the striking success of neighboring shrimp-farming areas, especially in Dam Doi and Nam Can districts, local farmers were eager to shift to shrimp farming at the same time. They applied extensive shrimp culture, which does not involve feeding shrimps.

As a result of the locally specific implementation of national policies, each locality has its own trajectory of transformation from rice to shrimp farming. In Long An, the shift has been gradual, while in Ca Mau, it happened rapidly. People raise mainly Black-tiger shrimps (*Penaeus Monodon*), although White-leg shrimps (*Penaeus Vannamei*) have been raised recently. The latter type of shrimps was introduced into Central Vietnam in 2000 and permitted to be raised in the Mekong Delta only in 2008. White-leg shrimps, with their lower prices, are favored in the market. Processed food from White-leg shrimps account for two thirds of the global market share (www.khuyenongvn.gov.vn, 23/2/2010). Although maturing more quickly and having a higher yield, White-legged shrimps usually cause extreme pollution to the ponds due to their higher waste release. Therefore, people often combine these two types of shrimp according to the specific conditions of their ponds. In my study site in Ca Mau, people raise mainly Black tiger shrimps; while in Long An, farmers raise both types of shrimp.

16.4 Ecological Change and the Locals' Socio-Economic Adaptations

16.4.1 Ecological Change

In Hoa My of Cai Nuoc and in Tan Chanh of Can Duoc, inhabitants' past living patterns involved rice cultivation during the rainy season and thus, mainly fresh-water ecology. Normally farmers grew local varieties of rice in a growing season of 5–6 months because these varieties were suitable to the low terrain and saline-affected land.

Living at the intersection of saline and fresh water, local inhabitants tried to build dams to prevent saline intrusion in the hope of having fresh-water ecological conditions to cultivate rice for the whole year. Due to the nature of each locality, such efforts produced different results.

In Tan Chanh, after the liberation of South Vietnam in 1975, the government tried its best to build dams to prevent saline water intrusion during the dry season and to allow rice cultivation all year around. However, this effort failed because, with the existence of dams, the soil contained more alum and there was insufficient fresh water for rice cultivation. Due to the uneven terrain, when the planting time came, all fields could not be supplied with enough fresh water through small water-inlet sluices. Consequently, farmers tried to get water for their rice fields in their own ways and damaged dams in the process.[1]

In Hoa My, dams were built when the first settlers came to reclaim the land. It was reported that, annually, when the dry season came, people were mobilized to reinforce dams to prevent the intrusion of saline water. In the rainy season, these dams were de-activated so that fresh water could intrude further inland for rice cultivation. After 1975, solid dams were built with permanent water-inlet sluices. With these saline-water-control dams and thanks to their abundant land, local inhabitants could cultivate rice and fruit trees and raise pigs and fish safely (without worrying about saline intrusion destroying the crops).

In Tan Chanh of Long An, most people just converted their rice fields into shrimp ponds. Some also enlarged the ponds by converting some of their limited garden and residential land. However, due to limited land, this type of conversion is not common in Tan Chanh. While in Hoa My, due to their large amount of land, people converted both rice fields and gardens into shrimp ponds. They hired contractors with excavation machines to do the conversions.

[1] Interviews with local authorities in 2009.

With the shift from rice to shrimps, local ecologies have experienced changes which require adaptation by their inhabitants. The local landscapes are no longer green rice fields but treeless shrimp ponds. Farmers in Hoa My describe the effects on their environment as well as their lives as follows:

> letting saline water into fields and ponds had major consequences. Gardens were damaged; vegetables and fruit trees died out…. In the past, during the dry season when we did not have enough water, we could rely on fresh water ponds. At present, we have no way to get fresh water for daily use. At present, we use well and rain water. When shifting to shrimp farming, we had to have drilled wells. When we let saline water into the fields, rice plants did not have grains. In the first few years, land was acidified and salinized. Since 2006, the environment has been getting better…
>
> In the past, there was no *rạm* (small crab living in brackish water). Since the day of transformation (the day farmers start to raise shrimps), there have been a lot of *rạm* which dramatically destroy rice plants. Previously, we used to bathe in canals. At present, we do not dare. Saline water also makes people "stunted"...
>
> Man, aged 74, Thi Tuong Hamlet
>
> Water inlet sluices and dams were destroyed in order to bring saline water for us to raise shrimps; however, the land was salinized… very affected. At first, all trees died out because of salinity. Previously, we had all kinds of vegetables. They are rare now. Two or three years ago, we raised dikes higher so that we could plant coconuts and bananas. In the past, fresh water fish were abundant. We could get tens of kilograms of them a day. It does not happen now.
>
> Man, aged 53, Thi Tuong Hamlet

Large-scale shrimp farming leads to water pollution in communities which, in turn, affects their production, as a farmer illustrates:

> At present, we have to take good care of water, shrimps, and their feeding. In the past, we did not. One of my neighbors bought small, raw shrimps to feed shrimps. He let food rot nastily on dikes, but he got a very good harvest. Shrimps were not susceptible to any disease. In the past, land was new and fertile which was good for shrimps. At present, land is polluted. Therefore we always suffer losses.
>
> Man, aged 65, Dinh Hamlet

This ecological change is a social process in which human beings play an active role in shaping their environment for the sake of their livelihoods. However, this is a multifaceted process. Shifting from a conventional pattern of living to a new one, humans have to adapt to the newly-created environment by changing their ways of life.

16.4.2 Changes in Living Patterns as a Local Adaptation to Environmental Changes

Tan Chanh inhabitants, due to their limited land and low earnings from one rice crop, have long engaged in other economic activities besides agricultural production for their livelihoods. This is possible because of the proximity of Tan Chanh to Ho Chi Minh City, the busiest city in the south of Vietnam, and its location as a bridge between the Mekong Delta and the eastern part[2] of southern Vietnam. Their common additional economic activities included trading by boat and labor migration to other agricultural areas. In the past, local people were famous for trading chinaware, earthen jars, pigs, household products, and firewood from the southeast to the Mekong Delta of Vietnam and vice versa.

In the customary gender division of labor, men did the heavy labor and served as bread-winners while women were oriented to family care. Both men and women engaged in rice cultivation because this kind of work requires much labor. After harvest time, men engaged in trading or went elsewhere to search for work. In the boat-trading business, each trip could last months. Their wives could accompany them, leaving the elderly and children at home. Local people did not do any agricultural activities during the six-month season of saline water. Other inhabitants, who could not afford trading boats, migrated to other places to find seasonal jobs. They could work as manual laborers (làm mướn), masons, and carpenters. Labor migration has long been a common phenomenon after harvest time in Khanh Hau, another rice-farming community in Long An (Hendry 1964, 135–137).

When people shifted to shrimp farming, although there was no difference in terms of seasonal rhythm with half a year of fresh water and half a year of saline water, local inhabitants' living patterns experienced an important change. They previously relied on fresh water for their agricultural production, and at present, on saline water. Saline water is recognized as a precious resource. At the beginning of the shift, the model of rice–shrimp rotation was introduced and encouraged as a sustainable system. Therefore, the structure of shrimp ponds included three components: rice field, ditch, and surrounding dikes. Rice fields were located at the center of the pond, occupying about 50% of the pond, where farmers were supposed to grow rice during the wet season. The ditch made up about 30% of the pond's area and encompassed the rice fields. Shrimps concentrated on the ditch bottom during daytime to avoid high temperatures. Surrounding dikes made up about 20% of the fields. Sludge and accretive silt from the ditch were moved to the dike surface every year. However, with the striking success of the first shrimp crops, people tended to neglect rice farming. They often had two shrimp crops per year instead of the shrimp–rice rotation, or they let the field vacant because rice yield was no longer considered as important as shrimps. Moreover, as many farmers did not grow rice, it was very difficult for other farmers to grow rice because of such crop destroyers as insects, birds, and rodents. Gradually, having limited land and wishing to increase shrimp production, farmers have removed the rice fields in their ponds to make the ponds deeper so that they can raise more shrimps. Consequently, long-existing rice fields have been totally converted to deep shrimp ponds. Local livelihoods became totally dependent on saline-water ecology. Local people raise shrimps in the dry season, and their leisure time is in the rainy season.

[2]The eastern part of southern Vietnam (miền Đông Nam bộ) is the part of southern Vietnam which is north, northeast, and northwest of Hồ Chí Minh city. It is also referred to as the southeast of Vietnam in this paper.

In rice cultivation, both men and women engage in agricultural work. In shrimp production, men provide the key labor because this kind of work involves much labor in water, complicated technology, and manual work such as dredging the ditch, operating pumps, and watching and feeding shrimps during the night. When shrimps are 1-month old, men sleep in temporary huts close to the ponds to watch and feed the shrimps at night. Men are usually considered the pillar of the family. Therefore, they are the persons who maintain the information and knowledge of shrimp production. However, due to the ecological change with a conversion from rice fields to shrimp ponds and repeated failures in shrimp production, along with the increase in the number of factories in the locality as well as in neighboring districts and cities, young men and women have decided to become workers in order to get stable monthly salaries instead of unpredictable and unstable incomes from shrimp production. Women usually become workers at garment and shoe factories, while the growing ship-construction yards in Tan Chanh attract many local men. They leave shrimp farming to their wives or parents. In 2009, in a shrimp-raising training course in Dinh Hamlet, Tan Chanh Commune, one witnessed a shift in labor division by gender. Because men went out to work, the majority of the course participants were women. Due to this change, women began engaging actively in shrimp production, including selecting shrimp fry, feeding, and watching shrimps. For other heavy tasks, they hired other people. However, the labor migration of men causes a shortage of labor in agricultural production. This leads to an increase in agricultural costs.

Although the remarkable profits of the first shrimp crops did help people to obtain larger incomes than ever had when cultivating rice; at present, due to the high risks in connection with price fluctuation, a changing environment, and shrimp diseases, this kind of production is no longer considered the main source of income for local inhabitants in Tan Chanh. A majority of farmers raise shrimps as one way to try their luck. In the past, due to limited land, rice yield could not meet their consumption for the whole year; therefore, they had to rely on other sources of income such as labor migration and trading. However, farmers could still subsist on fairly stable rice yields in those days. People often stored rice in their houses and only sold rice when they needed cash. Well-off farmers could buy rice at harvest time at low prices, store it, and sell it later at higher prices and for a profit. Other cash income covered their families' other expenditures such as ceremonies, health care, and education. In the shrimp-production period, all household needs, including rice as the staple food, are met mainly through the market.

Thanks to the big profits from shrimps, the infrastructure of the community is being upgraded. In the past, people mainly traveled in waterways in the rainy season and by road in the dry season. At present, people mainly travel by vehicles thanks to many new cement and asphalt roads.

In contrast to that of Tan Chanh, the seasonal cycle in Hoa My changed with the shift from rice to shrimp farming. Previously, local inhabitants had only a few small water canals. At present, many interlacing canals were enlarged in order to let saline water into all fields and to serve shrimp farming. These canals serve as the main transportation waterways for local people. At present, people mainly travel by motor boats. Sixty-nine percent of the households studied have their own motor boats. Those without motor boats can travel by service boats.

To get saline water for shrimp farming, saline water-prevention dams were destroyed to let water run naturally. The local landscape was totally transformed. The local subsistence economy was replaced. In the rice-farming period, local inhabitants had practiced a subsistence economy in which they produced rice, vegetables, fruits, pigs, poultry, and fish for their own consumption. They grew rice in the rainy season and did gardening, raising pigs, fishes, and poultry the whole year around. Moreover, thanks to the fresh-water ecology, people could catch wild fresh-water aquatic varieties, which were abundant in canals at that time. With abundant land, farmers could produce many more agricultural products than they could consume. Therefore, inhabitants could sell the excess volume of these products for cash. Labor migration was not a common phenomenon during that time because local agricultural work needed much labor.

In Hoa My of Ca Mau, at present, people mainly practice extensive shrimp culture and raise shrimps continuously for the whole year. In contrast to the shrimp culture in Tan Chanh, Hoa My farmers release shrimp fry (post larvae) every month or every 2 months and harvest shrimps on a weekly basis. As in Tan Chanh, previously both men and women engaged in rice production. At present, men provide the key labor in shrimp production. Men also maintain the information and knowledge on shrimp farming. Women are mainly involved in selling shrimps. A woman's explanation of her failure in shrimp farming illustrates a common perception of the role of men in shrimp farming:

> My family has no men. I do not have any knowledge of shrimp farming, no knowledge of shrimp fry, environment, and land. Men normally have more knowledge than us (women).
>
> Female shrimp farmer, aged 45, Thi Tuong Hamlet

Due to high alum content in the soil and the lack of technological information, Hoa My shrimp farmers, despite abundant shrimp land, did not get expected results. This is contrary to Tan Chanh where people were successful in raising shrimps from the beginning. In Hoa My, at the beginning, the majority of farmers suffered severe losses. While they invested a lot of money in pond building and shrimp fry (at the beginning, the price of shrimp fry was much higher than at present), they gained nothing from shrimp farming. Moreover, they could not grow rice because of water salinity, while they had no experiences in dealing with this. With this new livelihood, people have neither a stable source of income nor food. In the past, as rice production provided enough food for the whole year thanks to abundant land, the majority of farmers could sell the excess volume of rice to get cash and had other sources of cash such as gardening and fresh-water fish catching and raising. Due to repeated losses in shrimp production since 2001 and the excess of labor in agriculture, young people have chosen to migrate to other places to get jobs. They have become workers in various factories in other provinces such as Binh Duong, Dong Nai, and Ho Chi Minh City at an approximate distance of 400 km from Ca Mau, or have gone to get manual labor jobs in the same or neighboring provinces.

Shrimp farmers are mainly middle-aged and young people who are responsible for staying home to take care of their families.

Due to the abundance of land and scattered settlements, shrimp theft has become common in Hoa My. This phenomenon contributes to the losses suffered by local people, who often suffer from shrimp failures as well. Shrimp theft threatens people's livelihoods. Shrimps are much easier to be stolen than fish and other agricultural products. Local inhabitants have to watch their shrimp fields all night. However, because their fields are too large to be watched vigilantly, shrimps are often stolen. Thiefs often use pesticide to catch shrimps effectively and quickly. Pond owners not only lose their shrimps but also suffer such major environmental damage that it makes it hard for them to raise subsequent crops. This problem contributes significantly to the insecurity in people's material lives.

Thanks to the high value of shrimps, shrimp farming *generally* brings a much higher profit for farmers than rice in saline-water areas. However, along with this high profit, people have to face new kinds of risk which are associated with this new pattern of livelihood. A farmer makes an evaluation of the shift from rice to shrimps as follows:

> Fresh water ecology is very good. Trees are luxuriant. Aquatic varieties are abundant. At present, they have died out. In letting saline water into fields to raise shrimps, people have more difficult lives than in the past. Previously, we had various natural sources of income. Now, we can only rely on shrimps. Raising shrimps, despite its insecure earnings, is better than rice farming in terms of labor intensity and efforts. If we are lucky, we can earn much more money with less labor and effort than in cultivating rice.
>
> Man, aged 74, Thi Tuong Hamlet

In this new livelihood pattern, local inhabitants face an internal conflict in the community. That is a conflict among farmers regarding the use of water for shrimp farming. In Ca Mau, it is a conflict between extensive and intensive shrimp farmers. Although there are regulations for intensive farmers regarding the dredging of their ponds, it is hard to check their compliance. Extensive shrimp farmers often accuse intensive shrimp farmers of lack of compliance, leading to waste water from intensive ponds, polluting extensive ponds as a farmer provides details:

> Water is polluted. It's polluted because we do not have enough money to buy lime powder and fertilizer to treat it. Moreover, waste water and residues from intensive ponds greatly contribute to the pollution. When we get that water, we cannot raise shrimps. Recently, intensive ponds increasingly release waste into canals. Extensive farmers get that water without any treatment like in intensive farming. Their shrimps have consequently died out. Intensive farmers use many "toxic" chemicals to treat water, Shrimp food residues and various chemicals over months are then released to the environment. Other people get that water and their shrimps will certainly die.
>
> Man, aged 60, Thi Tuong Hamlet

In Long An, because people practice advanced shrimp culture, they have to invest a lot in shrimp food. Although there are regulations on draining water; due to individual benefits, people do not follow them. For example, when someone's shrimps die out because of diseases, according to regulations, he has to use chemicals to sterilize the pond before draining water out of the pond. However, people often do not do it because of high costs. If neighboring farmers use this affected water, their shrimps will likely catch diseases. For another example, the waste water from a farmer's shrimp pond can damage his neighbors' ponds, especially those which are farthest from canals. In both communities, the layout of shrimp ponds depends on individual households. Therefore, there are often no separate channels for fresh-water inflow and waste-water outflow. People get water from, and release water into the same canals. This condition is conducive to the outbreak of shrimp epidemics. The following narrative illustrates the situation:

> In raising shrimps, if water is polluted, shrimps will die. If everybody has a good conscience and self-awareness like me, I am sure the situation would be improved. When water in their ponds is dirty, farmers should release the water only when the tide is low so that dirty water is washed far away. However, they wait for high tide to change water. Therefore, waste water from their ponds runs further inland. Those who live downstream from the canals will be affected by released waste water.
>
> Man, aged 70, Dinh Hamlet

16.5 Rice-Shrimp Farming: An Environmentally Friendly and Sustainable Adaptation

Rice–shrimp rotation farming is considered to be more efficient and more sustainable than rice or shrimp monoculture system (Tran Thanh Be 1994, iv; Vuong and Lin 2001, 3). However, at first, the Mekong Delta farmers did not practice this system for the aforementioned reasons. Gradually, with cumulative experiences in shrimp farming and rice growing in saline-affected areas, the great support from local government, and some farmers' successful experiences with rice–shrimp rotation and their greater food security, shrimp farmers nowadays recognize the benefit of a rice–shrimp rotation. People do recognize that rice cultivation improves the environment for shrimps, and shrimp waste fosters paddy growth. In Long An, when the rainy season comes, farmers prepare their fields for growing rice. They use short-cycle rice varieties so that they can harvest rice before the shrimp-raising season. At the beginning of the rainy season, farmers practicing shrimp-rice rotation flush salinity out of their fields, using rain and fresh water from canals, in order to grow rice. Although in Long An, farmers do not raise shrimps in the rainy season; in Ca Mau, farmers do. The latter raise Black tiger shrimps which are familiarized with fresh water (*tôm hầm đất*) in the rice fields. Shrimps live in the ditch while rice grows in the field. Shrimp farmers perceive the advantage of rice–shrimp rotation as follows:

> Although growing rice is not as profitable as raising shrimps, rice cultivation is good for shrimps because it reduces the pollution resulting from the waste generated by shrimp-raising in the pond. Therefore, every year I try to grow rice regardless of yield and the rodent problem. Last year, in the 4,000 m^2 pond (with around 2,000 m^2 rice field), I only got five *giạ* (equivalent to 100 kg) of paddy. In the rice-growing time, 1,000 m^2 yielded 240 kg of paddy. This year I still keep growing rice.
>
> Man, aged 75, Dinh Hamlet

The reality of this rotation practice in the two communities studied shows that although farmers have chosen the right varieties of rice for saline-affected areas, the problems relating to irrigation water and consistently maintaining the rice–shrimp rotation have not been resolved. For example, in Hoa My of Ca Mau, in 2009, during rice cultivation from the eighth to the eleventh month of the lunar year, when the level of saline water rose unexpectedly, low surrounding dikes could not prevent saline-water intrusion into rice fields. Consequently, many households lost their rice crops. In Tan Chanh of Long An, people are reluctant to grow rice for the following reasons: (1) some have converted their rice fields to deep shrimp ponds and no longer have fields to grow rice; (2) some do not want to grow rice because of their limited land and because of low yield incommensurate with efforts; (3) some still practice rice growing, but they have to put much effort into protecting their rice paddies from rodents and birds for a very modest rice yield.

16.6 Conclusion

The shift from rice to shrimp leads to ecological changes. This shift shows that the environment is a social process in which humans constantly shape their environment and in turn have to adapt to this "created" environment by changing their socioeconomic lives. In the shift from rice to shrimps, when people cannot adapt to ecological change locally, labor migration becomes the optimal solution. Although it is the result of labor excess in the context of a reduced need for labor in shrimp farming, this phenomenon can also be seen as an indicator of agricultural unsustainability. In reality, rice-shrimp farming has been proven to be a sustainable practice. However, we need to pay more attention to measures to apply this practice, as demonstrated by this case analysis of shrimp-farming practices in the two communities studied in Vietnam.

References

Aquatic Information Center (2008) Xuất khẩu thủy sản chạm ngưỡng 4 tỷ USD (Aquatic export reaching 4 billion USD). http://www.fistenet.gov.vn. Nov 2008

Ca Mau People's Committee (2008) Hiện trạng và Giải pháp Phát triển Nuôi tôm Nước Lợ tại ĐBSCL (Present situation and solutions for brackish shrimp farming in the Mekong delta). http://www.camau.gov.vn. 12 Sept 2008

General Statistics Office (2006) Statistical yearbook of Vietnam. Thống kê publisher, Hanoi

General Statistics Office (2007) Statistical yearbook of Vietnam. Thống kê publisher, Hanoi

General Statistics Office (2008) Statistical yearbook of Vietnam. Thống kê publisher, Hanoi

Hendry J (1964) The Small World of Khanh Hau. Aldine Publishing Company, Chicago

Huynh PT (2008) "Đồng bằng Sông Cửu Long vào Cuộc Cạnh tranh Toàn cầu". Trong Đồng bằng sông Cửu Long đồng hành cùng WTO. Vietnam Economic News. www.mdec.vn (Website of The Mekong Delta Economic Cooperation)

Mai TT et al (2006) Đánh giá mối quan hệ tự do hóa thương mại, đói nghèo nông thôn, môi trường trong nghề nuôi tôm (An assessment of the linkages between trade liberalization, rural poverty, and environment in shrimp aquaculture in Ca Mau Province, Vietnam. Project abstract)

Nguyen PQ (2004) Thị trường lúa gạo Nam kỳ (1860–1945) (The Rice market of the south of Vietnam 1860–1945). Ho Chi Minh City Publishers, Ho Chi Minh City

Nguyen TP, Truong HM, Nguyen AT (2004) Tổng quan về các mô hình nuôi tôm sú ở Đồng bằng sông Cửu Long (An overview on farming types of Black tiger shrimp in the Mekong delta). Paper presented at the conference "The development of aquatic resources in the coastal areas", Ho Chi Minh City, 4 Aug 2004

Nhandan online (2008) Nông dân Cà Mau làm gì để giảm vay nợ? (What do the Ca Mau farmers do to alleviate debts). http:// www.nhandan.com.vn/ tinbai/?article =134650 & sub=55& top =38. 10 Nov 2008

Tran TB (1994) Sustainability of rice-shrimp farming system in a Brackish water area in the Mekong delta in Vietnam. Master of Science thesis, University of Western Sydney, Hawkesbury

Vietnam Economic News (2008) The Mekong delta: a big exporter. The Mekong Delta Rising with the WTO. www.mdec.vn. (Website of The Mekong Delta Economic Cooperation)

Vuong DQT, Lin CK (2001) Rice-Shrimp Farming in the Seawater Intrusion Zones of the Mekong Delta, Vietnam. ITCZM Monograph No. 6

Chapter 17
Relationship Quality in Fish Value Chains: Buyer–Supplier Management in the Pangasius Industry, Vietnam

Le Nguyen Doan Khoi and Nguyen Phu Son

Abstract This study deals with the importance of investment and trust in designing a high-quality, export-oriented fish value chain. It analyzes relationship quality in the Pangasius industry in order to improve product quality through closer supply-chain coordination. According to findings from several research streams, we argue that relationship quality must be conceptualized as a function of trust and investment. Consequently, we derived a conceptual model that links commitment and trust to relationship quality. The analysis brings together assumptions and concepts from a variety of sources ranging from Porter's value chain to transaction-cost economics to institutional economics.

To verify the applicability of the model, we conducted a survey in the Pangasius industry of Vietnam. The data were obtained from a sample of 120 Pangasius farmers and 5 Pangasius processing/export firms in the Mekong River Delta (MRD), Vietnam. Each interview made was taped and properly transcribed. We collected relevant information by interviewing the managers of four processing/export firms. In addition, we consulted secondary data by making use of documentary information, archival records from relevant organizations, and secondary sources. Hence, using different sources of evidence, we were able to triangulate our findings on the main issues under study.

Keywords Relationship quality • Trust • Transaction-specific investments • Value chain

---

Le Nguyen Doan Khoi (✉)
School of Economics and Business Administration, Can Tho University,
Campus 1 – 30/4 Street, Hung Loi Ward, Ninh Kieu District, Can Tho City, Vietnam
e-mail: khoihathanh@gmail.com

N.P. Son
Mekong Delta Institution, Can Tho University, Can Tho City, Vietnam
e-mail: npson@ctu.edu.vn

M.A. Stewart and P.A. Coclanis (eds.), *Environmental Change and Agricultural Sustainability in the Mekong Delta*, Advances in Global Change Research 45,
DOI 10.1007/978-94-007-0934-8_17,

17.1 Introduction

In recent years, many businesses have recognized the importance of commitment and trust in improving the performance of exchanging parties (Tsegai 2006; Boersma et al. 2003; Nooteboom 1999; Sako and Helper 1998; Forsgren et al. 1995; Morgan and Hunt 1994; Easton and Araujo 1994; Williamson 1993; Hakansson and Johanson 1992). As a result, a great deal of attention has been paid by both economists and sociologists to concepts relevant for studying investment and trust (Williamson 1985; Hakansson and Snehota 1995). The economic and sociological approaches differ in their theoretical assumptions and concepts. Generally speaking, economists approach these issues via transaction-cost analysis, while sociologists do so via networking theory. However, several efforts have been made to bridge the gap between these two perspectives, e.g. views of transaction cost economics and networking theory on the discussions of investment and trust (Johanson and Mattsson 1987; Nooteboom 1993). Their findings show that while transaction-cost approach focuses on the opportunistic behavior of exchanging parties and the risk associated with it, networking theory focuses on its correlate: trust. Networking theory also argues that trust minimizes transaction costs and that it can provide the basis for a viable governance structure in a dynamic network environment (Hakansson and Johanson 1993). Transaction-cost theory explains investment in terms of credible commitments or the reputation of the firm, and its discussion is limited to relationship-specific investments. Networking theory notes that investment is the outcome of mutual adaptation processes and provides a broader way of measuring investments made in a relationship.

In this chapter, we want to test the extent to which networking theory can integrate the main concepts of transaction-cost theory on investment and trust. Our approach seems in line with Williamson's view (1992) that transaction-cost economics needs to be refined and extended. It is also argued that such an integrated networking theory provides a better explanation of the problems of the seafood industry and other industries (Anderson et al. 1994; Dryer 1996; Jarillo 1988; Gulati 1995; Sven and Gronhaug 1995; Mitullah 1999; McCormick 1999). Based on this integrated networking theory, a comprehensive conceptual framework is developed.

In order to test the applicability of the model derived, we made a study on the Vietnam Pangasius industry by specifically analyzing the importance of investment and trust between fish suppliers and export firms, which groups both want to meet the demand of EU fish importers. In Vietnam, the largest production of aquaculture comes from the MRD, varying between 60 and 70% of total production during the last decade (MOFI 2008). The MRD in the southern part of Vietnam is known as the region for Pangasius farming. Pangasius has emerged as one of the key aquaculture species by value and volume in Vietnam. At the moment, Pangasius is exported to over 80 countries world-wide (VASEP 2008). In 2008, the EU was the largest importer of Pangasius. The EU imported 40% of the total Pangasius export volume. Expectations of future market demand in the Pangasius sector are robust. Consumption of white fish fillets is increasing, and wild stocks (especially in Europe) continue to decline (Globefish 2008).

17.2 Theoretical Approach

This section analyzes transaction-cost theory in relation to networking theory. The basic focus in transaction-cost theory concerns efficiency. In the case explored here, we found this theory inadequate for several reasons. However, the concept of efficiency can be integrated into networking theory. Networking theory makes a distinction between "transfer activities," which are related to efficiency, and "transformation activities," which are related to effectiveness (Hakansson and Johanson 1992). More specifically, the networking perspective proceeds under the assumption that instead of minimizing the cost of one transaction alone, the efficiency criterion should be based on a set of transactions between two parties or should be aimed at maximizing the joint transaction value of a given transaction among several value-system actors (Zajac and Olsen 1993). In this case, the unit of analysis concerns the relationship rather than one discrete transaction, which provides a strong basis for understanding the coordination of industrial activities in a broader context. Also, the network approach considers transaction costs as only one aspect of the total network relationships. To achieve an overall assessment of the network relationship, these costs must be compared with the total advantages of cooperation.

Transaction-cost theory claims that the choice of governance structure is determined by attributes of a given transaction and certain assumptions regarding human behavior (Williamson 1985). In order to guarantee the quality standards, vertical coordination between small-scale farmers and their chain actors are crucial (Ziggers and Trienekens 1999; Hobbs and Young 2001; Boger 2001; Schulze et al. 2006). Vertical coordination is important when examining ways to reduce transaction costs. In their study of agri-food supply chains, Hobbs and Young (2001) state that a reduction in transaction costs through vertical coordination is beneficial to both the firm and the farmers. The firm gets an assured and timely supply of the desired raw material. On the other side, the farmers get an assured market for their produce. Moreover, they gain more reliable access to production inputs, capital, technology, and market information (Han et al. 2006; Hobbs and Young 2001; Ruben et al. 2007). Therefore, smallholders can remain involved in using different strategies for improving vertical and horizontal coordination (Kaplinsky and Morris 2000; Henson et al. 2000; Key and Runsten 1999). Most work employing transaction-cost theory also shows that a high level of asset specificity leads to high-sunk costs. Furthermore, this implies that firms are likely to stick to a particular operating structure and therefore will not be able to respond to strategic changes in market expectations or competitive conditions. Due to the sunk costs, asset specificity can also be viewed as a variable that may have a negative influence on the development of long-term business relations. Note, though, that from the networking perspective, the concept of asset specificity is very closely related to the discussion of heterogeneity, mutual adaptation, power, and market assets (Hagg and Johanson 1982; Johanson and Mattsson 1987). According to the networking approach, investment is realized as a result of a mutual adaptation process and is positively related to the development of closer relationships. Firms in the network are engaged in exchange processes, and every transaction made is considered to be an investment. This investment concept is integrated in our conceptual model as one of the key features of relationships.

Frequent exchanges between partners may be the result of a gradual development of trust that helps partners to lower transaction costs by safeguarding against opportunism. The implications of the effect of trust on governance structures are generally ignored in transaction-cost theory. This limitation of social embeddedness of economic actions and trust is best dealt with in networking theory (Granovetter 1985; Uzzi 1997; Grabher 1993). From a networking perspective, opportunism is not considered as a basic characteristic of the actor. Instead, trust is an important concept in the networking approach. We share the view that informal networks reduce transaction costs because of the high level of trust in the relationships. A high level of trust enables firms to reduce negotiation costs; it helps to reduce transactional uncertainty; and it creates opportunities for the exchange of goods and services. Hence, our conceptual framework relies heavily on instruments that build trust. A detailed discussion on this issue is presented in later sections.

17.3 The Conceptual Framework

Based on the theoretical discussions, we derived two strongly inter-related elements: *investment* and *trust*, crucial for analyzing network processes.

17.3.1 Investment

If industry actors are to realize their objectives – such as getting access to resources or markets – each actor is expected to invest in relationships. The concept of investment in marketing and networking theories deserves special attention. Empirical studies – that of Easton and Araujo (1994), for example – show that Williamson's concept was a very narrow one, essentially concerning the bare minimum investment that a partner needs to make to sustain a relationship at all. They proposed a hierarchy of investments within buyer–seller relationships. Also, in Hagg and Johanson (1982) and Forsgren et al. (1995) three types of market investments are analyzed: general, market-specific, and relationship-specific investments. General market investment relates to overall investments made in a business. Market-specific investments refer to investments made for a specific market, product, or geographical region. Finally, relationship-specific investments are investments in which the value becomes zero if the relationship comes to an end.

In our conceptual model, we considered two types of investments, namely market-specific investments and relationship-specific investments, and we adopted some arguments from Hagg and Johanson (1982) in analyzing market investments.

To develop business relations, firms should make market-specific investments, which are flexible by nature and do not necessarily create sunk costs. For instance, flexibility is found to be one of the most important characteristics of successful firms in the seafood industry [2]. This is because the industry mainly relies on access to marine resources, predominantly fish, and the monthly catch fluctuates. At the same time, the demand for fish also changes from time to time. In such a situation, firms are expected to be flexible enough to re-adjust to such changes by adapting their organizations in terms of size, form of production, or technology. This implies that firms do not necessarily have to invest in assets that create sunk costs. It is also possible for industry actors to invest in relationship-specific investments. This especially holds true when exporters prefer to get a regular supply of fish: To make this possible they develop business ties with individual fish suppliers by providing credit or other services.

17.3.2 Trust

The literature clearly shows that trust creates stability and guarantees continuity in the relationships between industry actors, and it is the glue that holds such relationships together. Several studies consider trust as a central feature of business relationships and propose different ways of measuring trust. For instance, Sako (1992) identified three types of trust, namely, contractual, competence, and goodwill trust. Similarly, Shapiro et al. (1992) discussed deterrence-based, knowledge-based, and identification-based trust. Mishra (1996) focuses on competence, reliability, and openness in defining trust. According to Doney and Cannon (1997), the development of trust involves five processes: calculative, capability, predictive, intentionality, and transference [3]. Zucker (1986) defined trust as a set of expectations shared by all those involved in an exchange and identified three dimensions of trust, namely process-based, characteristics-based, and institutional-based trust. Such classification integrates most of the instruments of measuring trust as explained by other researchers and allows us to examine the concept of trust in broader perspective. Hence, in our study, we adopted Zucker's (1986) definition and classifications of trust, but redefined process-based trust as competence trust.

Competence trust is based on concrete experience concerning certain behavioral patterns. It results from the dynamics of past and future exchange processes, and it is influenced by the reputation of industry actors. Each party gathers information on past transactions on the basis of which they can evaluate the other partner's trustworthiness. As a means of evaluation, they consider both technical and managerial competences in living up to their promises. Competence trust combines the explanations provided by Sako (1992) on the same issue and the discussions of Shapiro et al. (1992) on knowledge-based trust. It is also consistent with the discussions of Mishra (1996) and Doney and Cannon (1997). *Characteristic-based trust* refers to the influence of social norms, religion, personal bonds, or friendship in the relationships between industry actors. This is similar to the discussions of

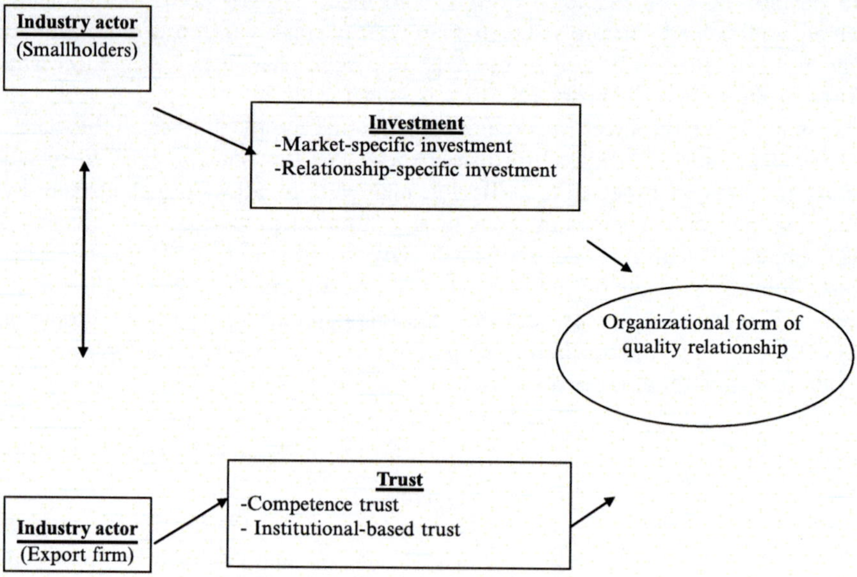

Fig. 17.1 Conceptual framework of the importance of investment and trust in developing business relations between fish farmers and export firms

Sako (1992) and Shapiro et al. (1992) on goodwill trust and identification-based trust, respectively. According to Williamson (2000), characteristic-based trust represents a level 1 form of institutional arrangement, which is characterized by informal institutions, customs, and norms.

Institutional-based trust concerns formal social structures, which are usually backed by sanctions based on the law. These include property rights, business contracts, formation of fish cooperatives, and fishery legislations. This type of trust incorporates the discussion by Sako (1992) on contractual trust and the implications of deterrence-based trust, as articulated by Shapiro et al. (1992). Similarly, according to Williamson (2000), institutional-based trust represents level 2 and level 3 forms of institutional arrangements, which include the rules of the game and actual playing of the game itself (Fig. 17.1).

17.4 Research Design

Let us say a few words about our research design. Obviously, there is no single perfect research design; the particular method or methods to follow depends on the research problem and its purpose (Yin 1994; Miles and Huberman 1992; Ghauri et al. 1995). Our research design is aimed at selecting a research method that is relevant to finding an answer to our research question: *What is the role of investment and trust in organizing an export-oriented fish supply chain in the Vietnam fish industry?* We make use of case studies. A case-study design depends on the unit of analysis. The unit of analysis may be an individual, a firm, a decision, or a program (Yin 1994). In choosing the case, the most important criterion is that we learn as much as possible from the case (Stake 1995). In our research, the unit of analysis refers to the business relationship that a firm has with other organizations. Accordingly, we developed a case-study protocol that is in line with the model, and each variable was properly classified and made operational. Each actor in the relationship is asked questions specifically related to investment and trust. In order to verify the applicability of the model derived, we studied business relations between fish suppliers and export firms in Vietnam.

There are different ways of collecting data. The case study and survey methods are two of the most frequently used research methods. In a survey method, samples are usually large, and the focus is not on an individual in a sample but rather on the general profiles or statistics derived from individual cases. Questionnaires, personal interviews, and telephone surveys are some of the methods typically used in the survey method. According to Yin (1994), a case study is defined as "an empirical enquiry that investigates a contemporary phenomenon within its real-life context, especially when the boundaries between the phenomenon and context are not clearly evident." In a case study, a single subject or phenomenon which is bound by time and activity (event, process) is explored. The case study's strength is its ability to deal with a full variety of evidence like documents, interviews, and observations.

In order to understand the view of fish suppliers, we carried out both case study interviews and survey research. A case study and survey methods may complement each other. In our research with fish suppliers, we found that it is useful to start with a case-study approach and then employ the survey method. By using both a case-study approach and survey research we managed to gather all relevant information, which allows us to present a compete case study on fish suppliers. In total, there are 11 villages in An Giang Province; and, after consultation with key fishery experts, we selected three fish villages for a case study. Accordingly, we carried out a case-study interview with 26 farmers. The respondents were asked about their views on the importance of investment and trust in developing business relations between themselves and export firms. Each interview was taped and properly transcribed. To complement the case-study results, we also administered a survey, through semi-structured questionnaires, with 120 fish farmers selected from six villages. This study examines questions relating to quality control and the quality-assurance system at the processing-firm level. Information regarding the five processing firms located in survey areas was obtained by interviewing the quality-control managers of these firms. The processing firms that were selected all produce for the international markets and include all three kinds of ownership forms in the region (joint stock, private, and state-owned firms). In the survey, the variables were classified; and the respondents were asked to rate the importance of each variable through a five-point Likert scale: (1) not important at all, (2) not important, (3) neutral, (4) important, and (5) very important. We presented the frequency, mean, and standard deviation of each response using an SPSS statistical package.

The results of both case-study interviews and the surveys were linked to each variable under investigation and were compared with the views of partner firms. In order to verify the views of both respondents, we also referred to financial records and reports prepared by respondents and other organizations. Hence, using different sources of evidence, we were able to triangulate our findings on the main issues under study.

17.5 Analysis

In this study, an analysis is made regarding the importance of investing in the relationship between fishermen and export firms and the role of trust in maintaining the relationship. The importance of flexible supply contracts as an organizational form of relationship is discussed as well.

17.5.1 Investment Made in Relationships

In order to gain access to credit and to obtain attractive fish prices from the export firm, fishermen have to show their willingness to make market-specific and relationship-specific investments. Regarding market-specific investments, we asked the respondents to rate two variables: the "willingness to invest in advanced quality assurance" and the "willingness to use new technology." According to the survey, 92% of the respondents said that a "willingness to invest in advanced quality assurance" was (very) important. Sixty-two percent of the respondents rated the "willingness to use new technology" as (very) important. In particular, they said that they were prepared to invest in waste-water treatment ponds and good aquaculture practices (Table 17.1). About 24% of the respondents answered that they did not need to invest in modern quality assurance for they already had these facilities.

Fish farmers are also expected to make commitments to the export firm in the field of relationship-specific investments. The respondents were shown four variables, and they were asked to rate each variable in terms of its importance. These variables were "willingness to deliver fish after every crop to the same processing firms," "commitment to deliver quality fish," "readiness to deliver fish in large volumes," and "adaptation to production schedules." According to the survey results, 86% of the respondents rated "willingness to deliver fish after every crop to the same processing firms" as being (very) important, and 84% agreed that a "commitment to deliver fish in large volumes" was (very) important. The respondents who rated "commitment to deliver quality fish" and "adaptation to production schedules" as (very) important made up 48% and 46%, respectively (Table 17.1). These findings show that fish farmers realize the importance of providing a regular supply of fish in large volume in order to develop relationships with export firms. However, the respondents found the delivery of quality fish to be less important for the relationship. According to the respondents, it requires more

Survey results on the views of fish farmers

Table 17.1 The importance of investment in the future relationship between fish farmers ($N=120$) and processing/export firms (*Source*: Survey data 2008)

Factors	Importance of investment: Very important	Important	Neutral	Not important	Not important at all	Mean	Standard deviation
Market-specific investments							
Willingness to invest in advanced quality assurance	73.0	19.0	4.8	3.2	0.0	4.62	0.73
Willingness to use new technology	25.4	36.5	14.3	20.6	3.2	3.60	1.17
Relationship-specific investments							
Willingness to deliver fish after every crop to the same processing firms	46.00	39.7	11.1	3.2	0.0	4.29	0.79
Commitment to deliver quality fish	12.7	34.9	25.4	22.2	4.8	3.29	1.09
Readiness to deliver fish in large volume	44.4	39.7	7.9	1.6	6.3	4.14	1.07
Adaptation to production schedules	15.9	30.2	27.0	19.0	7.9	3.27	1.18

of an effort and more costs to provide quality fish – and above all, fish farmers cannot predict what quality of fish they sell, for this is mainly a matter of luck. This often prevents fish farmers from delivering quality fish within a specific time schedule.

17.5.2 *The Role of Trust in Maintaining the Relationship*

The amount of investments to be made by both fish farmers and processing/export firms may change in time and will be influenced by the level of trust developed between the parties. According to our conceptual framework, two forms of trust explain the relationship between fish farmers and processing/export firms, namely competence trust and institutional-based trust.

17.5.2.1 Competence Trust

Competence trust concerns the managerial and technical ability of a fish farmer or an export firm in dealing with its promises and agreements. From four variables the

Table 17.2 The importance of trust in the future relationship between fish farmers and processing/export firms ($N = 120$) (*Source*: Survey data 2008)

	Importance of trust						
Factors	Very important	Important	Neutral	Not important	Not important at all	Mean	Standard deviation
Competence-based trust							
Attractive price	46.0	30.2	11.1	12.7	0.0	4.10	1.04
Promising to be a regular buyer	20.6	31.7	42.9	4.8	0.0	3.68	0.86
Getting access to credit	44.4	39.7	12.7	1.6	1.6	4.24	0.86
Providing market information	4.8	14.3	34.9	46.0	0.0	2.78	0.87
Institutional-based trust							
Written contracts	71.4	11.1	7.9	7.9	1.6	4.43	1.04
Unwritten contracts	11.1	19.0	27.0	25.4	17.5	2.81	1.25

respondents were asked to select the ones that best represent the competence of export firms in the future. The variables include attractive prices, the promise of being a regular buyer, the provision of credit, and the provision of market information. The survey results reveal that 84% rated "credit provision" as (very) important and 52% rated "attractive prices" as (very) important. On the other hand, 52% and 19%, respectively rated the variables "regular buyer" and "provision of market information" as (very) important (Table 17.2). This confirms that in order to win the confidence of fish farmers and to develop trust, processing/export firms should provide credit and set an attractive fish price. This may increase the number of fish farmers who regularly supply fish to the export firms.

Export firms stressed that the competence of a fish farmer is best evaluated by his reputation as a regular supplier and by his punctuality in meeting deadlines. However, export-firm respondents complained that fish farmers failed to exhibit the desired competence because they could not regularly supply, and also they did not adhere to quality requirements.

17.5.2.2 Institutional-Based Trust

Institutional-based trust is associated with property rights, laws, and mechanisms of enforcing laws that influence the relationship of fish farmers with the processing/export firms. So far, contractual agreements between fish farmers and export firms were almost non-existent, and the few agreements that did exist consisted of oral promises. The respondents were asked to evaluate the importance of written and non-written contracts in the future. Eighty-three percent rated "written contracts" and 30% "non-written contracts" as (very) important (Table 17.2). The managers of export firms think that written contracts are better than unwritten ones because they

Table 17.3 The importance of flexible supply contract in the future relationship between fish farmers and processing/export firms ($N = 120$) (*Source*: Survey data 2008)

	Importance of flexible supply contract						
Factors	Very important	Important	Neutral	Not important	Not important at all	Mean	Standard deviation
Gaining a reasonable profit margin	77.8	17.5	4.8	0.0	0.0	4.73	0.54
Fixing a specific contract in advance	17.5	44.4	19.0	11.1	7.9	3.52	1.15
Fish quality specification	9.5	25.4	36.5	25.4	3.2	3.13	1.01
Quantity specification	36.5	39.7	9.5	7.9	6.3	3.92	1.17
Just-in-time delivery	14.3	44.4	27.0	7.9	6.3	3.52	1.04

can be used as a reference document in case one of the parties refuses to act according to the agreement.

Fish farmers were shown five variables of importance in the design of a flexible supply contract, and they were asked to rate each variable in terms of its importance. The variables included realizing a reasonable profit margin, fixing the duration of the contract, quality specifications, quantity specifications, and just-in-time delivery. The results are shown in Table 17.3. They confirm that fish farmers are more interested in profitable contractual agreements, which allow them to supply fish in large volumes to the export firms.

The respondents from the export firms pointed out that apart from allowing a reasonable profit margin, the contract should also stress just-in-time delivery as well as quality specification. This also shows the preference for the delivery of quality fish over quantity, because the export firms prefer to buy specific species that are in high demand on the world market. The managers considered demand conditions, cost of fishing supplies, fixing the lifetime of the contract, and the exchange of market information as relatively less important. To conclude, the main finding is that both fish suppliers and exporters are willing to enter into a flexible supply contract that is profitable for them.

17.6 Discussion

The case study and survey results indicate a number of findings. More broadly, the findings show the importance of making investments to develop business relations between fish farmers and processing/export firms. The relationships between the buyers–suppliers may be used to find solutions for the smallholders participating in global markets. The responding fish farmers pointed out their willingness to invest in advanced quality assurance, to use new technology, and to deliver in large volumes to a single export firm. Smallholders generally lack some of the capabilities needed to comply with the high-quality requirements of processing firms. In order to solve this deficiency they can improve their position by horizontal coordination among themselves and vertical coordination with their buyers through farmers' groups. In addition, public and private intervention in providing market information, enforcement of agreements, establishment of quality-assurance systems, and the provision of appropriate financial services are crucial to facilitating the business relations between farmers and processors.

The size of the investment made by fish farmers and export firms through time is also influenced by the level of trust developed between the parties. The case study and survey results confirm that providing credit and setting attractive fish prices are not only the reasons for fish farmers to develop a relationship with an exporter, but the considerations are also major criteria to evaluate the competence of an exporter. According to the export firms, the competence of a fish farmer is best evaluated by his reputation as a regular supplier and his punctuality in meeting deadlines. It is also found that both fish farmers and exporters are willing to use a written flexible supply contract that is profitable to both of them. Such a contract can be realized by linking the main activities performed by fish farmers and export firms, and by developing commercial ties related to pricing decisions, credit ties that are concerned with sanctioning loans, and technical ties related to the adaptation of production processes.

17.7 Conclusions

This chapter attempts to bridge differences between the transaction-cost theory and networking theory by integrating the assumptions and limitations of transaction-cost theory into the discussion of networking theory. The theoretical discussions and the field research confirm the usefulness of networking theory in dealing with supply-chain problems. In order to verify the applicability of the framework derived, we considered the features of investment and trust in developing business relations between fish farmers and processing/export firms in a new business environment: the Vietnam Pangasius industry. In so doing, we hoped to contribute to the existing discussions on networking theory by exploring the extent to which such concepts could be applied in a totally different setting such as the Vietnam fish industry. The framework considers market-specific investment, relationship-specific investment, competence trust, and institutional-based trust. The case study and survey results demonstrated that the variables included in the framework were useful in analyzing supply-chain problems in the Vietnam fish industry. Industry actors acknowledged the importance of market-specific and relationship-specific investments in order to realize each other's objectives. Through time the relationship between both actors is influenced by the level of trust and extent of resource control exercised by exchanging parties. Managerial and technical competences of partners, compliance with formal and informal mechanisms of enforcing laws, and the existence of balanced bargaining power are also important factors for maintaining the relationship.

Public–private partnerships can play a key role in facilitating market linkages that can enable small-scale farmers to satisfy the market demands for food safety and quality while retaining a position in the supply chain. The government and the private sector can help farmers expand and upgrade their capabilities and practices to meet the quality requirements of global markets. There is a decisive role for the government to play in establishing an institutional order that guarantees the legal framework and defines transparent rules for conflict settlement. Farmers can only make the required investments to improve delivery frequency and quality when they can be relatively certain of available market outlets. The same concepts can also be applied in dealing with problems of the fish industry of other developing economies.

Acknowledgments The study is financed by SEARCA Seed Fund for Strategic Research and Training Program. The authors would like to express their deepest gratitude and special thanks to SEARCA for the research grant.

References

Anderson J, Hakansson H, Johanson J (1994) Dyadic business relationship within a business network context. J Marketing 58:22–38

Boersma M, Buckley P, Ghauri P (2003) Trust in international joint venture relationships. J Bus Res 56:1031–1042

Boger S (2001) Quality and contractual choice: a transaction cost approach to the polish hog market. Eur Rev Agric Econ 19:327–343

Doney PM, Cannon JP (1997) An examination of the nature of trust in buyer-seller relationships. J Marketing 61:35–51

Dryer JH (1996) Specialized supplier networks as a source of competitive advantage: evidence from the auto industry. Strategic Manage J 17:271–291

Easton G, Araujo L (1994) Market exchange, social structures and time. Eur J Marketing 28:72–84

Forsgren M, Hagg I, Hakansson H, Mattsson L (1995) Firms in networks: a new perspective on competitive power. Acta Universitatis Upsaliensis, Studia Oeceomiae Negotiorum, No. 38, Uppsala University, Uppsala

Ghauri P, Gronhaug K, Kristianslund I (1995) Research methods in business studies, a practical guide. Prentice Hall, Englewood Cliffs

Globefish (2008) Pangasius market report

Grabher G (1993) The embedded firm: on the socioeconomics of industrial networks. Routledge, New York and London

Granovetter MS (1985) Economic action and social structure: the problem of embeddedness. Am J Sociol 91:481–510

Gulati R (1995) Social structure and alliance formation pattern: a longitudinal analysis. Adm Sci Q 40:619–652

Hagg I, Johanson J (1982) Foretag I natverk [Firms in networks]. SNS, Stockholm

Hakansson H, Johanson J (1992) A model of industrial networks. In: Axelsson B, Easton G (eds) Industrial networks: a new view of reality. Routledge, London, pp 32–46

Hakansson H, Johanson I (1993) The network as a governance structure: inter-firm cooperation beyond markets and hierarchies. In: Graber G (ed) The embedded firm: on the socioeconomics of industrial networks. Routledge, London, pp 36–51

Hakansson H, Snehota I (1995) Developing relationships in business networks. Routledge, London

Han J. et al (2006) Vertical coordination, quality management and firm performance of the pork processing industry in China. In international agri-food chains and networks, management and organization, Wageningen, the Netherlands: Wageningen Academic Publishers, 319–32

Henson SJ, Brouder A-M, Mitullah W (2000) Food safety requirements and food exports from developing countries: the case of fish exports from Kenya to the European Union. Am J Agr Econ 82:1159–1169

Hobbs JE, Young LM (2001) Vertical linkages in agri-food supply chains in Canada and the United States. Research and analysis directorate policy branch, agriculture and agri-food Ottawa, Ontario, Canada

Jarillo J (1988) On strategic networks. Strategic Manage J 9:31–41

Johanson J, Mattsson LG (1987) Inter-organisational relations in industrial systems – a network approach compared with the transaction cost approach. Int Stud Manage Organ 17:34–48

Kaplinsky R, Morris M (2000) A handbook for value chain research, prepared for the IDRC. Sussex: institute of development studies

Key N, Runsten D (1999) Contract farming, smallholders, and rural development in Latin America: The organization of agro-processing firms and the scale of out grower production. World Development 27(2):381–401

McCormick D (1999) African enterprise clusters and industrialization: theory and reality. World Dev 27:1531–1551

Miles MB, Huberman M (1992) Qualitative data analysis: an expanded sourcebook, 2nd ed. Thousand Oaks, CA: Sage

Mishra AK (1996) Organizational response to crisis – the centrality of trust. In: Kramer RM, Tyler TR (eds) Trust in organizations: Frontiers of theory and research. Sage Publications, Thousand Oaks, pp 261–287

Mitullah W (1999) Lake Victoria's Nile perch industry: institutions, politics and joint action. IDS working paper 87, IDS, University of Sussex, Brighton, U.K.

MOFI (2008) Report of fishery production in Vietnam

Morgan RM, Hunt SD (1994) The commitment – trust theory of relationship marketing. J Marketing 58:20–38

Nooteboom B (1993) Transactions and networks: do they connect? In: Groenewegen J (ed) Dynamics of the firm: strategies of pricing and organisation. Edward Elgar, Aldershot, pp 9–26

Nooteboom B (1999) Inter-firm alliances: analysis and design. Routledge, London

Ruben R, Tilburg AV, Trinekens J, Boekel MV (2007) Linking market integration, supply chain governance, quality and value added. In Tropical Food Chains: Governance Regimes for Quality Management. Wageningen, the Netherlands: Wageningen Academic Publishers

Sako M (1992) Prices, quality and trust, inter-firm relations in Britain and Japan. Cambridge University Press, New York

Sako M, Helper S (1998) Determinants of trust in supplier relations: evidence from the automotive industry in Japan and the United States. J Econ Behav Organ 34:387–417

Schulze B, Wocken C, Spiller A (2006) Relationship quality in agri-food chains: supplier management in the German pork and dairy sector. J Chain Netw Sci 6:55–68

Shapiro D, Sheppard BH, Cheraskin L (1992) Business on a handshake. Negotation J 8:365–378

Stake R (1995) The art of case study research. Sage, London

Sven A, Gronhaug K (1995) Authority and trust in network relationships. In: Hakansson H, Snehota I (eds) Developing relationships in business networks. Routledge, London, pp 368–381

Tsegai M (2006) The importance of business relations in the fish supply chain and export market distribution channels: the case of the Eritrean fish industry. PhD thesis, University of Groningen, the Netherlands

Uzzi B (1997) Social structure and competition in inter-firm networks: the paradox of embeddedness. Adm Sci Q 42:35–67

VASEP (2008), Market report, Vietnam

Williamson OE (1985) The economic institutions of capitalism: firms, markets, and relational contracting. The Free Press, New York

Williamson OE (1992) Markets, hierarchies, and the modern corporation. J Econ Behav Organ 17:335–52

Williamson OE (1993) Calculativeness, trust and economic organization. J Law Econ 36:453–486

Williamson OE (2000) The new institutional economics: taking stock, looking ahead. J Econ Lit 38:595–613

Yin RK (1994) Case study research: design and methods. Sage, Thousand Oaks

Zajac E, Olsen CP (1993) From transaction cost to transactional value analysis: implications for the study of inter-organisational strategies. J Manage Stud 30:131–145

Ziggers GW, Trienekens J (1999) Quality assurance in food and agribusiness supply chains: developing successful partnerships. Int J Prod Econ 4:271–279

Zucker LG (1986) Production of trust: institutional sources of economic structure, 1840–1920. Res Organ Behav 8:53–111

Chapter 18
The Relationship Between Natural Conditions and the Formation and Development of Clam Grounds (*Meretrix lyrata*) in the Mekong Delta

Tong Phuoc Hoang Son and Nguyen Thanh Tung

Abstract Vietnam's overall mollusk production is estimated at 300,000–350,000 tons a year, of which clam production is estimated at about 60,000 tons. In the Mekong Delta, this species is concentrated mainly in six coastal provinces – Ben Tre, Tien Giang, Tra Vinh, Soc Trang, Bac Lieu, and Ca Mau.

In October 2009, the Ben Tre clam fishery received Marine Stewardship Council (MSC) certification, becoming the first fishery in Southeast Asia to meet the Council's sustainability and management standards. *Meretrix lyrata* – "Ben Tre" clam – became a famous commercial brand, a highlight of fresh aquaculture not only in Vietnam but also in the world.

This chapter focuses on assessing the relationship between natural conditions and the formation and development of clam grounds in Ben Tre Province as well as other parts of the Mekong Delta.

Statistical analysis based on survey data in 2007–2009, linked with the assessments of live server data and processed satellite image data, revealed the quantitative relationship between various environmental parameters and clam yield. The optimum environmental criteria set for the growth of clams have been established.

Based on the analysis of remote sensing data (aerial photographs, archived satellite images) in comparison with survey data, the evolution, formation, and development of clam grounds in these regions have been defined. Some recommendations on expanding the findings in the area studied to the whole Mekong Delta have been proposed, and the findings have been assessed in relation to various climate-change and human-interference scenarios.

T.P.H. Son (✉)
Institute of Oceanography, 01 CauDa Street, Nha Trang City, Vietnam
e-mail: tongphuochoangson@gmail.com

N.T. Tung
Institute of Economic and Fishery – Southern Branch of Vietnam,
731 Ly Thuong Kiet Street, Tan Binh District, Ho Chi Minh City, Vietnam
e-mail: ngthanhtung198@yahoo.com

M.A. Stewart and P.A. Coclanis (eds.), *Environmental Change and Agricultural Sustainability in the Mekong Delta*, Advances in Global Change Research 45,
DOI 10.1007/978-94-007-0934-8_18,

Keywords Ben Tre clam • Coastal geomorphology • *Meretrix lyrata* • Shore line change • Tidal flat

18.1 Introduction

Meretrix lyrata is a bivalve mollusk inhabiting the intertidal and shallow subtidal sand flat areas, typically in outer estuarine areas where the proportions of sand and mud are around 80% and 20%, respectively (Tuan 1999). The species is found in large parts of South East Asia, including Indonesia, the Philippines, Thailand, and Vietnam, but is densest in Taiwan and Vietnam. It commonly grows to 50 mm in length in intertidal area, but can be much larger in subtidal areas. In the Mekong Delta of Vietnam, this species is widely distributed in sand flat areas, and the highest density and production were recorded in Ben Tre province.

The Rang Dong Fishery Cooperative (in Thoi Thuan Commune, Binh Dai District), established in 1997, is a pioneering cooperative in clam fishing that is managed cooperatively by the community. Here, clam fishing is harmonized with the protection of the natural environment. Processing factories are also closely connected to the material bases. And, most importantly, profits from the clam fishery contribute greatly to poverty reduction in Ben Tre's coastal communities. In 2007, the Cooperative's profits reached 40 billion VND (approximately 2.2 million USD), which greatly improved the incomes of its members. Ten additional clam cooperatives have now been established, forming an alliance of cooperatives that covers all clam areas in Ben Tre. The cooperatives have 8,744 household members *in toto* and another 35 cooperating units involved in clam fishery, which are based mainly in the province's three sea-front districts of Ba Tri, Binh Dai, and Thanh Phu. The harvestable ground area involved is about 9,600 ha, while the total area where clams may be grown is more than 15,000 ha. The annual gross volume of clams (commercial and breeding) produced is between 8,000 and 27,000 metric tons. At the cooperatives, the surpluses from selling commercial clams help members in three ways: members share equally in profits, members derive income from clam harvesting work, and members benefit via welfare contributions. Nearby regions in coastal provinces of the Mekong Delta such as Tien Giang, Tra Vinh, Soc Trang, Bac Lieu, and Camau are also potentially ideal for clam production.

In October 2009, the Ben Tre clam fishery received Marine Stewardship Council (MSC) certification, becoming the first fishery in Southeast Asia to meet the council's sustainability and management standards. *Meretrix lyrata* – "Ben Tre" clam – quickly became a famous commercial brand, a highlight of fresh aquaculture not only in Vietnam, but also in the world.

This chapter focuses on assessing the natural conditions that affect clam resources in coastal provinces of Mekong Delta, especial in Ben Tre, Tien Giang, and Tra Vinh, where large river mouths exist, typically with funnel shapes and broad, sandy tidal flats.

To date, there have been scientific projects on the social, economic, and natural conditions in the Mekong Delta relating to the life of *Meretrix lyrata*. The publications arising from the environmental parameters affecting clam life, including nutrients, suspended matters, primary production, and phytoplankton (Phu 1999; An Nguyen and Nga 2001; Hung 2000); water temperature and salinity (An Nguyen and Nga 2001; Hung 2000; Tung 2007); rain fall and number of rainy days (Hao 2001; Tung 2007); geomorphology, altitude, and grain size of sediment in tidal flats (Tuan 1999; Tung 2007); the northeast monsoon's role in shore erosion and siltation on tidal flats and in raising mortality levels and constraining the development of clams (Tuan 1999; An Nguyen and Nga 2001; Tung 2007).

In their scientific report on *Meretrix lyrata* in the tidal flats of Tien Giang province, Tuan and Thom summarized three main parameters that affect the life and development of clams: the content of organic matters, the grain size of sediments, and the altitude of tidal flats (Tuan 1999).

The formation of parent clam grounds in Ben Tre strongly relates to the location of hydrological frontal zones in near-shore water, according to An Nguyen and Nga (2001).

According to Hao (2001), the main breeding season for clams relates directly to the rainy season, and breeding grounds in Ben Tre appear to form about 1–1½ months after the first rains. During this time, the rainfall is heavy enough and the warmer water temperature stimulated parent clams to reproduce and create great breeding grounds on tidal flats in Ben Tre.

As a result of the detailed study of biological features of clams such as food sources and reproduction characteristics, as well as the construction and analysis of a safety index and a "condition" index, researchers have established in a preliminary way the relationship between environmental and hydro-meteorological parameters, biological features, and clam life (Tuan 1999; An Nguyen and Nga 2001; Hao 2001; Tung 2007).

However, most studies on the environment and clams in the Mekong Delta do not attend to complex spatial and temporal variation in environmental factors affecting clams. Due to complex dynamic conditions in tidal flats, the distribution of environmental parameters is very complex. The hydrodynamic conditions in tidal flats are also very complex. Such conditions are related strongly to ebb- and -flood tidal phases, and much previous research has not accounted sufficiently for this fact.

Employing new approaches based on remote-sensing techniques and numerical and statistic modeling, this chapter attempts to explain in detail the quantitative relationship between natural conditions and the formation and development of clam grounds in the Mekong Delta (focusing on the tidal flats of Tien Giang, Ben Tre, and Tra vinh provinces).

The main goal of this study is to answer two key questions: When do clams (include breeding and parent grounds) appear? Where do they develop well?

Some recommendations on broadening the scale of the study to the tidal flats with clam grounds in the Mekong Delta as a whole are made, and a plan is proposed for selecting suitable sites for clam production in scenarios covering the next 10–20 years in the Mekong Delta, factoring in the impact of climate change (including the effect of sea-level rise and global warming).

18.2 Material and Methodology

18.2.1 Scale of the Study and Objectives

The study region focuses on tidal flats with clam grounds in three provinces: Tien Giang, Ben Tre, and Tra Vinh, within the geographical limits of 106°05′–106°50′E and 09°30′–10°25′N (Fig. 18.1).

Based on the results of previous studies, as mentioned above, and the applicability of remote-sensing techniques in marine study, we chose the following environmental parameters for an assessment of the relationship between the natural conditions and the formation and development of clam resource:

- Accumulated rainfall, sea-surface temperature (SST), water salinity (Sal), total suspended solids (TSS), chlorophyll-a (instead of food and nutrient sources).
- Hydrological frontal zones, alongshore and tidal currents, water level, and tidal characteristics.
- Geomorphologic features (topography and bathymetry) in tidal flats, history of evolution (shoreline and morphological changes), grain size of sediments, and

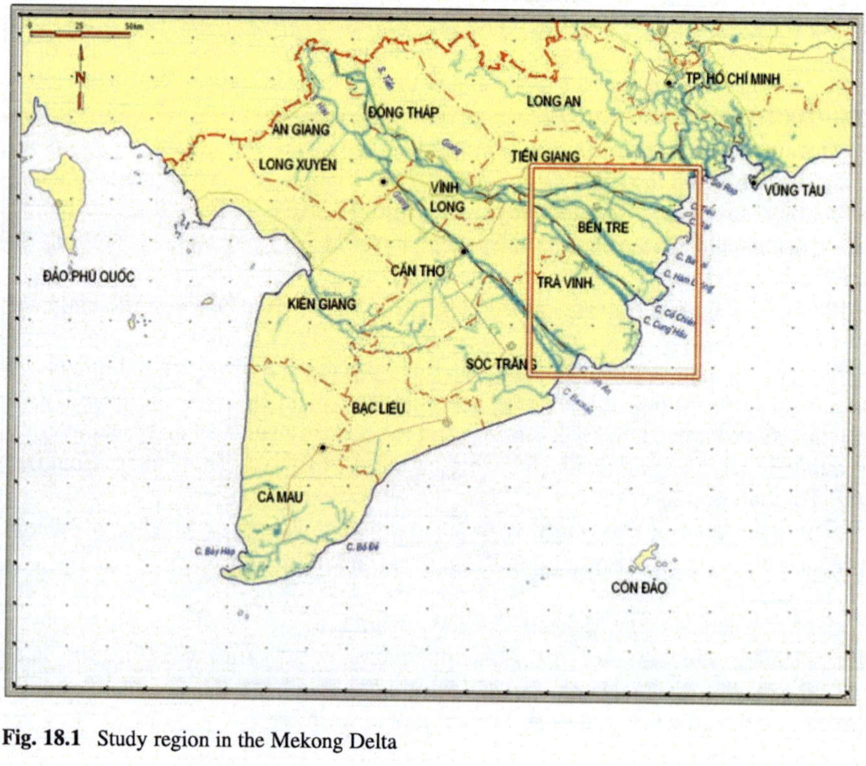

Fig. 18.1 Study region in the Mekong Delta

land use in land base and sea water (including distribution of mangrove forest, aquaculture, human settlement, paddy fields, garden trees, relict dunes, etc., and formation of underwater relief).

Each parameter group was processed by different methods and satellite imageries that will in part be presented below.

18.2.2 Material

18.2.2.1 Investigated Data

- Selected results from previous studies on clam life (including data on environmental, bio-resource, and geomorphologic features related to mollusk life) were collected from country reports (Tuan 1999; An Nguyen and Nga 2001; Tung 2007) and new results from our project (unpublished) in 2007–2009 were employed.
- Monthly meteor-hydrological serial data (2002–2008) on rainfall, wind, salinity, and air and water temperature at My Tho, An Thuan, Ben Trai meteorological stations were collected for purposes of calibration and validation of the processed data derived from satellite imagery.

18.2.2.2 Satellite Imagery and Related Data

A large number of satellite image scenes relating to the Mekong Delta from difference sources were employed, including:

- MODIS image dataset: including 166 daily scenes of MODIS aqua (Moderate Resolution Imaging Spectrometer) sensor (level 2B, 1 km resolution) which have been downloaded from the following website (http://oceancolor.gsfc.nasa.gov/) in 2007–2008. These scenes have been used to extract the spatial distribution field of SST, TSS, and Chlorophyll-a.
- The live assess serve data from the Giovanni website of NASA (http://disc2.nascom.nasa.gov/Giovanni/tovas/rain.GPCP.2.shtm) was used to extract data on the distribution and variation of rainfall, SST, and Chlorophyll-a. They were downloaded in grid 0.25°, ASCII format and include 96 monthly ASCII files (8 * 12) from 2002 to 2008.
- Multi-spectral datasets with high resolution imagery of USGS including: Landsat MSS – 60 m (1973–1977): 5 scenes, Landsat TM – 30 m (1988–1990): 4 scenes, Landsat ETM+ – 30 m (1999–2003): 9 scenes, Landsat ETM+–30 m (2007–2009): 8 scenes. These datasets were provided directly by the US Geology Survey of the USGS.
- Multi-spectral datasets with high resolution imagery of the Japan Aerospace Exploration Agency, including: JESR-MSSR1 – 60 m (1984–1988): 14 scenes, JERS-VNIR – 18 m (1997–1998): 8 scenes, and ALOS-AVNIR2 – 10 m (2007–2008): 4 scenes. They were provided directly by JAXA in ALOS-PI contract (ALOS PI 326).

18.2.3 Methodology

18.2.3.1 Analysis of the Spatial and Temporal Variation in Rainfall

Ninety-six monthly ASCII files on accumulation of rainfall were downloaded from the website of Giovanni – NASA and processed by serial time analysis via a special statistic technique – EOF (Empirical Orthogonal Function). The EOF technique allowed us to obtain and explain the main features of the temporal and spatial distribution of rainfall. In principle, this method is similar to the PCA method with varimax rotation (PCA is a technique used in multi-variable statistical analysis) (Reyment and Jvreskog, 1999).

We organized the data collected on monthly rainfall to create a numerical matrix $[N \times M]$, with $N = 86$ columns representing 86 months and $M = 121$ rows representing 121 data pixel of the 0.25° grid cover study region. A reverse matrix $[M \times N]$ was also used.

Statistica 7.0 software was used for processing.

18.2.3.2 Analysis of the Spatial and Temporal Variation of SST, TSS, Chlorophyll-a Derived from MODIS Sensor

One hundred and sixty six daily scenes (2007–2008) from the MODIS aqua sensor cover the region of the Mekong Delta. Daily level 2B LAC (local area coverage) data with 1 km resolution were obtained from the Distribution Active Archive Center Goddard Space Flight Center National Aeronautic Space Administration DAAC/GSFS/NASA (in website http://oceancolor.gsfc.nasa.gov/). Each image was geo-referenced, based on the WGS84 UTM zone 48 North coordinate system. Each data file contained the derived Chlorophyll-a, the sea surface temperature (SST), and the six water-leaving radiances (nLw, see http://oceancolor.gsfc.nasa.gov/product/SW_nLw.html).

The sea surface temperature (SST) images were extracted from the proto-MODIS infrared algorithm (Brown and Minnet 1999).

The Chlorophyll-a images were processed from the OC3M algorithm (Bailey and Werdell 2006).

The total solid suspended solid (TSS) images were extracted from the band ratio algorithm of nLw443/nLw553 (Murakami et al., 2004).

$$\text{TSS (mg/l)} = 10\text{\^{}}[0.38786 - (2.345 * R) + (1.0645 * R^2) - (1.03167 * R^3)]$$

where

- TSS: content of suspended solids in sea surface layer (mg/l).
- $R = \log_{10}[(\text{nLw}(443)/\text{nLw}(551))]$.

To reduce the effect of cloud cover, monthly composite chlorophyll-a was computed by monthly average (excluding missing data and clouds). The mask image of clouds and land from monthly chlorophyll-a images also was used for monthly SST and TSS images.

18.2.3.3 Application of Multi-Spectral High-Resolution Imagery

For extracting the geomorphologic features (topography and bathymetry) of tidal flats, history of evolution (shoreline and morphology changes), grain size of sediments, and land use in land base and sea water; some processing procedures for extracting geomorphologic features have been used as follows:

Detecting the topography on tidal flats: Based on color composite image (Landsat ETM+, MESSR, JERS-VNIR, AVNIR2,...) and by visual techniques and screen digitization, we were able to trace the water line and also the morphology of tidal flats in each different satellite image with different acquisition times.

According to time and day for acquisition image (restored in metafiles of raw images), we could get tidal height by WXTide32 software. The dataset of different water lines corresponds with different heights of tidal flats. Figure 18.2 is an example.

Tracing the shoreline and assessing the shore-line changes: The technique used for tracing the shore line at different (decadal) times was by the method of band ratio combined with threshold value of infrared bands (Van and Binh 2008). These techniques have been used for the assessment of the shore change over decadal periods as well as for the erosion–deposition process in areas where a lot of mangrove forests exist. The procedure for extracting shore-line changes has been carried out as in Fig. 18.3.

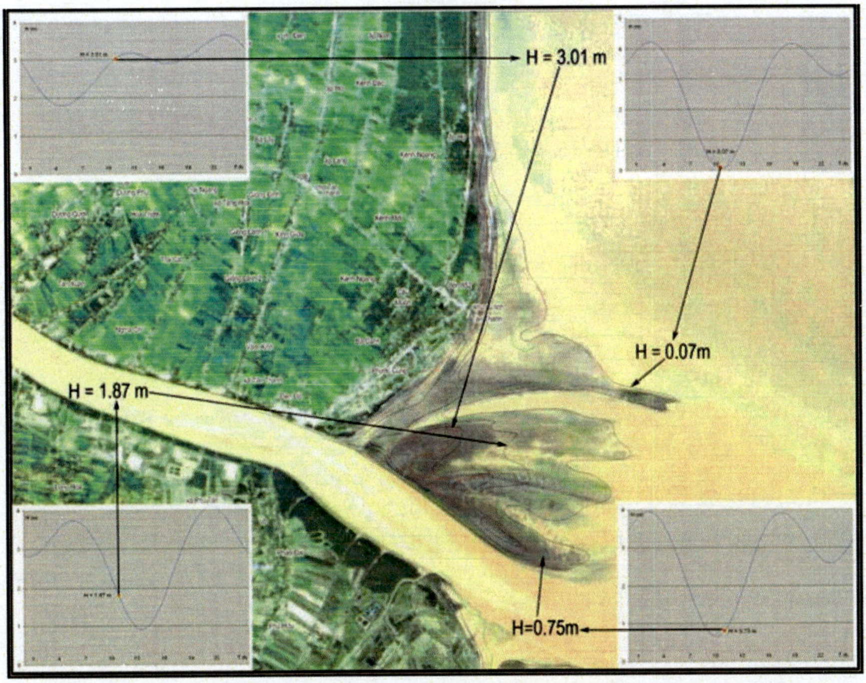

Fig. 18.2 Topography of tidal flat in Tien Giang in 2008 was extracted from four imagery scenes of Landsat ETM+

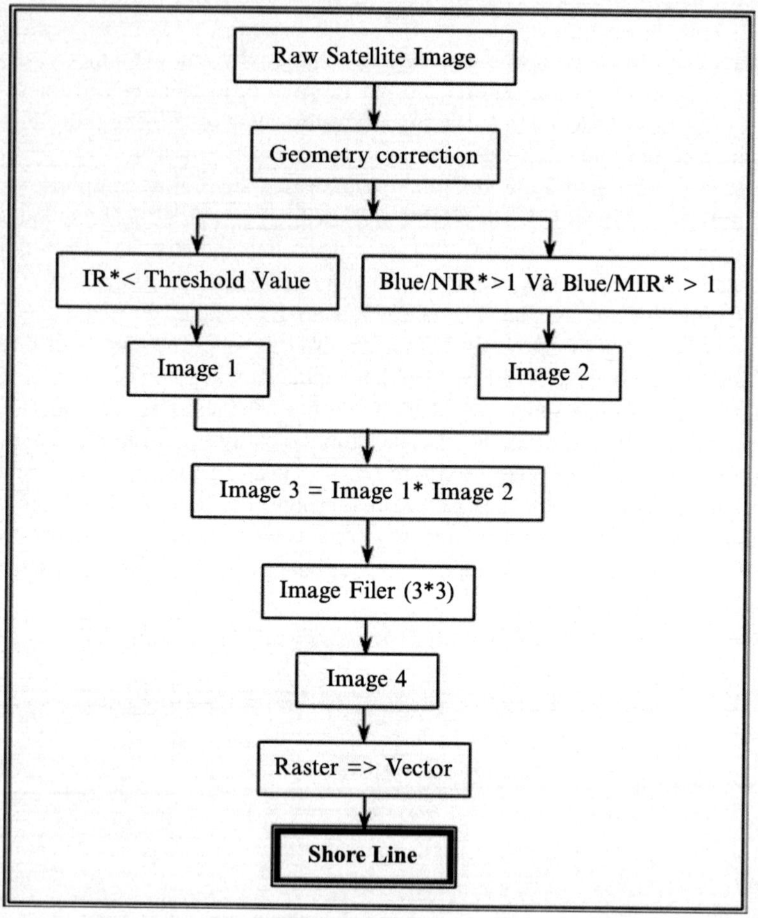

Fig. 18.3 Procedure for extracting the shore line

Land-use maps in coastal areas and tidal flats:

The results have been classified by the supervised classification method using the Maximum Likelihood algorithm based on ALOS-AVNIR2 images (10 m) during 2007–2008. The classified objects included were: tidal flat, sediment (sand, silt), mangrove forest, morphology in tidal flat, (sand wave, bar, channel, ...), river and water, shrimp pond, paddy field, garden tree, human settlement, relict dunes, etc. The training sites for classification were collected along the coastal area of three above named provinces in fields surveyed in 2008.

Hydrological fronts, tidal ebb and flood current, alongshore current, algae bloom, eutrophication events, etc. have been detected by visual interpretation and screening digitization based on color composite imagery of many kinds of sensors such as: Landsat, MESSR, JERS-VNIR, ALOS-AVNIR2.

18.2.3.4 Multi Variable Statistic Analysis

Used for the assessment of the universal relationship between environment parameters and clam yields; Cluster Analysis – CA is the main method used. Dataset criteria included: rainfall, SST, difference of sea surface temperature Δt; water salinity, chlorophyll, and production of clam (monthly data from 2002 to 2008 collected from Clam Rang Dong co-operative). Statistica 7.0 software was used for processing.

18.3 Results and Discussion

18.3.1 Meteorological Characteristics

The study region lies in the tropical monsoon zone where the air temperature is relatively high during the whole year with two distinct seasons: rainy and dry. The rainy season lasts from May to November, and the dry season is from December to April of the next year. The data from the Southern Branch of the Meteorological–Hydrological Office showed climate features in this region as follows.

18.3.1.1 Number of Sunny Hours

All regions studied (Tien Giang, Ben Tre, Tra Vinh) had almost the same number of sunny hours: 6.80–6.86 h/day. The number of sunny hours changed according to season. Almost all days with large numbers of sunny hours were concentrated from May to November. During this period, the sky was clear and blue, the cloud level was low, and the number of sunny hours fluctuated from 200 to 300 h/month.

18.3.1.2 Air Temperature

The temporal variation is clear. The minimum values for air temperature occurred during January (25.5°C). Values then increased quickly and reached a first peak during April (29°C). Values then decreased slightly to 28.8°C in May and to 28.0°C in June. They then decreased further, before becoming stable within a thermal range of 27.4–27.6°C.

18.3.1.3 Wind and Wave Regime

All regions studied were affected by monsoons. The SW monsoon blew during the rainy season with an average velocity of about 3.6 m/s and was strongest during August (4.5 m/s). The NE monsoon started in October (or a little later); the average velocity was 2.4 m/s and was strongest during February–April with a velocity between about 5–8 m/s. During this period the blowing wind was usually dry, hot, and when combined with local wind (brize wind), caused shore erosion, sea-level rise, high salinity, and also muddy depositions in some areas. This wind and wave regime badly affected clams.

18.3.1.4 Rainfall

Rainfall was distributed irregularly and slightly increased from Tien Giang to Ben Tre and Tra Vinh. There were two main peaks of rainfall: one in May and the other in October. Between these two peak periods were dry periods.

Analyzed results of EOF based on series data from the rainfall dataset from 2002 to 2008 determined the main modes of time and spatial EOFs. Four first EOFs predominated, accounting for about 50% of explained information in the dataset: 21.5%, 13.4%, 9.5%, and 6.1%, respectively, as shown in Figs. 18.4 and 18.5.

EOF1 explained the information regarding the two yearly rainfall peaks in May and August. As seen in spatial EOF1, during these periods, high rainfall was mainly concentrated in Tien Giang and Tra Vinh (where totals reached to 250–300 mm): while in coastal areas of Ben Tre, totals only reached a medium level of about 150–200 mm. In upstream districts of BenTre (Chau Thanh, Cho Lach), rainfall was slightly greater (220–250 mm). The distribution map of rainfall during May and August for 2008 typified the spatial distribution of rainfall during the above mentioned periods (Fig. 18.6a).

These distributional features, combined with characteristics of the alongshore current from the N-S direction (presented below), create conditions that delay the reproduction of clam breeding in Ben Tre about 1–1½ months compared to Tien Giang and TraVinh.

EOF2 shows the distribution of rainfall in October (the end of the rainy season). During this period, rainfall in upstream areas is relatively higher (250–300 mm), while in the coastal zone the rainfall decreases sharply and only reaches about 150–170 mm. In inshore waters there are usually some areas with low rainfall (60–80 mm), as in Travinh and in Ben Tre. The distribution map of rainfall during October of 2008 (Fig. 18.6b) typifies this weather pattern.

In ecological terms, the medium rainfall totals are enough to provide food with high organic matter from river water and also from wetlands with mangrove forests. In addition, the warmer water of terrestrial origin stimulates the development of mature clams during this period.

EOF3 illustrates the distribution of rainfall in June. At that time, rainfall reached relatively high levels (200–250 mm). The June rains were the last rains before the weather changed to the "Ba Chan" drought. The rainfall distribution map during June of 2006 (Fig. 18.6c) is typical of this weather pattern.

EOF4 shows rainfall distribution in June, the middle of the dry season. At that time, in land-based areas, the rainfall was minimal; while thundershowers brought modest amounts of rain (25–30 mm) in inshore waters (Fig. 18.6d). The appearance

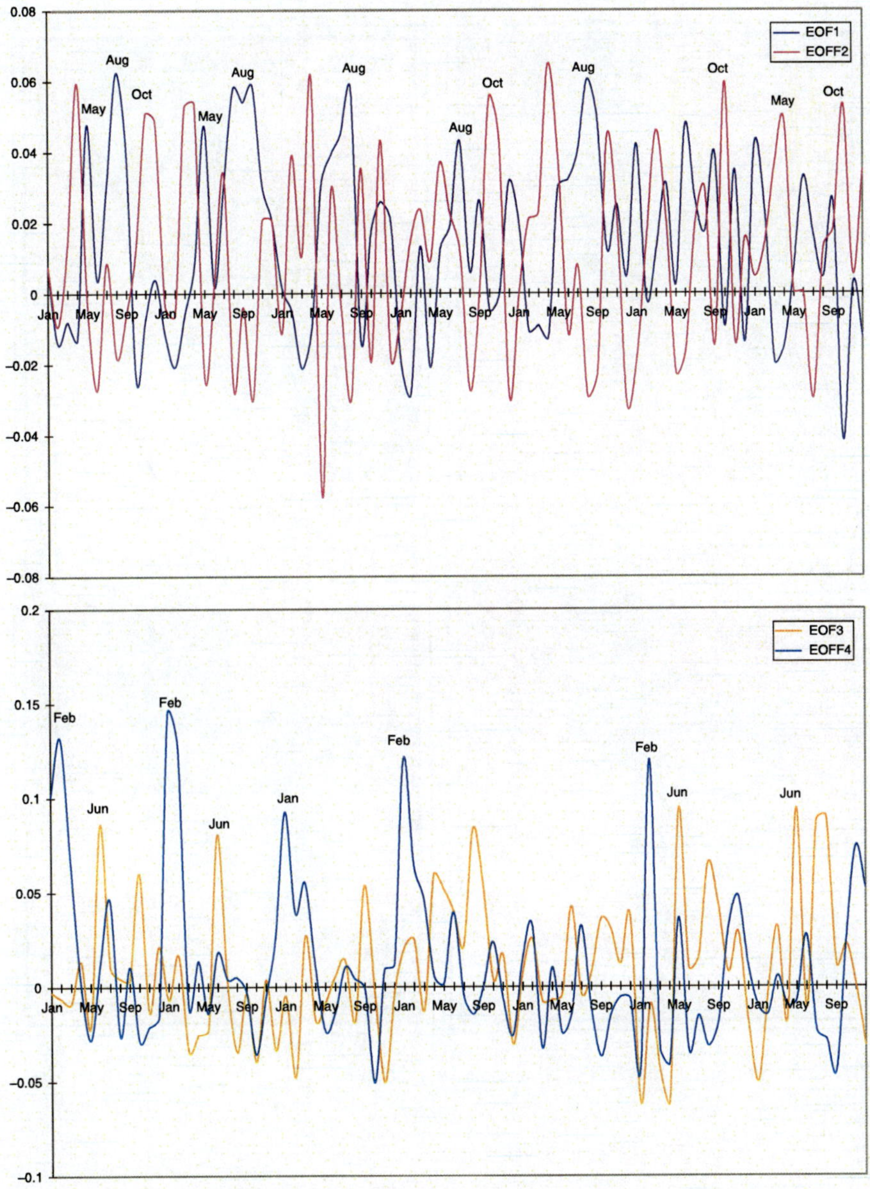

Fig. 18.4 Four first temporal EOF modes related to rainfall

of such rains created sudden changes in salinity and harmful algae blooms of Dinophyceae and Diatom. As a result of east winds, together with flood tidal currents, these algae blooms landed onto tidal flats and created an additional food source for clams during a food-deficient period.

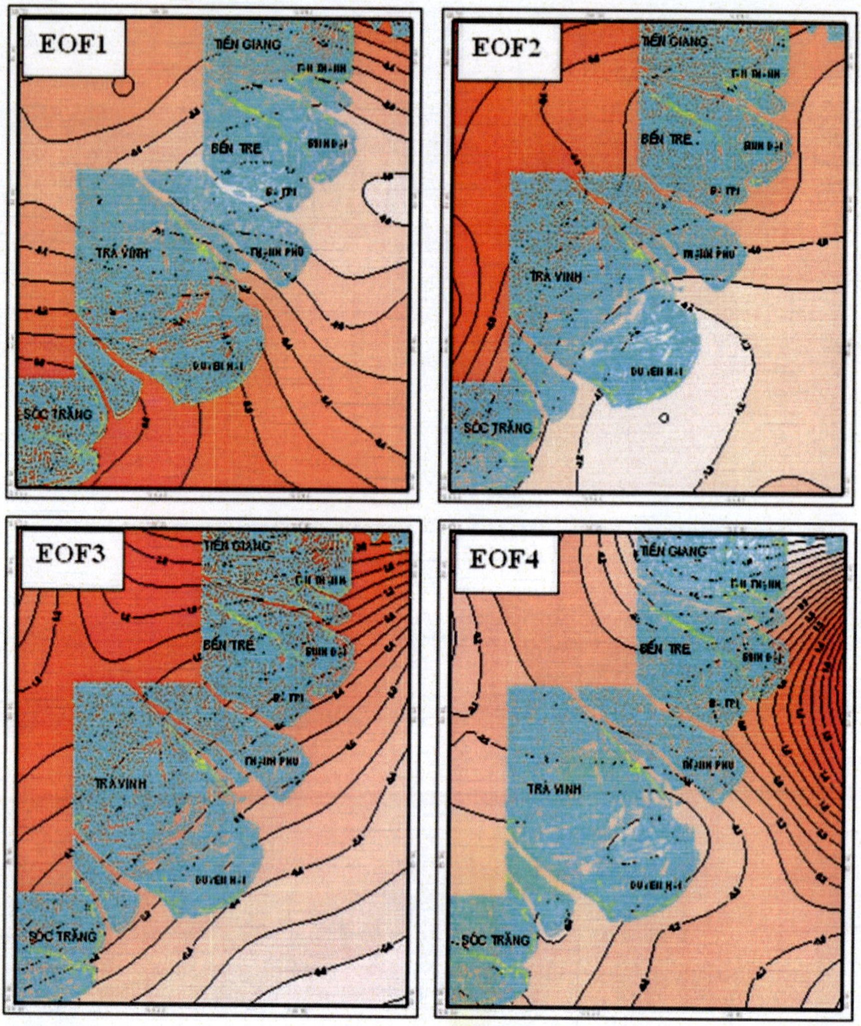

Fig. 18.5 Four first spatial EOF modes related to rainfall

We sometimes saw harmful algae blooms during our field surveys. In some cases, when algae patches were too dense and large, they caused eutrophication and serious depletion of oxygen. Moreover, pH increased to very high levels (pH>9) due to the impact of photosynthesis. This phenomenon usually occurred on Tan Thanh beach (Tien Giang Province). Local people call the algae blooms "water flowers" or "water bubbles," with their yellow, brown circular grains, which settle on the bottom when they were dead. The algae seriously affect clam life in January–February in this region and in the nearby region of Thua Duc and Thoi Thuan grounds (Ben Tre Province).

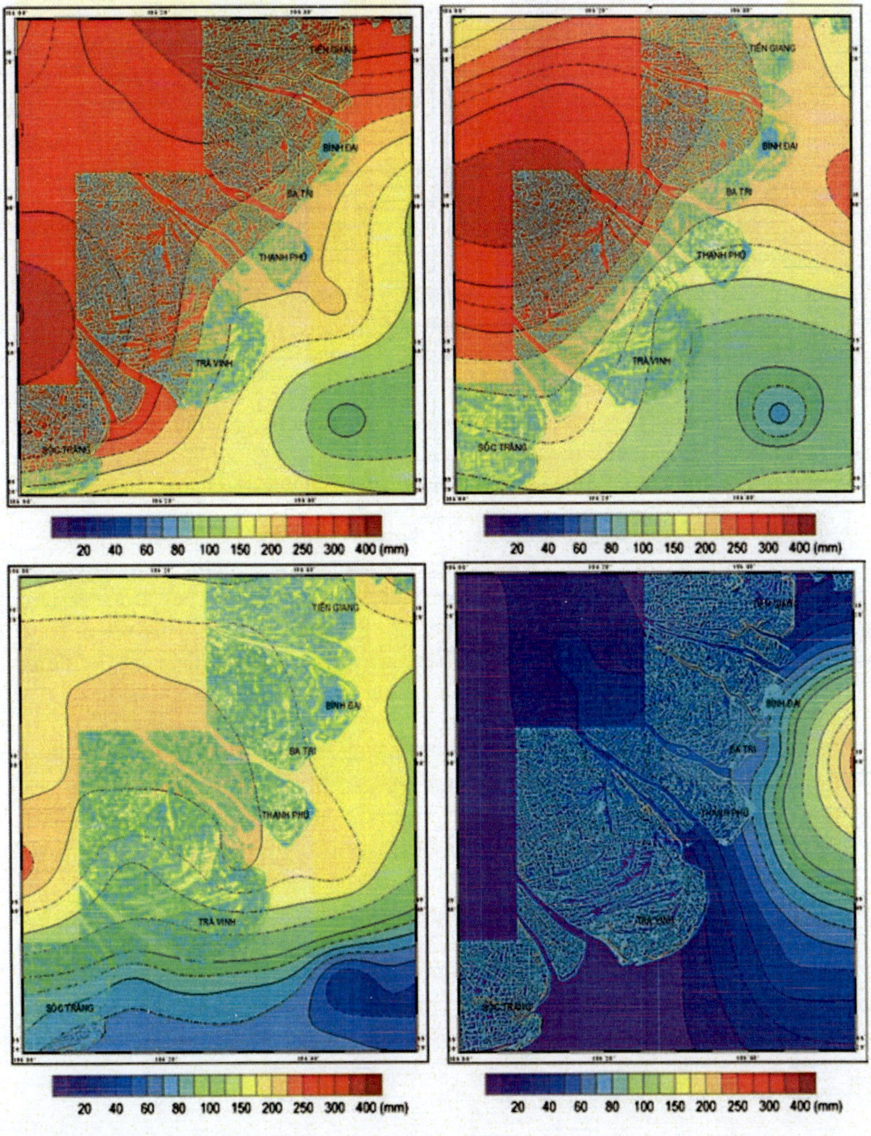

Fig. 18.6 Monthly distribution of rainfall during May, October, June, and January 2008 typical for EOF1, EOF2, EOF3, and EOF4

18.3.2 Hydrological Characteristics

18.3.2.1 Tidal Regime

The tide regime in the region studied was assessed by means of WXTide32 software in typical stations (in Vam Kenh, Binh Dai; Ben Trai, and Dinh An). The tide regime in this region is a semi-irregular diurnal one with two flood tides and two ebb tides. Tidal amplitude changed from 2.5 to 3.5 m during spring tide days and 0.8 to 1.0 m during days of neap tide.

The tidal amplitude tends to decrease from north (Tieu River mouth) to south (Dinh An River mouth) because the dominant direction of the tidal current and the alongshore current is from North East to South West, which causes erosion–deposition that affects clams.

Tidal amplitude was usually higher in November, December, and January. This coincided with the period of strongly blowing NE winds and the dry season, causing salinity intrusion inland. Lower tidal amplitude occurred from April to September. Calm winds during the rainy season combined with a weak tidal current (due to low amplitude) and smaller waves brought about favorable conditions for the formation of breeding grounds for clams during this period.

Investigations in 2008–2009 demonstrated the strong temporal and spatial variation of sea temperature, which fluctuated in the range between 29.6°C and 31.5°C. The SST dataset in 2007–2008 (Fig. 18.7), derived from MODIS imagery, showed that:

- SST changed to the range of 26.1–30.0°C (day) and 25.5–29.5 °C (night).
- SST during the rainy season was higher in comparison with the value in the dry season. Clearly, the effect of warmer terrestrial water plumes from in-river explains the big difference of SST between the two seasons. This warm water plume (especially during the immediate season) was an important condition that stimulated the reproduction and development of clam larvae during July–August.
- The value of day–night SST difference during the rainy season was two to three times greater than the values during the dry season.
- SST tends to decrease from North to South. The SST is closely related to along-shore current features (running North to South) and tends to increase from coastal to offshore areas in relation to warmer terrestrial water from interior rivers.
- During the rainy season, the warmer plume from interior rivers created the thermal fronts in river mouths. There was a relative coincidence between location of parents' clam grounds (from investigated data) and location of above fronts.

18.3.2.2 Total Suspended Solids (TSS)

Processed data regarding Total Suspended Solids (Fig. 18.8) derived from MODIS images show that, generally speaking, the coastal current carried a lot of food sources for clams, predominantly from NE to SW. In addition to nutrient sources from mangrove forests, algae patches were carried onto the coastal shore and into river mouths by waves in the investigated area. The nutrient and food sources carried from outside included food sources from the Xoai River mouth transported toward the southern side, which supplied clams during the dry season (November to April), and food sources transported toward the north from Batsac and Tran De estuaries (May to October). Note that the rate of the transported sediment in the latter direction was weaker.

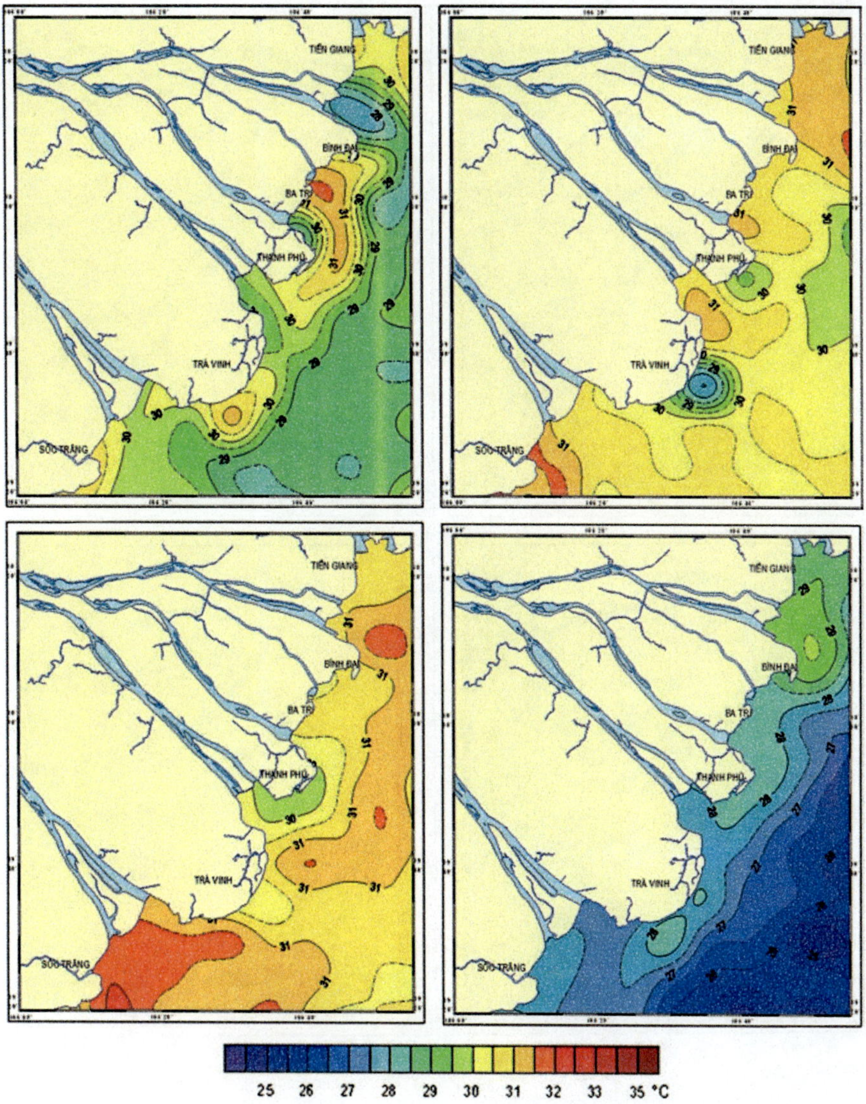

Fig. 18.7 Monthly distribution of SST (°C) in Jan, Apr, July, and Oct 2008

18.3.2.3 Plant Pigment Concentration (Chlorophyll-a)

The concentration of chlorophyll-a is one of the elements affecting primary productivity, phytoplankton, and water quality. Concentration of chlorophyll-a is the main factor creating high biomass. This is a way for clams to identify food sources.

Processed data of the content of chlorophyll-a (Fig. 18.9) derived from MODIS images show that:

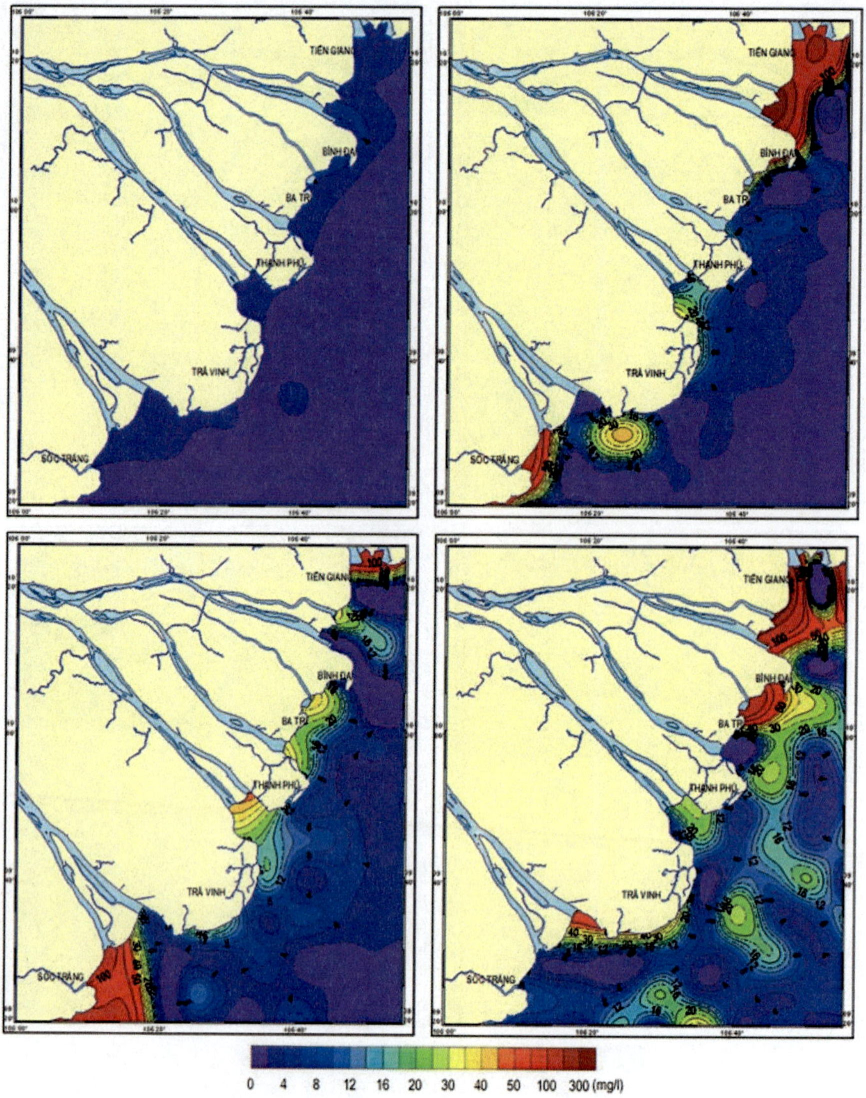

Fig. 18.8 Monthly distribution of TSS (mg/l) in Jan, Apr, July, and Oct 2008

Chlorophyll-a patterns changed in complex ways over time, but as a general rule, the content of chlorophyll-a in the dry season was higher than that in the rainy season. Its peaks often appeared about 1 month into the transition period from rainy to dry season (December, January) and from dry to rainy season (July).

In this period (December and January), solar radiation made the sea-surface temperature increase, and organic substances decomposed quickly. This created favorable conditions for the development of algae and then created algae patches on

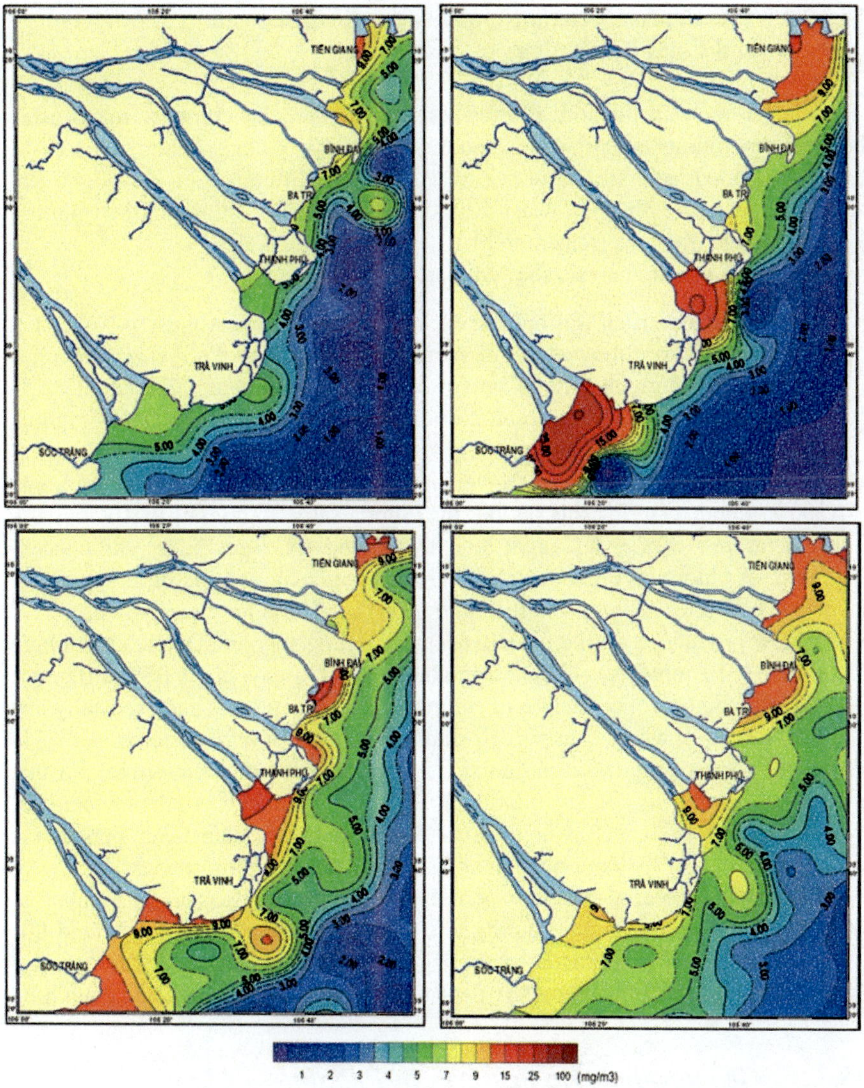

Fig. 18.9 Monthly distribution of Chlo-a (mg/m³) during Jan, Apr, July, and Oct 2008

the surface water. In addition, the effect of NE waves carried materials onto the shore, causing muddy deposits of organic substances and food. In general, the coastal water was high in nutrients in this period but not appropriate for the development of clams.

At the second peak in July, the content of chlorophyll-a was lower, but it had ecological significance for clam life. The early rains caused the water to be warmer, stimulating the reproduction and growth of clams in this area.

18.3.2.4 The General Relationship Between Environmental Parameters and Clam Production

The cluster-analysis method allowed us to ascertain the complex relationship between environmental parameters and the production of clams. The criteria in the dataset included: rainfall (mm/day), SST (°C), the difference of day–night sea-surface temperature ΔT (°C), water salinity (‰), content of chlorophyll-a (mg/m^3), and production of clams (tons/month).

The results of our analysis showed that:

- Clam production often was highest during the period from August to November (350 tons/month), followed by the period from March to April (220–250 tons/month). The optimum criteria set for clam production was: SST about 28.5–30.0°C, salinity >10‰, concentration of chlorophyll-a about 0.5–0.8 mg/m3, and the medium rainfall <10 mm/day.
- In some periods (September–November), production was still relatively high (200 T/month). During this period, the criteria set was: rainfall (>10 mm/day), temperature (>30°C), day–night SST difference >4°C (very high), and concentration of chlorophyll-a >2 mg/m3. These criteria limited the development of clams, but water salinity was still high enough (>10‰) for clam growth.
- The low production period usually occurred during July and October, with high rainfall (>10 mm/day), high temperature (>30°C), day–night SST difference >4°C (too high), concentration of chlorophyll-a >1.7 mg/m3, and especially low salinity (<8‰), which caused high clam mortality in several periods.
- Another lower production period was in the rainy season (May–June). At that time the rainfall was at a medium high level of 7.5 mm/day, but the temperature was too high (>30.5°C), which restricted the growth of clam. This period was only suitable for the reproduction and formation of breeding sources.
- Still another lower production period was in the dry season (November to February). The temperature during this period was very low (<26°C), and this was the most important factor preventing the growth of clams. In addition, at this time, the algae bloom and muddy deposition processes badly affected clam life.

18.3.3 Geomorphologic Features in Relation to Clams

18.3.3.1 Dynamic Geomorphologic Features in Tidal Inlets of the Mekong Delta

According to Davies (1964), tidal inlets in river mouths have been affected by both river dynamics and the tidal regime. Tidal inlets can be divided into three groups based on tidal amplitude.

- Micro tidal inlets: These exist in areas where tidal amplitude was lower, less than 2 m. The wave action and sand bodies formed by storms were the most important factors creating these tidal inlets. Typical morphological characteristics for this type of inlet were linear long sand bodies and narrow lagoons. Tu Hien

Fig. 18.10 River mouths with funnel shapes in the Mekong Delta

inlet and Tam Giang lagoon (Hue), Oloan, and DaNong inlets (Phu Yen) in North Central of Vietnam belong to this type.

- Meso-tidal inlets exist in areas where tidal amplitude was about 2–4 m. Ebb-tidal deltas, flood-tidal deltas, as well as meso sand waves in tidal flat were characteristic features in tidal inlets of this type. This type exists in the southern side of Central Vietnam, places such as Phan Ri, Phan Thiet, Ham Thuan Nam (Binh Thuan), and Vung Tau.
- Macro-tidal inlets exist in areas where tidal amplitude was higher than 4 m. In such areas, the river mouth was funnel-shaped and linear sand bodies move along in the direction of the ebb-tidal current. The tidal inlets in the Mekong Delta belonged to this type (Fig. 18.10).

In tidal flats of the Mekong Delta, there are several characteristic relief features, including main ebb channels, channel margin linear bars, ripple marks, sand waves, spillover channels, swash bars, and terminal lobes such as those described in the geomorphology models of Hayes (1973, 8), Bruun and Gerristsen (1959, 5), and Boothroyd (1985, 3, Fig. 18.11).

In general, during ebb times, the trend of the water current was to coincide with the axis of the main channel. Meanwhile, the flood-tidal current was close to the shore and along the marginal flood channel and the spillover channel.

As the ebb flood-tidal current reaches the shore, along the spillover channel, it creates a “slime muddy layer” which was conducive for clam-breeding grounds in the tidal flat.

18.3.3.2 Erosion–Accumulation and Shore-Line Change Processes in the Mekong Delta

By means of the band-ratio method (Van and Binh 2008), the shorelines of the Mekong Delta in different periods (1973, 1988, 1998, and 2008) have been extracted. The temporal shoreline changes in Mekong Delta are presented in Fig. 18.12.

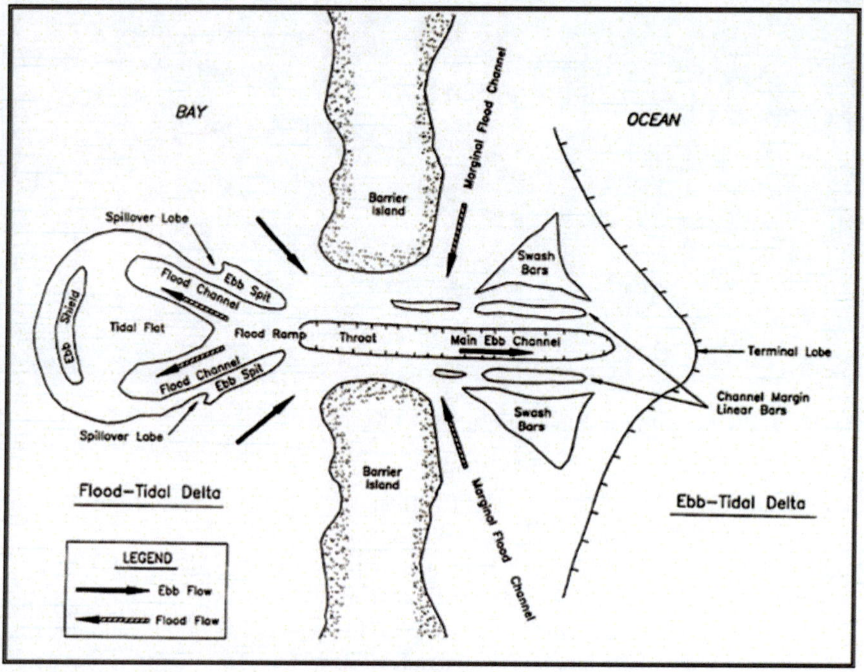

Fig. 18.11 Conceptual dynamic geomorphology model of a Tidal Inlet (Boothroyd 1985, 3)

The results show that:

- The erosion–deposition processes in tidal flats occurred alternately. The erosion process occurred in NE tidal flats; meanwhile, the deposition process occurred in SW tidal flats.
- Sand dunes moved from NE to SW in accordance with the alongshore direction.
- Flood-tidal deltas inside the river increased in size closer to the sea; and along the islets, flood-tidal deltas formed not only agriculture land but also clam grounds.

18.3.3.3 Geomorphologic Features in Relation to the Formation and Appearance of Clam Grounds

From historical satellite images linked to surveyed data and some basic knowledge from dynamic geomorphologic models (such as Boothroyd's (1985), Oertel's (1988), and Fitzgerald's (1988)), we were able to interpret the formation and appearance of clam grounds in relation to hydrodynamic regimes.

- Distribution of clam grounds on tidal flats of Tien Giang Province

The tidal flat in Tan Thanh was 6 km long, 5 km wide, and covered an area of about 1,500 ha. This is one of the main Potential Culture Zones (PCZ) for clams

a

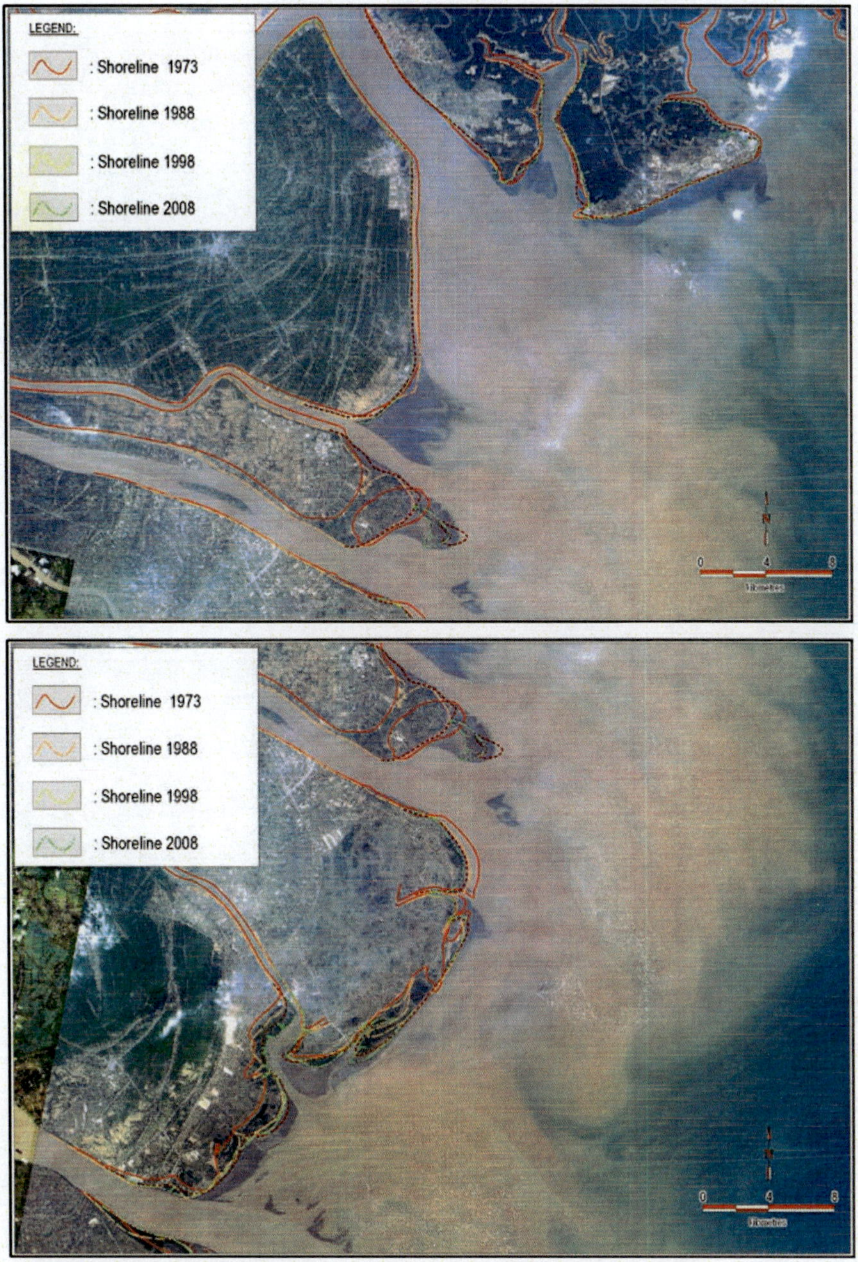

Fig. 18.12 Shoreline change in the Mekong Delta detected from satellite imageries 1973 (*red lines*), 1988 (*orange lines*), 1998 (*yellow lines*), and 2008 (*green lines*)

b

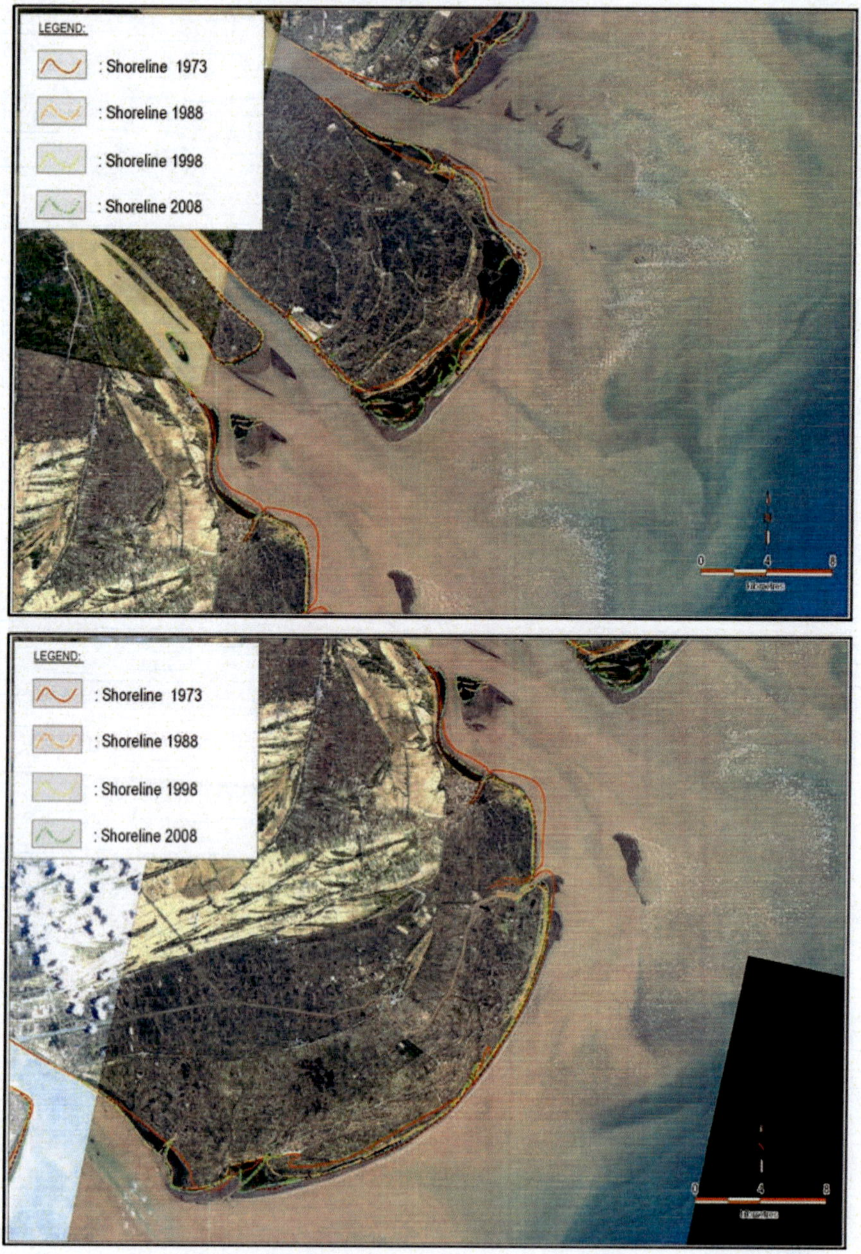

Fig. 18.12 (continued)

in Tien Giang. The tidal flat in the Tan Thanh area was "hand"-shaped with fingers (sand bars interlaid with troughs covered by fine materials) and a spillover channel where the breeding grounds usually appeared.

Ngang Island (Con Ngang) was considered a flood-tidal delta of the tidal inlet. It is located between the mouths of Cua Tieu and Cua Dai rivers, where clam-breeding grounds existed (Fig. 18.13).

The 12 km-long coastal section from Ap Cho (Vam Lang) to Cau Muong (Tan Thanh) (Fig. 18.14) was the area that was affected both by N-S alongshore and tidal currents and where the erosion process was dominant. The tidal flat was sloped, narrow, and unsuitable for the formation and development of clam grounds.

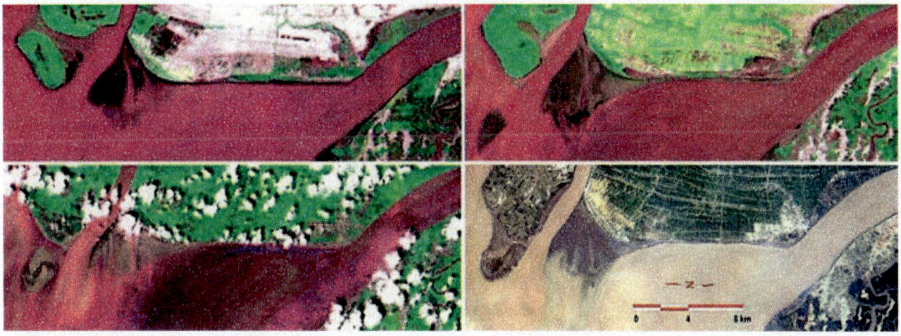

Fig. 18.13 The historical evolution of tidal flats in Tien Giang province (*upper-left*: 1973, *upper-right*: 1988; *lower-left*: 1997; and *lower-right*: 2008) (Note: The map direction was rotated 90° clockwise)

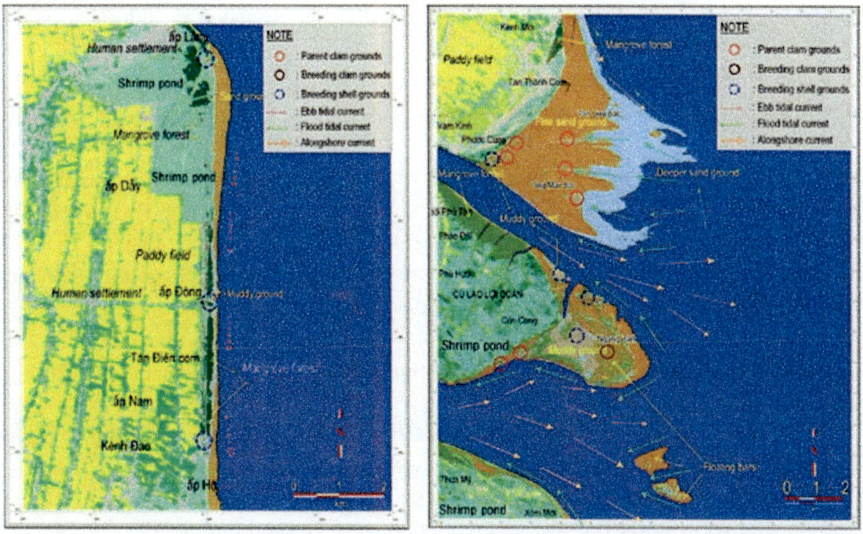

Fig. 18.14 The location of Potential Clam Grounds in Tien Giang in relation to geomorphology and dynamic conditions

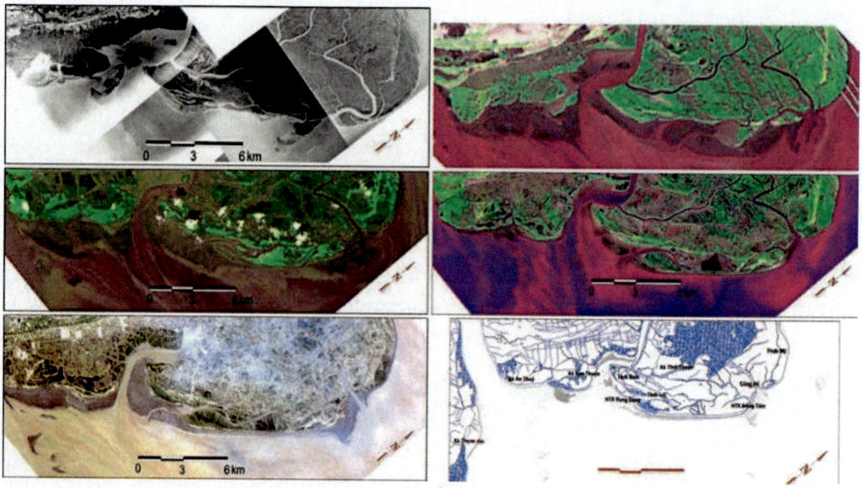

Fig. 18.15 The historical evolution of the tidal flat in Binh Dai and Ba Tri district (Ben Tre province) from 1968–1973–1988–1998–2008 (rotated maps to 45° clockwise)

- Distribution of clam grounds on the tidal flats of Ben Tre:

The historical evolution of tidal flats in Binh Dai and Ba Tri from multi-temporal imageries (Fig. 18.15) showed that:

In general, the geomorphology of the tidal flat in these regions changed greatly over time. The sand spit in Binh Dai moved in an alongshore direction (7 km/35 years). It narrowed the Ba Lai River mouth and created a large tidal flat in Thoi Thuan Village, where the biggest "clam mine" in the Mekong Delta existed.

The process of breaching the sand spit in the Cong Be River mouth during the period of 1968–1973 formed a second clam ground in Thua Duc Village.

To generalize, the biggest clam grounds in the Mekong Delta were located close to regions where there were small river mouths such as the mouths of the Ba Lai and Cong Be.

The clam grounds were concentrated in two main areas: one on tidal flats (PCZ=620 ha) in Thoi Thuan Village (Rang Dong co-operative; Fig. 18.16) and the other in Thua Duc Village (PCZ=260 ha – Dong Tam cooperative; Fig. 18.17).

The parents' grounds usually were located along the channel marginal linear bars at 4–5 m deep. The formation mechanism of these bars related to tidal-water circulation as mentioned above. The bottom relief was flattened, and the velocity of the flood current was medium (60–80 cm/s), which created a favorable habitat area for parent clams.

On the tidal flat in these areas there usually appeared a breeding ground of high density and consistency. The location of the breeding grounds was close to small channels, near the mangrove forest (mainly *Sonneritia alba*). The ground's morphology was lightly hollow, and well-sheltered by sand dunes. The permanent existence of a "slime muddy thin layer" was a typical characteristic of breeding grounds in this area.

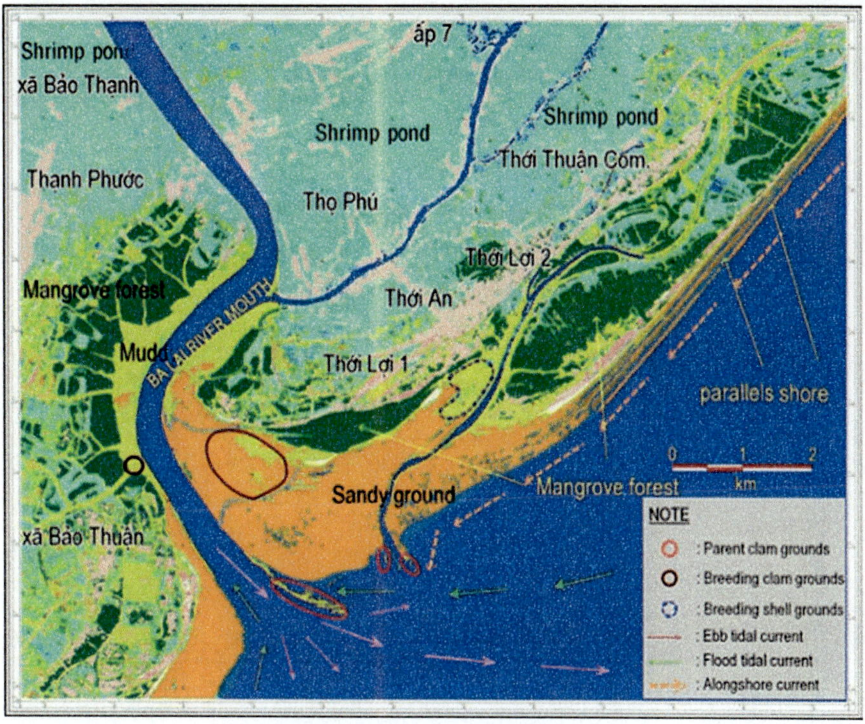

Fig. 18.16 The location of Potential Clam Grounds in Thua Duc – Binh Dai district (Ben Tre province) in relation to geomorphology and dynamic conditions

The coastal area of Ba Tri District included PCZ for clam culture in Bao Thanh, An Thuy, and Tan Thuy communes (Fig. 18.18). Breeding grounds sometimes appeared in the tidal flat of Tan Thuy but were of small size and inconsistent. The results from satellite images also found that "channel marginal linear bars" lie nearby the main channel axis, a typical indication of the predominant tidal current (Oertel 1988). The predominant tidal current in the Ham Luong River mouth is stronger, the result of which is that parent clams are not able to inhabit there and that breeding grounds rarely exist in nearby areas.

The historical evolution of tidal flats in Thanh Phu District revealed the same movement in the alongshore direction of the sand spit, as was the case in the tidal flat in Binh Dai District (Fig. 18.19). The breeding grounds also usually appeared in Thanh Phong and Thanh Hai villages.

The coastal area of Thanh Phu District included PCZ for clam culture in Thanh Hai (PCZ = 50 ha) (Fig. 18.20) and Thanh Phong communes (PCZ = 100 ha) (Fig. 18.21). There was a series of "linear bars" 5–6 m deep in the coastal waters of Thanh Hai. The parent grounds sometimes appeared there. A rough estimate, assuming an ebb-tidal-current velocity of about 80 cm/s, is that after 6 h, the materials in the water could overcome this bar and then rotate to return by flood tidal

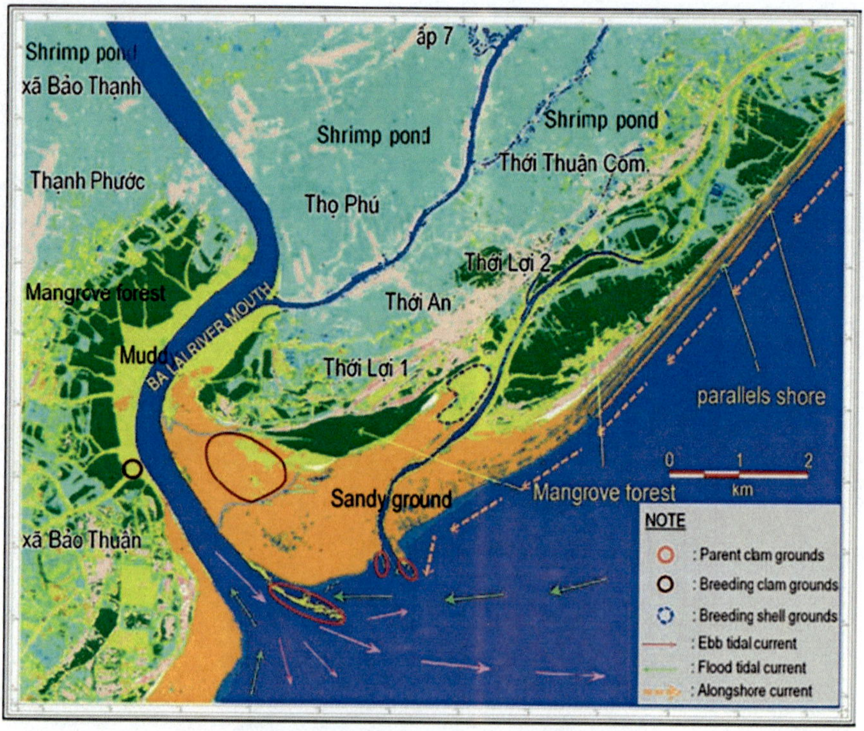

Fig. 18.17 The location of Potential Clam Grounds in Thoi Thuan commune – Binh Dai district (Ben Tre province) in relation to geomorphology and dynamic conditions

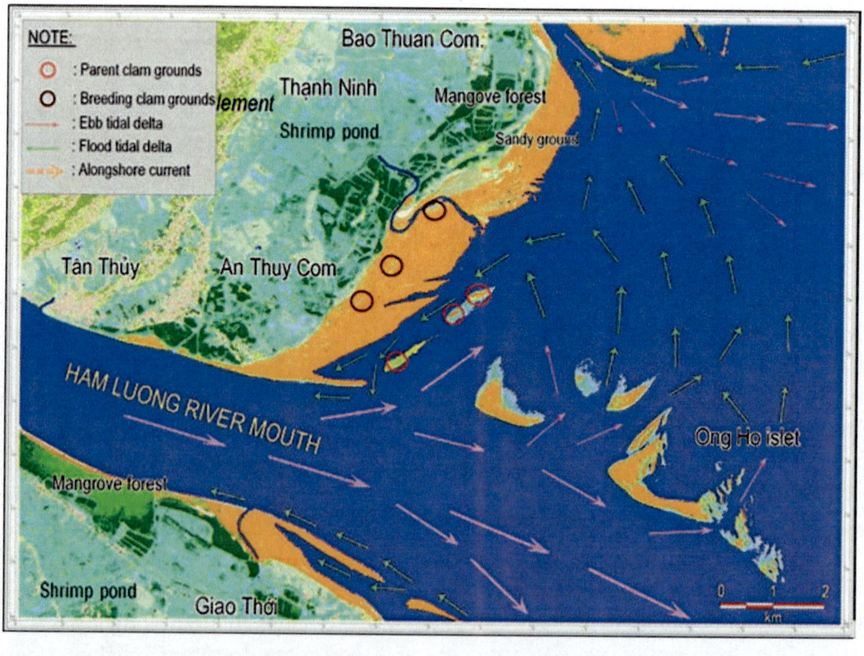

Fig. 18.18 The location of Potential Clam Grounds in Ba Tri district (Ben Tre province) in relation to geomorphology and dynamic conditions

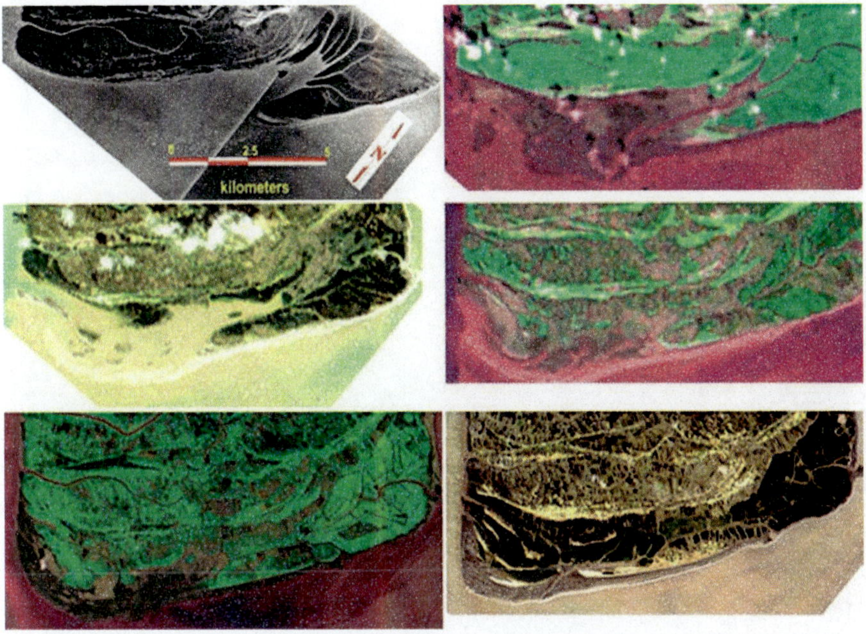

Fig. 18.19 The historical evolution of tidal flats in Thanh Phu district (Ben Tre province) from 1968–1973–1988–1998–2003–2008 (rotated maps to 50° clockwise)

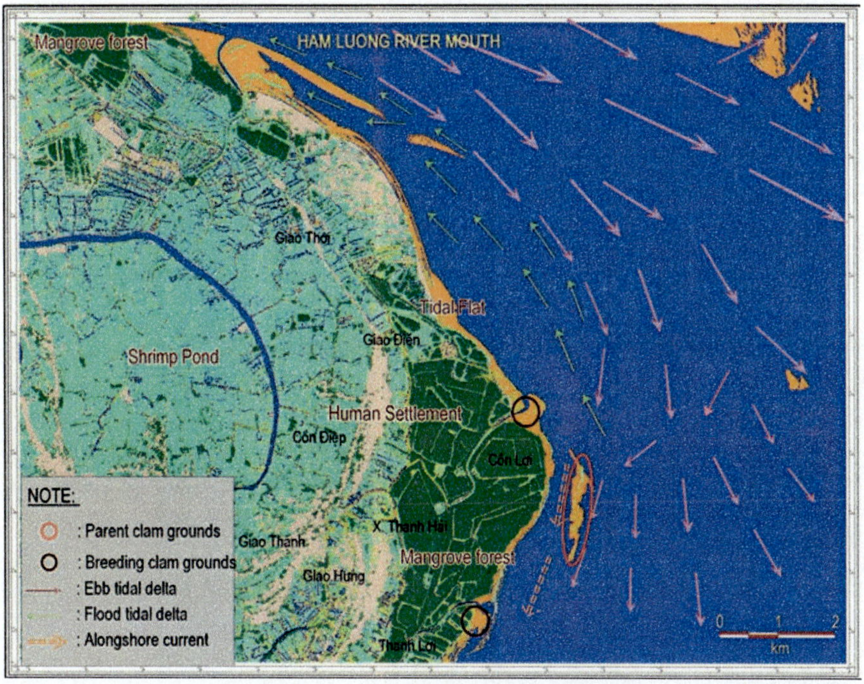

Fig. 18.20 The location of Potential Clam Grounds in Thanh Hai commune – Thanh Phu district (Ben Tre province) in relation to geomorphology and dynamic conditions

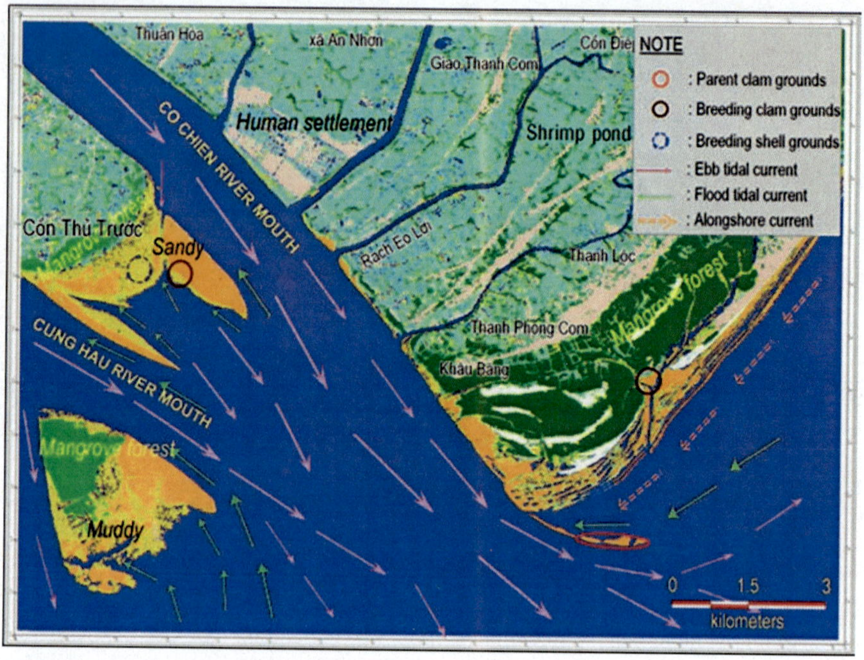

Fig. 18.21 The location of Potential Clam Grounds in Thanh Phong commune – Thanh Phu district (Ben Tre province) in relation to geomorphology and dynamic conditions

current (due to characteristics of semi-diurnal tide in the Mekong Delta). On the other hand, there exists a "stand water" area in this region that was suitable for a parent ground. This clam ground could form the basis of a breeding ground in nearby tidal flats. However, because of of the sloping littoral and weaker shelter, the erosion process usually occurred, resulting in breeding grounds that existed for a short time (sometimes days) and then disappeared.

- Distribution of clam grounds on the tidal flats of Tra Vinh.

The historical evolution of tidal flats in Tra Vinh Province exhibited the same trend as the tidal flat in Ben Tre (Fig. 18.22), moving in the alongshore direction of the sand spit toward North East–South West.

The processed data image results (Fig. 18.23) allowed us to explore the question of the relative similarity between dynamic-geomorphologic conditions and the formed locations of clam parents' grounds and breeding grounds in the two regions of Ben Tre and Tra Vinh. Our results were:

- Linear bars in on outside boundary of the "stand water" area were the forming mechanism: these bars existed in inshore waters close to Nha Mat Village (Truong Long Hoa Commune) where the parents' ground also appeared.
- On the tidal flat of Nha Mat and Ba Dong villages there were sometimes breeding grounds derived from linear bars in deeper waters.
- There was a Bar-through system which paralleled the shore.

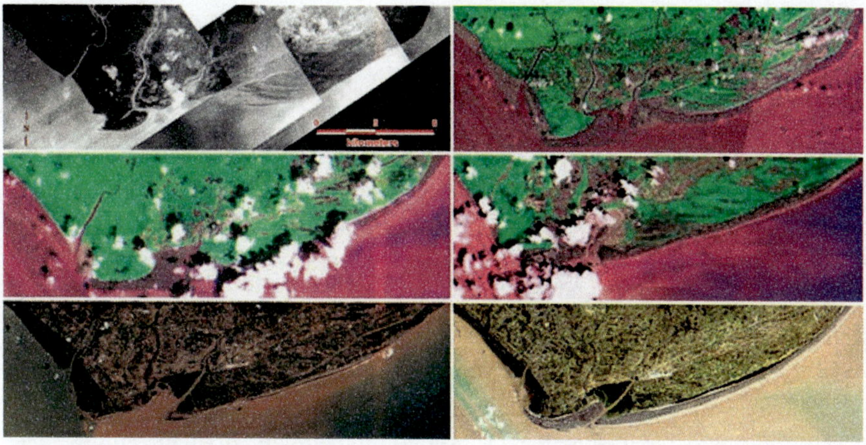

Fig. 18.22 The historical evolution on tidal flats in Tra Vinh province from 1968–1973–1988–1998–2002–2008 (map is not rotated)

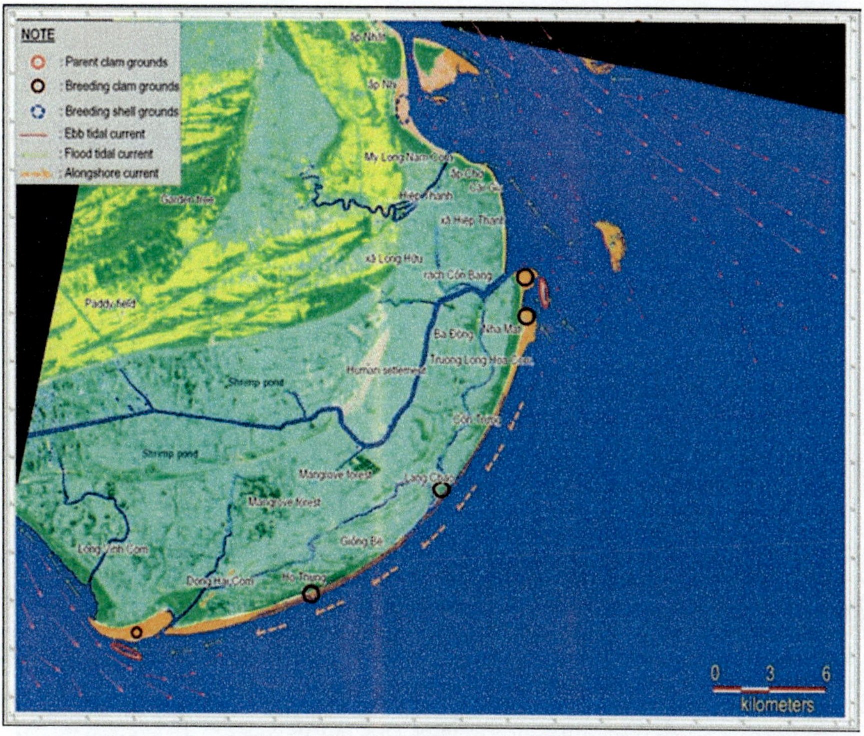

Fig. 18.23 The location of Potential Clam Grounds in tidal flats of Tra Vinh province in relations to geomorphology and dynamic conditions.

- The existence, appearance, position, and morphology of sandy, muddy grounds on Thu Truoc Island in the Cung Hau River were relatively similar to the morphology and bio-resources units on Dung Island in the Dinh An River mouth.

18.4 Conclusion

- The application of the remote-sensing technique based on multi-spectral imagery allowed us to determine qualitative and quantitative relationships between environmental parameters and clam life in the Mekong Delta.
- The rainfall, sea-water temperature, sea-water salinity, concentration of Chlorophyll-a, wind-wave regime, tidal regime, water circulation, and geomorphologic features in the Mekong Delta were major factors affecting clam life.
- The optimum environmental criteria for the growth of clam culture was found out.
- The suitable locations of parents' grounds and clam-breeding grounds, as well as areas of Potential Culture Zone for clam culture in Tien Giang, Ben Tre, and Tra Vinh provinces, were determined.

18.5 Recommendation

- There is a need to enlarge the study region so as to include the whole Mekong Delta (Can Gio, Soc Trang, Bac Lieu, and Ca Mau provinces as well) and also to study animal life forms (including oyster, clam, and other bivalve species), so that clams from areas other than Ben Tre achieve MSC certification and so that fresh marine culture in the whole Mekong Delta achieves commercial-brand status.
- There is a need for additional (and more detailed) studies of bio resources, community structure, the environment, hydro-dynamic regimes, and geomorphological features in relation to mollusk life. There is also a need to predict the effects of humans on mollusk life in the area in the future.
- Similarly, there is a need for a study to select suitable sites for the cultivation of clams (and other bivalve species) in the Mekong Delta, factoring in the effects of climate change (sea level rise, global warming) in various scenarios for the next 10, 20, and 50 years.

Acknowledgments The authors thank the national project 2008–2010 "The building of models for protection and development of the resources of Clam (*Meretrix lyrata*) and blood Shell (*Andara granosa*) on tidal flats in coastal areas of Tien Giang, Ben Tre, and Tra Vinh provinces – Mekong Delta." They thank Professor Mart Steward and Professor Peter A. Coclanis for their constructive comments and for editing the manuscript. They also thank the Japan Aerospace Exploration Agency (JAXA) for supporting the ALOS-AVNIR2 satellite imageries under ALOS-PI 326 project, the United State Geological Survey (USGS) for LandSat ETM$^+$ imagery, and the National Aeronautics and Space Administration (NASA) for the MODIS imageries which contributed in this study.

References

An Nguyen Tac, Nga TTT (2001) Scientific basis of solutions for protection and development of Clam and blood Shell resources in tidal flats in BenTre provinces (in Vietnamese). Project final report. Institute of Oceanography, Nha Trang

Bailey SW, Werdell PJ (2006) A multi-sensor approach for the on-orbit validation of ocean color satellite data products. Remote Sens Environ 102:12–23

Boothroyd JC (1985) Tidal inlets and tidal deltas. In: Davis RA (ed) Coastal sedimentary environments. Springer, New York, pp 445–532

Brown OB, Minnet PJ (1999) MODIS infrared sea surface temperature. Algorithm theoretical basis document version 2.0. University of Miami, Miami, FL 33149–1098, Under Contract Number NAS5-31361

Bruun P, Gerritsen F (1959) Natural bypassing of sand at coastal inlets. J Waterw Harbors Div ASCE 85:401–412

Davies JL (1964) A morphogenetic approach to world shoreline. Z Geomorphol 8:127–142

Fitzgerald DM (1988) Shoreline erosional-depositional processes associated with tidal inlets. Hydrodynamics and sediment dynamics of tidal inlets. In: Aubrey DG, Weishar L (eds) Lecture notes on coastal and estuarine studies, vol 29. Springer, New York, pp 186–225

Hayes MO (1973) Morphology of sand accumulation in estuary. Proceedings of 2d international estuarine research

Hao NV (2001) The characteristics on biology and production of Clam (*Meretrix lyrata*) in Mekong Delta. (in Vietnamese) Proceedings of scientific conference "Bien Dong" 2000, pp 219–230

Hung ND (2000) Study on ecological and environmental conditions effect to culture procedure in coastal areas of Clam *Meretrix lyrata* (Sowerby,1851) in Tien Giang and Ben Tre. Master's Thesis. National Science University, Ho Chi Minh City (in Vietnamese)

Murakami H, Sasaoka K, Hosoda K et al (2004) Version up of near real-time high level processing of MODIS data around Japan. In: MODIS near real time data

Oertel GF (1988) Processes of sediment exchange between tidal inlets, ebb deltas, and barrier Islands. Hydrodynamics and sediment dynamics of tidal inlets 29. In: Aubrey DG, Weishar L (eds) Lecture notes on coastal and estuarine studies. Springer, New York, pp 297–318

Phu TQ (1999) Study on some features on biology, biochemistry and culture technique of clam – *Meretrix lyrata* in coastal areas of Tien Giang, Ben Tre provinces. PhD Dissertation (in Vietnamese), The College of Aquaculture and Fisheries, Can Tho University

Reyment RA, Jvreskog KG (1999) Applied factor analysis in the natural sciences. Cambridge University Press, Cambridge and New York

Tuan VS (1999) Investigation on ecological – environmental conditions and developing ability of Clam *(Meretrix lyrata)* in Go Cong district – Tien Giang province. Project final report (in Vietnamese). Institute of Oceanography, Nha Trang

Tung NT (2007) Study the solution for protection and development of Clam (*Meretrix lyrata*) Sowerby, 1851 in Mekong Delta. Project final report (in Vietnamese). Southern Branch of Institute of Economic and Fishery, Ho Chi Minh City

Van TT, Binh TT (2008) Shore line change detection to serve sustainable management of coastal zone in Cu Long Estuary. In: Proceedings of international symposium on geoinformatics for spatial infrastructure development in earth and allied sciences, Hanoi

Chapter 19
Compost Potential from Solid Waste: Toward Sustainable Agriculture and Mitigation of Global Warming in the Mekong Delta, Vietnam

Nguyen Phuc Thanh and Yasuhiro Matsui

Abstract This study was conducted to evaluate the positive effects of compost application in agriculture and the benefit of the composting process for solid-waste treatment in the Mekong Delta region in southern Vietnam (comprised of 12 provinces and 1 centrally city). A simple prediction model for waste generation based on population – and employing *time-series analysis* – was constructed to assess and forecast waste generation to identify the potential of waste for composting and discharging. The greenhouse gas baseline emission from biodegradable components and the greenhouse gas reduction emission for alternative composting options were also calculated in g(CO_2 eq.) per capita per day. Scenarios regarding different composting applications for the Mekong Delta region were defined. The estimations focused on such matters as environmental impacts, greenhouse-gas emission and reduction, resource consumption, economic benefits, and the application potential of the composting process. The principal result was that organic-waste composting was beneficial not only for mitigating the waste burden in landfill sites but also for reducing greenhouse-gas emissions compared to baseline emission and the attendant costs. In addition, the compost product was identified as having strong potential to displace chemical fertilizers in agriculture. It was shown that the composting process was a sound alternative for moving toward sustainable development in waste treatment and agriculture.

Keywords Agriculture sustainability • Composting • Development sustainability • Greenhouse gas • Mekong Delta • Mitigation • Solid-waste management sustainability

---

N.P. Thanh (✉)
Graduate School of Environmental Science, Okayama University, 3-1-1 Tsushima-naka, 700-8530 Okayama, Japan
and
College of Environment and Natural Resources, Can Tho University, Can Tho, Vietnam
e-mail: npthanh@ctu.edu.vn; phucthanhn@gmail.com

Y. Matsui
Graduate School of Environmental Science, Okayama University, 3-1-1 Tsushima-naka, 700-8530 Okayama, Japan

M.A. Stewart and P.A. Coclanis (eds.), *Environmental Change and Agricultural Sustainability in the Mekong Delta*, Advances in Global Change Research 45,
DOI 10.1007/978-94-007-0934-8_19,

19.1 Introduction

Solid-waste management is facing challenging problems in many countries. Two forms of waste in particular – kitchen waste (food waste) and yard waste – account for major shares of total waste. Composting of organic waste not only reduces the amount of waste taken to landfills but also yields a relatively low-cost product that can be used in agriculture (Jakobsen 1995; Wolkowski 2003). Several authors have assessed the positive repercussions of compost application in agriculture and the benefits of the composting process; among them, Martínez-Blanco et al. (2009) found the use of compost in horticulture to be a treatment with fewer negative environmental impacts than mineral fertilizer, and no differences were observed in terms of agricultural performance.

Global warming is one of the most serious matters of public concern in recent years, and organic-waste composting is useful not only for the mitigation of the waste burden in landfill sites but also for the reduction of greenhouse-gas (GHG) emissions and their attendant costs. Ayalon et al. (2001) found that compared to solid-waste treatment alternatives, composting was the most cost-effective option, although its efficiency of CH_4 reduction (90%) was lower than incineration or anaerobic digestion methods (100%). Moreover, the authors of that study highlighted the potential for developing Clean Development Mechanism (CDM) projects in the solid waste field (IPCC 2006).

Vietnam, like other developing countries in Southeast Asia, uses open dumping as its main disposal method (Idris et al. 2004). However, municipal solid waste contains a large proportion (60–80%) of easily degradable organic waste (World Bank et al. 2004), which has great potential for composting applications (Byer et al. 2006). For that reason, this study was conducted to evaluate the positive effects of compost application in agriculture and the benefit of the composting process for solid-waste treatment. A simple prediction model was developed to assess and forecast waste generation in order to identify the potential of waste for composting and discharging. Moreover, GHG baseline emission from the predicted waste generation was calculated and warned. The authors also evaluated GHG emission mitigation for various options of composting application via scenario analysis. The evaluations concentrated on such matters as GHG emission, production of the compost product, displacement of chemical fertilizer for agriculture through employment of the waste-compost product, required capacity for treating (landfill), and costs and benefits. In addition, the authors reviewed and integrated into their analysis the positive effects and benefit of the composting process on GHG emission and reduction, energy consumption and generation, investment and operation cost, and the potential use of compost in agriculture.

19.2 Methodology

19.2.1 Study Area and Current Status

This study estimated the waste generation of the Mekong Delta region (including 12 provinces and 1 centrally city) with an estimated population of 17,524,000 in 2007 (GSO 2007). The Mekong Delta spreads over approximately 40,000 km^2, occupying about 12% of the total natural area of the country (Fig. 19.1). The Mekong Delta is a vital agricultural zone in the country. With a tropical monsoon climate and favorable weather conditions, the area lends itself to the growth of paddy and a wide range of plants and vegetables all year round. Today, a fourth of the Delta is under rice cultivation, making this area one of the premier rice granaries in the world.

Fig. 19.1 Map of the Mekong Delta region within Vietnam (Source: http://en.wikipedia.org/wiki/Mekong_Delta (Edited by the authors))

The Mekong Delta is one of the areas in the world that is most vulnerable to the effects of climate change, because this region is on average a few meters above sea level. Furthermore, this region faces great challenges in solid waste management, including waste collection, transportation, final disposal, and especially the need to introduce sustainable waste-treatment alternatives. In the Mekong Delta region, generally, waste is only collected in central urban areas, where the collection rate is estimated by the Urban Environment Company to be about 70–80%. Meanwhile, no waste collection is undertaken in rural and suburban areas (European Commission 2009).

19.2.2 Waste Generation Prediction for the Mekong Delta Region

In the prediction model the amount of waste generated was calculated based on the estimated population, and per-capita waste generation rate. Population was estimated until 2020 via the *time-series* method that used a database containing information on population levels by provinces and cities in the period 1995–2008 (GSO 2009). The per-capita waste generation rate was 0.7 and 0.3 kg/cap/day for urban areas and rural areas, respectively (World Bank et al. 2004). The quantity of waste generation and waste collection was calculated for two groups: the residents in rural areas and the residents in urban areas. It was assumed that the per-capita waste-generation rate would not change in the estimation period.

The linear regressions of the 14-year time-series (1995–2008) are shown in Table 19.1 with the linear equation and the coefficient of determination (R^2), respectively. The model of linear function used is represented in the following equation:

$$Y_i = aX_i + b \tag{19.1}$$

Table 19.1 Population trend of each province and city in the period of 1995–2008

No.	Name of provinces and city	Area (km²)	Population trend according to time (years)			
			Urban areas		Rural areas	
			Linear equation	R^2	Linear equation	R^2
1	Long An	4493.8	$Y=4.613X+185.8$	0.857	$Y=10.80X+1,051$	0.966
2	Tien Giang	2484.2	$Y=5.125X+191.5$	0.963	$Y=7.796X+1,361$	0.962
3	Ben Tre	2360.2	$Y=2.916X+98.05$	0.935	$Y=3.768X+1,172$	0.875
4	Tra Vinh	2295.1	$Y=4.592X+96.52$	0.944	$Y=5.036X+825.4$	0.912
5	Vinh Long	1479.1	$Y=2.102X+134.5$	0.878	$Y=4.105X+847.7$	0.948
6	Dong Thap	3375.4	$Y=7.609X+168.4$	0.940	$Y=7.090X+1,301$	0.882
7	An Giang	3536.8	$Y=24.33X+311.5$	0.958	$Y=6.156X+1,605$	0.734
8	Kien Giang	6346.3	$Y=13.11X+266.3$	0.983	$Y=12.55X+1,104$	0.977
9	Can Tho	1401.6	$Y=9.825X+540.6$	0.915	$Y=3.59X+555.6$	0.906
10	Hau Giang	1601.1	$Y=4.717X+72.6$	0.961	$Y=3.208X+627.1$	0.814
11	Soc Trang	3311.8	$Y=3.608X+194.0$	0.935	$Y=8.820X+932.2$	0.959
12	Bac Lieu	2585.3	$Y=3.688X+165.0$	0.961	$Y=5.713X+529.3$	0.973
13	Ca Mau	5331.6	$Y=5.941X+176.2$	0.961	$Y=10.29X+859.6$	0.994

b is the intercept and it indicates the mean value of the response variable when = 0; *a* is the slope and it indicates the average change in the year *i* ($i = 1...n$, started from 1995 ($i = 1$)), when the random variable rises. Y_i is the population rate (persons) in the year *i*, X_i is number of years started from 1995.

Therefore, the predicted population, based on those linear equations, is used to calculate waste generation by specific areas.

19.2.3 Estimation of GHG Emission Reduction

19.2.3.1 GHG Baseline Emission

The GHG baseline emission from waste is the amount of methane calculated in tons of carbon dioxide equivalent (tCO_2 eq.) that would be generated from disposal of waste at a solid waste-disposal site in the absence of the proposed composting project. GHG baseline emission was determined via the "*2006 IPCC Guidelines for National Greenhouse Gas Inventories*" (IPCC 2006). For the sake of simplicity, for calculation and prediction of the baseline emission from solid-waste disposal sites, we used a spreadsheet model for *Estimating Methane Emissions from Solid Waste Disposal Sites* (*IPCC Waste Model*) developed by the IPCC (2006). In this study, the GHG baseline emission was calculated using the default emission factors in the guideline of the 2006 IPCC tier 1 approach.

19.2.3.2 GHG Emission and Reduction

GHG emission reduction is the difference between GHG emission in the composting project compared to baseline emission (IPCC 2006; Plöchl et al. 2008; UNFCCC 2008). Generally, emission reduction in the year *y* is calculated according to the following equations:

$$ER_y = BE_y - PE_y \pm L_y \quad (19.2)$$

$$PE_y = PE_{processing,y} + PE_{energy\ consumption,y} - PE_{energy\ generation,y} \quad (19.3)$$

Where ER_y is the emission reduction in the year *y*; BE_y is the baseline emission in the years; PE_y is the project emission in year *y*; and L_y is the leakage emission in year *y*. The emission was considered within the project boundary as defined in the Glossary of CDM terms version 01 (UNFCCC 2007).

The project emission (PE_y) in this study was defined so as to include the emission from energy consumption ($PE_{energy\ consumption,y}$), energy generation ($PE_{energy\ generation,y}$), and processing of project activities ($PE_{processing,y}$), which is expressed in the Eq. (19.3).

This study concentrated only on composting, which has no energy generation, but energy consumption was 30 kWh/t waste inputs (McDougall et al. 2001). In this study, it is assumed that electricity consumed was imported from the grid. Tuyen and Michaelowa (2004) reported that the baseline emission factor for the Vietnam electricity system was 0.585 tCO_2/MWh by 2008.

GHG emission from processing of project activities ($PE_{processing,y}$): The quantity of emissions of GHG (CH_4 and N_2O) from composting depends on the amount of waste treated and the related emission. For the present study, the emission was calculated using the default emission factors of the 2006 IPCC tier 1 approach (IPCC 2006).

The above-mentioned theories and explanations were used to calculate the GHG emission and reduction from waste generation and composting.

19.2.4 Economic Estimation

Regarding the economic benefits from the project: These were calculated on the basis of the value of useful products, such as compost products or GHG reduction credits (Certified Emission Reductions – CER).

Regarding compost products, we are thinking of products such as fertilizer for soil supplementation. Good-quality compost accounted for approximately 30% of the final compost product; if supported by suitable nutrient components, it can sell on the market at prices as high as 30 US\$/t. By reducing GHG emissions from alternatives, "carbon credits" or "CERs" are generated, that is, climate credits issued by the CDM – Executive Board – UNFCCC. These credits can be sold in international markets to developed countries. The trading price of carbon credit units (US\$/t CO_2 eq.) depends on the world market. In this study, the trading price of "carbon credits" is based on the current market price in Vietnam, 8–12 US\$/t CO_2 eq. (Michaelowa et al. 2009) and the price of 10 US\$/t CO_2 eq. was used as an estimate.

19.2.5 Application Potential of Compost in Agriculture

To estimate the potential for employing compost in agriculture, we needed to calculate and evaluate the *land area equivalent of compost production.* This was defined as the land area "gained" through composting solid waste. Estimates of the land area equivalent by compost feeding were based on the amount of compost production and the appropriated supplying rate of compost employed in agriculture, calculated according to Eq. 19.4. In the study, the use of the compost product (3R-HN 2009) focused on three kinds of trees: agriculture trees, food trees, and fruit trees. The supplying rate by kind of tree is shown in Table 19.2.

$$\text{Land area equivalent by compost feeding (ha)} = \frac{\text{Compost production (ton)}}{\text{Appropriated supplying rate (ton/ha)}} \qquad (19.4)$$

Table 19.2 Rate of compost employment in agriculture

Kinds of trees	Unit	Range of application	Average value
Agricultural trees	kg/ha	300–400	350
Food trees	kg/360 m^2	20–30	25
Fruit trees	kg/360 m^2	30–40	35

Source: 3R-HN (2009)

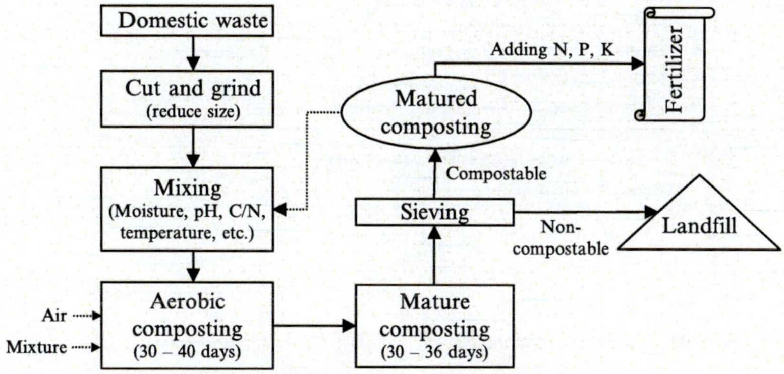

Fig. 19.2 The typical composting technology applying in Vietnam

19.2.6 Waste Composting Process

Regarding composting, the final composting product accounted for 50% of input loading; the loss in mass was due to evaporation and biodegradation of organic fraction (McDougall et al. 2001). Furthermore, the authors assumed that 10% of input loading for composting was removed by the sieving process after composting process; this part would be discharged to landfill. Above -mentioned methods were utilized to calculate the composting product and residual amount after composting which went to landfill. In this study, the composting process and the derivative product (considered in terms of quality and quantity) were adapted to the real situation in Vietnam. Composting of organic waste has been applied rather generally and the technology differs little. The typical technology for organic solid-waste composting is presented in Fig. 19.2.

19.2.7 Description of Scenarios

The scenarios in the study are based on alternative composting waste-treatment plans. The authors used simple assumptions and the specific boundaries and technologies presented in Table 19.3. The study estimated waste generation environmental effects, and benefits for the period of 2009–2020 in all of the scenarios, with the exception of the baseline scenario, for which estimates run until the year 2075.

Table 19.3 Scenario description

Scenarios	Definition	Note
0	Baseline: Non-collection and treatment	It is assumed that residual amount of waste generation will go to open dumping site
1	Assume that collection rate is 82 (following the "Resolution No. 23/2008/QH12" (VNA 2008)), and 100% of collected organic waste is treated by composting	
2	A-part composting: 25% Organic waste is treated by composting	
3	A-part composting: 50% Organic waste is treated by composting	
4	A-part composting: 75% Organic waste is treated by composting	
5	Complete composting: 100% Organic waste is treated by composting	

19.3 Results and Discussion

19.3.1 Prediction of Waste Generation and Composition

Figure 19.3 presents the waste generation rate by provinces and cities in the Mekong Delta region, including waste generation from both urban and rural areas. This figure was derived from statistical data (14 years, from 1995 to 2008) and forecasted data (11 years, from 2009 to 2020) from the linear equations in Table 19.1. The figure also shows that the prediction models are rather linear; except in the cases of Can Tho city (—o—) and Hau Giang province (—◆—), where the linear trend changed in 2003. These cases are explained by the fact that Can Tho Province was separated into two units in 2003, with Can Tho City now included in Hau Giang Province.

The results presented in Fig. 19.4 relating to patterns of waste generation in the Mekong Delta region show that most waste is generated in rural areas (around 90%). Urban areas were marked by low collection efficiency, with collection and treatment in such areas averaging only 70–80%. This point demonstrates that most of the solid waste generated has not been collected and treated in the Mekong Delta region. This finding is quite alarming because of the serious environmental consequences resulting from waste generation that has not collected and treated yet.

Regarding waste composition, mainly physical compositions of typical provinces and cities have been shown in Table 19.4. For the estimation purpose in this paper, the representation value of main compositions of waste generation for the case of the Mekong Delta region was also calculated. It was average value of typical cases, shown in the last row of Table 19.4. This representation value was used to calculate various aspects in this study.

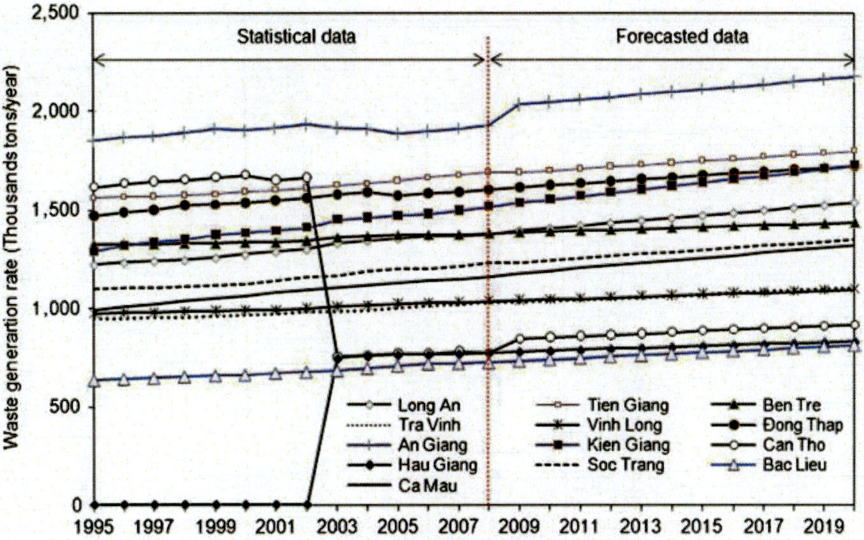

Fig. 19.3 Waste generation by provinces and cities in the Mekong Delta region

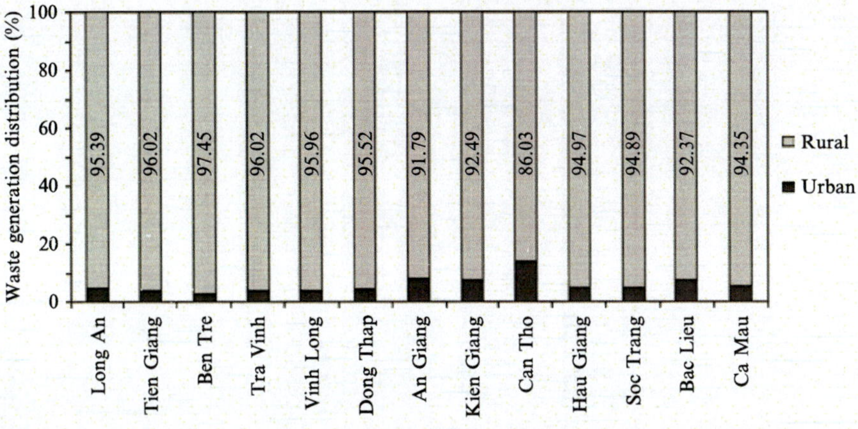

Fig. 19.4 Waste generation distribution in the Mekong Delta region

19.3.2 GHG Baseline Emission

Figure 19.5 shows the GHG baseline emission from waste generation in the Mekong Delta region in the period 1995–2075 (80 years), including the loading period (1995–2020) for waste disposal calculated from Sect. 19.3.1, and the residual waste decomposition period – the period wherein residual waste emits pollution at the disposal site – over the next 55 years (2020–2075). The baseline emission was presented in terms of the amount of GHG (CO_2 eq.); it showed the same

Table 19.4 Estimation of waste composition (%) for the Mekong Delta region (*Source*: European Commission 2009)

Provinces	Paper	Glass	Metals	Plastics	Organic	Hazardous	Bricks and ceramics	Rubber, leather, textiles	Miscellaneous
Long An	5.10	0.70	0.37	13.63	76.30	0.15	2.68	–	1.08
Tien Giang	3.89	0.21	0.23	6.37	77.53	0.06	2.14	–	9.57
Ben Tre	6.50	0.85	1.75	3.40	72.80	0.25	1.60	1.35	6.30
Vinh Long	11.50	4.00	0.55	9.45	66.25	–	0.75	6.50	1.00
Kien Giang	6.38	1.64	1.27	7.69	72.52	–	7.49	1.29	1.72
Can Tho	2.79	1.53	0.70	9.57	79.65	0.03	3.10	1.68	0.76
Hau Giang	1.80	0.90	0.40	5.70	82.60	4.00	1.60	1.50	1.50
Soc Trang	4.12	0.66	0.78	7.24	70.35	–	9.63	3.11	4.11
Bac Lieu	4.51	4.91	4.59	4.44	53.34	2.78	10.81	7.69	6.92
Ca Mau	4.50	0.50	0.10	6.10	57.30	–	2.10	1.40	28.00
Representation	5.11	1.59	1.07	7.36	70.86	0.73	4.19	2.45	6.10

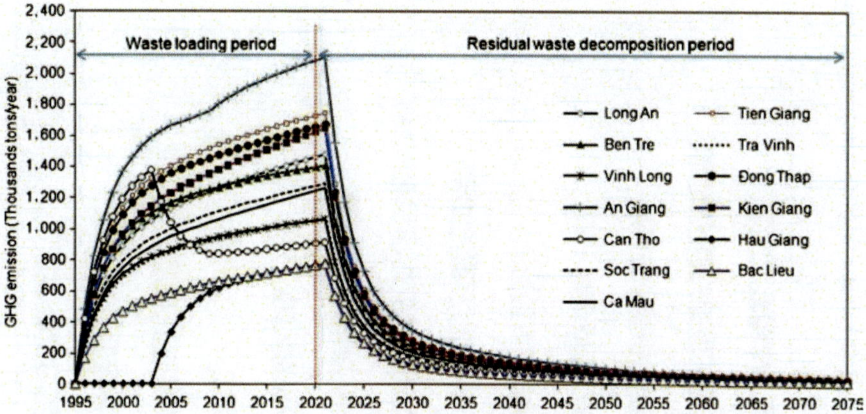

Fig. 19.5 GHG baseline emission by provinces and cities in the Mekong Delta region

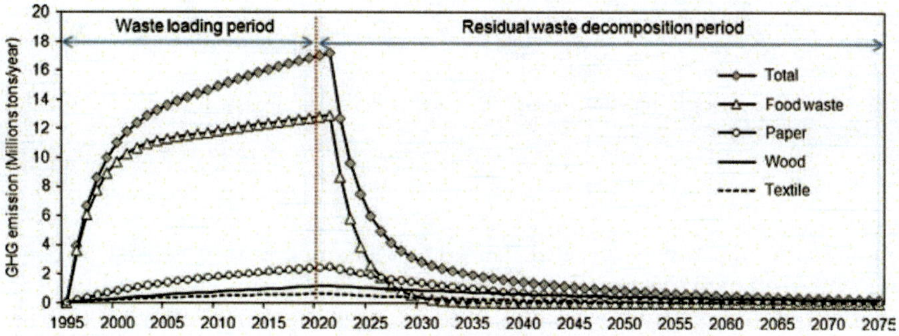

Fig. 19.6 GHG baseline emission by waste components

gradually increasing linear trend for provinces and cities in the Mekong Delta region. Emissions also gradually increase in line with the time and amount of disposal waste. Emissions accumulated to reach a maximum at the end of the waste-loading period by 2020, then are reduced in a gradual slope in the following next period.

The amount of emissions mainly depended on waste composition, as can be seen in Fig. 19.6. The major GHG emissions from waste were caused by the food component. In comparison to other components, the baseline emission of the food component quickly dropped drown after waste-loading stopped. This suggests that the GHG emission in the food component occurred in a short-time period after the disposal of waste.

Figure 19.7 shows the total GHG baseline emission in the estimation period 1995–2075. This figure also presents the distribution of GHG baseline emissions among the Mekong Delta provinces and cities. The figure demonstrates conclusively that GHG emissions in rural areas accounted for the larger share of emissions.

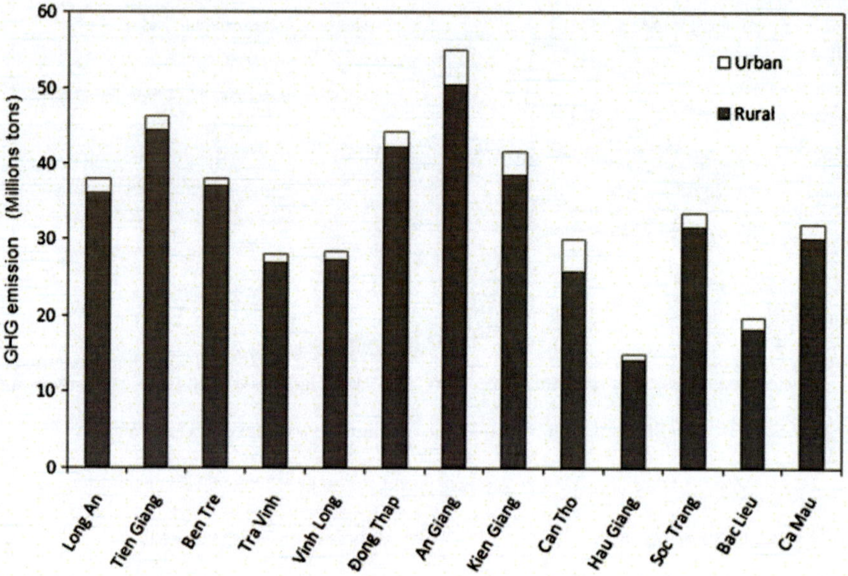

Fig. 19.7 GHG baseline emission during the period (1995–2075) by provinces and cities in the Mekong Delta region

Furthermore, the results revealed that GHG emissions were generated in large amounts in all of the Mekong Delta's provinces and cities. However, the severity of emission problems was due mainly to the fact that most emissions could not be controlled in rural areas.

19.3.3 Estimation of Waste Treatment Alternatives

The estimation of environmental effects, resource consumption, investment, cost of operation, and the benefit of waste-treatment alternatives were encoded in qualitative scales as illustrated in Table 19.5. A positive symbol (+) indicates that the process in question had a positive effect and vice versa with the negative symbol (–). The comparative importance of the effects of the five disposal processes is represented by the number of symbols.

Regarding GHG emissions, Thanh and Matsui (2009) found that the highest emissions came from open-dump landfills, followed in descending order by emissions from sanitary landfills, composting, anaerobic digestion, and incineration, as shown in Table 19.5. In addition, this table expresses the GHG emission-reduction of each alternative; open-dump landfills resulted in the lowest emission reduction, followed by sanitary landfills, composting, and aerobic digestion. Incineration was identified as the greatest GHG emission-reduction alternative.

Table 19.5 Estimation of waste-treatment alternatives

Alternative methods	Energy		GHG (CO_2eq.)		Cost		
	Consumption	Generation	Emission	Reduction	Investment	Operation	Benefit
Open-dump landfill	0	0	– – – – –	0	–	–	0
Landfill with energy recovery	–	+ +	– – – –	+	– – – –	– – – –	+ + +
Composting	– –	0	– – –	+ +	– –	– –	+ +
Aerobic digestion with energy recovery	– – –	+	– –	+ + +	– – –	– – –	+
Incineration	– – – –	+ + +	–	+ + + +	– – – – –	– – – – –	+ + + +

Energy content was a very important environmental effect relating to GHG emission and resource consumption (Thanh and Matsui 2009). The estimation of treatment methods based on energy content is presented in Table 19.5; the incineration process resulted in the largest amount of energy recovery, although this method also consumed the most energy. Meanwhile, composting was non-energy producing, and its energy consumption was limited.

The investment and operation costs to reduce 1 ton of CO_2 eq. by various alternatives was reported by Ayalon et al. (2001), and the benefit from various treatment processes was examined by Thanh and Matsui (2009). These results are presented in Table 19.5. Note that in terms of investment and operation costs, composting was the cheapest alternative.

Regarding the estimations of investment, operation cost, and benefit; the treatment processes ranked in descending order as follows: composting, anaerobic digestion, landfill with energy recovery, and incineration. Moreover, for the estimations of environmental effects and resource consumption, the treatment processes ranked in descending order as incineration, aerobic digestion, composting, and landfill with energy recovery. In sum, if it is impossible to separate the organic waste from the waste stream, composting seems the most economically and environmentally sound treatment process for the current situation of the Mekong Delta region.

19.3.4 Compost Potential for Sustainable Agricultural Development

19.3.4.1 Waste Compost Products Applied as Fertilizer in Agriculture

Several authors have assessed the positive repercussions and benefits of applying compost products in soil. Hargreaves et al. (2008) reported that organic-waste composting had potential as a beneficial recycling tool for waste management and treatment. This use not only reduced the volume of waste going in to landfills but also was a source of safe agricultural fertilizer.

Table 19.6 Quality of compost product at Thuy Phuong composting plant (*Source*: TSN 2009)

Characteristic	Units	TSN (1-0.5-1)	TSN (3-3-3)
Moisture	%	<30	<25
pH	–	6–8	6–8
Organic matter content	%	≥20	≥20
Total nitrogen content	%	1.0	3.0
Phosphorus pentoxide content (P_2O_5)	%	0.5	3.0
Potassium oxide content (K_2O)	%	1.0	3.0
Total nitrogen fixing bacteria	CFU/g	1×10^6	1×10^6
Total phosphorus decomposing bacteria	CFU/g	1×10^6	1×10^6
Total cellulose decomposing bacteria	CFU/g	1×10^6	1×10^6

CFU/g Colony forming units per gram

A primary benefit of composted waste is the high organic matter content and low bulk density (He et al. 1995). Moreover, the change in soil structure was found to persist 9 years after initial application (Garcia-Gil et al. 2004). Repeated application of composted waste consistently increased soil organic matter content and soil C/N ratio to levels greater than those of unimproved soil (Garcia-Gil et al. 2004; Perucci 1990; Montemurro et al. 2006). Composted waste was found to have a high water-holding capacity because of its organic matter content, which in turn improved the water-holding capacity of the soil (Hernando et al. 1989). Furthermore, application rates of 30 and 60 Mg ha^{-1} of waste compost increased the aggregate stability of soil through the formation of cationic bridges, thereby improving the soil structure (Annabi et al. 2007; Hernando et al. 1989). In addition, the fiber content of the compost product also had positive effects: when used for mulching, it has proven important for preventing erosion of the topsoil by irrigating water (Jakobsen 1995).

At present, there are many composting plants in Vietnam; among them the Thuy Phuong composting plant (Hue City, Vietnam) is the most successful exemplar for employing the composting process to convert waste to fertilizer. This plant yields two kinds of products: TNS (1-0.5-1) and TSN (3-3-3). The quality of these compost products is summarized in Table 19.6. These fertilizers are sold to highland areas for feeding coffee trees and tea plants.

19.3.4.2 Compost Products: Limitations and Modes of Improvement

Waste composting is generally safe for use in agriculture. However, sometimes pesticides, heavy metals, and salt have been found in soils where composted waste has been applied. The quality of waste compost depends on many variables, including the design of the composting facility, feedstock source, composting procedure, and length of maturation. In addition, the effects of composted waste differ when applied to different types of field soils and when used for different plants.

The best method for reducing these problems and improving the quality of waste compost is early-source separation, perhaps requiring separation to occur before or at curbside collection. Sewage sludge should not be added to the composting process at any point since it will raise the metal content of the compost (Richard and Woodbury 1992). In addition, factors such as feedstock selection, aeration, and maturity should be optimized and controlled to increase nutrient availability, specifically nitrogen, in waste compost.

19.3.5 Scenario Analysis

Various scenarios were analyzed to estimate the environmental impact, economic benefits, and application potential of the composting process. The analyses and estimations are presented in Fig. 19.8. Table 19.7 presents the comparison among scenarios during the period 2009–2020, considered within the specific boundaries and technologies, presented earlier in Table 19.3.

The results show that, when the amount of organic waste increased in composting applications, the amount of compost product, benefits, GHG reduction, and landfill-area reduction were also increased in step-wise fashion from scenario 1 to scenario 5, respectively. The results also demonstrate that the composting process is a sound, sustainable-development tool for both waste treatment and agriculture application.

This study also evaluated the potential economic benefits from waste composting for the Mekong Delta region. The estimates included the benefits of the compost product and those emanating from the sale of CER credits. These benefits were

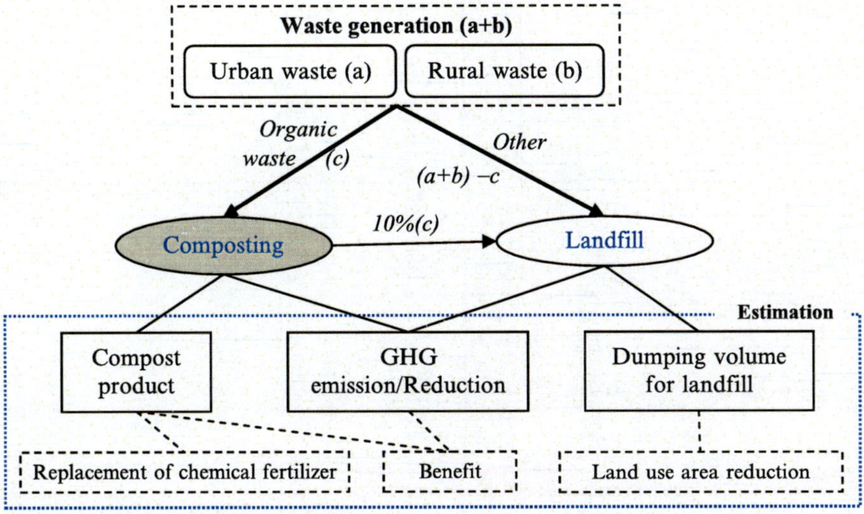

Fig. 19.8 Outline of scenario analysis

Table 19.7 Scenario analysis

Scenarios	Composting product (ton)	Land equivalent by compost feeding (ha)			Dumping volume for landfill		GHG	Benefit (composting and CER) (1,000 USD)		
		Agriculture trees	Food trees	Fruit trees	Loading (Gg)	Reduction (%)	Reduction (Gg)	Compost product	CER	Total
0										
1	991,125.9	2,831,788.5	11.1	7.8	182,461	3.5	8,560	29,734	85,600	115,334
2	4,480,773.6	12,802,210.5	49.8	35.7	159,196	15.8	38,754	134,423	387,539	521,962
3	8,961,547.5	25,604,421.3	99.6	71.1	129,325	31.6	77,508	268,847	775,079	1,043,926
4	13,442,321	38,406,631.8	149.4	106.8	99,453	47.4	116,262	403,270	1,162,618	1,565,888
5	17,923,095	512,088,423.6	199.2	142.2	69,581	63.2	155,016	537,693	1,550,157	2,087,850

found to relate positively to the amount of waste and the amount of compost applied. Moreover, composting distributed the large share of GHG reduction and mitigated the burden of waste volume dumped into landfills. In addition, the authors estimated the amount of land that could be fertilized by compost feeding of three general kinds of trees (see Table 19.7). This shows potential effects on agriculture of replacing chemical fertilizer with compost product from waste.

19.3.6 Political Implications of the Results

Some useful remarks can be made in this regard based on the results of this study. In the Mekong Delta region, biodegradable waste accounted for a very high percentage of all solid waste, around 70%. This share had the most GHG emission potential among all components of total solid waste (see Fig. 19.6). Due to the current status of cities of the Mekong Delta region, the best disposal option for biodegradable waste is composting. This method is useful not only for the reduction of GHG emissions and the mitigation of the waste burden in and cost of landfill sites, but also for the organic-waste fertilizers used in agriculture. Furthermore, the results show the potential for developing CDM projects in municipal solid waste fields.

However, the Mekong Delta region is a very large area where most of the waste generated has not been collected and treated, especially in rural areas. This has become a serious environmental problem. Organic composting seems an appropriate option for treating waste, although this treatment method has not been generally applied. In order to combine efficient solid-waste management with environmental and agriculture sustainability in the future, the Vietnamese government and environmental authorities should combine with municipal decision-makers, authorities, and planners to devise and put into effect the strategic plans outlined in Table 19.8.

Table 19.8 Strategic actions for encouraging recycling activities with biodegradable waste in the Mekong Delta region

Strategic actions and plans	
Urban areas	Rural areas
Increasing the coverage of waste collection system	Encouraging application of individual composting at each household
Promoting the waste separation at source for effective utilization of recyclable materials	Encouraging utilization of compost replacing chemical fertilizer for agriculture purposes
Improving and rehabilitating of existing composting plants	
Researching and developing the effectively centralized composting facilities	
Establishing sufficient market for consuming of the composting product	
Subsiding for recycling activities, especially biodegradable waste	
Approaching CDM to credit CER for saving money	
Developing community-based waste management system	

19.4 Conclusion and Recommendation

1. This study found that most waste generated in the Mekong Delta region has not been collected and treated. Waste generated from rural areas accounted for approximately 90% of total waste generation; however, in urban areas waste was collected and treated at low rates (average 70–80%).
2. GHG baseline emission was generated in large amounts in all Mekong Delta provinces and cities; however, it was really a serious problem because emissions of GHG could not be controlled in rural areas.
3. In the study area, municipal solid waste contains a large percentage (70.86%) of easily degradable organic waste, which has a high potential of GHG emission. Results suggest that composting is a preferable option for promoting waste reduction and recycling from this source.
4. Regarding estimations of environmental impacts, resource consumption, economic benefits, and application potential of the composting process; organic waste composting was beneficial not only for mitigating the waste burden going to landfill sites but also for reducing GHG emissions compared to baseline emissions and their attendant costs. Besides, the compost product was found to have strong potential to displace chemical fertilizer in agriculture. It was shown that the composting process was a sound alternative for promoting sustainable development in waste treatment and agriculture.
5. For easer composting, waste should be separated into at least two portions: compostable waste and other. Compostable waste should be collected and transported to composting plants; this procedure prevents hazardous matters from getting into the compost product. To minimize the amount of untreated waste and GHG emissions discharged from urban and rural areas, the waste-collection system should be expanded as much as possible. Moreover, in rural areas very far from cities, programs and projects should be implemented to encourage and train farmers to compost their organic waste to prevent negative environmental impacts and to obtain the local benefits for agriculture of composted waste.

References

3R-HN (2009) Instruction for use of compost fertilizer, Cau Dien composting plant. Obtained through the Internet: http://3r-hn.vn/Phanhuucocaudien/eguide.html

Annabi M, Houot S, Francou C, Poitrenaud M, LeBissonnais Y (2007) Soil aggregate stability improvement with urban composts of different maturities. Soil Sci Soc Am J 71:413–423

Ayalon O, Avnimelech Y, Shechter M (2001) Solid waste treatment as a high-priority and low-cost alternative for greenhouse gas mitigation. Environ Manage 27:697–704

Byer PH, Hoang CP, Nguyen TTT et al (2006) Household, hotel and market waste audits for composting in Vietnam and Laos. Waste Manage Res 24:465–472

European Commission (2009) Handbook: integrated waste management modules for different courses of graduate studies, INVENT project (Asia-Link Programme). Obtained through the Internet: http://homepages.hs-bremen.de/~office-ikrw/invent/Teaching%20Modules.htm

Garcia-Gil JC, Ceppi S, Velasca M et al (2004) Longterm effects of amendment with municipal solid waste compost on the elemental and acid functional group composition and pH-buffer capacity of soil humic acid. Geoderma 121:135–142

GSO (2007) Statistical yearbook. General Statistical Office – Statistical Publishing House, Ha Noi

GSO (2009) Statistical information of Vietnam: population and employment. General Statistical Office – Statistical Publishing House, Ha Noi, Vietnam. Obtained through the Internet: http://www.gso.gov.vn/default_en.aspx?tabid=467&idmid=3. Accessed Oct 2009

Hargreaves JC, Adl MS, Warman PR (2008) A review of the use of composted municipal solid waste in agriculture. Agric Ecosyst Environ 123:1–14

He X, Logan T, Traina S (1995) Physical and chemical characteristics of selected U.S. municipal solid waste composts. J Environ Qual 24:543–552

Hernando S, Lobo M, Polo A (1989) Effect of the application of a municipal refuse compost on the physical and chemical properties of soil. Sci Total Environ 81(82):589–596

Idris A, Inanc B, Hassan MN (2004) Overview of waste disposal and landfills/dumps in Asian countries. J Mater Cycles Waste Manage 6:104–110

IPCC – Intergovernmental Panel on Climate Change (2006) Guidelines for national greenhouse gas inventories. In: Eggleston HS, Buendia L, Miwa K, Ngara T, and Tanabe K (eds), Prepared by the National Greenhouse Gas Inventories Programme. IGES, Hayama

Jakobsen ST (1995) Aerobic decomposition of organic wastes 2. Value of compost as a fertilizer. Resour Conserv Recycl 13:57–71

Martínez-Blanco J, Pere Mu˜noz P, Antón A, Rieradevall J (2009) Life cycle assessment of the use of compost from municipal organic waste for fertilization of tomato crops. Resour Conserv Recycl 53:340–351

McDougall FWP, Franke M, Hindle P, Procter G (2001) Integrated solid waste management: a life cycle inventory, 2nd edn. Blackwell Science, Oxford

Michaelowa A, Perspectives, GmbH., Wucke A, GTZ (2009) CDM Highlights 74: Monthly newsletter of the GTZ Climate Protection Programme (CaPP). A project carried out on behalf of the German Federal Ministry for Economic Cooperation and Development. Obtained through the Internet: http://www.gtz.de/en/themen/umwelt-infrastruktur/umwelt-politik/14317.htm

Montemurro F, Maiorana M, Convertini G, Ferri D (2006) Compost organic amendments in fodder crops: effects on yield, nitrogen utilization and soil characteristics. Compost Sci Util 141:14–123

Perucci P (1990) Effect of the addition of municipal solid-waste compost on microbial biomass and enzyme activities in soil. Biol Fertil Soils 10:221–226

Plöchl C, Wetzer W, Ragoßnig A (2008) Clean development mechanism: an incentive for waste management projects? Waste Manage Res 26:104–110

Richard T, Woodbury P (1992) The impact of separation on heavy metal contaminants in municipal solid waste composts. Biomass Bioenergy 3:195–211

Thanh NP, Matsui Y (2009) Evaluation of the alternative treatment methods for GHG emission mitigation from municipal solid waste management: case study of Ho Chi Minh City, Vietnam. As J Energy Environ 10:35–52

TSN (2009) Introduction of product: TSN compost fertilizer, Tam Sinh Nghia investment-development join-stock company. Obtained through the Internet: http://phanbonvietnam.com/index.php?choose=productdetails&catp_id=2&pro_id=13

Tuyen TM, Michaelowa A (2004) Discussion paper: CDM baseline construction for Vietnam national electricity grid. Hamburgisches Welt-Wirtschafts-Archiv (HWWA)

UNFCCC – United Nations Framework Convention on Climate Change (2007) Glossary of CDM term Version 01. Obtained through the Internet: http://cdm.unfccc.int/Reference/Guidclarif/index.html

UNFCCC – United Nations Framework Convention on Climate Change (2008) Tool to determine methane emission avoided from dumping waste at a solid waste disposal site, Version 02, EB39. UNFCCC-CDM Executive Board

VNA – Vietnamese National Assembly (2008) Resolution no. 23/2008/QH12: on the 2009 socio-economic development plan, dated 6 Nov 2008, The National Assembly of the Socialist Republic of Vietnam. Obtained through the Internet: http://vbqppl2.moj.gov.vn/law/en/2001_to_2010/2008/200811/200811060001_en

Wolkowski R (2003) Nitrogen management considerations for land-spreading municipal solid waste compost. J Environ Qual 32:1844–1850

World Bank, Vietnam Ministry of Environment and Natural Resources, Canadian International Development Agency (2004) Vietnam Environment Monitor: Solid Waste. Obtained through the Internet: http://www-wds.worldbank.org/external/default/WDSContentServer/WDSP/IB/2005/07/28/000012009_20050728112421/Rendered/PDF/331510rev0PAPER0VN0Env0Monitor02004.pdf

Chapter 20
Biogas Production of Pig Manure with Water Hyacinth Juice from Batch Anaerobic Digestion

Nguyen Vo Chau Ngan, Le Hoang Viet, Nguyen Dac Cu, and Nguyen Huu Phong

Abstract Anaerobic digestion, a series of processes in which micro-organisms break down biodegradable materials in the absence of oxygen, is normally used for wastewater treatment and energy production. Anaerobic digestion has been applied to hog – wastewater treatment and biogas production in the Mekong Delta of Vietnam, where a high percentage of the population works in the agricultural sector. However, the supply of pig manure for operating the anaerobic digesters is not always available. Therefore, to stimulate the application and operation of anaerobic digesters that are very significant to water-pollution prevention in the Mekong Delta, there is a need to find supplemental inputs to pig manure to help maintain the operation of the digesters.

The aim of our research was to find out the yield of biogas produced from batch anaerobic digestion by using water hyacinth (WH) – a plant commonly grown in the canal networks in the Mekong Delta – as a supplemental source to pig manure in biogas digesters. We conducted experiments using 50 L plastic bags of pig manure (PM) mixed with three different types of WH (C1, WH juice after 2 days of hydrolysis; C2, WH juice after 2 days of hydrolysis+chopped WH; and C3, WH juice after 2 days of hydrolysis+crushed WH). Based upon the organic dry matter (ODM) of input sources, we set up 13 experiments (with three replicates for each experiment) of 100% PM; 75% PM+25% WH; 50% PM+50% WH; 25% PM+75% WH; and 100% WH. The produced biogas volume was recorded daily while other parameters of pH, buffer capacity, biogas components (CH_4, CO_2, O_2, and H_2S) were recorded once daily for 7 consecutive days during the 35 day period.

According to our experimental conditions, the various preparations of WH showed differences in gas production. The gas production from experiments with C2 and C3 was not significantly different. Meanwhile, the gas production from experiments with C1 and C2 was significantly different.

N. Vo Chau Ngan (✉), Le Hoang Viet, and N.D. Cu
College of Environment and Natural Resources, Can Tho University, Can Tho, Vietnam
e-mail: nvcngan@ctu.edu.vn

N.H. Phong
VIE/020 Hyacinth Project, Can Tho University, Can Tho, Vietnam

M.A. Stewart and P.A. Coclanis (eds.), *Environmental Change and Agricultural Sustainability in the Mekong Delta*, Advances in Global Change Research 45,
DOI 10.1007/978-94-007-0934-8_20,

With respect to C2 treatments, on the one hand, the total gas production tended to be higher in the treatments with high percentages of WH (231 L, 315 L, 392 L, 358 L, and 439 L in the treatment of 100% PM; 75% PM+25% WH; 50% PM+50% WH; 25% PM+75% WH; and 100% WH, respectively). On the other hand, total biogas yield per kg ODM of input materials reached the highest value in the treatment of 50% PM+50% WH (122 L, 148 L, 176 L, 153 L, and 171 L in the treatment of 100% PM; 75% PM+25% WH; 50%PM+50% WH; 25% PM+75% WH; and 100% WH, respectively).

These results strongly confirm that the farmers in the Mekong Delta of Vietnam can use water hyacinth as a potential supplement to pig manure in anaerobic digestion in case they do not have enough pig manure for their anaerobic-digestion treatment systems.

Keywords Batch-type system • Biogas production • Mekong Delta • VACB farming system • Water hyacinth

20.1 Introduction

The VACB farming model combining gardening, fishery, pig husbandry, and biogas plants (V: garden, A: fish pond, C: pigsty, and B: biogas plant) is very popular in the Mekong Delta of Vietnam. This farming model has proved to be very beneficial to aqua-agricultural production and improved sanitation in the region. Research has shown that in the same area the annual income through VACB farming is three to five times greater (and sometimes as much as ten times greater) than that derived from growing two rice crops per year (Nguyen Van Man 2008). In the VACB system, the biogas digester is a treatment module not only for pig manure but also for human excreta. Therefore, after treatment in the biogas digester, the waste will be sanitarily safe to apply in the fish ponds and gardens.

Up-to-date biogas digesters have proven not only to be a safe treatment for livestock waste but also an efficient source of gas supply for household cooking, lighting, etc. Nevertheless, pig manure in the region is often in short supply. Most farm households in the Mekong Delta have small-scale piggeries with four to ten animals, and in case the market price of pigs is low or the cost of livestock fodder is high, the households tend to decrease the number of pigs or even to stop raising pigs. In such cases, the biogas digester processes discontinuously due to the shortage of pig manure, the base input of biogas digesters in the Mekong Delta. Consequently, the investment return from installing biogas digesters is less profitable. This impedes the broad diffusion of biogas digesters in the Mekong Delta and subsequently affects the pollution-prevention program in the area. Because of the shortage of pig manure in biogas digesters in the Mekong Delta, there is a need to find materials to supplement pig manure if biogas digesters are to come into wide usage in the area.

In seeking supplements to pig manure for biogas digesters, we found that there may be some potential materials such as agricultural and aquacultural residue that can be used. Many years ago Ottmar Philipp et al. (1983) found that water hyacinth is suitable for the process of anaerobic fermentation. Malik et al. (1990) reported that some plant wastes (including WH) and other agricultural wastes, along with poultry litter and dung, could be used for biogas production. Chanakya et al. (1992) found that fresh WH, dried WH, and urban garbage collected immediately after dumping were amenable to solid-phase fermentation. Other researchers have tested a two-stage, rumen-derived, anaerobic digestion process for the conversion of WH shoots and a mixture of the shoots with cow dung into biogas (Kivaisi and Mtila 1998). And Ganesh et al. (2005) have developed a procedure for extracting volatile fatty acids from WH using simple and inexpensive equipment of the type commonly available in rural households.

Based on these references, we found that water hyacinth, *Eichhornia crassipes*, is potentially a supplemental input for biogas digesters. WH is a very common plant growing everywhere throughout the year in a dense system of canals in the Mekong Delta. Thanks to the widespread availability of WH in local areas of the Mekong Delta, if WH can be used as a supplement to pig manure in biogas digesters, it would make the broad diffusion of biogas digesters much more feasible in the Mekong Delta. Our study aims to test whether or not WH can be effectively used to supplement pig manure in biogas digesters in real situations of the Mekong Delta by measuring gas production from batch anaerobic digesters with the input materials of PM mixed with WH. The result of this study is expected to encourage the owners of biogas digesters to consider using WH as an additional input material in their biogas digesters when animal manure is in short supply.

Our experiments were expected to determine:

- Levels of gas production according to differential treatments of WH; and
- The impact of the differential mixing ratio of WH to PM on gas production.

20.2 Materials and Methodology

20.2.1 Research Location and Time

The experiments were conducted at Hoa An Research Center of Cantho University, while some physical–chemical analyses were done at the Environmental Engineering Laboratory, College of Environment and Natural Resources, Cantho University, Vietnam. We carried out our experiments from November 6 to December 16, 2008.

20.2.2 *Input Materials*

20.2.2.1 Experimental Set-up

We installed the apparatus needed for 39 airtight digestion units. Each consisted of a 50 L plastic bag (0.5 mm thickness) connected via soft joints to ensure air-tightness and water-tightness to one polyvinyl chloride (PVC) input pipe, one PVC output pipe, and one gas-collection pipe. For the safety of the test, only 40 L of materials were loaded into the 50 L digester. This quantity of input materials was estimated in proportion to a typical 4-m^3 biogas tank commonly used in the Mekong Delta.

The aluminum bags (at maximum containing up to 30 L) were connected directly to the digestion bags through the gas-collection pipe in order to collect gas samples (Fig. 20.1).

20.2.2.2 Collection and Treatment of Input Materials

Input materials loaded for the anaerobic process were prepared as follows:

- PM collected from the different piggeries at Hoa An Research Center was dried out at ambient temperature (about 25°C±4°C) for 1 week before use, and then the dried PM was smashed and mixed together until it became a homogeneous mixture.
- WH was collected from the canals around Hoa An Research Center. The WH (without the roots) was chopped into 0.5–1.5 cm pieces by hand. Then the WH was air-dried at ambient temperature (about 25°C±4°C) up to unchanged weight. Once completely dry, the WH pieces were mixed together until they became homogeneous (Fig. 20.2).

Fig. 20.1 Set-up of batch experiments

Fig. 20.2 Prepared input materials

To speed-up the gas production, the WH material was hydrolyzed by the inoculum for 2 days before being loaded into the digestion bags. The hydrolyzed WH then underwent three different types of pre-treatments - one with only the juice of WH (C1); one with WH juice and WH residue (C2); and one with WH juice and crushed WH residue (C3). We prepared the WH pre-treatments differently in order to test which type of preparation of WH would give an optimal result in gas production.

- To shorten the time for gas production, we used the residue manure taken from an activated biogas tank as an inoculum. Note that there is a 100-m^3 biogas tank at Hoa An Research Center that we used for treating the piggeries' wastewater.

20.2.2.3 Mixing Ratio of Input Materials

The ratio of PM to WH in the mixture was based on the ODM in these materials. Before conducting the experiments, we carried out sample analysis of organic dry matter in the PM and WH to determine an appropriate input-material loading rate.

For each type of WH pre-treatment, we set up five different treatments, with mixing ratios between PM and WH as follows:

- R1, 100% PM+0% WH: 1 experiment with C0 (control treatment).
- R2, 75% PM+25% WH: 3 experiments with C1, C2, and C3, respectively.
- R3, 50% PM+50% WH: 3 experiments with C1, C2, and C3, respectively.
- R4, 25% PM+75% WH: 3 experiments with C1, C2, and C3, respectively.
- R5, 0% PM+100% WH: 3 experiments with C1, C2, and C3, respectively (Table 20.1).

Table 20.1 The arrangement of all 13 experiments (with triplicate treatments of each)

WH treatments Ratio of PM to WH	C0			C1			C2			C3		
R1: 100% + 0%	C01[a]	C01[b]	C01[c]									
R2: 75% + 25%				C12[a]	C12[b]	C12[c]	C22[a]	C22[b]	C22[c]	C32[a]	C32[b]	C32[c]
R3: 50% + 50%				C13[a]	C13[b]	C13[c]	C23[a]	C23[b]	C23[c]	C33[a]	C33[b]	C33[c]
R4: 25% + 75%				C14[a]	C14[b]	C14[c]	C24[a]	C24[b]	C24[c]	C34[a]	C34[b]	C34[c]
R5: 0% + 100%				C15[a]	C15[b]	C15[c]	C25[a]	C25[b]	C25[c]	C35[a]	C35[b]	C35[c]

[a, b, c] are three replicates for each experiment

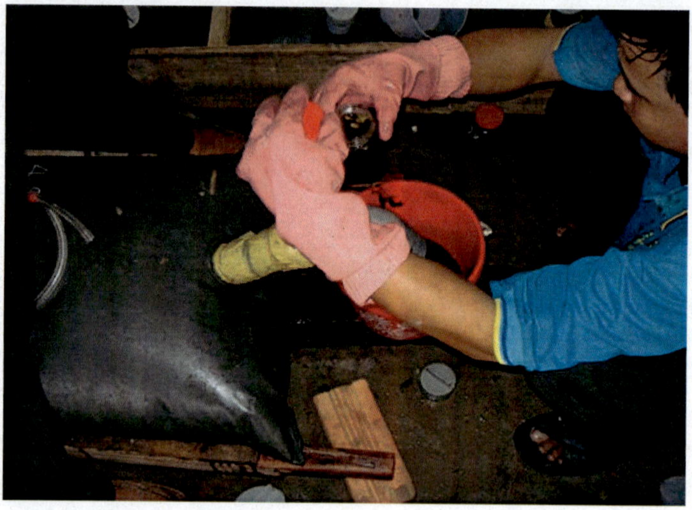

Fig. 20.3 Collecting wastewater samples

20.2.3 Analyzing Parameters

The target analysis was conducted according to standard methods.

- pH: pH meter Orion model 230.
- DM, ash, ODM, C: Memmert UI 40 oven; heating 550°C; electronic balance Sartorius CP 324.
- N: Tecator digestion system; Gerhart distiller.
- Buffer capacity: Biogaspro instrument.
- CH_4, CO_2, O_2, H_2S: Geotechnical 0–100/100GA94.
- Gas volume: RITTER meter (with smallest scale of 20 mL) (Figs. 20.3 and 20.4).

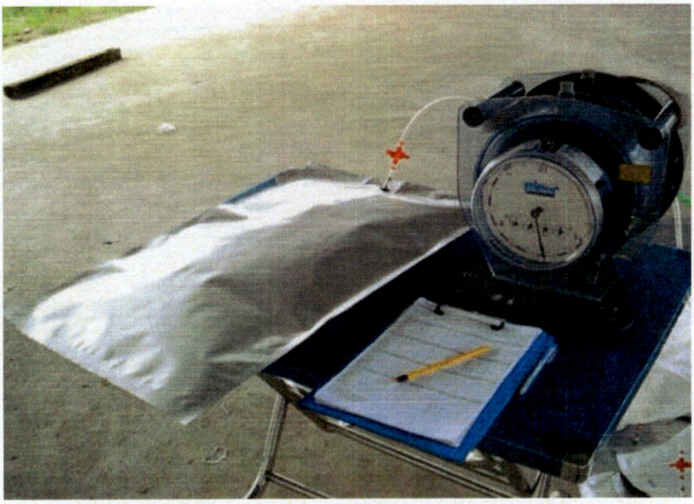

Fig. 20.4 Measuring gas volume

20.3 Results and Discussion

After using statistical tools to analyze output results, we found that gas production in experiments for C2 and C3 was not significantly different. For that reason, in this paper we only present and discuss the output results of experiments for C1 and C2.

20.3.1 Biogas Volume

The biogas volume was collected by the aluminum bags and recorded every day. The daily results were shown according to each type of WH treatment.

As described in Fig. 20.5, the biogas volume reached maximum value in the 2nd week; and between the 2nd and the 5th week, biogas volume reduction was very slow; this observation showed that the material decomposed most in the first 2 weeks. This result suggests that the gas-produce period of wastewater in C1 treatments possibly lasts approximately 30 days.

With respect to the different ratios of WH to PM in the mixture, we found that the treatments with higher percentages of WH in the mixture resulted in less biogas volume. The biogas volume of treatments mainly comes from the decomposition process of PM, but the WH used in the treatments was not completely hydrolysized because 2 days is not long enough to decay all organic materials in the WH. This possibility may have a negative impact on biogas volume in C1 treatments.

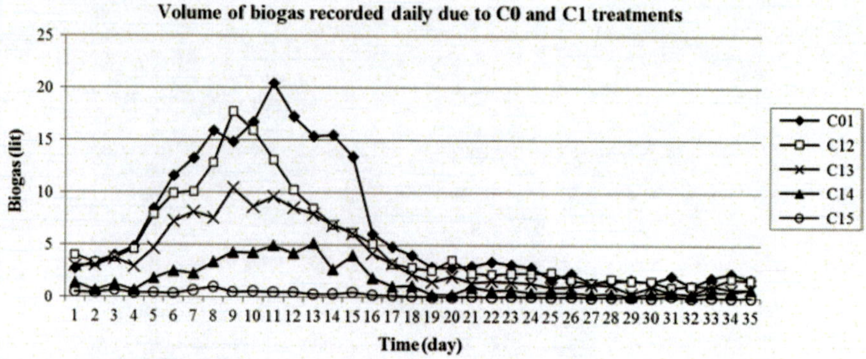

Fig. 20.5 Biogas volume of C0 and C1 treatments

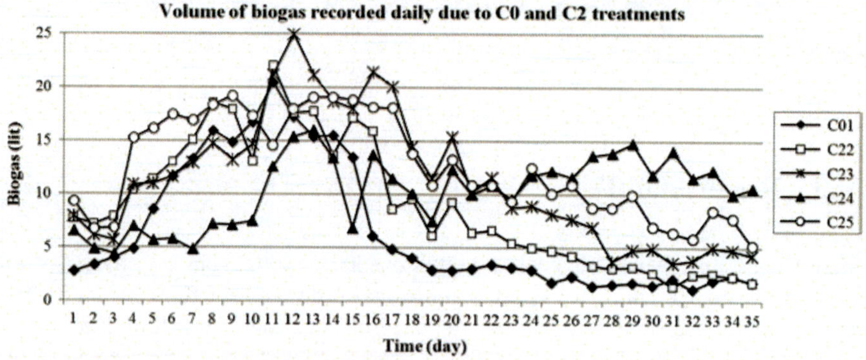

Fig. 20.6 Biogas volume of C0 and C2 treatments

As shown in Fig. 20.6, the biogas volume of each C2 treatment reached maximum value at close range of 110–130 L after 2 weeks (except for the C24 treatment in which biogas volume increased slowly from week 2 to week 5 and got maximum value in week 5). Compared to the biogas produced from C1 treatments, in the last 3 weeks the biogas yield of C2 treatments also slowed down but was still higher than C1 treatments. This result suggests that in case of application of C2 treatments for wastewater, the gas-produce period in the biogas digester could last up to 40 days.

The biogas volume of the treatments reached the maximum value in the 2nd week and then gradually decreased in the last 3 weeks. This result of our experiments is consistent with previous findings. In our opinion, the biogas volume could reach its maximum value in the second week due to the use of wastewater as buffer water in the experiments and the stimulating effects of shaking the digester bags. The inoculum water was taken from an activated 100-m^3 biogas tank, and the biogas digester bags were shaken daily to mix the materials completely.

20.3.2 The Main Composition of Gas

In addition to the measurement of biogas volume produced in each treatment, the main gas components (including CH_4, CO_2, O_2, and H_2S) were collected daily and stored in bags for 7 days before being measured. Every day, we extracted a certain amount of gas from the aluminum bags and stored the extracted gas in bags. The extraction work was carried out either by a self-made piston (in the case of an aluminum bag with much gas) or by a 50-mL syringe (in the case of the one with less gas). In the first 3 weeks, the extracted gas was 300 mL/bag*day, but in the last 2 weeks, we could get 500 mL/bag*day due to the large amount of gas production (Table 20.2).

The weekly records of the main gas components show that the biogas quality was rather high (due to the weekly average value of the CH_4 of around 51.63–63.18%). The CH_4 reached the highest value of 70.73% in the C14 treatment; the O_2 reached a high of 4.7% in the C15 treatment, while in the other treatments the O_2 did not exceed 1.45%. Among our treatments, only the C23 treatment could reach the highest value of H_2S with 0.0316% (316 ppm). Meanwhile, the gas collected from the activated 100-m^3 biogas tank at Hoa An Center (where we took inoculum water) could reach the H_2S value of 1,000–3,000 ppm. The difference in H_2S yield has two possible explanations:

- First, our experiments were conducted in only 35 days while the 100-m^3 biogas tank is stable and in operation for a long time.
- Second, the biogas samples were collected and stored in the aluminum bags for 7 days before the gas composition was recorded. These aluminum bags were re-used for multi-records, which might have caused water condensation inside these bags. Available water vapor (H_2O) and oxygen (O_2) remaining in the gas container itself will reduce the H_2S by the following reaction:

$$H_2S + O_2 \rightarrow 2S + H_2O$$

Also, NH_3 is present in biogas composition:

$$NH_3 + H_2O \rightarrow NH_4^+ + OH^-$$

$$H_2S + OH^- \rightarrow HS^- + H_2O$$

Of the main gas components in biogas, the average value of CH_4 is presented in Fig. 20.7.

In the 2nd week, all CH_4 outcomes of the C1 treatments were greater than 50%, increasing with higher percentages of WH loads and attaining the highest value in the C15 treatment (69.5%). By contrast, all C2 treatments produced CH_4 yield in lower levels (less than 50%) with the lowest found in the C24 treatment (32.25%).

In the first 2 weeks, almost all of the CH_4 weekly values in the C1 treatments were higher than those in the C2 treatments, but over the last 3 weeks of the experiment, the

Table 20.2 The weekly records of the main gas components (%)

	CH_4			CO_2			O_2			H_2S		
Treatment	Min	Ave	Max	Min	Ave	Max	Min	Ave	Max	Min	Ave	Max
C01	53.70	60.63	64.43	36.10	38.10	39.40	0.00	0.43	1.20	0.0004	0.0042	0.0128
C12	57.70	60.04	62.85	36.85	39.35	43.45	0.00	0.21	0.50	0.0000	0.0023	0.0074
C22	46.27	57.50	61.00	38.17	41.82	51.50	0.00	0.22	0.65	0.0000	0.0013	0.0044
C13	57.53	60.13	63.15	33.57	35.71	37.15	0.00	0.56	1.37	0.0000	0.0018	0.0058
C23	39.15	54.74	62.57	36.63	43.86	55.25	0.00	0.16	0.65	0.0011	0.0092	0.0316
C14	57.60	63.18	70.73	28.73	31.39	34.00	0.00	0.12	0.60	0.0022	0.0103	0.0260
C24	32.25	53.15	62.17	40.90	46.35	61.55	0.00	0.00	0.00	0.0008	0.0017	0.0026
C15	25.45	51.63	69.50	14.05	20.90	26.35	0.00	1.45	4.70	0.0010	0.0066	0.0153
C25	38.10	52.01	60.55	38.95	44.86	56.85	0.00	0.39	0.90	0.0016	0.0020	0.0028

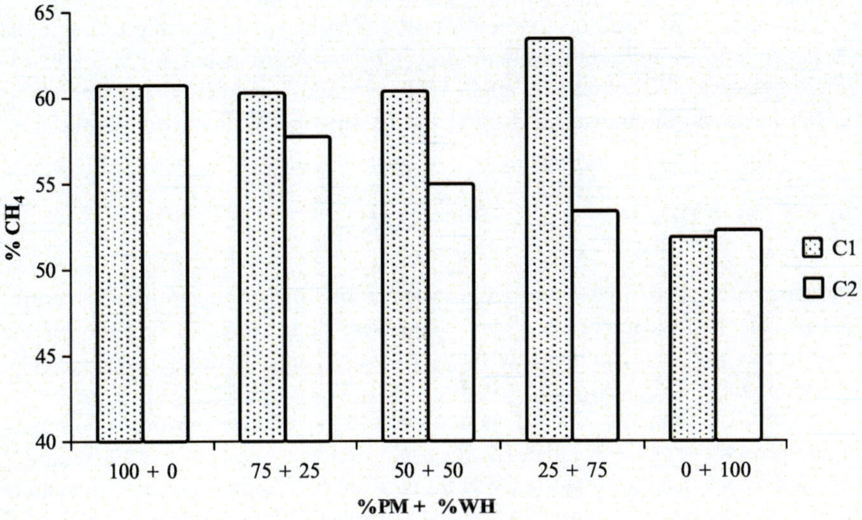

Fig. 20.7 The average value of CH_4 by treatments C1 and C2

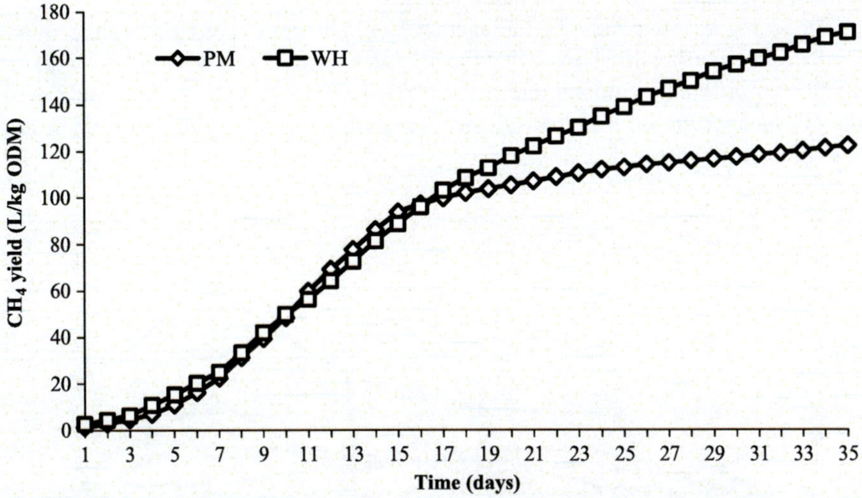

Fig. 20.8 Comparison of cumulative CH_4 between 100% PM and 100% WH treatments

CH_4 weekly values in the C1 treatments were lower than those in the C2 treatments. This demonstrates that the loaded WH juice keeps producing methane in the first 2 weeks. Therefore, to be able to apply the C1 treatments successfully in reality, the hydrolysis time of WH should last more than 2 days so that the organic matters of WH could be totally decomposed, and the buffer water would be highly qualified.

As shown in Fig. 20.8, CH_4 cumulative values of 100% PM and 100% WH treatments seem to be similar in the first 2 weeks; but from the third week, the CH_4 cumulative value of 100% WH treatment is much increased compared to the 100%

PM treatment. This result clearly confirms that WH had good decomposition in the anaerobic digester and produces quite good CH_4 yields compared to PM. In reality, biogas owners could completely apply WH as an additional material input in their biogas digesters, and biogas production would still yield well.

20.3.3 pH Value

The parameter of pH plays a key role in anaerobic-digester operation. To control this parameter, measurements of pH value were done from the first day of the experiments and then continuously recorded every 7 days during the experimental period (35 days) (Figs. 20.9 and 20.10).

In the C1 treatments, the pH values were higher due to the treatments with higher percentages of WH input. However, this tendency was in contrast in the C2 treatment. As a result of inputting WH into the biogas digester, the organic matters from WH will be decomposed, which creates an acidification phase. For the C2 treatments, both WH juice and WH residue were loaded into the biogas digester, so more organic matters will be decomposed, and then the acidification phase will remain longer and make the pH value lower.

Regarding a comparison of the average pH values produced among the treatments: Our experiments showed that the pH value ranged from 6.87 to 7.60 with no significant differences among the treatments ($p<0.05$). This range is suitable with the pH value proposed for biogas digester operation 6.8–7.2 (optimum from 7.0 to 7.2) (Gerardi 2003).

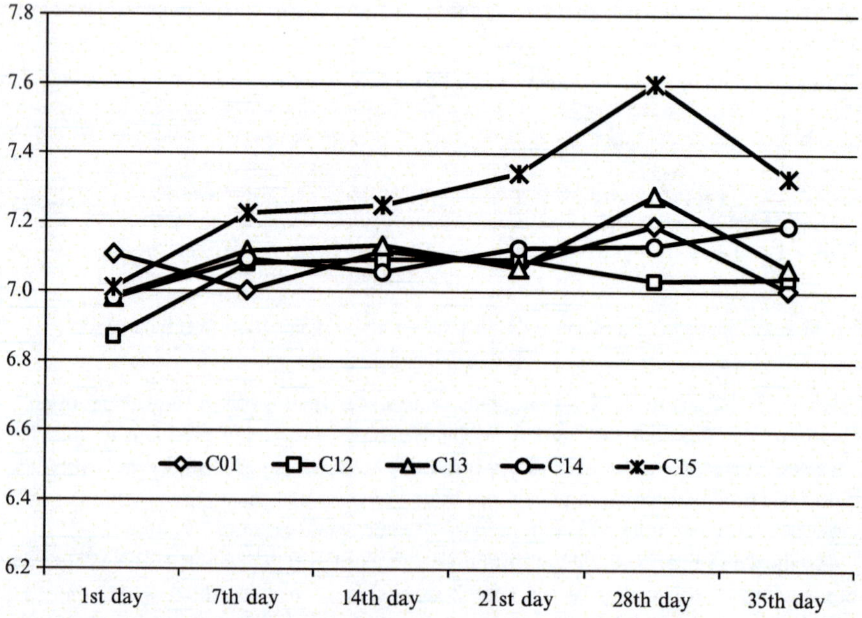

Fig. 20.9 pH value measured from C0 and C1 treatments

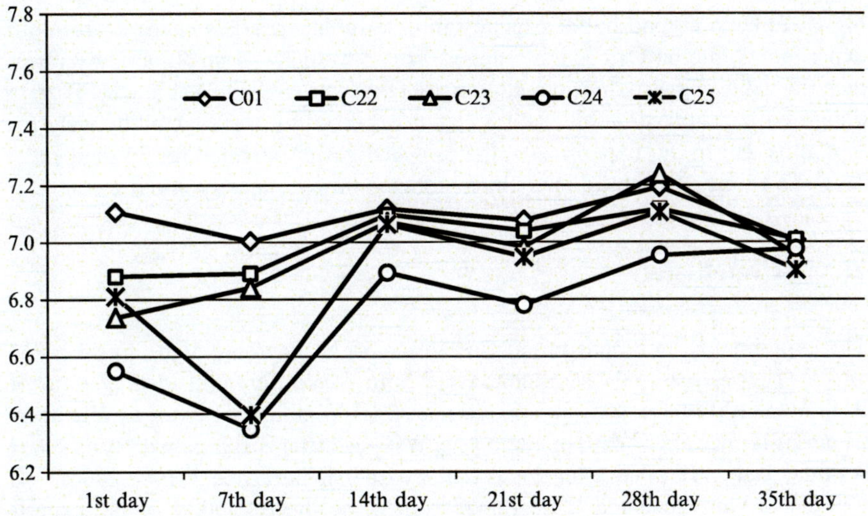

Fig. 20.10 pH value measured from C0 and C2 treatments

Table 20.3 The weekly average value of alkalinity (mL/L)

Treatment	Week 1[a]	Week 2	Week 3	Week 4	Week 5	Average
C01	–	500	733	900	867	750
C12	–	1,017	667	833	867	846
C22	–	850	1,033	767	767	854
C13	–	850	1,200	550	900	875
C23	–	1,133	1,150	933	1,050	1,067
C14	–	1,017	783	767	650	804
C24	–	983	533	717	817	763
C15	–	700	567	517	817	650
C25	–	717	767	850	900	808

[a]Not recorded due to non-working equipment

20.3.4 Buffer Capacity

The weekly buffer capacity were measured parallel to pH measurements. However, owing to problems with the equipment in the first experiment week, we could record buffer results only for the last 4 weeks (Table 20.3).

The buffer capacity were recorded by Biogaspro equipment that acts based on the principle of using acid (HCl) to respond with $CaCO_3$ (the major component of buffer) to produce CO_2. The CO_2 volume was measured in order to determine the concentration of $CaCO_3$ in the wastewater as a result of the following reaction:

$$2HCl + CaCO_3 \rightarrow CO_2 + CaCl_2 + H_2O$$

According to Joachim (2008), buffer is a parameter to ensure stability of gas produced. The buffer capacity should be maintained at the level of 1,000 mL/L or higher to keep a good buffer capacity in case input materials change. From our experiments, the buffer capacity ranged from 500 to 1,150 mL/L, slightly lower than the value suggested for maintaining the digester well in a stable way. To gain a high buffer capacity, a biogas digester is expected to operate continuously for 2 months, but in our experiments, the biogas digester operated for 35 days only. This short time of operation may have affected buffer capacity in our test.

20.4 Conclusion

The experiments on biogas production in the 50 L anaerobic digesters with PM and WH input materials were conducted with five differential loading ratio of PM + WH (100 + 0, 75 + 25, 50 + 50, 25 + 75, 0 + 100) and three types of WH treatments (WH juice after 2 days of hydrolysis; WH juice after 2 days of hydrolysis + WH residue; and WH juice after 2 days of hydrolysis + crushed WH residue). All treatments were repeated three times to increase the reliability of the experiments, i.e., the total number of experiments was 39. Our principal finding after 5 consecutive weeks of experimental work was that water hyacinth is potentially a supplement to pig manure in biogas digesters appropriate to the local conditions of the Mekong Delta of Vietnam. Regarding the method of WH pre-treatment, our findings concluded that to get the optimal yield of gas production in digesters, before the WH juice and WH residue are loaded into biogas digesters; fresh WH should first be cut into small pieces, dried for 2 days, and then hydrolysized in inoculum for 2 days.

According to our findings, the biogas yield was 14.3 L/kg of fresh-water hyacinth. Therefore, in case pig manure for the biogas digesters is in short or no supply, we can use water hyacinth not only as a supplement but also as an alternative to pig manure. A daily cooking-energy demand for one household of four to six persons is normally 1,200 L of biogas per day. A 4-m^3 biogas digester with six to ten 90-kg pigs can produce this amount of gas to meet such a household's daily cooking-energy demand. In case pig manure is in short supply due to the decrease of pig herds, 10 kg of fresh-water hyacinth is equivalent to the amount of manure produced by a 90-kg pig per day. This means that if there is one pig absent, we can use 10 kg of fresh-water hyacinth to maintain the normal yield of the biogas production for household cooking-energy consumption. In case of no pig manure supply, a biogas digester whose only input is about 84 kg of fresh-water hyacinth per day can produce 1,200 L of biogas per day.

We conclude that water hyacinth is a potential supplementary source in biogas digesters suitable to the local circumstances of the Mekong Delta where water hyacinth is highly available. This finding is important for promoting the broader application and operation of biogas digesters in the Mekong Delta and for helping to prevent water pollution in the area. In so doing, the development of biogas digesters will bring financial benefits to farmers and greater aqua-agricultural sustainability to the region.

Acknowledgments This study was kindly supported by the project "Production of feedstuffs and renewable energy from water hyacinths in Vietnam, VIE/020-WH" funded by the Ministry of Foreign Affairs of Luxembourg. Thanks to this project, we have more opportunities for further research on the use of WH as an additional input material for biogas digesters.

We would like to thank all members of the Department of Environmental Engineering, Cantho University, who helped with field sampling and lab work, including N. T. Thanh and H. L. Toan. Our thanks should go to Ms. Catherine Mertz and Ms. Kieu for their help with the administration work of the VIE/020-WH project. We are also indebted to Dr. D. N. Quynh and V. H. Nam for fruitful discussions which gave us a good chance to clarify our thinking regarding the interpretation of outcomes of the biogas experiments.

References

Chanakya HN, Borgaonkar Sushama, Rajan MGC, Wahi M (1992) Two-phase anaerobic digestion of water hyacinth or urban garbage. Bioresour Technol 42:123–131

Ganesh PS, Ramasamy EV, Gajalakshmi S, Abbasi SA (2005) Extraction of volatile fatty acids (VFAs) from water hyacinth using inexpensive contraptions. Biochem Eng J 27:17–23

Gerardi MH (2003) The microbiology of anaerobic digesters. Wiley, Hoboken

Joachim C (2008) Control process – pH, buffer capacity and high added as buffer substances. Reports, Cantho University, Cantho, Vietnam

Kivaisi AK, Mtila M (1998) Production of biogas from water hyacinth (*Eichhornia crassipes*) (Mart) (Solms) in a two-stage bioreactor. World J Microbiol 14:125–131

Malik MK, Singh UK, Ahmad N (1990) Batch digester studies on biogas production from Cannabis sativa, water hyacinth mixed with dung and crop wastes and poultry litter. Biol Waste 31:315–319

Nguyen Van Man (2008) Vuon – Ao – Chuong: the Traditional Vietnamese Farm. http://permaculture.org.au/ 2008/10/04/vuon-ao-chuong-the-traditional-vietnamese-farm/. Accessed 22 April 2009

Philipp O, Koch W, Köser H (1983) Utilization and control of water hyacinth in Sudan. German Agency for Technical Cooperation (GTZ), Eschborn

Chapter 21
Community-Scale Wind-Powered Desalination for Selected Coastal Mekong Provinces in Vietnam

Ha T. Nguyen and Joshua M. Pearce

Abstract Global climate destabilization is exacerbating water problems in Vietnam, most acutely in the South and Central regions where the majority of the inhabited area lies in the low elevation coastal zone. Off-grid community-scale reverse osmosis desalination powered by small wind turbines offers a solution to this problem for the coastal fringe of Vietnam's Mekong Delta. Using a geographical information system (GIS) platform, a wind resources atlas developed by the Asia Sustainable and Alternative Energy, and projected rural population available from Columbia University's Center for International Earth Science Information Network, this chapter explores this potential. The GIS analysis estimated that in the absence of all other water supply facilities, off-grid wind desalination could provide clean water to 5.4 million rural residents living in 18,900 km^2 of the Mekong Delta coastal provinces at the rate of 60 l/person/day. In addition to providing clean water, the use of wind-powered desalination in the region would have educational benefits to combat poverty and unemployment and ease water-related conflicts, and it has been shown to improve environmental and agricultural sustainability. Thus this technology was found to represent a decentralized and community-based method to adapt to and mitigate climate change in the coastal fringe of the Mekong Delta.

Keywords Renewable energy • Desalination • GIS • Sustainability • Mekong Delta

Ha T. Nguyen
Department of Mechanical and Materials Engineering, Queen's University, 60 Union Street, Kingston, ON, Canada K7L 3N6
e-mail: nguyenh@me.queensu.ca

J.M. Pearce (✉)
Department of Mechanical and Materials Engineering, School of Environmental Studies, Queen's University, 60 Union Street, Kingston, ON, Canada K7L 3N6
e-mail: pearce@me.queensu.ca

M.A. Stewart and P.A. Coclanis (eds.), *Environmental Change and Agricultural Sustainability in the Mekong Delta*, Advances in Global Change Research 45,
DOI 10.1007/978-94-007-0934-8_21,

Abbreviations

ASTAE	Asia Sustainable and Alternative Energy Program
CFD	Computational Fluid Dynamics
CIESIN	Center for International Earth Science Information Network
CPRGS	Comprehensive Poverty Reduction Growth Strategy
LECZ	Low Elevation Coastal Zone
ERS SAR PRI	European Remote Sensing Synthetic Aperture Radar Precision Image
IRDC	International Research and Development Council
NRCWSS	National Rural Clean Water Sanitation and Supply Strategy
NREL	National Renewable Energy Laboratory
RO	Reverse Osmosis
SAR	Synthetic Aperture Radar
UNFAO	United Nations Food and Agriculture organization
UTM	Universal Transverse Mercator
VLSS	Vietnam Living Standard Survey
WPD	Wind Powered Desalination

21.1 Introduction

Although it has always been a serious problem, providing a clean water supply for the population of Vietnam is one of the major concerns for the government in the face of looming global climate destabilization (Soussan et al. 2005). In 1999 the total renewable freshwater supply of the country, not the actual fresh water supply, was 891 km^3/year, but this supply of course varies spatially.[1] The spatial non-uniformity of groundwater resource distribution was documented by University of Bern, Switzerland.[2] Small towns in the coastal fringe of the Mekong Delta suffer the lowest groundwater quality nationwide, are home to the worst habits in water sanitation (Soussan et al. 2005), pay higher water tariffs, enjoy few alternatives, and sometimes cannot afford connection fees to relatively sanitary supplies (World Bank 2000). In some areas, over-exploitation has resulted in falling water tables, which contributes to further land subsidence and salinity intrusion, a problem of particular concern in the Mekong Delta.[3]

To correct these problems, the National Rural Clean Water Sanitation and Supply Strategy (NRCWSS) 2020 has aimed to have 85% of the rural population consume clean water at 60 l per capita per day in 2010 (Ministry of Construction and Ministry of Agriculture and Rural Development 2000); the demand in large urban

---

[1] http://www.worldwater.org/data.html

[2] http://www.cde.unibe.ch/Regions/Mekong_Proj_2_Rs.asp#DISS_EP PI

[3] http://www.wepa-db.net/policies/state/vietnam/overview.htm

centers already ranged from 50 to 180 l per capita per day (Vo 2007). The target for 2020 is even more ambitious yet unlikely to succeed as there are no clear terms of action required to meet these goals (Soussan et al. 2005).

Vietnam's 192 water supply plants are deteriorating (Soussan et al. 2005). In the Mekong Delta the popular means of water supply is groundwater extraction by pumping. The number of pumps has increased dramatically within the last several years from 75,000 pumps in 1990 to around 800,000 pumps by 1999. The current rate of groundwater extraction is estimated at 430,000 m³/day serving about 4.5 million people out of the Delta's population of 14 million (Eastham et al. 2008). This is in part due to the fact that the existing uniform promotion of piped water is inappropriate and inefficient with as much as 30–55% of the water pumped out of the country's treatment plants lost through the distribution network (Soussan et al. 2005).

Despite these challenges, those living in the Mekong Delta have ample access to water along the coast, which only needs to be desalinated. Yet, there is no sea water desalination in the coastal areas of Central and South Vietnam at the municipal level (SunLift 2001). This can largely be attributed to the high cost of extending the transmission network (World Bank 2006a) and the cost barrier of economical distributed energy technologies able to provide power for rural households, let alone for desalination (Nguyen 2006). Recent technical advances and the scaling up of renewable energy sources such as wind systems have reduced the energy requirement as well as improved efficiency, making it possible to drive down the costs (SunLift 2001; García-Rodríguez 2004; Fritzmann et al. 2007), and thereby offer the potential to provide this power for desalination to develop municipal-scale clean water systems.

This chapter will explore the potential for wind-powered desalination (WPD) for the coastal fringe of Vietnam's Mekong Delta. Specifically, this chapter considers off-grid medium- to small-scale reverse osmosis desalination run with electricity from small wind turbines. First, aspects of the water problems for Vietnam will be reviewed in light of the probable effects of climate change during the twenty-first century. Next, regions with the highest potential and most urgent demand will be identified and quantified. Finally, the introduction of wind desalination in rural Mekong Delta will be discussed in the context of sustainability and technical and financial viability.

21.2 Background

21.2.1 Threats to Vietnam's Clean Water Supply

Anthropogenic factors play a major role in bringing about the water supply crisis and impeding sanitation improvement in the Mekong Delta. Described by Kakonen, the dual nature of human attitudes toward the river flow regime and the Delta's environment is characterized by human intervention, control, and adaptation;

the former of which has lately dominated the choice of actions, which have escalated in terms of administration level and a number of public construction works (2008). Despite becoming very productive and export-oriented, the Delta has been facing worsening water quality, increased demand for water, and increased saline water intrusion. The costs and drawbacks of the engineering works have largely fallen on poor farmers and landless people.

The coastal areas experienced shifts first from brackish water to freshwater and then back to brackish water systems. Although the conversion of large areas into intensive shrimp farms adds value to brackish water, experience from Thailand has demonstrated the difficulty in making the shrimp business a socially-equitable and environmentally-sustainable livelihood source (Kakonen 2008). It has been argued that more social and environmental sustainable options be explored to maintain the Mekong Delta's enormous productivity and mitigate the increasing production costs in terms of strengthening saline intrusion, deteriorated water quality, declining biodiversity, and increased social differentiation (Pretty 2008).

The aggravation of the water problem by global warming is and will be significantly seen in the South and Central regions (Chaudry and Ruysschaert 2007), most of the inhabited area of which lies in the low elevation coastal zone[4] (LECZ) (SunLift 2001). First, sea level rise contributes to saline water intrusion (Soussan et al. 2005) and limits the water available for irrigation and aqua-farming (Nguyen 2007a). In the Mekong Delta salt water infiltration into groundwater and freshwater shortages, being already very common, will be increased through prolonged droughts and reduced river discharge. The damage implicates 1.5 million hectares of cultivated land out of the total 2.4 million hectares currently employed for agriculture in this national "rice-bowl" (Le and Wyseure 2007). In 2000 irrigation accounted for approximately 84% of water demand nationwide (World Bank 2006b) and grew to between 85% and 90% in 2003 (Le and Wyseure 2007). Between 2000 and 2010 the volume of irrigation water will likely rise from 76.6 to 88.8 km^3 (Soussan et al. 2005; Eastham et al. 2008). To put this in perspective, 95% of groundwater reserves in Vietnam remain untapped (World Bank 2003, 2006b), of which the total potential exploitable reserve of the country's aquifers is estimated at nearly 60 km^3 per year. Unless better management of water resources occurs there is a serious future food-security threat in the region.

Water supply, secondly, is connected to poverty in Vietnam (SunLift 2001, Nguyen 2007). In the Mekong Delta basin, the four farthest downstream provinces, which are most severely affected by salinization, have 80% of their water from drilled tube wells whereas 70% of the Central coastal communities get drinking water from wide diameter open wells. As these areas are most vulnerable to flood and storm surges (Soussan et al. 2005), the practices (1) lower both quality and availability of the water resources, (2) force the population to accept low quality and substandard consumption of water (e.g., An Giang and Dong Thap in downstream Mekong) (Berg et al. 2006), and (3) lead to water borne diseases

[4] http://sedac.ciesin.columbia.edu/gpw/lecz.jsp

(Soussan et al. 2005). Being water-poor is equivalent to poverty and to low-health security (Soussan et al. 2005).

Finally, ecology and habitat security is endangered by the intensified and less predictable seasonal patterns of drought and rainfall (Government of Vietnam 2002; Nguyen 2007; Soussan et al. 2005). Ninh Thuan province provides an example of accelerated desertification hurting the local livelihoods. More still needs to be compiled and done for drought data in Vietnam (Soussan et al. 2005), but it is clear that this severe problem is likely to be aggravated by climate change this century.

21.2.2 *Reverse Osmosis*

Common desalination technologies include reverse osmosis, electrodialysis, multi-stage flash, multi-effect distillation, and vapor compression distillation (SunLift 2001; Einav et al. 2002; Blank et al. 2007; Karagiannis and Soldatos 2008). In a previous study of sea water desalination in the coastal areas of Central and South Vietnam, reverse osmosis (RO) was found to be the most economical means of providing desalinated water in many cases due to its lower energy consumption, leading to lower unit water costs (SunLift 2001). Einav et al. presented a general layout of a reverse osmosis desalination plant, which consists of a bundle of membranes placed in a pressure chamber, a high pressure pump, a turbine for recovering energy from the high concentration brine that is discharged from the plant, and a system for the pretreatment of the feed water and the product water. Seawater enters a pretreatment system, which contains sand filters, micron filters, and a system for chemical dosing. The purpose of this pretreatment system is to protect the membranes from fouling by dirt, biological, or chemical deposits. The feed pump drives seawater through the membrane system. The desalinated water then receives a final treatment, which includes the adjustment of its reactivity ratio, the reduction of its corrosiveness, and its disinfection. The discharged brine passes through the turbine, which recovers 30–40% of the energy invested by the process pump and is then returned to the sea. A secondary system used for periodical cleaning of the membranes is installed in each reverse osmosis plant (2002).

RO unit water costs are primarily determined by the membrane life and energy costs (Einav et al. 2002; Avlonitis et al. 2003). RO systems can be designed for intermittent or continuous operation (Thomas 1997) according to fluctuating water demand and benefit only slightly from economies of size. This makes them ideal for intermittent renewable energy sources as long as there are adequate water reservoirs for storage. RO has tremendous energy advantages over other technologies under the following conditions: a 1% salt passage can be tolerated, steam is not available, and good quality sea water is available, which exists in the South China seawater. Further, when compared to thermal processes, RO has: (1) higher conversion ratios due to the increased strength of new composite membrane modules to higher operating pressure levels, (2) improved plant availability due to the enhancement of membrane lifetimes, (3) simplicity and modularity of plant design and operation, (4) reduced installation space, (5) shorter delivery times, and (6) lower environmental impacts (Kiranoudis et al. 1997; Dore 2005).

The US Department of Energy estimates the capital cost to be between $1,600 and $2,000/m^3/day of capacity for developing country applications (Thomas 1997), which is higher in terms of upfront investment compared to thermal processes (SunLift 2001) and explains the decreasing market share of the technology (Kiranoudis et al. 1997). One solution is to reduce the operational cost of the entire plant, which constitutes more than half of the total annual cost of the plant via the use of alternative (renewable) energy sources for direct production of electricity. The most promising alternative source of energy under consideration is electricity produced by wind turbines in regions with high wind potential, because they reduce the unit cost of clean water up to 20% provided that the regional wind mean velocities are higher than 5 m/s (Kiranoudis et al. 1997). Kaldellis and Kondili (2007) confirmed that wind desalination is most suitable for small (1–50 m^3/day) and medium (50–250 m^3/day) scale plants and that wind energy can be used efficiently provided that the average wind velocity is above 5 m/s.

Energy requirement varies with the type of feed-in water and the sources of the study. For sea water it can be as high as 17 kWh/m^3 (Thomas 1997) or between 5 and 7 kWh/m^3 (SunLift 2001; Carta et al. 2003). High efficiency pumps designed for renewable energy power RO desalination can lower energy consumption to between 3.5 (García-Rodríguez 2004) and 2 kWh/m^3 (Fritzmann et al. 2007). Many reverse osmosis systems now use energy recovery to increase the efficiency of the system further. However, energy recovery does not increase the operational complexity of RO systems (Thomas 1997), but it can significantly increase the costs of the system (Kalogirou 2005).

When assessing the environmental impacts of the RO desalination technology, Einav et al. (2002) considered the use of land, the impact on groundwater, the impact on the marine habitat by way of the discharge of the brine, noise pollution, and the intensified consumption of energy. To minimize the environmental impact of a desalination plants, the siting of the plants and the consequential use of long pipes are advised to avoid salinization of the local water system. Attention must be paid to the effects of discharging the concentrated brine on the marine environment. Although examples in the literature were drawn from plants by the Red Sea, the sensitivity of mangrove and coral reefs is worth noting given the presence of mangroves around the Ca Mau Peninsula (Binh et al. 1997) and that of coral reefs offshore of the southern Central Vietnamese coast (Wilkinson 2005). Reverse osmosis by itself is not a significant greenhouse gas emission source (Fritzmann et al. 2007).

In Vietnam, even though several attempts at demonstrating the benefits of desalination systems have occurred recently, the diffusion of desalination in Vietnam is minimal (SunLift Systems 2001; Pankratz 2008). Previous literature on desalination in Southeast Asia is dominated by large scale nuclear powered desalination; while in the Mediterranean region (Greece), the Middle East (Iran, Oman), and North Africa (Egypt, Algeria) it is abundant. This is not helped by the fact that the only desalination plant in a comparable geographical region is the SingSpring project in Tuas, Singapore, which can provide over 100,000 m^3 per day or 10% of the country's needs.[5] Following standard definitions (Buitenhuis et al. 2010; Pearce et al. 2008) of appropriate technologies, there is a clear investigative gap to fill as large-scale centralized sources of desalination are not appropriate for the distributed rural areas for the Mekong Delta. A report by SunLift in 2001 considered diesel/wind and wind/hydrogen hybrid desalination systems and found that in light of the increasing prices, significant cost benefits will be achieved for wind/diesel systems even in locations with medium wind conditions (SunLift 2001). Nearly a decade has passed since the report was produced and the first landmark for the NRCWSS goal has arrived; there is a clear need for a new cost evaluation and comparison for renewable (wind) powered desalination in Vietnam.

21.2.3 Wind Energy Modeling and Estimation

Landberg et al. provide an excellent theoretical base for tackling wind-energy related problems and present eight methods to estimate energy yield: (1) "Folkore," (2) measurement only, (3) measure-correlate-predict (MCP), (4) global databases (most utilized is the NASA Surface Solar Energy SSE), (5) wind atlas methodology (CFD models), (6) site data-based modeling, (7) meso-scale modeling, and (8) and combined meso/microscale modeling (2003). Manwell et al. (2002) provide an additional introduction to the popular Weibull and Raleigh distributions, which in the absence of station measurements will allow the fitting of available data and predicting and classifying the wind at a specified location. A resolution of 1 km or less requires tremendous amounts of computing although increasing speeds of computers has allowed the order of a few kilometer (Landberg et al. 2003). Recent work by Pryor et al. in 2004 gives an intensive mathematical discussion of the uncertainties associated with distribution fitting methods.

To deal with the coastal wind, data from either synthetic aperture radar (SAR) or daily scatterometer (QuickScat) are used. Wind mapping from the former requires imaging and signal interpretation algorithms (Christiansen et al. 2006) and suffers from steep commercial costs for each ERS SAR PRI and Envisat ASAR PRI scene covering an area of 500 km by 500 km. Wind energy calculations can be carried out using the WasP (Wind atlas analysis and application Program) (Hasager et al. 2004). In the case of QuickScat, free tools are available to interpret the level 3 QuickScat .hdf files into a GIS platform.[6] However the data suffers from low resolution.

Khan and Iqbal (2004) used the MCP method, which correlates surface data from short measurements or sparsely located stations with the global Surface Solar

[5]http://www.water-technology.net/projects/tuas/

[6]http://comlmaps.org/how-to/layers-and-resources/oceanographic-data/quikscat-wind

Energy dataset and found it to be an excellent and user-friendly source for preliminary study purposes. Contours plotted by Matlab[7] using this database were found to indicate similar trends with those from the NCEP/NCAR (National Center for Environmental Predictions/National Center for Atmospheric Research) Reanalysis and from the COADS (comprehensive ocean–atmosphere data set). Ground data extrapolation for turbine heights was done using the power law.

The Wind Energy Resource Atlas of Southeast Asia resulted from the MesoMap system, which is an integrated set of atmospheric simulation models, global weather, and geographical databases. Simulations were run on a representative set of days picked from 1984 to 1998 on a 1-km grid scale. Error assessed for the case of the Philippines was 8% within the true mean wind velocity. It was found that good to excellent wind resource areas concentrate partly into the coastal areas of southern Vietnam, where northeast monsoon winds converge with offshore wind and accelerate round the end of the Southeast Asia peninsula. Areas classified as fair or better for small wind turbines include large sections of coastal southern and Central Vietnam (TrueWinds 2001).

21.3 Methodology

21.3.1 Definitions

A distinction between rural and urban population density was necessary for this chapter, since defining urbanism and urban population is country-specific and effort-consuming (United Nations 2010; Long et al. 2001) and no uniform definition exists for the world (United Nations 2010). For Vietnam such a definition was not found in the National Statistics Office. Hence rural population density for Vietnam was taken to be lower than 2,000 people per km^2, which is the minimum requirement for urban density in the lowest class (class V) in the government's scheme of urban center classification (Coulhart et al. 2006).

By 2001 an estimated 53% of the urban population and 30% of the rural population had access to potable water or 65 and 40 l per capita per day, respectively (SunLift 2001). It should be noted that 40 l/person/day is the World Health Organization standard water consumption per urban resident (United Nations 2007). By 2004, 99% of urban population and 80% of the rural population (85% of the total population) in Vietnam had access to safe drinking water.[8] However according to the Vietnam Living Standard Survey (VLSS) (General Statistics Office of Vietnam 2006), 62.2% of urban households and 6.3% of rural households rely on tap water for their main source of drinking water. The Comprehensive Poverty Reduction and Growth Strategy (CPRGS)'s 2003 mandate that 60% of rural and 80% of urban

[7] Scilab provides an free and open source platform for numerical computation http://www.scilab.org/
[8] http://www.worldwater.org/data.html

population have access to clean and safe water by 2005 (Government of Vietnam 2003) signals some optimism, at least by coverage. However, it also raises a question by caution: by what standard or definition is the water "safe and clean"? Conflict can exist among the definitions of safe and clean water used by different institutions.

Very limited information about the groundwater resource size, use, sustainability, and quality in this region is available from the literature (Eastham et al. 2008). Acknowledging the problem of incompatible data sets for water supply and sanitation coverage (World Bank 2002), this study will reconstruct maps for water treatment coverage based on the Vietnam Living Standard Survey (General Statistics Office of Vietnam 2002). The maps for the nationwide NRCWSS coverage were based on definitions of clean water supply, which encompasses clean water, drilled wells, rainwater, mountain springs with filters, constructed hand dug wells at least 7 m away from pollutant sources, and filter/chemical-treated water from unprotected sources; and that of improved water sources, which include tap water, drilled wells, rain water, mountain spring with filters, and all constructed hand dug wells (Soussan et al. 2005).

21.3.2 Data Inputs

As GIS was intended to be the platform for calculation and visualization of the results, all project data had to be acquired and imported in a suitable and usable format for the program used, which was ArcGIS version 9.3, and conform to the time line and definitions discussed above. Data ranged from geographic (to define the geographical constraint of the study) to demographic (to reflect the human central focus of the study) and technical (for the wind potential).

- VN's country and province boundaries, in Universal Transverse Mercator zone 49N (UTM 49N) and World Geographic System 1984 datum (WGS 1984), developed by Ha T. Nguyen.
- VN's population grids for 2010, in Lambert Azimuthal Equal Area projection, at 6 km resolution, developed at Columbia University's Center for International Earth Science Information Network (CIESIN).[9]
- ASTAE wind map: UTM 48N at 1 km resolution and at 30 m elevation.

21.3.3 Yield of Desalination by Wind Energy

The conceptual methodology of the chapter can be summarized in the following steps:

- Spatially map the rural population density of the region for 2010
- Spatially map the rural population within the feasible wind areas

[9] http://sedac.ciesin.columbia.edu/gpw/global.jsp

- Estimate local water demand corresponding to each feasible wind point based on population density
- Estimate required wind speed to meet local demand at each feasible wind point
- Calculate the difference between available potential supply and local demand for water at each feasible wind point

For a resolution of 4.16 km by 4.16 km there is a cut-off of 22,210 people per cell (e.g., population densities lower than this are not considered). A reclassification scheme using this value as a cut-off produced a density Boolean grid, g_{db}. Multiplying this grid with g_d the original population density grid of 2010 gave g_{rd10} the rural population density grid of 2010 with the low density areas removed:

$$g_{rd10} = g_{db} * g_d \tag{21.1}$$

A zonal statistics gave the projected 2010 rural population density in each province, p_{prd10}, which was multiplied with area and divided by cell size to give p_{prt10} the province's rural total population in 2010, just to be verified with 2008 government statistics:

$$p_{prt10} = p_{prd10} * S_{prov} / S_{cell} \tag{21.2}$$

The unsupplied population p_u that has no access to clean water is

$$p_u = p_{prt10} * (1 - c) \tag{21.3}$$

where c is the percentage of clean water coverage obtained during the VLSS 2002.

Only areas with wind greater than or equal to 5 m/s for 30 m were selected based on recommendations in previous studies (TrueWinds 2001; Kiranoudis et al. 1997; Carta et al. 2003). The resultant point shapefile is named selected_wind, which was converted into raster – $g_{wind.position}$ – at the same grid size as the density grid, in order to eliminate cells with no technically viable wind potential:

$$g_{wind\ times\ pop} = g_{wind\ position} * g_{dr10} \tag{21.4}$$

The product has the same extent (coverage) as the area with wind prospects and of smaller extent than the population density raster. Thus only the areas of rural population and good wind potential are counted. Next the grid that contains population density as cell values was given by:

$$g_{wind\ pop} = g_{wind\ times\ pop} / g_{wind\ position} \tag{21.5}$$

The new raster has the highest population density of 21,290/cell compared to 21,601 for the whole coastal region. It should be noted that Eq. 21.5 is used to restore units in a different coverage.

$g_{wind.pop}$ was then transformed into a new raster with integer values. The integer raster has an attribute table, which enables further eliminating cells with population density lower than 50 people per cell (of approximately 20 km^2 each). The cut off was decided based on an estimation that if 50 people only consume 20 l/day for drinking, cleaning, and cooking (Ayoub and Alward 1996) they will need a desalination unit with daily output of 1 m^3 which would represent prohibitive costs to the users. The command Extract values to points was run on the newly filtered $g_{wind.pop}$ to assign the appropriate population size to the applicable wind points.

A bottom-up approach was used to project the demand of local residents, and hence the size of desalination units, against the available wind resources. Reverse osmosis technology was chosen because of the advantages discussed above and compatibility to small and medium scale systems (Kalogirou 2005). A conservative estimate took 7 kWh/m^3 as the energy requirement for reverse osmosis, i.e., excluding energy recovery, and 60 l per capita as the 2010 goal of the NRCWSS strategy for rural water consumption, the daily rural water demand per cell is thus:

$$d_{cell} = p_{cell} * 60/1{,}000 (\text{l/day}) \quad (21.6)$$

The current best wind turbines can harvest up to 45% of the available wind resources (Manwell et al. 2002), hence the energy consumption (e_{cell}) needed for such daily water demand is:

$$e_{cell} = d_{cell} * 7 / 0.45 (\text{kWh/day}) \quad (21.7)$$

A simple equation to calculate the wind power per unit area was formed under three assumptions: (1) the standard density of air, ρ, is 1.225 kg/m^3; (2) power from the wind is proportional to the swept area by the rotor; and (3) the wind power density is proportional to the cube of the wind velocity (Manwell et al. 2002). The wind turbine chosen for the study is Fortis Alize wind turbine with a pole height of 30 m, a blade diameter of 3.5 m, and a power rating of 10 kW.[10,11] Electricity generated by the wind is used for pumping water and pressurizing the brine. From equation 2.2.8 in Manwell et al. (2002) the average daily wind speed required to operate such system is

$$V_{req} = [(e_{cell} * 2 * 1000)/(\rho * S_{swept})]^{1/3} (\text{m/s}) \quad (21.8)$$

where 2 results from taking the inverse of kinetic energy and 1,000 is the conversion factor between Watts and kiloWatts.

Since each cell encompasses multiple wind sites, the required wind speed for a whole 20 km^2 area may at first seem to overwhelm the available wind resources. A sum of all the available wind speed can give by proxy estimation of the potential desalination volume and better compare what is required and what is "available" for

[10] http://www.fortiswindenergy.com/products/wind-turbines

[11] http://www.solarenergyalliance.com/fortisturbines.htm

each population size. The *summarize* command was run for population density against the required wind speed and population density against the total available site wind speed, the output of which was further handled in a spreadsheet by applying:

$$\text{If} \quad V_{\text{req}} < \text{or} = V_{\text{avail}} \quad \text{then returns 1 else 0} \tag{21.9}$$

where 1 means that the cell is included for the area of feasible wind powered desalination and 0 otherwise. A table joint enabled matching these boolean values to the appropriate population density, where all wind sites in the "counted" cells were extracted as a new shapefile: *ws_ final*. This shapefile was then converted to a raster whose all values were reclassified as 1 and got multiplied by $g_{\text{wind.pop}}$. The product is a raster of selected population density that is fit for wind powered desalination: $g_{\text{selected.wind}}$ and plotted in Sect. 21.4.

21.4 Results

Figure 21.1 shows the position of the feasible wind powered desalination areas within Vietnam's coastal Mekong Region, where the region that can be supported by wind powered desalination was overlaid on the low elevation coastal zone.

Figures 21.2–21.9 provide a detailed enlarged view of this region on a per province basis and an inset perspective to the inland and sea boundaries. The grey scale gradient corresponds to low (white) to high (black) population density, while column height signifies the difference between required wind velocity and available wind velocities for each area of approximately 20 km^2. On the figures some representative net wind resource values (velocity available minus velocity demanded to supply water production) are shown. It should be noted that these are aggregates of the net velocity over the grid, which is (17.3 km^2). Hence the darker and taller the column the more desirable a desalination location because it represents a demand and an adequate margin between what is required and what is available in terms of mean wind energy. But even if the columns are of medium height, but dark, it will still make more sense to put a desalination system there than in areas with pale, tall columns because those represent very sparsely inhabited communes and even fields and mangroves. The ellipses in the figures denote areas that should be targeted first for WPD.

Figures 21.2–21.9 present excess wind resources against rural population for each province of Ba Ria Vung Tau, Bac Lieu, Ca Mau, Ben Tre, Kien Giang, Soc Trang, Tien Giang, and Tra Vinh, respectively. Dark tall columns (high population and favorable wind velocities) concentrate on the provinces of Ba Ria Vung Tau (Fig. 21.2), Tien Giang (Fig. 21.8), Ben Tre (Fig. 21.5), and inland Tra Vinh (Fig. 21.9). The same ideal conditions are only present around towns and provincial capitals of Bac Lieu (Fig. 21.3), Kien Giang (Fig. 21.6), Soc Trang (Fig. 21.7), and Ca Mau (Fig. 21.5). These results correspond to the provincial economic standings. Ba Ria Vung Tau (Fig. 21.2) is projected to be an international port and center of the

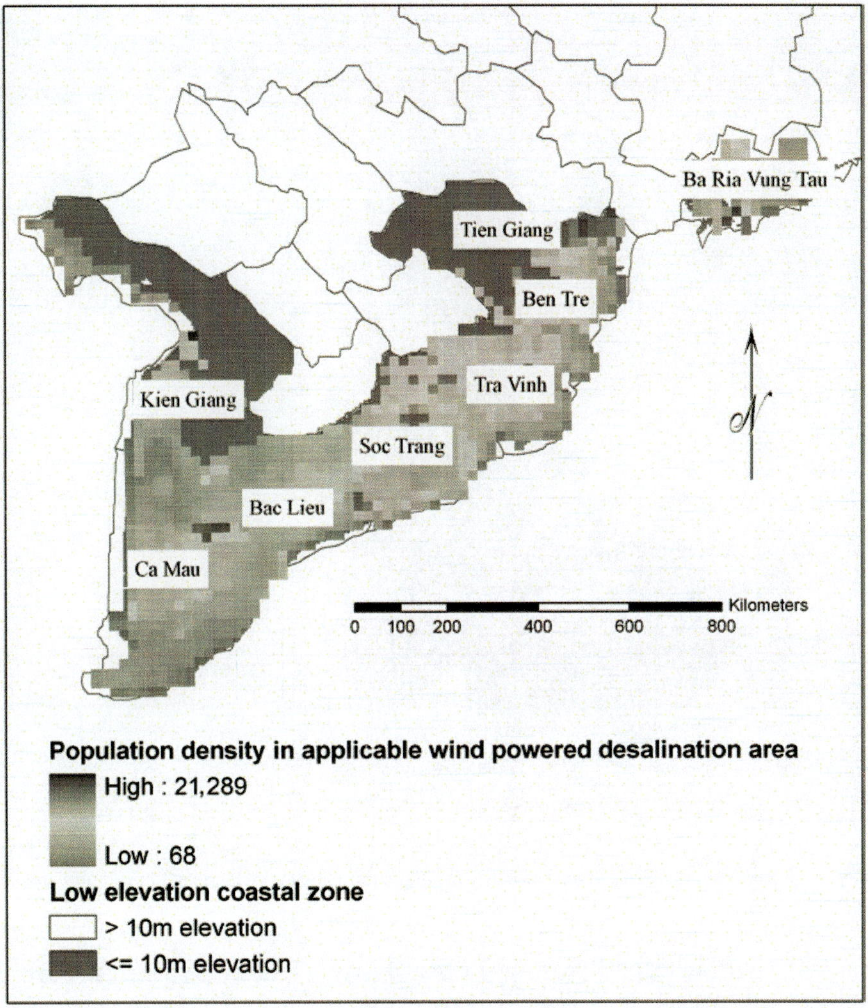

Fig. 21.1 Position of the feasible wind desalinated area in coastal Mekong

petroleum and tourism industries.[12] Tien Giang occupies the center point in the waterway systems of the Mekong Delta and prospers on aquaculture and fruit agriculture.[13] In a similar standing with Tien Giang, Ben Tre guards the waterway and highway access to Ho Chi Minh City and western Mekong.[14] Advantages in transportation access and industrial development attract more population, which in turn raises an immediate need for clean water supply.

[12] http://www.baria-vungtau.gov.vn/

[13] http://www.tiengiang.gov.vn/xemtin.asp?idcha=960&cap=2&id=1132

[14] http://www.tiengiang.gov.vn/xemtin.asp?idcha=960&cap=2&id=1132 n&id=6&Itemid=46

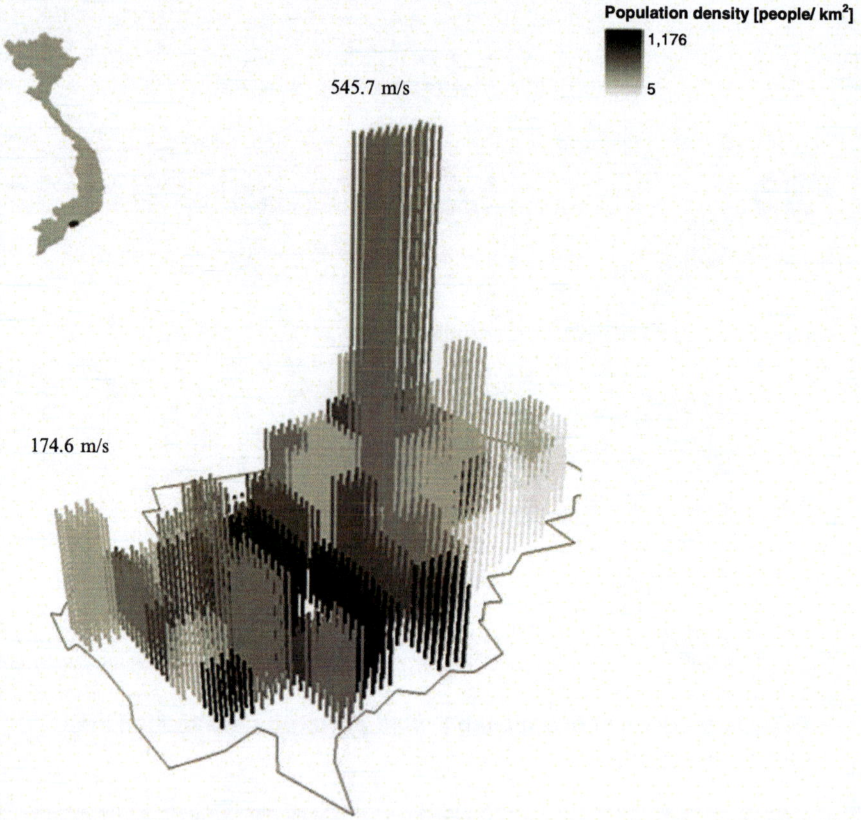

Fig. 21.2 Net wind resources (wind velocity available minus velocity demanded to supply water production over 17.3 km^2) in Ba Ria Vung Tau

For the remaining provinces water demand largely reflects the patterns of agricultural and civil activities. There is a special case in Fig. 21.9 of Tra Vinh because of sand dunes running parallel with the coastline and complicating the topology of the province. In addition its vulnerability to salinization explains the deep inland occurrence of human settlement and the importance of fresh water channeling here.[15] The majority of land use in Soc Trang (Fig. 21.7) and Bac Lieu (Fig. 21.3) is dedicated for aqua farming and rice and hence the observation of tall, pale columns representing a large marginal difference between demand and supply in areas of low inhabitance and high wind resources, which may well be open aqua farming water bodies and rice paddies. For Ca Mau (Fig. 21.4) the dominant presence of these columns agrees with the substandard poverty and education levels and minimal infrastructure or else the vast mangroves of U Minh.[16] In Kien Giang (Fig. 21.5)

[15] http://www.travinh.gov.vn/

[16] http://www.camau.gov.vn/#intro_details/8920/1

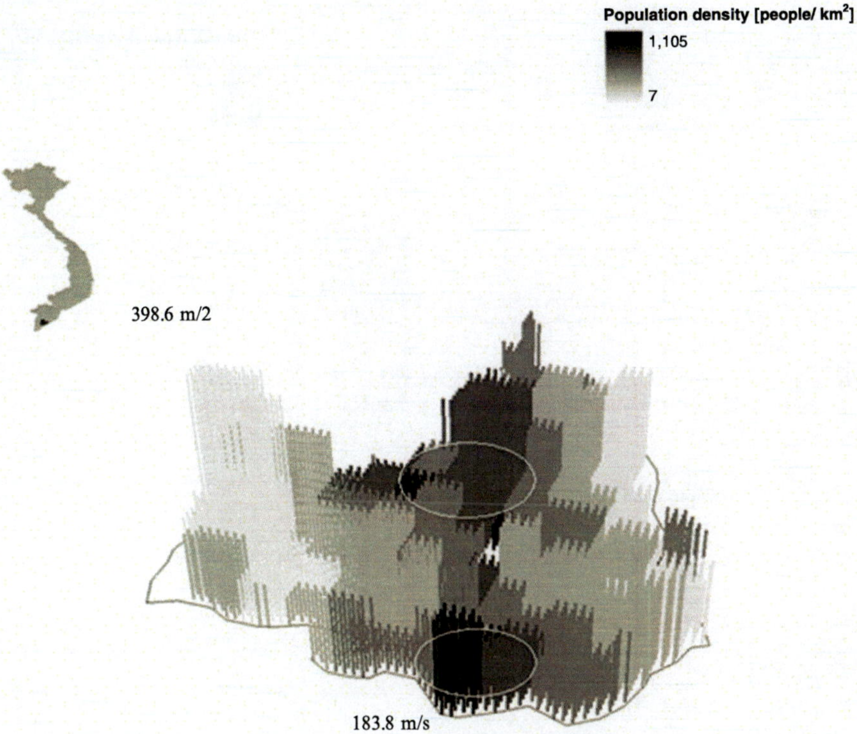

Fig. 21.3 Net wind resources (wind velocity available minus velocity demanded to supply water production over 17.3 km^2) in Bac Lieu

the threat of salination has directed the local livelihood toward fishing, aqua farming, tourism, and very recently a few heavy industries, which are to be developed, located near new ports.[17] But overall the column profile of Fig. 21.5 for Kien Giang is relatively flat and short, illustrating only fair wind potential (TrueWinds 2001) and relatively few demand centers.

It was deemed more appropriate to measure the potential of small wind turbine applications via the proportion of rural population that could be served by small turbines (TrueWinds 2001). In addition, villages located in very good or excellent wind areas are very rare globally except in Vietnam. It was calculated that available wind resources can provide for 5.4 million people in applicable areas shown in Fig. 21.1 at the rate of 60 l per capita per day, in the absence of any other facilities. This is 5.4 million out of a total of 8.7 million predicted for the 2010 rural population, spreading over 18.9 thousand km^2 of 27.2 thousand km^2 total area of the coastal provinces.

[17] http://www.kiengiang.gov.vn/index2.jsp?menuId=275&articleId=11012

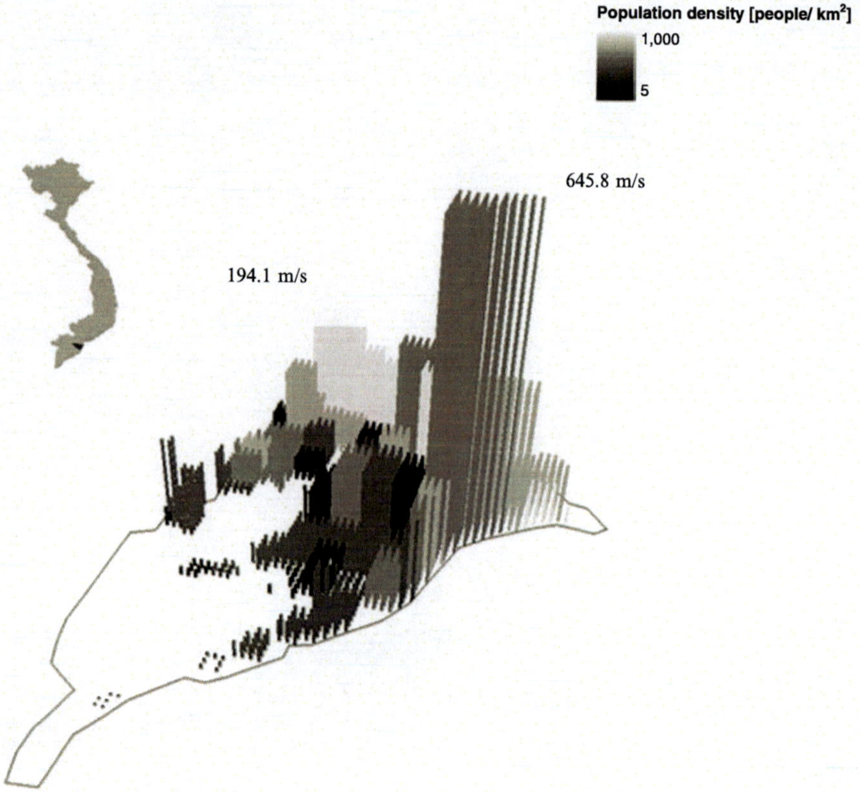

Fig. 21.4 Net wind resources (wind velocity available minus velocity demanded to supply water production over 17.3 km^2) in Ben Tre

WPD system size was taken as directly proportional to the demand of the local populations and hence to the population density at the point wind speed measurement is available. The number of wind sites that are technologically and financially feasible in providing electricity for desalination were found to concentrate in regions with moderate density i.e., up to 1,400 people per km^2. This agrees with Nguyen (2007b) that small wind turbines appear to be one of the best options for isolated rural areas, particularly when there is a coupling of wind turbines with desalination. This is not entirely surprising because the higher the density is at a site, the more human activities take place; hence, more complex terrain can be created. In addition the natural topology and land-use characteristics of the coastal Mekong are already perceived as dynamic and complex (Kakonen 2008), the effect of which on the energy output from a turbine may be significant enough to force the economics of the whole project to depend on proper siting. In this case the more wind sites are available for each cell of 20 km^2 coverage the more flexible and more decentralized it becomes for system siting and design. However it would be premature to

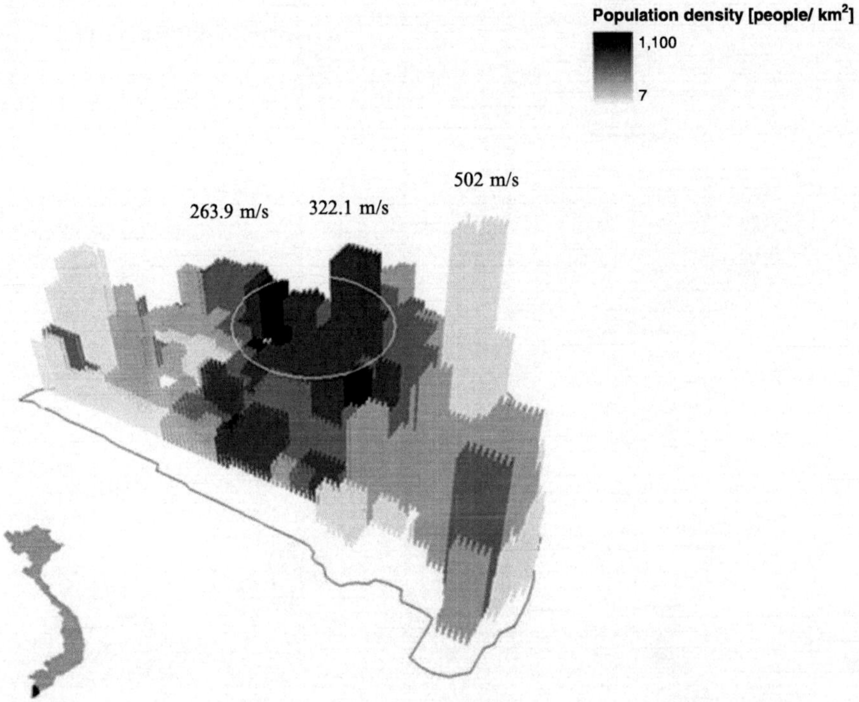

Fig. 21.5 Net wind resources (wind velocity available minus velocity demanded to supply water production over 17.3 km²) in Ca Mau

exclude those locations with higher density without further investigation, being that larger population means greater demand and a more urgent need of access to clean water.

21.5 Discussion

21.5.1 Desalination for Employment, Conflict Reduction, and Sustainability

Wind energy is still very new in Vietnam and despite its perceived potential it has not been well received on both household and utility scales (Nguyen 2007b), which can be extrapolated to the public attitude that wind powered desalination may encounter in the coastal Mekong rural communes. Demonstration, promotion, and propagation activities need to be put in place to allow developers to evaluate their projects and users to appreciate the reliability and cost competitiveness of the technology. The appreciation can be more profound and long lasting if it is seen through

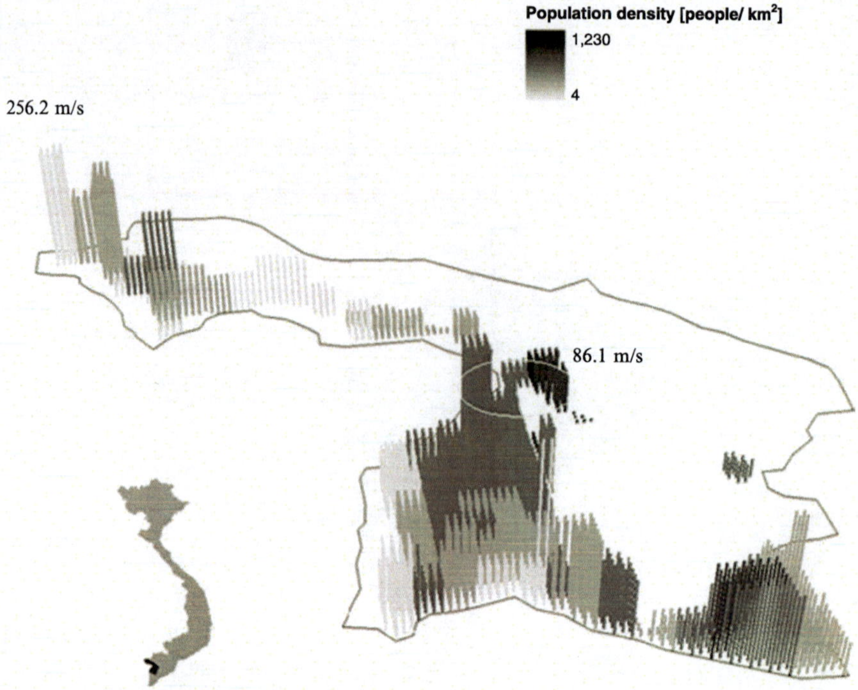

Fig. 21.6 Net wind resources (wind velocity available minus velocity demanded to supply water production over 17.3 km²) in Kien Giang

the lens of (1) skill acquisition and educational benefits to combat poverty and unemployment, (2) easement of water-related conflicts, and (3) the context of environmental or agricultural sustainability. In the context of agricultural sustainability, it offers social, environmental, and technical improvements for the region and hence represents a community-based (and decentralized) method to adapt to and mitigate climate change, which has attracted little focus in Vietnam (Shaw 2006).

WPD has the potential to improve the living standard and local infrastructure. Many implementations of WPD in the area would not only lead to improved water quality but also a demand for and presumably an inevitable improvement in local technical knowledge to accommodate maintenance. This is a significant benefit, because despite massive regional productivity the Mekong Delta ranks nearly the lowest nationally in terms of school enrollment and literacy levels (Taylor 2004, Nguyen 2007a). As WPD promises new jobs and an incentive to improve the education of local residents, the demand for the landless in particular to seek hard manual and low paid jobs in faraway urban centers will be reduced. Note that Ca Mau and Soc Trang, two of the provinces under consideration, are home to the highest proportion of the poor in the labor force (72% and 74.6%, respectively) (AUSAID 2004). It should be pointed out that initially it is likely that seawater desalination projects in Vietnam will be mostly proposed and carried out by foreign

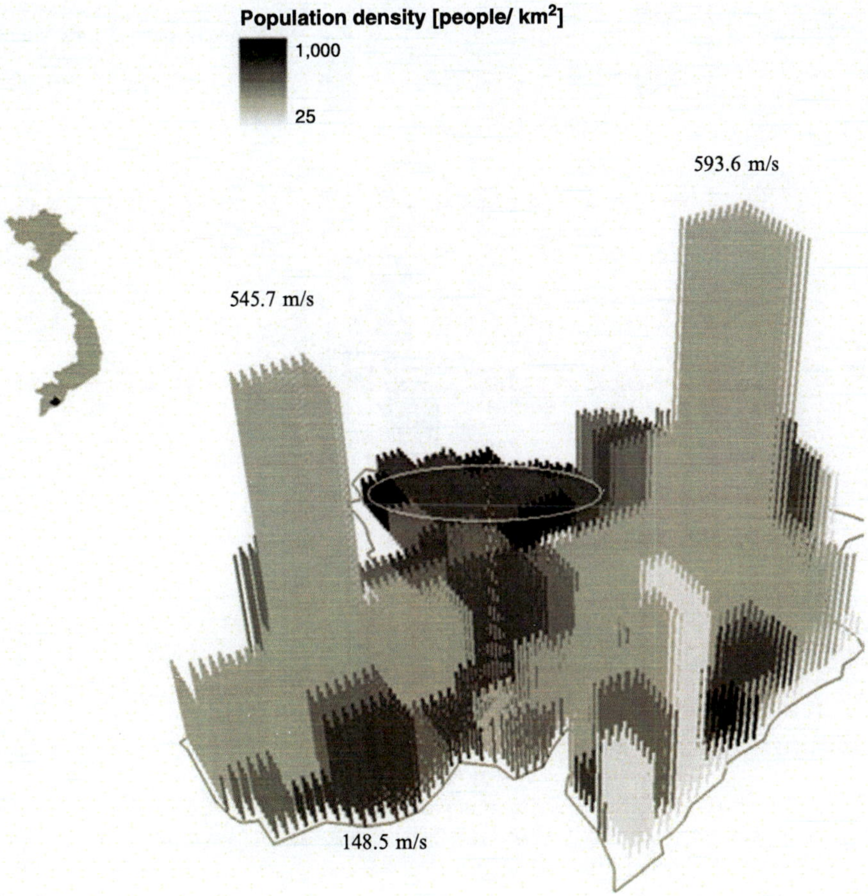

Fig. 21.7 Net wind resources (wind velocity available minus velocity demanded to supply water production over 17.3 km²) in Soc Trang

companies, who possess more skill and experience than companies in Vietnam even at the national level. It has been cautioned that only small scale plants (less than 1,000 m³ per day) have a good chance of being approved by the government (SunLift 2001). This last note should be considered with caution as it could quench any ambitious plan to gain entry into the desalination market of Vietnam and jump start the industry.

Just as in the US, poverty in Vietnam often follows racial and ethnic lines. The Khmer appears to make up the majority of the poor/landless in the Mekong Delta or at least their rate of escaping poverty is much lower than the Vietnamese and Chinese. When it comes to clean water and environmental sanitation the percentage of Khmer households accessing sources of clean water is lower than that of the other less isolated, richer ethnic groups (AUSAID 2004). Based on the spatial distribution

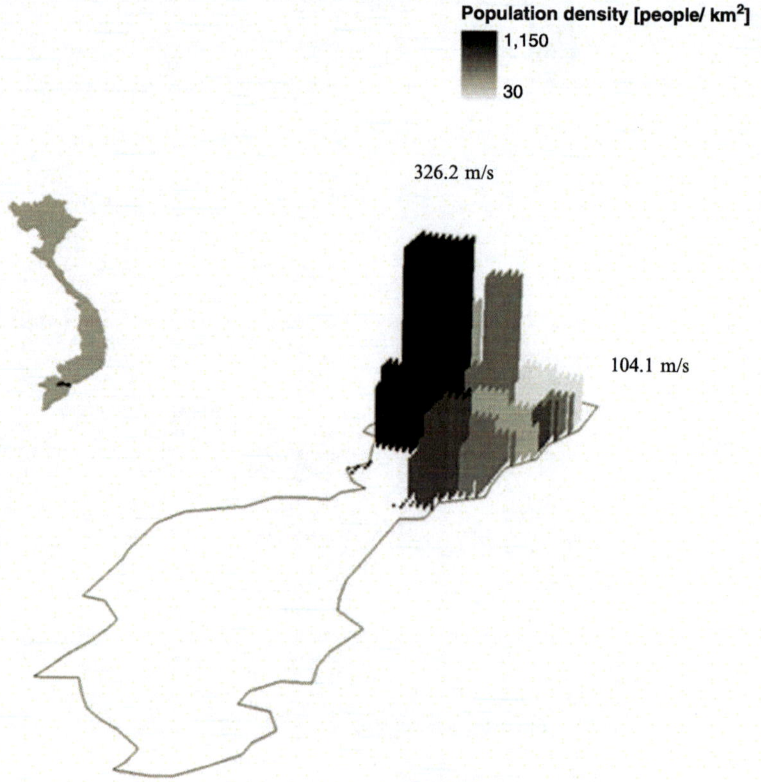

Fig. 21.8 Net wind resources (wind velocity available minus velocity demanded to supply water production over 17.3 km^2) in Tien Giang

of ethnic communities in the Mekong Delta, decentralized WPD has potential to address this under-served group. Less vulnerability to poverty leads to a lower rate of internal migration, which is a compounding problem for rural development and poverty reduction policies (Phuong et al. 2008). These are only among the most obvious benefits that WPD projects can directly bring to the poor in this region.

WPD also holds promise to reduce the water-related conflicts in the region, where conflict is understood as interest incompatibility or livelihood loss among various water users as a result of access to water of inadequate quantity and quality (Dang et al. 2007). As aqua-farming and agriculture are dominant among Mekong rural livelihoods, there is a potential for water conflicts between rice and shrimp farmers and between upstream and downstream users. The water conflicts relate to intensive water abstraction for rice farming in the upper delta, effluent discharges from aquaculture farming, acidity and metals released from land reclamation in acid sulfate soil zones, and water competition among crop culture, shrimp farming, mangrove forests, and fishing in coastal zones. An injection of clean water supply

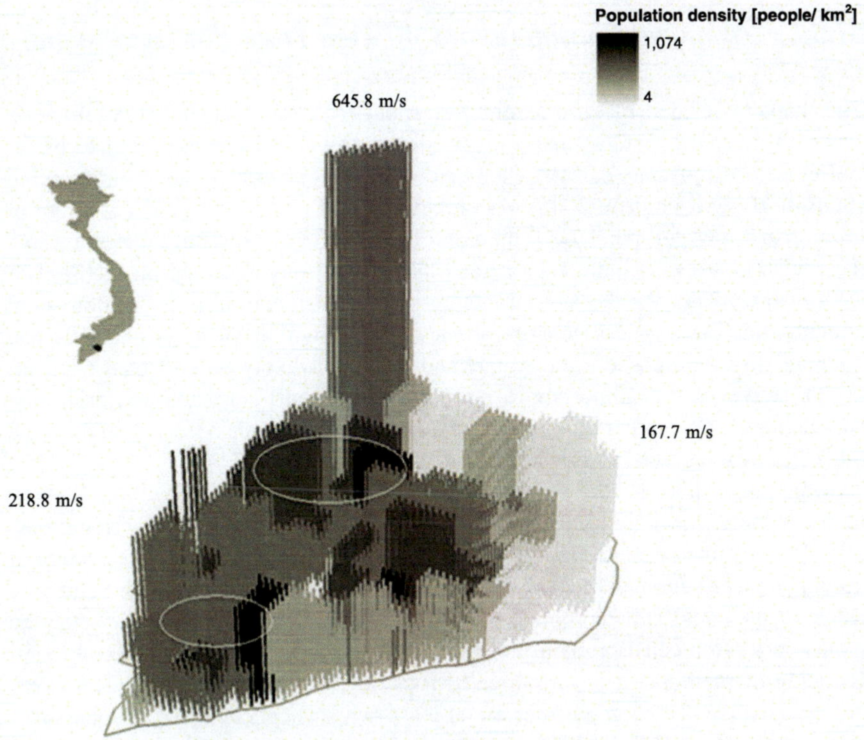

Fig. 21.9 Net wind resources (wind velocity available minus velocity demanded to supply water production over 17.3 km²) in Tra Vinh

via desalination can clearly help to alleviate some of the tension in the existing interdependencies.

Within the defined region WPD can alleviate the consequences of hitherto unsustainable practices by providing water in a non-carbon-intensive and environmental-friendly way. The material input to generate electricity from wind emits about 10 t of CO_2 per GWh while that of coal is between 830 and 920 t per GWh on a life cycle basis (Akermann and Soder 2002). Decentralized desalination is totally devoid of large-scale waterway construction that has changed the ecosystem in massive and not necessarily sustainable ways (Kakonen 2008) and relatively unaffected by dissolved contamination in local channel and groundwater (Pretty 2008; Dang et al. 2007) since the feedstock is seawater. Coupled with the above social benefits wind-powered desalination also satisfies the criteria of agricultural sustainability as outlined by Pretty (2008). In light of these potential contributions WPD targets many important sustainable development indicators for Vietnam (Dinh et al. 2006) and therefore is worth being considered among poverty-reduction strategies.

21.5.2 Technical Considerations of Wind Power for Desalination

The chapter considers off-grid medium- to small-scale reverse osmosis desalination run with electricity and/or mechanical energy provided by small wind turbines. Before the design stage can start, the quality of seawater input should be assessed (SunLift 2001; Kalogirou 2005). As energy demand increases linearly at a rate of more than 1 kWh/m^3 per 10,000 ppm and likewise for high salinity concentration, the quality of seawater in the region governs the suitability of such projects (SunLift 2001; Dore 2005). As the system design under consideration is for remote rural communities there is less concern about pollution within the river channels and openings to the sea where the water feed will most likely be collected (SunLift 2001). However, it is important to point out that salinity intrusion and brine discharge might concern the proximity of the catchment for such systems and the local biota (Einav et al. 2002; Laspidou et al. 2010).

Membranes play a very important role in determining the cost and lifetime of the system, the final quality of the product (water), and the eco-friendliness of the plant (Einav et al. 2002; Avlonitis et al. 2003). Membranes separate water of varying concentrations under high pressure and come into contact with solutes of different chemical and biological properties; hence maintenance of desalination plants very often consists mainly of the cleaning and replacement of membranes. Cleaning of membranes occurs three to four times a year and replacement every 5–6 years since discharged brines are at least twice as concentrated in substances as seawater (Einav et al. 2002; Blank et al. 2007). Although membrane costs have been reduced by a factor of 10 during the last 30 years, pretreatment of the feed water, which affects the lifetime of the membranes, and replacement cost are still the most expensive items in the membrane system (Blank et al. 2007; Karagiannis and Soldatos 2008). Avlonitis et al. recommended effective pretreatment of the feed water to maximize the efficiency of the process and lengthen membrane life by minimizing scaling formation, fouling deposition, and membrane degradation (2003).

Wind energy is a highly variable energy source in terms of both velocity and frequency. When wind energy is used for electricity generation the variation of the wind source can be balanced by the addition of battery banks, which are charged when wind is available and discharged to the load (desalination plant) when required. In the case of mechanical energy production from wind, the desalination plant can operate only when there is wind. If so the desalination plant is usually oversized with respect to water demand, and instead of storing the energy, the water produced when wind is available is stored. This same methodology can also be used in the case of electrical generation from wind when used for desalination. Thus, a water reservoir is used in place of a battery to reduce system complexity and costs. As discussed previously the wind turbine size can also be minimized by including energy recovery technologies in the system. The omission of energy recovery calls for a trade off between lower energy consumption and up-front investment (Fritzmann et al. 2007; García-Rodríguez 2004).

One concept that has gained attention recently is the use of a modular desalination system as a deferrable load in order to increase the wind penetration potential in a wind-diesel mini-grid. The modules of the desalination unit would be brought on-line as power from the wind became available to increase utilization of available wind power, decreasing the amount of excess output which must be "dumped" and reducing the amount of diesel run-time, and to allow the hybrid system to operate optimal loads. RO is particularly suited for such systems, because RO is a modular technology, with typical installations containing several RO membrane/pump modules. However, the only operational system that could be identified by a literature search is located at Fuerteventura in the Canary Islands (Carta et al. 2003; Thomas 1997). This is because the system is still in the process of being optimized (Thomas 1997) in connection to a variable power system. For instance, the reverse osmosis system has to cope with the sensitivity of the membranes regarding fouling, scaling, as well as unpredictable phenomena due to start–stop cycles and partial load operation during periods of oscillating power supply (Kalogirou 2005).

The sizing of a WPD system to the demand is also a very important aspect of system design. The absence of a less than random distribution for the population density made it impossible to arrive at a typical size for a desalination system in the coastal Mekong Delta region. Sizing a desalination plant depends on multiple criteria, with one of the primary variables being the geographical conflicts with local trade network and tourism. Moreover as there is but one desalination plant in the region (Tuas, Singapore) and wind desalination is still relatively new there was a lack of local experience and comparison to draw from for estimations made here. Finally since this does not require utility scale wind farms for generation or aim at urban centers, no such constraint outlined by Nguyen (2007b) will have to be considered.

21.5.3 Financial Estimates

The reported price of RO systems varies widely because: (1) a large number of RO systems specialized are for various applications, (2) the cost of pretreatment varies depending upon the feed water quality, (3) the energy price, which is one of the main parameters for cost evaluation (Avlonitis et al. 2003; Blank et al. 2007), is volatile on its own, let alone the prices of renewable energies, (4) different assumptions and calculations by different researchers (Blank et al. 2007; Karagiannis and Soldatos 2008), and (5) geography makes a plant different from another at a different location.

Karagiannis and Soldatos provide the most updated and detailed analysis of the cost of desalinated water (2008). Most relevant to the case study in this chapter, a seawater desalination plant of capacity between 1,000 and 5,000 m^3 per day costs between 0.68 and 3.78 US$/$m^3$. If that uses wind energy, each cubic meter costs between 1.2 and 6 US$. Using RO each cubic meter costs between 0.68 and 1.66 US$.

Any smaller size will result in socially unacceptable price (Blank et al. 2007) and be prohibitive for energy recovery, the concept of which is to lower energy consumption and hence lower cost in the first place (Avlonitis et al. 2003). However, any larger size will not be appropriate to the rural communities of interest. In addition, including energy and water production in one complex and ruling out the use of pipes offer the least expensive compromise between sustainable development, economic evaluation, and innovation (Karagiannis and Soldatos 2008; Blank et al. 2007; Pang 2006; Zejli et al. 2002). The unseen benefit, which is the avoidance of piping, grid involvement, and access to off-grid electricity is challenging to quantify, but should be appreciated.

Another value-added benefit, which has only been partially employed elsewhere (Einav et al. 2002), is the opportunity to direct the discharged brines to salt production instead of treating it as waste and discharging it into the sea. Since brine discharge is the biggest problem for desalination plants (Einav et al. 2002; Laspidou et al. 2010), this option presents an advantage to the additional reprocessing of the brines through the membranes, thereby increasing the salinity of the discharged water. Surrounding communities receive additional employment and generate an additional commodity, which is salt to be used not only in daily life and in local food industries but also for export.

The importance and advantage of economies of scale should not be ignored. A more detailed analysis of the costs for WPD RO systems has been made where considerations of economy of scale have been taken into account (Nguyen and Pearce 2009). It should be noted that the per capita usage being considered is only 60 l for individual needs, which effectively limits an equally important need of water for agriculture. Technically, the plant size can be larger, if the availability of wind power allows it. Water can be continuously treated, stored, and with much larger systems other means of water transportation can be arranged by use of the extensive canal system in the Mekong Delta. However, this approach may not be appropriate particularly in the short term as the benefits of decentralization to poverty reduction and local sustainable autonomy might not be achieved. Future work is necessary to evaluate the economics of an optimized WPD system in the Mekong Delta.

21.5.4 Limitations and Future Work

The main limitation of the study is that it was based on numerical simulations, without testing at a local (pilot) system and thus experimental wind and output data. Without a regional plant within range capacity it is hard to draw more conclusions from previous studies that are site-specific. In the absence of a pilot system a bottom-up estimation using population density as the indicator of demand was the best approach available. The coverage of wind desalination in the low elevation coastal zone of Mekong Delta also suffered from the lack of detailed land use or zoning maps. The availability of such a pilot plant with wind turbines and land use GIS data will allow a life cycle analysis to be carried out to more realistically assess the feasibility of wind powered desalination, benefits, and negative effects (if any) of such projects in coastal Mekong. If smaller turbines are preferred in order to provide a distributed clean water system appropriate for the distributed rural populations, the power law exponent needs to be investigated further as additional wind speed measurements need to be taken as wind characteristics in Vietnam have only focused on 30 and 65 m heights (TrueWinds 2001, Nguyen 2007). Three popular techniques were introduced by Manwell et al. (2002) which could be investigated: (1) correlation of the power law exponent as a function of velocity and height, (2) correlation dependent on surface roughness, and (3) correlations based on both surface roughness and velocity.

Autonomous wind-powered desalination is only one potential solution. Hybrid systems of wind and bio-diesel could be implemented to take advantage of agricultural waste in the region. Additional hybrid systems of wind and solar photovoltaic (PV) cells may better utilize the energy potential of the region. The cost of PV technology has been dropping quickly with scale up of the industry, and industrial symbiosis within the thin film PV manufacturing sector is predicted to aggressively reduce the cost of solar electricity (Pearce 2008). Work here could be expanded to include recent developments in the use of GIS to predict PV potential (Nguyen and Pearce 2010) for hybrid desalination systems. Thus, future work is needed to determine the optimal renewable energy mix for the region and if some form of small electrical storage is necessary to smooth out fluctuations for the RO system. Future work should also consider mechanical vapor compression (MVC). Although MVC consumes more energy than RO, it presents fewer problems due to the fluctuations of the energy resource. MVC systems also have an advantage in remote areas as they are more robust and need fewer skilled workers and fewer chemicals than RO systems. Although as noted above, this may detract from one of the educational and economic development benefits of RO systems. In addition, MVC do not need membrane replacements and in case of polluted waters, the distillation ensures the absence of micro-organisms and other pollutants in the product (Kalogirou 2005).

Although a detailed cost estimate for wind desalination was beyond the scope of this chapter, the economic viability is very dependent on both scale of the plant, the scale of the industry for both desalination and wind power in Vietnam, and government policy. Wind desalination here was only presented as one technically viable option to provide clean water for the selected coastal Mekong provinces in Vietnam, which is but a subset to the much bigger problem of water governance in Vietnam. Where wind desalination is too expensive, a combination of options should be investigated including: (1) demand side management and conservation, (2) rainwater harvesting, and (3) less water-intensive crops and methods of irrigation.

21.6 Conclusions

The chapter considered the potential of off-grid medium- to small-scale reverse osmosis desalination systems powered by small wind turbines to provide safe drinking water to the residents of the Mekong Delta in Vietnam. The GIS-based numerical analysis revealed that in the absence of all other water supply facilities, off-grid wind-powered desalination could provide 5.4 million rural residents living in 18.9 thousand km^2 (or 62% of the predicted 2010 rural population and 70% of the total area of the coastal provinces) with clean water at the rate of 60 l per capita per day. In the context of agricultural sustainability, it offers social, environmental, and technical improvements for the region and hence represents a decentralized and community-based method to adapt to and mitigate climate change.

Acknowledgments The authors would like to thank R. Andrews for helpful discussions and support from the Natural Sciences and Engineering Research Council.

References

Akermann T, Soder L (2002) An overview of wind energy – status. Renew Sust Energ Rev 6:67–128

AUSAID (2004) Mekong Delta poverty analysis. AuSAID final report

Avlonitis SA, Kouroumbas K, Vlachakis N (2003) Energy consumption and membrane replacement cost for seawater RO desalination plants. Desalination 157:151–158

Ayoub J, Alward R (1996) Water requirements and remote arid areas: the need for small-scale desalination. Desalination 107:131–147

Berg M, Stengel C, Pham TKT, Pham HV, Sampson ML, Leng M, Samreth S, Fredericks D (2006) Magnitude of arsenic pollution in the Mekong and Red River Deltas – Cambodia and Vietnam. Sci Total Environ 372(2–3):413–425

Binh CT, Phillips MJ, Demaine H (1997) Integrated shrimp-mangrove farming systems in the Mekong Delta of Vietnam. Aquac Res 28:599–610

Blank JE, Tusel GF, Nisan S (2007) The real cost of desalted water and how to reduce it further. Desalination 205:298–311

Buitenhuis AJ, Zelenika I, Pearce JM (2010) Open design-based strategies to enhance appropriate technology development. In: Proceedings of the 14th annual national collegiate inventors and innovators alliance conference: Open, 25–27 Mar 2010, pp 1–12. Available http://nciia.org/sites/default/files/pearce.pdf

Carta JA, Gonzalez J, Subiela V (2003) Operational analysis of an innovative wind powered reverse osmosis system installed in the Canary Islands. Sol Energ 75:153–168

Chaudry P, Ruysschaert G (2007) Human development report 2007/48: climate change and human development in Vietnam. UNDP, Vietnam

Christiansen MB, Koch W, Horstmann J, Hasager CB, Morten N (2006) Wind resource assessment from C-band SAR. Remote Sens Environ 105(1):68–81

Coulhart A, Nguyen Q, Sharpe H (2006) Urban development strategy: meeting the challenges of rapid urbanization and the transition to a market oriented economy. Vietnam's infrastructure challenge volume. World Bank

Dang KN, Be NV, Trung NH (2007) Water use and competition in the Mekong Delta, Vietnam (Chapter 4). In: Tran TB, Bach TS and Miller F (Eds), Challenges to Sustainables Development in the Mekong Delta: Regional and National Policy Issues and Research Needs

Dinh ND, Tuyen NT, Cuong TT, Chung NK (2006) Sustainable development in Vietnam – a socio-cultural approach. Hanoi Publication, Hanoi

Dore MHI (2005) Forecasting the economic costs of desalination technology. Desalination 172:207–214

Eastham J, Mpelasoka F, Mainuddin M, Ticehurst C, Dyce P, Hodgson G, Ali R, Kirby M (2008) Mekong river basin water resources assessment: impacts of climate change. CSIRO: water for a healthy country national research flagship

Einav R, Harussi K, Perry D (2002) The footprint of the desalination processes on the environment. Desalination 152:141–154

Fritzmann C, Löwenberg J, Wintgens T, Melin T (2007) State-of-the-art of reverse osmosis desalination. Desalination 216:1–76

García-Rodríguez L (2004) Desalination by wind power. Wind Eng 28:453–466

General Statistics Office of Vietnam (2002) Vietnam living standard survey 2002

General Statistics Office of Vietnam (2006) Vietnam household living standard survey 2006

Government of Vietnam (2002) Vietnam national action program to combat desertification

Government of Vietnam (2003) The comprehensive poverty reduction and growth strategy (CPRGS)

Hasager CB, Nielsen M, Christiansen MB (2004) RWT tool: offshore wind energy mapping from SAR. In: Proceedings of the 2004 Envisat & ERS symposium, Salzburg, 6–10 Sep 2004

Kakonen M (2008) Mekong Delta at the crossroads: more control or adaptation? http://enviro.lclark.edu/resources/vietnam/Mekong/MekongDelta.pdf

Kaldellis JK, Kondili EM (2007) The water shortage problem in the Aegean archipelago islands: cost-effective desalination prospects. Desalination 216(1–3):123–138

Kalogirou SA (2005) Seawater desalination using renewable energy sources. Prog Energ Combust Sci 31:242–281

Karagiannis IC, Soldatos PG (2008) Water desalination cost literature: review and assessment. Desalination 223:448–456

Khan MJ, Iqbal MT (2004) Wind energy resource map of Newfoundland. Renew Energ 29(8):1211–1221

Kiranoudis CT, Voros NG, Maroulis ZB (1997) Wind energy exploitation for reverse osmosis desalination plants. Desalination 109(2):195–209

Landberg L, Myllerup L, Rathmann O, Petersen EL, Jørgensen BH, Badger J, Mortensen NG (2003) Wind resource estimation – an overview. Wind Energ 6(3):261–271

Laspidou C, Hadjibidos K, Gialis S (2010) Minimizing the environmental impact of sea brine disposal by coupling desalination plants with solar saltworks: a case study for Greece. Water 2:75–84

Le AT, Wyseure G (2007) Water environmental governance in the Mekong River Delta, Vietnam. http://www.wepa-db.net/pdf/0712forum/paper27.pdf

Long JF, Rain DR, Radcliffe MR (2001) Population density vs. urban population: comparative GIS studies in China, India, and the United States. In: Proceedings of the international union for the scientific study of population conference, Brazil, 18–25 Aug 2001

Manwell JF, McGowan JG, Rogers AL (2002) Wind energy explained: theory design and application. Wiley, London

Ministry of Construction and Ministry of Agriculture and Rural Development (2000) Natural rural clean water supply and sanitation strategy up to year 2020

Nguyen QK (2006) Alternatives to grid extension for rural electrification: decentralized renewable energy technologies in Vietnam. Energ Policy 35(4):2579–2589

Nguyen HN (2007a) Human Development Report 2007/ 53: Flooding in Mekong River Delta. UNDP, Vietnam

Nguyen QK (2007b) Wind energy in Vietnam: resource assessment, development status and future implications. Energ Policy 2:1405–1413

Nguyen HT, Pearce JM (2009) Renewable powered desalination in coastal Mekong Delta. In: Proceedings of the 4th ASME conference on energy sustainability, Phoenix, 17–22 May 2009, USA ES2010-90224

Nguyen HT, Pearce JM (2010) Estimating potential photovoltaic yield with r.sun and the open source geographical resources analysis support system. Sol Energ 84:831–843

Pang E (2006) The potential of wind powered desalination in water scarce countries. Master thesis, the Fletcher school, Tufts University. http://repository01.lib.tufts.edu:8080/fedora/get/tufts:UA015.012.DO.00144/bdef:TuftsPDF/getPDF

Pankratz T (2008) Water desalination report. http://www.menarec.org/resources/wdr2008-10.pdf

Pearce JM (2008) Industrial symbiosis for very large scale photovoltaic manufacturing. Renew Energ 33:1101–1108

Pearce JM, Grafman L, Colledge T, Legg R (2008) Leveraging information technology, social entrepreneurship and global collaboration for just sustainable development. In: Proceedings of the 12th annual national collegiate inventors and innovators alliance conference, 201–210

Phuong NT, Tam TNTM, Nguyet NT, Oostendorp R (2008) Determinants and impacts of migration in Vietnam. DEPOCEN working paper series no. 2008/01

Pretty J (2008) Agricultural sustainability: concepts, principles and evidence. Phil Trans R Soc Biol Sci 363:447–465

Pryor SC, Nielsen M, Barthelmie RJ, Mann J (2004) Can satellite sampling of Offshore wind speeds realistically represent wind speed distributions? http://ams.allenpress.com/archive/1520-0450/43/5/pdf/i1520-0450-43-5-739.pdf

Shaw R (2006) Community-based climate change adaptation in Vietnam: Inter-linkages of environment, disaster, and human security. In: Sonak S (ed), Multiple dimensions of global environmental changes. The Energy Research Institute, TERI Press, pp 521–547

Soussan J, Do MC, Patrick KJ, McGrath T, Le THH, Nguyen VT, Nguyen TN (2005) Joint GoV – donor sector review: rural water supply, sanitation and health in Viet Nam

SunLift Systems GmbH (2001) Asia eco best project report: sea water desalination in the coastal areas of Central and South Vietnam

Taylor P (ed) (2004) Redressing disadvantage or rearrange in-equality? Development interventions and local responses in the Mekong Delta (Chapter 7). In: Social equality in Vietnam and the change to reform. Institute of Southeast Asian Studies

Thomas KE (1997) Overview of village scale renewable energy powered desalination. National Renewable Energy Laboratory, Golden

TrueWinds Solutions (2001) Wind energy resource atlas of Southeast Asia

United Nations (March 2007) Indicators of sustainable development: guidelines and methodologies, 3rd edn. United Nations, New York

United Nations Statistics Division (2010) Principles and recommendations for population and housing censuses. http://unstats.un.org/unsd/demographic/sconcerns/densurb/densurbmethods.htm

Vo LP (2007) Formulation of an integrated approach to sustainable water management in Hochiminh City, Vietnam. http://digital.library.adelaide.edu.au/dspace/bitstream/2440/49486/2/01front.pdf

Wilkinson CR (2005) Global and local threats to coral reef functioning and existence: review and predictions. Mar Freshwater Res 50:867–878

World Bank (2000) Global small towns water and sanitation initiative: Vietnam case study. The world bank rural water supply and sanitation thematic group and the water and sanitation programme East Asia Pacific (WSP-EAP)

World Bank (2002) Ensuring environmental sustainability

World Bank (2003) Vietnam environment monitor 2003

World Bank (2006a) Vietnam's infrastructure challenge workshop edition: power strategy managing growth and reform

World Bank (2006b) Vietnam environment monitor 2006: water quality in Vietnam

Zejli D, Bouhelal OK, Benchrifa R, Bennouna A (2002) Applications of solar and wind energy sources to sea water desalination – economic aspects. In: Proceedings of international conference on nuclear desalination: challenges and options, Marrakech, 16–18 Oct 2002

Chapter 22
Promotion of Organic Cocoa in Mixed Farming System in the Mekong Delta Region: A Preliminary Analysis

Vo Van Phong, Daniel Valenghi, and Nguyen Lam Giang

Abstract Over the past decades, the Mekong River Delta has become a strategic region for the socio-economic development of Vietnam. While the region is currently producing the majority of the exported products (mainly rice) of the country, the sustainability of its conventional production system is challenged. It is known that organic agricultural systems are more sustainable compared to conventional, chemical-based agricultural systems. Farmers growing crops organically not only can improve their soil fertility and structure with the result of stable yields and better product quality, but also sell their products at market prices plus premium. Organic agriculture also yields lower levels of green house gas emissions than conventional agriculture. To enhance sustainability and fight poverty, Helvetas, the Swiss Association for International Cooperation, in collaboration with residents of selected provinces of Tien Giang and Ben Tre and with Nong Lam University (NLU) in Ho Chi Minh City, has initiated an organic and fair trade cocoa program for a 5-year period. In that period of time, a total of 2,000 selected cocoa cum coconut farmers (men and women) in the two provinces of Tien Giang and Ben Tre are expected to produce certified organic and fair trade cocoa for the international market. For those purposes, the program is aimed to support all key aspects of the cocoa value chain: the development of appropriate techniques, the provision of training for farmers and fermenters, the guidance for qualifying and obtaining key certifications, and the linking of the farmers' organizations to the market. In 2009, the first 22 pilot farms were converted to organic agriculture by adopting measures such as composting and the biocontrol of pests and diseases. The report of these first positive experiences has been used to convince other farmers to convert to organic cocoa. As a result, there are now 154 farmers in this area registered to grow cocoa organically.

Vo Van Phong (✉), D. Valenghi (✉), and N.L. Giang
Program for the Development of Organic and Fair Trade Cocoa in Vietnam, Helvetas Vietnam, Helvetas Swiss Intercooperation, Hanoi, Vietnam
e-mail: vo.phong@helvetas.org; daniel.valenghi@helvetas.org

M.A. Stewart and P.A. Coclanis (eds.), *Environmental Change and Agricultural Sustainability in the Mekong Delta*, Advances in Global Change Research 45,
DOI 10.1007/978-94-007-0934-8_22,

Keywords Sustainability • Organic agriculture • Organic and fair trade cocoa • Organic certification • Farmer organization

22.1 Introduction

Vietnam's Mekong Delta, covering an area of some 40,000 km^2, and home to some 18 million people, has become a strategic region for the socio-economic development of the country. It accounts for 50% of the nation's total food production, 95% of the rice exports, 65% of the fisheries production, and 70% of the fruit production (Truong 2009). However, over the past decades, agricultural production in the Delta has used roughly a million tons of chemicals – mainly pesticides and fertilizers – a year (Chanh 2009). This has begun to have discernible negative impacts on local environments and adversely affect the sustainability of agricultural production in the region. The recent and recurrent outbreaks of rice brown plant hoppers are a typical consequence of the improper and sometimes futile use of pesticides. The improper use of agro-chemicals not only pollutes water resources but also soils. In general, a crop effectively uptakes only 30% of the total amount of fertilizers applied and the remaining 70% is either washed off or accumulated in the soil (Truc et al. 2006).

For developing the agricultural sector in a more sustainable manner the government of Vietnam has made great efforts to encourage the application of more environmentally sound farming techniques. Since the early 1990s, with support from the international community, the government has promoted and successfully applied integrated pest management (IPM) techniques on rice, vegetables, and other crops in 61 provinces nationwide; 97% of the villages in the Mekong Delta have participated (Community IPM 2001). Recently, good agriculture practice (GAP) has been introduced in some crops such as rice, star apple, and pineapple; GAP in star apple production in Tien Giang Province is a typical example of this direction. At the same time, organic agriculture has begun to be practiced in Vietnam, and some 22 ha are now under organic production.

In this context, Helvetas Vietnam, the Swiss organization for international cooperation, which has substantial experience in rural development in Vietnam, sees the combination of organic farming and fair trade approaches as a key strategy for poverty alleviation and sustainable natural resources management in the country. It has initiated an organic cocoa production project with the development of this strategy in mind. This project is based on the two-fold premise that organic agriculture can help smallholding farmers improve their income and that cocoa is a new crop in Vietnam, thus making it easier to introduce organic farming techniques and to set up a value chain for organic cocoa products at the same time.

In organic agricultural production, small scale farmers are enabled to have better opportunities to access agricultural value chains and stable markets with premium prices (FAO 2005, 2007). Also, participation in this sector will enable small scale farmers to become less dependent on chemical inputs that become expensive when factoring in the high transportation costs for these items to reach rural areas and the higher unit costs for small volumes (IFAD 2003; United Nations 2008).

Cocoa has successfully been introduced to Vietnam since 2000 after several failures in the past. At present, cocoa is a new crop in Vietnam that is experiencing a steady increase in planting area and productivity. By 2009, the total area under cocoa cultivation was about 12,000 ha, a 19% increase over 2008. The total productivity of some 1,500 tons of dried cocoa beans accounted for a very tiny fraction of the world cocoa productivity of some 3.8 million tons in 2009. However, with favorable climatic conditions and diligent farmers it is believed that cocoa will provide a good opportunity for income generation for smallholding farmers.

Based on those strong foundations, a 5-year program for the development of organic and fair trade cocoa in Vietnam has been developed by Helvetas Vietnam, with funding from several international donors including SECO, Rabobank, and Helvetas. This program aims to increase the living standards of the rural population in the southern regions of Vietnam by way of improved social, economic, and environmental conditions related to agriculture (Helvetas 2008b). The project is to be implemented through a multi-stake-holder approach: participants and supporters include the private sector (Ritter Sport, Cargill), farmers' organizations, and local public institutions (Nong Lam University in Ho Chi Minh City, Ben Tre Province through the Department of Agriculture and Rural Development, and Tien Giang Province through the Department of Science and Technology).

This paper aims to present the rationales for the program interventions and an analysis of preliminary results of the program; though the project is still in the early stages of development, information about its origins and first stage may serve as a useful source of information for any organic initiatives that are planned for the Mekong Delta in the future, as well as for projects that seek to develop niche organic products in Southeast Asia.[1]

22.2 Description of Program Rationales for Promoting Organic Agriculture in the Mekong Delta

The world organic agriculture sector increased substantially in recent decades (United Nations 2008). Generally speaking, organic agriculture is an agricultural production system that sustains the health of soils, ecosystems, and people by combining site-appropriate traditional farming techniques and scientific innovations that rely on non-chemical inputs and ecologically sustainable techniques such as crop rotation, green manure, compost, and biological pest control (UNEP-UNCTAD CBTF 2009). Chemical substances are mostly prohibited in organic production systems, and inorganic fertilizers are replaced with organic nutrient management

[1]See Chap. 23.

practices such as crop rotation and the use of compost, mulches, green fertilizers, and nitrogen fixation crops. This is done in order to conserve and raise the level of organic material and fertility in the soil (Conseuga 2003). In fact, unlike conventional systems, organic production does not focus on the agro-products per se but rather on the whole system employed to produce and deliver the product to the end user (FAO 2005). It should be noted that not all agriculture systems without external inputs can be considered as organic agriculture according to international organic standards. This is because purchased inputs such as mineral fertilizers or synthetic pesticides are not used in some agriculture systems simply because the farmers cannot afford them and not because of a commitment to organic methods; in these cases, the consequence is often low and declining productivity and eventually environmental degradation (FAO 2005).

With the principles for successful promotion of organic cocoa in the Mekong Delta provinces of Tien Giang and Ben Tre in place, the program has identified five key areas to intervene: the development of site-appropriate organic cocoa farming techniques; the provision of training for local cocoa farmers; training in the necessary steps needed to obtain organic certification; the marketing of organic cocoa; and enabling in general a friendly policy for organic agriculture.

22.2.1 Development of Organic Farming Techniques Suited to Local Conditions

Helvetas has identified the development of site-appropriate organic farming techniques as the first and foremost important intervention for promoting organic production in the Mekong Delta. International experience in organic farming promotion has demonstrated that it is difficult to develop a one-size-fits-all organic farming model. The principles of organic agriculture serve only as guidelines for organic practitioners to formulate organic farming practices suited to individual farming locations and specific social, technical, and climatic conditions of these sites (IFAD 2005; CSIRO 2006). Therefore, before designing and implementing the program, Helvetas conducted a feasibility study on promoting organic cocoa to identify the most suitable regions, and perhaps more importantly understand the opportunities and challenges the program might likely face. The study revealed that Tien Giang and Ben Tre Provinces are the most suitable areas for the program, because of the low input requirements of local cocoa–coconut farming systems, the availability of manure and green materials for compost making, and a high level of commitment from local authorities (Helvetas 2008a).

For conducting experiments to define organic farming techniques suited to local conditions, Helvetas has collaborated with the Nong Lam University to conduct studies on the management of cocoa pests and diseases and nutrient management using compost made from local resources. Instead of using agro-chemicals to control pests and disease, Helvetas has directed local farmers to use weaver and black ants to control cocoa mirids and to apply cultural methods such as canopy management to manage cocoa black pod disease (*Phytophtora palmivora*). At the same time, with the support of the program, local farmers have used locally available resources such as cow dung, water hyacinths, and coconut husks to make compost to fertilize their cocoa instead of using NPK inorganic fertilizers. The research activities to formulate site-appropriate organic farming techniques have not been one off, but rather on-going. Helvetas acknowledges the need for several years of trials to define organic farming practices that are technically, socio-culturally, and economically suited to local conditions. This is practically accepted in the sense that it usually takes time for local farmers to test and learn newly introduced farming technologies (IFAD 2005). It would therefore be desirable to promote organic agriculture to a local area gradually, at least during the initial years (IFAD 2003).

22.2.2 Provision of Training to Farmers and Other Value Chain Actors

Helvetas has prioritized training activities to make sure local authorities and local farmers have a sound understanding of organic requirements and principles. This is because organic agriculture as it is practiced according to international standards is new to Vietnam. Also, Helvetas needs to make sure that participating farmers fully understand the benefits of organic agriculture and not just the requirements and standards of organic agriculture. International experience shows that the likelihood of success in organic production is greater if farmers are highly motivated, by recognition of ongoing economic advantages and also by health and environmental concerns (IFAD 2003).

For the purpose of training, once the organic cocoa farming techniques are developed and defined by the research team, local cocoa farmers and local extension service providers will be trained in those techniques. The training activities of the program will be accomplished using farmer to farmer extension approaches, where key farmers will be selected and trained to become farmer trainers. These farmer trainers will train their fellow farmers. It has been proven that these approaches will be more cost-effective in general and more sustainable after the program ends because local people are still in their community to act as a local resource and continue the work.

22.2.3 Organic Certification

For organic cocoa to be recognized in the international market, the cocoa produced in the program area must be certified by internationally recognized organizations. Certification provides confidence to buyers that an organic product has been produced and processed in accordance with international organic requirements and standards. Certification also enables small scale farmers to benefit from premium prices (FAO 2007). Organic certification systems were developed in the early 1970s; and today 70 countries have domestic certification organizations, and a dozen internationally active organizations offer organic certification services in virtually all countries in the world (United Nations 2008).

In the Mekong Delta, organic certification has not been accomplished before, and thus certification is also a key program intervention for Helvetas. International experience shows farmer organizations play a key role in organic production, as working in groups enable individual farmers to take advantage of collective marketing, the management of larger volumes, and perhaps most importantly, to accomplish increased efficiency in negotiating and implementing contracts with buyers and suppliers (IFAD 2003).

Farmer groups also provide mechanisms for setting up and running internal control systems to ensure the full compliance to organic regulations by participating farmers. Fraud can be a big challenge to small-farmer organizations producing organic products; group members may be tempted to obtain premium prices without complying fully with organic methods of production (IFAD 2003). Group certification is equally important, as it will reduce the cost of certification per farmer. Organizing farmers for this purpose is crucial, as it is financially impossible for small scale Mekong farmers to individually achieve certification. Since farmers in the Mekong Delta operate on a small scale and are not very well organized in the first place, they need to group themselves in organizations to maximize their advantage in developing organic cocoa production. The program needs to have a minimum number of farmers participating in the first place in order to cover the costs associated with organic certification (IFAD 2003; FAO 2005). Also, from the marketing perspective, local farmers need to have a minimum volume of organic products for a buying company to set up a buying system. For the program to thrive, Helvetas needs to motivate at least 2,000 small scale farmers to convert to organic cocoa.

22.2.4 Marketing Support

Over the past decade the worldwide organic trade has experienced substantial growth (FAO 2007). For Mekong cocoa growers, the development of stable markets for organic agro-products is crucial if organic agriculture is to be successfully promoted. For this purpose, organic agriculture developers need to prioritize marketing activities to introduce the organic products from the Mekong Delta to international buyers. This can be achieved by participation in international fairs or other marketing means.

Also, for the program intervention to be sustained, farmer organizations need to build up their capacities in marketing and contract negotiation to ensure that farmer groups are enabled to have sufficient bargaining power in negotiating with buyers during the program life and long afterward. International experience indicates that capacity-building at the farmer level (local farmers associations, local training, and advisory services) should be a central aspect of any strategy aimed at using organic agriculture as a tool for poverty alleviation (IFAD 2005). Also, the marketing of organic products through farmer organizations with direct contacts with buyers is a key to obtaining better prices (IFAD 2003). It may be too ambitious to turn a farmer into a trader; however, a network of organizations can also improve bargaining power and substantially lower transaction costs (IFAD 2005).

22.2.5 *Enabling Policy for Organic Agriculture*

As with other development initiatives, for the effectiveness and sustainability of organic program interventions, an enabling policy environment plays a pivotal role in promoting organic agriculture in the target area. International experience reveals that for developing countries, an enabling environment is required to support institutional development and to set up norms and standards to create favorable conditions for organic agriculture and markets to grow (FAO 2007). Equally important, enabling national laws and regulations will help reduce certification costs, either by stimulating foreign certification bodies to open local offices or by supporting the development of local service providers (United Nations 2008; IFAD 2003).

22.3 Analysis of Program Preliminary Results

To successfully promote organic agriculture in the Mekong region, Helvetas has prioritized the development of organic cocoa farming techniques which technically, socially, and economically are suited to local conditions. They have done this with research and on-farm demonstrations. Helvetas has collaborated with NLU to conduct a number of studies on the management of cocoa pests and diseases and on compost making, and also with provincial partners to implement on-farm demonstration plots to introduce and test organic farming. For the effective implementation of on-farm demonstration, Helvetas is working with three partners: NLU, the Department of Agriculture and Rural Development (DARD) in Ben Tre Province, and the Department of Science and Technology (DOST) in Tien Giang Province. Meetings with these collaborators were held to reach consensus on criteria for demonstration holder selection and to identify locations for demonstration plots and procedures for demonstration implementation.

As a result, 22 cocoa farmers have been selected for managing demonstration sites: ten in Ben Tre Province and twelve in Tien Giang Province. On the average, each demonstration site holder has 300 cocoa trees intercropped with mature coconuts. Nearly half of them practice animal husbandry and have cows or pigs. For the successful implementation of the demonstration plots, the demonstration holders were trained in compost-making and in the management of pests and diseases with organic methods. The introduction of organic farming techniques to local farmers is formulated and evaluated using three major factors: technical feasibility, social acceptance, and economic viability.

22.3.1 Technical Feasibility

For the effective management of cocoa black pod disease (*Phytophthora Palmivora*), the Trichoderma fungus was tested. Trichoderma is a fungus antagonistic against the cocoa black pod disease-causing fungus. With a significant presence, this antagonistic fungus can suppress plant pathogens. *Trichoderma* specimens have been isolated from cocoa gardens in the program area, multiplied in the NLU laboratory, and tested in the demonstration plots. *Trichoderma* have then been sprayed on cocoa trees to test against the disease. However, the effectiveness of *Trichoderma* against the black pod disease has not yet been clear, as it may take more time for *Trichoderma* populations to build up to a level that can sufficiently defeat the pathogen causing the disease. In the meantime, since fungicides are prohibited in organic production, the conversion process should be started at the beginning of the dry season (November) when the threat of black-pod disease is lower.

Experimentations by NLU technicians and on-farm demonstration activities also reveal that effective management of cocoa mirids is crucial in the successful conversion to organic production. In farms with a low weaver ant population, the level of damage by this pest is very significant. Experience shows that rearing weaver ants during the rainy season is very difficult and therefore it is highly recommended that the conversion should be started during November at the outset of the dry season.

Investigators have also learned from an analysis of compost samples from the demonstration farms that for 1 kg of dried beans, farmers need to apply about 15 kg of finished compost during the first 3 years after conversion. After a few years of organic practices and once the soil acquires balance and microbiological and nutrient richness, the amount of compost applied can be expected to be much lower. In the first years of conversion, local farmers may find it challenging to collect and chop sufficient amounts of green materials for compost making. One option for overcoming this challenge is to introduce nitrogen-fixing crops that are suitable for shade conditions in the cocoa gardens and that are intercropped with coconuts in the program areas. International experience shows that the use of purchased organic matter, organic nutrient sprays, and green manure in combination with traditional techniques such as crop residues and compost can solve the problem of insufficient nutrients for crops when converting to organic production.

This is important because yield capacity depends partly on the application of organic fertilizers in qualities and quantities that compensate for nutrients extracted by crops (IFAD 2003). Another solution is to select the most suitable time to start conversion. For the ease of compost-making, the program experience shows that the conversion should be started at the beginning of the dry season (November) when cocoa husk is abundantly available and activities such as material collection, chopping, and compost piling are easier. If compost making is started in November, the converting farmers will have sufficient finished compost to fertilize their cocoa once the rainy season comes (usually at beginning of April of the following year). In addition, the demonstrations also revealed that without animal of their own to supply manure, local farmers may face financial difficulties in buying manure such as cow dung for compost making once the financial support from the program is no longer provided. Therefore, organic cocoa should be initiated with farmers for whom cattle or pigs are part of their farm operations.

22.3.2 Economic Viability

As with any rural development and poverty reduction effort, cost-benefit analysis is one of the prioritized program activities; however, due to time constraints the Helvetas program has not yet done this systematically. In principle, once they are converted to organic production organic farms will have lower, equal to, or higher yields compared to conventional farms; depending on farm conditions, the knowledge and skills of the farmers, and the extent to which synthetic inputs were used before conversion (FAO 2005; IFAD 2003; Conseuga 2003). This last point is clearly illustrated in the context of this program. The preliminary results indicate that farmers can gain similar yields as they accomplished before conversion if manures had been used predominantly to enrich soil before conversion, but lower yields are found on the farms which had used a lot of inorganic fertilizers before conversion.

At this point, the lack of sound understanding about the implications of organic agriculture for food production and food security in Vietnam makes it impossible to fully evaluate its importance (Conseuga 2003; FAO 2005). Meaningful comparisons of the performance of organic and conventional agriculture systems need to be conducted over a long period of time; farmers may experience a decline in yields after converting to organic production, and then experience a significant yield increase after the agro-ecosystem is restored and organic management systems are fully implemented (FAO 2005; IFAD 2005).

Moreover, it may not be easy to evaluate the benefits of organic agriculture from environmental and health perspectives. Organic production can help to mitigate the environmental pollution associated with animal husbandry. Also, using water hyacinth for compost-making may help local authorities save hundreds of millions VND otherwise expended to clear local canals of this invasive species. In addition, organic agriculture can help to reduce the exposure of famers to agro-chemicals and contribute to better health. According to a United Nations study (2008), organic production can help prevent the recurrence of the estimated 3 million cases of acute severe pesticide poisoning and 300,000 deaths that result from agrochemical use in conventional agriculture every year worldwide.

22.3.3 Social Acceptance

The program has also learned that gender issues and social networking via group formation needs to be taken into consideration for successfully promoting organic production in the Delta. As with other rural development activities gender issues play an important role in the promotion of organic agriculture. This is because the conversion of a farm to organic practices influences all facets of farm operation, including labor demand, social structures, and decision-making processes (FAO 2005). Once joining the organic program, the farm owner needs to convert his whole farm to organic farming, and not just the target crop. Both wife and husband need to fully understand and comply with the organic principles, and both need to avoid the use of inorganic fertilizers on their crops. Women can also be actively involved in the process of compost-making as well as in cocoa pod harvesting (Helvetas 2010).

Moreover, the experience gained until now has proven the importance of social networking and of peer monitoring and motivation-building by famers. One of the initial 22 demonstration farmers, for example, attempted to use inorganic fertilizers. The unusual performance of his cocoa crop (very big cocoa leaves) drew the attention of group members, and after several meetings for discussion the farmer had to admit to his peers that he had introduced inorganic materials to his farm and promise not to do that again. Group motivation to comply with organic requirements and regulations is higher than individual motivation. Organic production is more labor-intensive than conventional production, since organic farmers need to spend more time making and applying compost. In recent months, some farmers have felt discouraged by the effort needed to make sufficient recommended compost. However, with the support of group members, they have successfully continued the recommended organic farming techniques such as compost-making and weaver ant rearing.

22.3.4 Stable Markets for Organic Products

The program has learned that for promoting organic production in the region, the development of a stable market for locally produced organic products is a key factor. In the beginning, Helvetas made it clear to local farmers at the outset that they would be able to connect farmers through organic production to stable market conditions. Local farmers have had bitter experiences with bumper crops and consequent lower market prices and other fluctuating and uncertain market conditions, and providing them with the assurance of more market security is an important incentive to converting to organic production. When designing the program, Helvetas identified potential buyers of organic cocoa such as Ritter Sports and Cargill. Recently, Helvetas has signed a purchasing agreement with Cargill Vietnam to buy conversion cocoa, beginning December 1, 2009. Without some market security, local authorities as well as local farmers are reluctant to join the program.

22.3.5 Sound Understanding of Organic Agriculture by Local Authorities and Farmers

Over the course of the program, the importance of a sound understanding of organic agriculture principles and practices by all stakeholders has been demonstrated. To ensure common understandings of organic cocoa and organic certification procedures as well as of the requirements and standards of organic production, the program has contracted a service provider to organize a 2-day training workshop on organic cocoa production for key program stakeholders. However, due to ineffective course delivery, local partners and cocoa farmers did not fully understand the principles and requirements of organic cocoa growing. Consequently, their negative and passive attitudes about organic cocoa production have hindered the program's progress and expansion in some localities, particularly in Ben Tre Province. As with other development initiatives, all program stakeholders must fully understand the principles and the requirements of organic agriculture. International experience shows that for the successful promotion of organic production, local authorities need to understand the long-term benefit of organic agriculture with due consideration of the pros and cons thereof (IFAD 2003; CSIRO 2006).

22.4 Conclusion

Until recently, the program has made great efforts to accomplish numerous activities, making good progress toward the achievements of its goals. Generally speaking, after 1 year of conversion, organic cocoa appears to be technically feasible in the sites where it has been tried. Farmers have been able to apply recommended organic cocoa farming techniques such as composting and using ants to control cocoa mirids instead of using agrochemicals. The yields of the conversion farms vary depending on the farming techniques practiced before conversions and the farmers' ability to make compost during the conversion period. More importantly, the benefits of using compost compared to NPK fertilizers have been acknowledged by the demonstration farmers. Applications of compost help the cocoa tree better cope with drought and saline conditions. Also, at the start of the conversion, participating farmers were worried about a significant reduction in coconut productivity. The reality, however, proves otherwise. If farmers have been able to use compost as recommended, coconut productivity has been even higher than before.

Moreover, the program has learned that key factors for successfully promoting organic agriculture in the Mekong Delta include, but are not limited to: the sound understanding of local authorities about organic agriculture in order to motivate local farmers to join the program; the development of site-appropriate organic farming techniques; farmer organization for peer monitoring and motivation; the development of stable markets for locally produced organic products; and the important roles of women in organic cocoa production.

More specifically, the program has learned that the most suitable time for conversion to organic cocoa production is from November to January of the following year because of the abundant availability of local materials and favorable conditions for compost-making, lower damage by the black pod disease, and favorable conditions for building up weaver ant populations. In the years to come, the program should give attention to local cocoa farmers who are also practicing animal husbandry to join the program as the likelihood of successful conversion for these farmers is higher. For a successful cost-benefit analysis, the program needs more time to collect information. It should at the same time be kept in mind that organic practitioners need to encourage local authorities as well as local farmers to understand that the meaningful comparisons between organic systems and conventional ones will require an effort and a commitment over a long period of time.

References

Chanh DT (2009) Rural areas are polluted with chemicals: living with agro-chemicals (Làng quê đang bị đầu độc: Ăn, ngủ cùng phân hóa học, thuốc trừ sâu). Vietnam Agriculture News dated 9/7/2009

Community Integrated Pest Management (Community IPM) (2001) Vietnam country report. http://www.communityipm.org/Countries/vietnam.htm. Accessed 4 Jan 2010

Conseuga NP (2003) Organic agriculture: foundations for the ecological management of pests. HIVAS, Havana, Cuba

FAO (Food and Agriculture Organization) (2005) Organic agriculture, environment and food security. Environment and Natural Resources Series No. 4, FAO, Rome

FAO (Food and Agriculture Organization) (2007) Organic certification schemes: managerial skills and associated costs. FAO, Rome

Helvetas (2008a) Feasibility study for organic and fair trade in Vietnam. Helvetas, Helvetas Vietnam Swiss Association for International Cooperation, Hanoi

Helvetas (2008b) Program for the development of organic and fair trade cocoa in Vietnam. Program document. Helvetas Vietnam Swiss Association for International Cooperation, Hanoi

Helvetas (2010) Gender assessment for the development of organic and fair trade cocoa in Vietnam. Helvetas Vietnam Swiss Association for International Cooperation, Hanoi

IFAD (International Fund for Agriculture Development) (2003) The adoption of organic agriculture among small farmers in Latin America and the Caribbean. Thematic evaluation. IFAD, Rome

IFAD (International Fund for Agriculture Development) (2005) Organic agriculture and poverty reduction in Asia: China and India focus. Thematic evaluation. IFAD, Rome

Kristiansen P, Taji A, Reganold J (2006) Organic agriculture: a global perspective. CSIRO, Collingwood

Truc NTT, Trinh BV, Danh VT (2006) Agricultural development and natural resources degradation: an environmental review. School of Economics and Business Administration. Cantho University, Vietnam

Truong TV (2009) Integrated water resource management in Vietnam's Mekong Delta. Workshop on scientific and rational exploitation of water resources – our common responsibility in China. http://www.vncold.vn/Modules/CMS/Upload/13/Documents/TVTMekong1Ew_25_10_08/Truong%20PPT%20paper3.pdf. Accessed Mar June 22

UNEP-UNCTAD Capacity Building Task Force on Trade Environment and Development (UNEP-UNCTAD CBTF) (2009) E-learning course on successful organic production and export. UNEP-UNCTAD CBTF

United Nations (2008) Best practices for organic policy: what developing country governments can do to promote the organic agriculture sector. United Nations, New York

Chapter 23
The Transition from Conventional to Organic Rice Production in Northeastern Thailand: Prospect and Challenges

Natedao Taotawin

Abstract This chapter examines the process of market integration and its impact on farmer livelihoods and the environment and agriculture of northeastern Thailand. Using an anthropological approach and fieldwork conducted in the northeast of Thailand during 2008–2009, the chapter shows the way in which northeastern Thai farmers have integrated their practices into the ethical niche markets of fair trade and organic food. This chapter also illustrates the impacts of the buyer-driven food chains on social–nature relations in the northeast of Thailand and the changes that have occurred as a result. Competition in the organic trade market depends significantly on the capacity of farmers to comply with international regulations, so this paper will also focus on the distinctiveness of socio-economic and ecological conditions in northeastern Thailand, especially those which have contributed considerably to competition in fair trade products and organic rice commodities.

The chapter argues that conversion to organic agriculture is a strategy employed by northeastern Thai farmers in northeastern Thailand to cope with the problems of environmental deterioration and increasing production costs, yet the emergence of niche markets has brought both opportunities and challenges. The farmers participating in these niche markets can maintain their farmland and gain benefits from the value-added production they engage in, but they confront difficulties related to intensive labor-use, tighter standards and controls, and increasing production costs. In addition, the emergence of niche markets in northeastern Thailand has been an uneven process, as the majority of farmers have been excluded from the rise of niche markets and have not been able to convert to organic agriculture because of a lack of technical knowledge and finances.

N. Taotawin (✉)
Department of Sociology and Anthropology, Faculty of Social Science,
Chiang Mai University, Thailand
e-mail: tnatedao@gmail.com

M.A. Stewart and P.A. Coclanis (eds.), *Environmental Change and Agricultural Sustainability in the Mekong Delta*, Advances in Global Change Research 45,
DOI 10.1007/978-94-007-0934-8_23, © Springer Science+Business Media B.V. 2011

23.1 Emergence of Export-Oriented Organic Rice Production in the Northeast of Thailand

The world rice market is an unstable commodity market, in which the availability of supplies is uncertain and prices are highly variable. The pre-Green Revolution period, from 1950 to 1964, was characterized by high and stable prices. Then during the Green Revolution period, from 1965 to 1981, when modern fertilizer-responsive varieties were adopted in many countries, a period of high and unstable rice prices began. The immediate post-Green Revolution period, from 1982 to 1984, then marked a short transition to low and stable prices (Dawe 2002).

For the most of the twentieth century and since, the major rice exporters in the world have been the nations of mainland Southeast Asia, including Thailand, Burma, Cambodia, and southern Vietnam. Because a large share of their production is exported to world markets, these countries have needed to be commercially oriented and participate actively in the world market as reliable rice suppliers. Not only is rice an important share of the economy for these countries, but it is also a key source of foreign exchange earnings and government revenue. In Thailand, taxes on rice exports consistently accounted for more than 10% of all government revenue in the period from 1950 to 1965, and the share occasionally reached more than 25%.

In 1957 Asian per capita production fell by 4%. Aggravating matters, production for the two major exporters, Burma and Thailand, dropped sharply. In 1965, the situation changed considerably, as rice exports from Burma decreased, and in Thailand, revenue from rice export taxes fell to just 6% of total government revenues by 1967 and to just 1% by 1971, since the Thai Government was no longer so reliant on a stable domestic price. As a result, during the world price spike in 1967, Thailand raised its rice premium to a level more than double the previous high. Thailand's rice exports fell to just 10% of domestic production between 1973 and 1975, reaching their lowest point in the post-war period.

By the middle to late 1980s, the situation had changed considerably for the better, as Thailand's commercial orientation had increased steadily, with exports typically accounting for 40% of domestic production. The presence of Thailand and Vietnam as commercially oriented rice exporters was a major factor in stabilizing the world market in 1998; Thailand was by this time much more commercially oriented compared to the mid-1970s. Between 1961 and 1993, world trade fluctuated between 3.5% and 5% in terms of world production. Since 1994; however, this fluctuation has exceeded 5% every single year, and the share traded has averaged 6.1% (Dawe 2002).

There was a downward trend in the volume and value of non-organic rice exports after the 1970s. However, the importance of the agricultural sector to the Thai economy had not diminished, as a large proportion of the population still worked within the agricultural sector, so the sector had the capacity to absorb surplus labor in rural areas. The presence of neo-liberal economic policies in Thailand by the 1980s affected competition on the world markets within the agricultural sector, and as a result, some scholars suggested that the Thai Government should adopt an agricultural restructuring policy, focusing on a shift to the production of high-value food products (Siamwala 1989).

The adoption of an agricultural restructuring policy by the Thai Government during the late 1990s, one focused on high-value food production, was an attempt to enhance the competitiveness of the Thai agricultural sector within a neoliberal context (Manarangsan and Suwanjindar 1992). Within this move, the conversion to organic agriculture reflected an attempt on the part of the Thai Government to shift to high-value food production (Department of Agricultural Extension 2004). Since then, organic agriculture for export has been promoted by government bodies as a weapon to be used to eradicate rural poverty, improve the nation's economy, and create food security and environmental sustainability.[1]

In 2008, the planted area of organic crops was estimated to be approximately 140,963 rai[2] (2254.08 ha), of which 80% of the total planted area was used for organic agriculture. In Thailand, the Thai Government wanted to increase the area planted with organic crops by 40%, to reach 200,000 rai (33313.33 ha) by 2010 in order to meet the increased demand for organic food products on the world market. The small area planted in Thailand put it in 71st place out of 85 countries growing organic produce, earning only around one billion baht[3] (US$28.57 million) in export revenue. The figure represented less than 0.1% of the total value of organic products around the world, estimated to be 1.3 trillion baht in 2006. World market demand for organic produce has been rising by 20% a year, especially in wealthy countries in Europe and North America and in Japan, and the major organic product exporters are Argentina, Mexico, Brazil, India, and China.

To capture this growing niche market, the Thai Ministry of Agriculture and Cooperatives allocated a budget of 1.7 billion baht (US$48.57 million) to improve organic plantations during the year 2005–2009. The fund was allocated to improve land, to support research and development initiatives, and to enhance marketing. The Ministry had allocated a budget of 354 million baht (US$10.11 million) for a Mentor Program, in which selected experts were hired to advise farmers on organic farming. The 40 "mentor centers" set-up proved successful in disseminating information and advising farmers on the sufficiency economy and organic agriculture. By 2009, the Thai Ministry of Agriculture had established 130 of these centers, with each one supporting at least 200 farmers, therefore, the Ministry had set a goal to turn around two million rai (51.20 thousand ha) of chemical-use land into organically farmed land.[4]

In particular, the Thai Government has implemented a policy to promote organic *jasmine* rice production. In 2000, it allocated a budget of 1.7 billion baht (US$4.86

[1] http://www.ldd.go.th/link_fertilizer/home.htm

[2] 6.25 rai is equivalent to 1 ha.

[3] One US Dollar is equivalent to 35 Baht.

[4] Bangkok Post Newspaper, Jan 9, 2008, http://www.biothai.org/cgi-bin/content/news/show.pl?0518

million) to increase the areas used for organic jasmine rice production to 120 billion tons. The Ministry of Agriculture, in conjunction with other governmental bodies, also set up a national rice strategy, implemented between 2004 and 2008, in order to encourage the production of safe foods, to improve environmental management, and to improve the environment as a whole.

The Thai Government considers the standardization of organic foods as a pre-condition for the exporting of organic agricultural products on to the world market. The standardization of organic farming in Thailand was initiated through the institutionalization of organic standards, marked by the establishment of the Office of Organic Agriculture Certification Thailand (ACT) in 1998, the establishment of formal ACT standards in 1999, and the improvement of ACT standards to fit International Federation Organic Agriculture Movement (IFOAM) standards in 2000.[5] The Government established the National Bureau of Agricultural Commodity and Food Standards (ACFS) in 2002 to consolidate all national organic certifications under a single authority. Although organic agriculture was promoted as a national agenda, the growth of organic food supplies in Thailand resulted from the engagement of agribusiness in organic production; the majority of smallholders did not convert to organic agriculture because of a number of barriers (Panyakul 2004).

23.2 Methods and Data Collection

This study adapts anthropological approach methodology to understand the position of farmers who grow fair trade and organic rice under contract, and the role of contract farming in re-shaping labor processes within the organic rice scheme. Ethnography was the key methodology employed to observe and gather data in the field, together with a farmer household survey and in-depth interviews with key informants.

The field sites studied were several districts in Ubon Ratchatani Province along the Thai-Laos border, which are important sites for export-oriented organic rice production. A household survey was conducted at these field sites in order to collect data about the socio-economic status of farm households engaged in organic agriculture. This survey collected data on demographics, education, land-use patterns, sources of inputs and supply, labor arrangements, water management, migration, and sources of income and debt information. In-depth interviews with key informants were conducted in District "A" of the Province, which was selected as a field site for the ethnographical research.

[5]Initially, the ACT was named Alternative Agriculture Certification Thailand and had four main objectives, one of which was to assure consumers of the distinctive quality of organic foods produced by alternative agriculture. After the ACT had run for a few years, it changed to focus on certification alone. In 1998, the Alternative Agriculture Certification Thailand was renamed to Organic Agriculture Certification Thailand, and its standards were also revised to meet those of IFOAM in 2000 (http://www.actorganic-cert.or.th/standard.html).

23.3 Case Study

According to the fair trade standards, most producers who enter into fair trade markets are smallholders and are united through the presence of associations, cooperatives, and producer groups. The organic jasmine[6] rice scheme in northeastern Thailand was initiated by a local NGO in 2002 through a local development scheme. To be able to export organic jasmine rice to the fair trade markets, the farmers have to be united as an organic rice producer's group. The initial mobilization of farmers for growing organic rice was based on their conversion from standard to organic agriculture, and the certification of organic rice took 3 years after the initial conversion. The initial inspection by internal inspectors has been followed by an annual inspection by external inspectors accredited by the certifying organizations. Initially, the producer's group had a membership of less than 100 farmers, however, its membership had reached 500 farmers by 2007.

It was expected that the long-term linkage between the fair trade scheme and the producer's group would improve the well-being of smallholders, with a fair trade premium paid to the farmer's organization in addition to the guaranteed rice price, and that this would support locally-initiated development projects. The producer's group used the fair trade premium to begin to focus exclusively on organic rice production. From 2003 to 2008 the group received a fair trade premium of approximately 1.5 million baht (US$24,857). The premium was employed to support a variety of activities, such as training the farmers in the technical knowledge required for organic rice farming, paying certification fees, and setting up a fund to dispense group loans for farm improvements.

In theory, it is correct to say that the fair trade and organic rice scheme was meant to open a space for all potential farmers to produce high quality organic rice products. However, in practice the integration of smallholders into the global organic market has occurred through the mobilization of conventional farmers who have been induced to grow organic jasmine rice for export. The local NGO has linked smallholders with global niche markets through contract farming. The local NGO imposes criteria for the selection of land and labor in order to produce the organic jasmine rice commodity. The farmers were originally informed that they were to apply to be members of the producer's group through their personal connections with former farmer members, NGO staff, and the committee of the producer's group. The NGO organization had extension service staff who constantly visited the farmers in order to convince them to apply for membership of the producer's group.

A survey conducted by the local NGO in 2007 indicates that those farmers engaged in growing organic jasmine rice for export come from different sized farms. From a total of 553 farm households, the farmers can be classified into three categories: (1) large-scale farmers, (2) medium-scale farmers, and (3) small-scale farmers or smallholders. Smallholders refers to those peasants who own land

[6]Jasmine rice or jasmine rice is called the 'Khao Dawk Mali 105' (KDML 105) (in Thai script, it is written as **'ข้าวดอกมะลิ'**).

Table 23.1 Classification of farm size by land holders

Farm sizes	Number of farm households	% of farms
Small-scale farmer (0–15 rai, or 0–2.4 ha)	250	45.21
Medium-scale farmer (16–30 rai, or 2.5–4.8 ha)	208	37.61
Large-scale farmer (larger than 31 rai, or 4.9 ha)	95	17.179
Total (11302.16 rai, or 1834.48 ha)	553	100

between 0 and 15 rai (0–2.4 ha), of which there are 250 households, comprising 45.20% of all farms. The medium-scale farmers are those who owned land of between 16 and 30 rai (2.5–4.8 ha), of which there are 208 households, comprising 37.61% of all farms. Finally, the large-scale farmers refers to farmers that own land over 31 rai (4.9 ha), of which there are 95 households, comprising 17.17% of all farms. The number of farm households and size of farms is shown in Table 23.1.

The reasons for the farmers' decision to grow organic rice for export are fairly complex. The farmers believe that the conversion to organic rice farming benefits them in several ways. For instance, it allows them to cope with cost-price squeeze problems and to gain a premium price. Once the farmers have been integrated into the global niche markets, they can sell their produce at a minimum guaranteed price, as set by the FLO – which is also a premium price.

Although economic reasons are an important factor in the conversion to organic jasmine rice production, social and cultural reasons are also involved. The farmers consider organic agriculture to be safe, healthy, and environment-friendly, and they believe that organic agriculture improves land fertility and generates safe food both for their own family members and consumers; at the same time it can be seen as value-added production.

Export-oriented organic rice production in northeastern Thailand is a response to emerging niche markets and to national agricultural restructuring policies. Moreover, the export-oriented organic jasmine rice scheme in northeastern Thailand is a buyer-driven commodity chain.[7] It is a commodity chain in which large retailers, brand-named merchandisers, and trading companies play a pivotal role in setting up decentralized production networks in exporting countries, typically located in the so-called Third World. These chains allow buyers and branded merchandisers to act as strategic brokers in linking overseas factories and traders with evolving product niches in their main consumer markets. The flexible specialization of these chains gives them a greater capacity to deal with the changing conditions of doing business in a post-Fordist world (Gereffi 1994).

The creation of flexible specialization takes place through changes across three main domains: (1) volume and differentiation of the farmers' product, (2) farm

[7]Buyer-driven commodity chains are identified by reference to three main dimensions: (1) the input–output structure i.e., a set of products and services linked together in a sequence of value-adding economic activities, (2) territoriality i.e., a spatial dispersion or concentration of production and distribution networks, comprised of enterprises of different sizes and types, and (3) a governance structure i.e., authority and power relationships that determine how financial, material, and human resources are allocated and flow within a chain (Gereffi 1994).

management and a mix of internal factors, and (3) the farmers' relations with other firms, such as suppliers or buyers (Raynolds 1994). For example, the Kenya flower commodity chain is driven by the force of retailers and the force of abstraction coming from international regulations (Hughes 2004). Similar to the Kenya flower commodity chain, the organic jasmine rice commodity chain is based on the re-arrangement of the conventional rice commodity chain in accordance with the demands of retailers and the forces of abstraction coming from international regulations.

Figure 23.1 illustrates the conventional rice commodity chain, which is locally embedded and is facilitated and controlled by local actors. Local rice mills and local rice merchants play an important role in collecting rice from the farmers to sell at both local and international markets.

In contrast, the organic jasmine rice commodity chain is facilitated and controlled by global actors. Although the production system is organized by local actors, such as the local NGO through initiation of the local development program, the organic jasmine rice commodity chain is a vertical network. Contract farming is employed as a major mechanism by which local farmers are linked to global niche markets. Contract farming allows global actors to gain control over farmers and farming practices from a distance (Young 2006). The use of contract farming in the organic jasmine rice scheme allows the global actors such as retailers, certification bodies, and inspection organizations to control both the quantity and quality of organic jasmine rice produced by the farmers. The farmers who produce under the contracts are orchestrated by the global actors via regulations; as contract farming is employed by capitalists to create flexible specialization, the role of the contract within the organic jasmine rice commodity chain is vital.

Figure 23.2 illustrates the complex web of social relations that exist between firms, NGOs, and farmers within the organic jasmine rice commodity chain. The farmers are organized together as a producer's group, according to the fair trade standards. Milling, processing, and packaging processes are operated by a local rice mill. Audit inspections are conducted by independent inspectors under the control of international certifying bodies such as FLO-CERT and IMO, and the organic jasmine rice is exported by another export firm, one which has market ties to importers overseas. These importing firms are responsible for milling the brown rice into white rice and for the processing of the organic jasmine rice. Following this, the organic jasmine rice is packaged by different brand-name companies and sold through ethical marketing chains overseas, through fair trade networks, and organic niche market retailers.

The organic jasmine rice scheme in northeastern Thailand has been certified by six international standards systems, the Fair Trade,[8] NOP,[9] EEC 2092/91,[10] BIO-SUISSE,[11] JAS,[12] and ISO 22000: 2005 standards[13]; the production system has

[8]The fair trade standards.

[9]The American USDA National Organic Program (NOP).

[10]The Council Regulation of the European Union (EEC) No 2092/91 on organic production of agricultural products.

[11]The Swiss Organic Regulation.

[12]The Japanese JAS standard.

[13]The International Organization for Standardization.

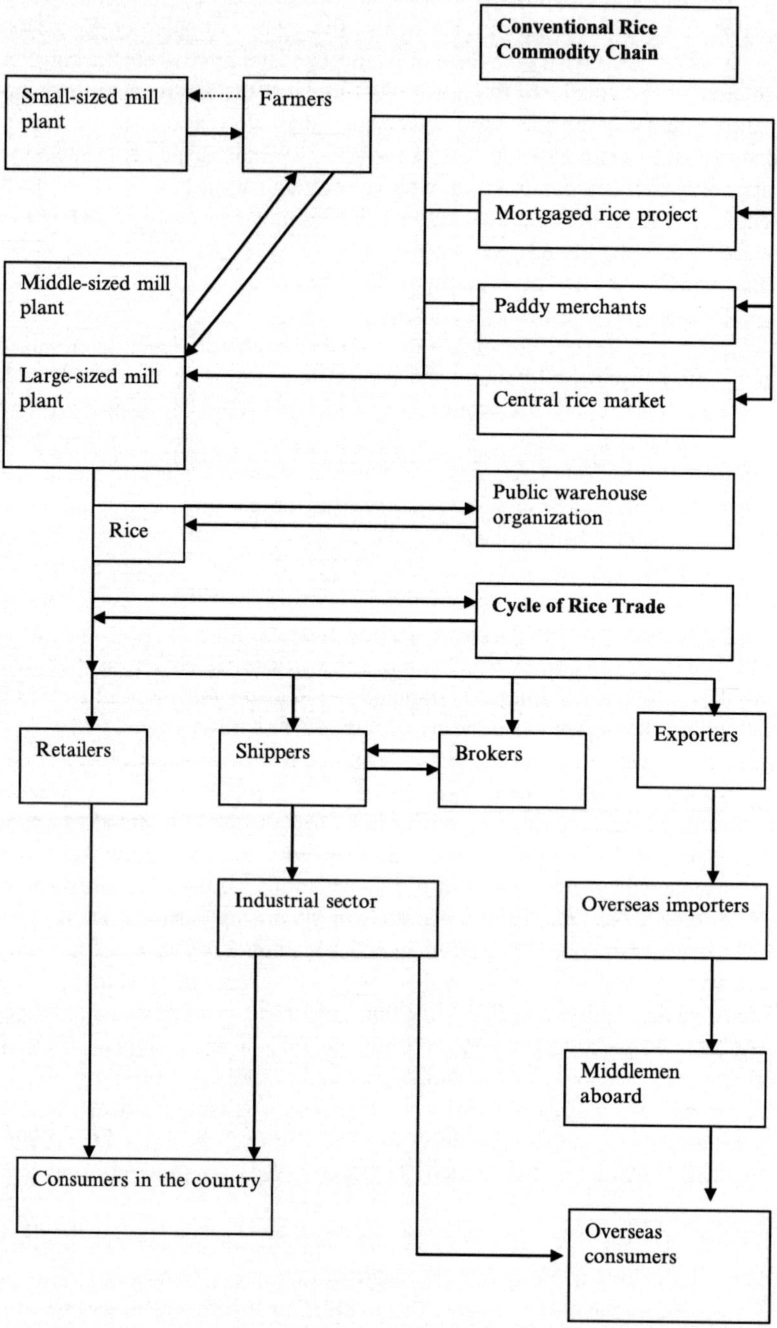

Fig. 23.1 Conventional rice commodity chain

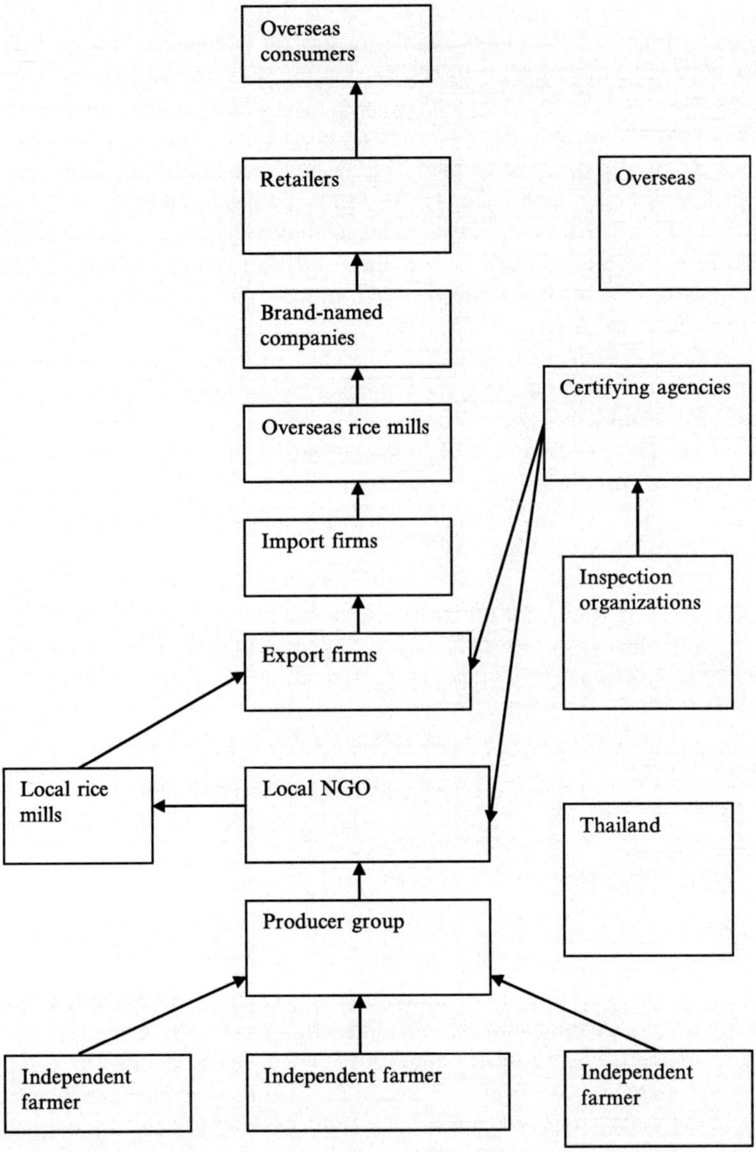

Fig. 23.2 Organic jasmine rice commodity chain

been re-arranged to comply with fair trade and the organic standards. The fair trade standards are mainly concerned with an improvement in the social conditions of production by paying the producers a guaranteed minimum price, one which covers overall production costs. In addition, the buyers are required to pay an additional social premium to support local development schemes initiated by producer organizations. For producers to get the fair trade premium, they are required to coordinate with a producers group, an association, or a cooperative. The group must consist of small farmers who depend on family labor to produce the fair trade products. Additionally, the group must be organized and operated along democratic lines and have democratic decision-making processes.[14] Recently, the Fair Trade Labeling Organization International (FLO) has recognized the importance of ecological well-being. Consequently, the fair trade producer group is required to reduce negative environmental impacts during the production, post-harvesting, and processing stages (Rice 2001). Hence, the regulating spaces that exist under fair trade standards are essentially social conditions of production, rather than ecological conditions of production.

The organic agriculture movement is founded on an improvement in soil fertility, the prohibition of synthetic agro-chemical use, and the minimizing of off-farm resource use (Darnhofer 2005). From this perspective, the regulating spaces set by organic standards are ecological conditions of production rather than social conditions of production. Consequently, organic regulations play a critical role in determining the legal definition of organic and in providing guidelines for the practice of organic agriculture (Guthman 1998).

Since the competiveness of export-oriented organic jasmine rice production in the northeast of Thailand is constituted through an ability to comply with five systems of certification, key factors to be considered are: (1) the mechanisms through which the international regulations are enforced on the ground, (2) the ways in which the production system is re-arranged to comply with international regulations, and (3) the degree to which the farmers are able to comply with the international regulations.

23.4 The Creation of Flexible Specialization in Export-Oriented Organic Jasmine Rice Production

This study argues that competition in the organic jasmine rice scheme is created by the ability to manage production time and the ability to integrate together different firms and organizations across different geographical spaces in order to match production with purchase orders and to control the quality of products (Schoenberger 1994). In addition, socio-economic and ecological conditions in northeastern Thailand contribute to competition in the organic jasmine rice scheme.

The northeast of Thailand is known as a region which has insufficient infrastructure, deteriorated natural resources, low education levels, and a lack of capital. Rice farming in the Northeast depends mostly on rainfall and has relatively low yields when compared to other regions of the country. The level of rainfall in the Northeast is not markedly lower than that of other regions, but the poor quality of the local soil reduces its effectiveness; the land has lower levels of nutrients and water-retention

[14]Food Safety Management Systems.

capacity and drought is thus a common occurrence. The poor quality of the soil and the drought are obstacles to agricultural intensification and expansion of agricultural territory, and they underpin the problem of rural poverty in the Northeast.

Socio-economic and environmental conditions in the Northeast have limited the impacts of the Green Revolution. As a result, the attempts to modernize agriculture in northeastern Thailand did not meet with total success. The Green Revolution could be identified as a "package approach" used to increase productivity, and it included widespread use of high yielding varieties (HYVs), intensified cropping, and a high level of reliance on irrigation systems, machinery, and agricultural chemicals. Some research has argued that the Green Revolution was only effective if the package approach was undertaken. From this perspective; therefore, the introduction of HYVs alone was not enough to create the beneficial Green Revolution effects, because water control, high fertilizer inputs, and the prevention of disease and insect management were also important (Pingali et al. 1997).

The impacts of the Green Revolution were obvious in central Thailand, as it improved rice yields per rai. However, a variation in rice growing successes across the different regions of the country was reported. Many writers on the impacts of the Green Revolution state that variations in the yields cultivated across different regions of the country were the result of variations in the environmental conditions under which rice was grown, rather than variations in technological access (Silcock 1970; Hseith and Ruttan 1967). A closer examination of the regional yield data in Thailand highlights the impact of environmental conditions on the yields, including the proportion of paddy land under irrigation in different regions and the occurrence of floods or droughts which damaged harvested areas and reduced yields.

Central Thailand had fairly high yields, due to a larger proportion of planted areas with access to irrigation systems. In contrast, rice yields in northeastern Thailand were only about two-thirds of those in the central plains, due to poor weather and soil conditions (Silcock 1970). At the early stages of the Green Revolution in the Northeast, an important report contended that northeastern Thai farmers had no confidence in the effects of chemical synthetic fertilizers, for two reasons. First, the farmers could not access micro-credit, so they could not afford expensive synthetic chemical fertilizers. Second, the farmers were not convinced of the efficacy of synthetic chemical fertilizers in their paddy fields, because in general chemical fertilizers did not generate dramatic yield increases and profitability in that area, as was frequency claimed (Kolseus 1972).

The statistics from the 3-year period 2003–2005 confirm the studies mentioned above. Central Thailand had fairly high yields over this period, at about 571, 584, and 594 kg per rai, respectively, but yields in northeastern Thailand were only 333, 329, and 340 kg per rai for the same period.[15]

Moreover, northeastern Thailand has few major rivers and an insufficient water supply, factors which generally lead to severe drought and placed a limitation on the effects of the Green Revolution. The HYVs depend on a combination of adequate synthetic fertilizer and insecticide usage. They also

[15] Office of Agricultural Economics.

require an adequate water supply, in order to allow the rice to mature and ripen; hence, an insufficient irrigation system in northeastern Thailand directly affected the yields.

However, northeastern Thai farmers do not rely on rainfall alone; they do try to channel water into their paddy lands. Some use short-life pumps to draw water from channels or underground water supplies into their paddy fields so as to prevent their fields from drying up. At the start of the rainy season, when the rainfall is unreliable, paddy fields that do not have nursery crops often serve as catchment areas for those that do. Each paddy field has the characteristics of a small reservoir, but even then water may be insufficient. In places with a limited catchment area, or where the level of underground water is low and the soil is highly porous, farmers' efforts bring little reward.

Without using the package approach of the Green Revolution, this study argues that the introduction of one single technological element of the Green Revolution into the traditional cultivation system, at the personal discretion and initiative of the farmers and under certain limited environmental and socio-economic conditions, only had a partial effect.

The partial effects of the Green Revolution in northeastern Thailand have led to a comparative advantage in terms of the conversion to organic agriculture, as the potential risk of cross-contamination in the Northeast is perceived as less than in other regions. jasmine rice[16] is a photo-period sensitive rice variety and can be grown in the wet season in every region of the country. However, northeastern Thailand is the most suitable place for growing jasmine rice, due to its rainfall pattern and soil salinity (Isvilanonda and Seiichi 2005).

The socio-economic and environmental conditions in northeastern Thailand have also contributed to its competitiveness in terms of organic jasmine rice production. The Northeast is known as a region in which a large number of residents live below the poverty line. In 2004, the average income of residents in the Northeast was estimated to be about 1,078 baht per month, the proportion of those considered as poor was estimated to be about 17.16%, and the number of poor was estimated to be 3,650,000 people. In contrast, average income in central Thailand at that time was estimated to be 1,339 baht per month, the proportion of poor people was estimated to be 5.09%, and the number of poor was estimated at 757,000 people.[17] Moreover, another report said that 24.53% of farmer families were landless and had problems in terms of a lack of stable land tenure. In addition, at that time about 4.86 million farmer families (or 86.4%) were indebted to an average

[16]Jasmine rice is known for specific characteristics, particularly its aroma and low amylase content, which makes the cooking of jasmine rice different from other rice grains. These special characteristics can be distinguished among Asian consumers and help generate a premium price. The aroma and grain quality of jasmine rice grown in different geographical regions are not uniform, due to the production environment, soil nutrient levels, and cultural management practices (Isvilanonda and Seiichi 2005).

[17]A survey conducted by the Office of National Statistics, calculated by Office of the National Economics and Social Development Board in 2004.

Table 23.2 Indicates the poverty line, proportion of poor people, and the number of poor people in 2004 (*Source:* A survey by the Office of National Statistic, calculated by Office of the National Economic and Social Development Board)

Region	Poverty line (baht/person/month)	Proportion of poor people (%)	Number of poor people (thousand)
Bangkok	1,853	1.64	108.40
Central plain	1,339	5.09	757.40
Northeast	1,078	17.16	3,650.80
North	1,131	16.24	1,907.40
South	1,164	7.82	655.00
Total	1,242	11.25	7,079.00

amount of 24,700 baht (US$642.7) per family.[18] The poverty line, plus the proportion and actual number of poor people found in northeastern Thailand in 2004, are identified in Table 23.2.

Initially, most farmers who engaged in export-oriented organic jasmine rice production were smallholders. A 2007 survey pointed to the fact that wealthy farmers were the smallest group of farmers engaged in export-oriented organic rice production, though they had relatively high potential in terms of access to labor, land, and capital, and that the smallholders were the biggest group, although they had relatively less potential in terms of access to labor, land, and capital. Hence, the introduction of export-oriented organic jasmine rice production in the Northeast of Thailand has involved a complex and uneven process.

As only a minority of smallholders have been integrated into niche markets, in some way the emergence of niche markets in northeast Thailand has created a new rural differentiation regime, between the farmers who have been integrated and those who have been excluded from the niche markets. Moreover, rural differentiation has also occurred among farmers integrated into the fair trade and organic jasmine rice scheme but who have different capacities in terms of access to land, labor, and capital.

The use of contract farming in the fair trade and organic jasmine scheme has created flexibility for the capitalists to access suitable land owned by the smallholders and utilize labor which is uneven across the different seasons in terms of availability. The farmers are considered as a flexible specialty labor; they are poor, vigorous, patient, honest, dutiful, and flexible. The conversion to organic agriculture as a livelihood strategy employed by the farmers to cope with vulnerability, to seek security in the context of a weakening peasant movement in Thailand, as well as to deal with the internationalization of high-value food, has been a common strategy used. Within the last few decades, rural development in the Northeast has focused particularly on the improvement of productivity to attain self-sufficiency,

[18]The Bank of Agriculture and Cooperatives reported in 2004.

to resolve the problems of environmental destruction and social inequality resulting from economic development (Phatharathananunth 2002). Nevertheless, we are still left with the question as to whether this strategy has helped reduce vulnerability and poverty, particularly among poor farmers.

Rural diversification is an important strategy employed by rural households in developing countries to reduce vulnerability and enhance well-being (Rigg 2001), and migration has been a livelihood strategy employed by younger generations in northeastern Thailand to gain more income and to cope with the problems of vulnerability (Mills 2001). Therefore, livelihood diversification, or the process by which rural families construct a diverse portfolio of activities and social support capabilities in order to survive and to improve their standard of living, is vitally important (Ellis 1998). Empirical evidence shows that rural families in developing countries rely significantly on migration and non-farm livelihoods for their survival (Rigg 2006). Nevertheless, the impacts of migration and remittances on local economies and the changes that have taken place in the agricultural sector are complex, because agrarian transition in developing countries has not been linear but has appeared along multiple trajectories (Hart 1997).

This study confirms the argument of Rigg that non-farm livelihoods are important for the survival of rural families of northeastern Thailand. I argue that the conversion to organic agriculture in northeastern Thailand has been the result of complex drivers. Farmers now combine on-farm and non-farm activities in order to make a living; they often grow glutinous rice for their own consumption and organic jasmine rice for commercial production. They have a number of options open to them: (1) to persist with conventional rice farming plus carry out wage labor or other non-farm activities; (2) to shift to organic rice farming plus carry out wage labor or other non-farm activities; (3) to try a combination of conventional and organic rice farming, plus carry out wage labor or other non-farm activities; (4) and, to abandon rice farming altogether and turn to wage labor. When considering the rural households' portfolios as mentioned above, the conversion to organic agriculture helps them to persist with their on-farm livelihoods. Non-farm income also plays a significant role in the way that the farm households make a living, as the farmers usually spend their incomes derived from non-agricultural activities to invest in agriculture.

It is worth noting that only a minority of farmers in the rural communities of northeastern Thailand have been able to convert to organic agriculture; the majority of their fellow farmers have continued with conventional agriculture. This indicates that the process of market integration is complex and uneven. For the farmers, the decision to convert to organic agriculture is not a freely-made choice, but it is influenced by their economic and social contexts. Some of the main determinants of diversification are seasonality, differentiated labor markets, risk strategies, coping behavior, credit market imperfections, and inter-temporal saving and investment strategies (Ellis 1998). Key factors that shape choices to diversify have to be examined.

23.5 Examining Capacities and Challenges in Recent Trend of Export-Oriented Organic Rice Production

To export organic jasmine rice to global niche markets, the production system and agro-ecological environment must be organized in compliance with international regulations. The ability to compete in the trade of organic jasmine rice depends significantly on the capacity of individual farmers to arrange their time and space to comply with international regulations. The organic jasmine rice scheme in northeastern Thailand has been certified by five systems of certification, yet international regulations have a certain set of rules to follow. The complexity and difference in international regulations when compared to domestic standards can be regarded as a non-tariff trade barrier for the export of products from developing countries to international markets (Henson and Loader 2001). Therefore, differences between the fair trade system of certification and the organic trade system of certification have created practical constraints to exports (Browne et al. 2000).

The differences in international regulations make their enforcement on the ground nearly impossible for both development practitioners and local farmers. As a result, the international regulations have had to be transformed by the relevant development programs into 20 locally-established rules. The primary reason for simplifying international regulations to 20 locally-established rules is to make the regulations practicably enforceable and to make possible the monitoring and auditing of farming practices.[19]

Contract farming is the key mechanism through which international regulations are enforced on the ground. Contract farming in organic and fair trade agriculture plays two key roles, the first being to re-arrange agricultural practices to comply with international food regulations, and the second to assure the availability of disciplined labor to work on the farms. Although the role of contract farming in organic and fair trade agriculture is important, a full discussion of it is beyond the purpose of this paper. This study will focus on the re-arrangement of agricultural practices at place in contract farming so that it complies with fair trade and organic regulations.

The chronology of this process is: (1) producers are required to attend training; (2) farmers negotiate the size of the paddy fields they want to use for the contract production of organic rice; (3) the contract between the project manager and the producer is created on the basis of the individual farmer involved; (4) the project manager sends out a standard contract to each producer on a take-it-or-leave-it basis; (5) the producer signs the contract, specifying the paddy fields to be used for rice growing; (6) the project manager inspects the paddy fields, choosing the contract producers based on this inspection; (7) the project manager decides on the production plan; (8) the farmers prepare the soil, sow the rice seeds on the nursery plots, and transplant rice crops from the nursery plots to paddy fields; (9) the farmers nurture and protect the organic rice; (10) the farmers harvest the rice crops, dry the cultivated paddy fields, and deliver the rice to the project; (11) and finally, the farmers are paid.

[19] Interview with a development programmer on 28 December 2009.

Since accurate planning is important within contract farming, the transplanting activities must be synchronized to match the capacity of the harvest and threshing laborers, taking into account transportation time and the ripeness of the rice. Therefore, decisions regarding the harvest period are made by the development project.

The 20 locally-established rules include: (1) the farms must completely convert to organic agriculture across all plots, including subsistence food plots. The transition to a fully organic situation takes 3 years; (2) mixing animal rearing and organic rice production is forbidden – if animals are kept on the farm, some measures are required in order to ensure animal welfare; (3) farmers are required to set aside at least 7% of their total agricultural area to conserve "natural space"; (4) the farmers must apply for membership in the organic jasmine producer's group and must sell all their organic paddy to the project; (5) the farmers must sign the contract; (6) it is compulsory to use certified organic rice seeds; (7) the use of chemical synthetic fertilizers and pesticides is prohibited; (8) all production inputs must be organic, must not be chemically treated, and must be approved by the development project; (9) use of genetically modified organisms (GMOs) is prohibited; (10) farmers are required to grow multi-annual rotation crops to improve soil fertility; (11) the prevention of airborne and waterborne contamination is required; (12) rice must be handled in such a way as to avoid mixing organic jasmine rice with other types of rice produced by non-certified producers during the harvesting, threshing, rice storage, and rice containing stages; (13) harvesting and post-harvesting processes must follow the instructions specified in the guidelines; (14) farmers are required to always participate in the annual training program, otherwise their status as members of the organic producer's group will be rescinded; (15) farm records[20] must be clear, and the original receipts for purchasing and selling items must be kept ready for checking by internal and external inspectors; (16) farmers must allow internal and external inspectors to carry out spot checks on the fields, the house, rice storage areas, the warehouse, and other crop checks. The inspections can be carried out at any time, with no need to inform the farmers in advance; (17) farmers must attend the annual meeting. If the farmers cannot attend the meeting for whatever reason, they have to assign a family member who engages in organic rice production to replace them, otherwise their status as a member of the organic jasmine rice producers group will be rescinded; (18) during the transition period, the farmers cannot sell their produce at the minimum guaranteed farm-gate price set by the FLO, and the farmers must be verified by internal and external inspectors at least once a year during the transition period; (19) the delivery of organic rice to the rice mills and the costs of delivery are the responsibility of the farmers; and (20) the NGO must be informed of any adjustment of production and post-harvesting activities, as well as control of pest epidemics. The development programmer claims that these 20

[20]Farm records cover the recording of farm history, including the agricultural area, land ownership, land use patterns, types of crop, chemical usage, water resources used for cultivation, management of agricultural inputs, the harvesting and selling of crops, and total production and income from product sales. Importantly, farm records record the field history of the farm for the previous 3 years.

locally established rules cover all five systems of certification and both fair trade and organic standards.

These locally-established rules represent codes of practices for the on-farm management activities carried out by the farmers and also represent codes of practices for the inspectors. These codes of practices for organic certification are mainly concerned with: (1) careful on-farm management; (2) proper documentation; (3) a crop protection strategy; (4) the prohibition of chemical pesticide and insecticide use; (5) the application of organic inputs[21] and the methods used; (6) enhancing animal welfare; (7) and protection of the environment. The codes of practices for fair trade certification are mainly concerned with: (1) careful on-farm management; (2) proper documentation; (3) decision-making along democratic lines; (4) and the protection of workers' rights. These codes of practices follow the ideals of the fair trade and organic regulations, aimed at changing the way in which export-oriented rice farming is handled.[22]

To achieve the ideals promoted by the fair trade and organic standards, the on-farm management and labor processes have to be re-arranged according to these codes of practice. As a result, farmers have to spend more time working in the fields, particularly during the transplanting and harvesting periods. Transplanting and harvesting organic rice depends entirely on manual labor. The transplanting and harvesting must be synchronized and must be handled in a way that meets the contract schedule, as any delay in the production process, such as during transplanting and harvesting, will prevent the rice from growing properly and from generating a fruitful yield, thereby affecting both the quantity and quality of the yield.

Across the other seasons, organic rice farming[23] still requires labor provisions to be consistent throughout the year. In the summer, which is between January and April, most of the rice is harvested. However, due to the inadequate water supply at this time, paddy fields are often left without crops. As a result, farmers who pursue conventional rice farming often seek non-farm income,[24] but those farmers who

---

[21] Products used in organic production and processing, e.g. fertilizers, soil conditioners, plant protectants, including additives and processing aids used in organic processed products.

[22] Handling refers to wind-drying, sun-drying, cleaning, cutting, sorting, packing, storage, and transportation of the product (IFOAM).

[23] Includes all agricultural land holdings (for crops or animal rearing) under the management of the same person; it also includes land rented from others for farming purposes, where the farmer is not the owner.

[24] The primary sources of income for rural households are categorized as farm, off-farm, and non-farm sources. Farm income includes livestock as well as crop incomes and comprises both the consumption of own-farm outputs and cash income from the outputs sold. Off-farm income typically refers to wage or exchange labor on other farms. It also includes labor payments in kind such as harvest-share systems and other non-wage labor contracts. Non-farm income refers to non-agricultural income sources and includes: (1) non-farm rural wage employment, (2) non-farm rural self-employment, (3) property income, (4) urban-to-rural remittances arising from within national boundaries, and (5) international remittances arising from cross-border or overseas migration (Ellis 1998).

undertake organic agriculture are not allowed to do this; they have to work on their farms to ensure that they are properly managed and in order to comply with all the international rules.

Though the farmers intend to follow the regulations, some find it quite difficult to comply with the international regulations. At a meeting, on March 12, 2008, farmers specified the difficulties they encounter: (1) the provision of non-qualified rice seeds by the project causes low yields and contamination problems; (2) no compensation is provided to farmers during the conversion period,[25] even if their yields are significantly decreased; (3) conversion to organic rice farming causes increased production and transaction costs; (4) social conflicts associated with land and resource management sometimes occur between farmers in pursuit of non-organic and organic rice farming; (5) the degree of risk arising from insect epidemics is high; (6) the prevention of diseases, weed control, and pest management is difficult and costly; (7) the lack of a water supply prevents the rice from growing properly; (8) there is insufficient animal manure; (9) tensions arise from the tighter controls and rules; (10) the minimum guaranteed rice price is relatively low compared to the non-organic rice price outside the contract; (11) there is sometimes no available organic rice market (from participatory observation during a meeting on, March 12, 2008).

The field study confirmed the difficulties encountered by the farmers. Moreover, the farmers complain that they "get more tired" than those who practice conventional agriculture. Such complaints reflect the fact that organic agriculture requires an extended time and intensive labor, more so than conventional agriculture. In all, the process of market integration has affected different scales of smallholders in different ways.

Evidence derived from my 2007 survey also shows that the use of contract farming in organic and fair trade agriculture allows the capitalists to use dispersed land for growing organic jasmine rice for export. Under the contract, paddy land is managed by individual growers, even if these lands are owned only through customary rights. My survey in 2007 also found that farmers receive a total average income derived from off-farm jobs of about 7,730 baht (US$220.85) per year. The farm households use some of their income derived from off-farm activities to invest in their farms. The average number of family members within each family is five people, but those who work on organic rice farming have less than five per family. The average age of farmers engaged in organic rice farming is 49 years, so since the number of family laborers that work on the farm is insufficient and those that do live there are old, the problem of insufficient labor is severe. To solve this problem, the households rely on hired labor to meet their labor demand in some seasons, when intensive labor is needed.

[25] "Conversion period" means the period from starting to carry out organic agriculture in accordance with IFOAM standards, until the product has been certified as organic. For organic rice, this conversion period takes 3 years.

The process of market integration has affected the different sized smallholders in different ways. My survey found that smallholders depend mostly on their family's own labor to work on farms; the majority of large-scale farmers depend entirely on hired labor to work on their farms, as they earn most of their income from off-farm activities and have more time to work off-farm. It is common to find that larger-scale farmers are wealthy, using the income derived from their off-farm activities to hire laborers to work on their farms. Because of their significant dependence on hired labor and less time devoted to farm activities, wealthy farmers are in effect capitalists acting as farm managers.

The wage labor problem stems from the fact that the demand for agricultural labor is uneven, and laborers have to be available all at the same time. Additionally, another problem of hiring workers is not only that they must be hired in the first place, but also that they must be made to work. The task of supervision is especially problematic in agriculture, because the labor process is carried out over an extensive geographic area, and under such circumstances, farmers find it difficult to gain constant exertion from the hired laborers. Although special care is needed for many tasks, the uneven labor demand created by the seasonality of agricultural production means that those who are hired for particular tasks are those who cannot get full-time jobs and may be neither skilled nor inspired to work hard by the low pay and difficult working conditions.

Contract farming shifts both the risks and hazards of seasonally uneven labor requirements onto the contract farmer. The farmer is paid according to how much she or he delivers, which effectively skirts the need for supervision by making the farmer the agricultural laborer on a piece work rate. The costs of supervision are not eliminated entirely, however, since the farmer may have reasons for reducing the amount of labor devoted to a contracted crop, at the expense of receiving a slightly lower income. Labor costs generally form a substantial part of total production costs. In the organic jasmine rice scheme, total production costs in 2008 were estimated to be about 3,210 baht (or US$91.71) per rai, with labor costs estimated to be about 2,365 baht (or US$67.57) per rai, comprising 73.99% of total production costs. The farmers pursuing organic agriculture need to pay labor costs for seedling and weed management, while their fellow farmers who pursue conventional agriculture do not need to pay for such labor costs. Because of this fact, the labor costs for organic agriculture are relatively high when compared to conventional agriculture. Uneven labor requirements are the farmer's biggest problem, since the buyers pay for produce only upon delivery; therefore, if the production procedures require intensive labor, the contract ensures that the farmer exploits himself or herself.

Self-exploitation is a crucial advantage of small-scale peasant farming over other forms of agriculture. For the small-scale farmer, labor may include both family labor and labor hired on advantageous terms, plus that which only a peasant can obtain – labor hired in kind, not for cash. Sanctions against child labor, whether legal or based on public disapproval, do not apply to sharecroppers or to contract growers who make use of their own child's labor. Contract farming offers access to a hidden peasant labor market with economies not available to the large-scale hirer of wage labor, such as labor exchanged among peasant relatives and neighbors within the peasant community.

When compared to the farmers who pursue conventional agriculture, the famers who pursue organic agriculture pay relatively higher production costs, with the production costs of organic agriculture calculated to be 3,535 baht (or US$101) per rai in 2008, while the production costs of conventional agriculture were calculated at 3,210 baht (or US$91.71) per rai in the same year. A comparison between the production costs of the organic jasmine rice scheme and that of a non-organic jasmine rice production scheme in the Northeast of Thailand is shown in Table 23.3.

Furthermore, the farmers who pursue organic agriculture have to pay opportunity costs, due to the locally-established rule specifying that the land must be used for growing organic crops only. After cultivating organic rice crops, rotation crops are required to be grown in order to enhance land fertility, and as a result, the land cannot be used for growing multiple crops of rice. Following the rice harvest, the conventional farmers usually grow cash crops such as cassava, which tolerates dry weather and can be grown even on infertile land. Moreover, cassava does not need intensive labor and takes only a few months of cultivation; however, cassava does require a certain amount of synthetic chemical fertilizers to produce commercially viable yields. As a mixture of organic and non-organic crops is prohibited in the organic jasmine rice scheme, the farmers who pursue organic agriculture cannot grow cassava for commercial purposes.

Among many difficulties encountered by the farmers, the period of conversion to organic agriculture, which takes 3 years, is identified as the cause of greatest hardship. The yields dramatically decrease in the first year of the transition to organic agriculture, but there is no compensation provided to farmers for this loss. Some farmers claim that the yields from organic rice plots in the first year of transition decrease by about a half over the yields gained from conventional agriculture.

Table 23.3 Shows production cost of organic jasmine rice production and production cost of non-organic jasmine rice production per one rai estimated in the Northeast of Thailand in 2008

Items	Organic jasmine rice (baht)	Percentage	Non-organic jasmine rice (baht)	Percentage
Plowing (labor cost)	250	7.78	250	7.07
Secondary plowing (labor cost)	400	12.46	400	11.32
Seeds	85	2.65	85	2.40
Seedling	140	4.36	–	–
Transplanting (3 persons/1day/1 rai)	600	18.69	600	16.97
Fertilization compost (250 kg/3 baht) and EM (0.5 l) Labor cost	800	24.92	1,300	36.76
Weed management (1 day/1 rai)	35	1.09	–	–
Harvesting (2 persons/1 day/1 rai)	250	5.79	250	7.07
Threshing (1 day)	500	15.58	500	14.14
Packaging and delivery (1 day)	140	4.36	140	3.96
Transportation	10	0.31	10	0.28
Total	3,210	100	3,535	100

Yet, if the farmers expect to be certified as organic, they also need to pay for increased production and transaction costs, plus the cost of risk management arising from exposure to insect epidemics. Even though yields cultivated in the second year of transition improve, one cannot say that the yields from organic rice farming can compete with yields from conventional rice farming, even in the third year of conversion to organic rice farming.

As organic rice farming requires careful attention to be paid to the overall pre-production, production, and post-production procedures, the requirement for intensive and consistent labor in organic rice farming is a crucial factor. However, a survey I conducted in 2007 found that the average number of family members within the small-, medium-, and large-sized farms is 5, 6, and 5 persons, respectively. The majority of laborers within the small-sized farms work on organic rice farming. In contrast, most of the laborers in the large-sized farms earn their money from off-farm activities. In addition, my survey found that the average age of farmers engaged in organic rice farming is 49 years, and since there is not a sufficient pool of labor, the family farms in this category depend a considerable amount on hired labor to work on the farm. Moreover, farmers who pursue organic agriculture need to hold back at least one family labor member in order to handle the cattle throughout the year, because livestock manure is crucial to making organic compost; an allocation of labor for animal husbandry is thus important in order to ensure that sufficient animal manure is available.

The demand for intensive manual labor for transplanting and harvesting activities is vital, forcing the farmers to seek hired labor to meet their labor requirements, and it has not been uncommon in recent years for labor to have been recruited from Lao PDR to work on these organic rice farms. The replacement of Thai laborers by those from Laos is a strategy employed by the contract farmers to cope with the labor shortage problems that occur in the agricultural sector, as it helps the contract farmers to reduce production costs and to cope with the problem of uneven labor requirements.

It is the household income derived from non-farm wages and remittances that allows the contract farmers to hire laborers to cope with the problem of uneven labor demand. My survey in 2007 found that average household income derived from remittances was approximately 34,339 baht (or US$918.11) per year, and that the farmers depend significantly on remittances to invest in organic rice production. However, an increasing reliance on hired labor has led to higher production costs and a higher risk of monetary loss. One year, Lao seasonal labor was disallowed by the Thai Government, and confronted with this difficulty the farmers faced significant competition for wage labor, plus rising labor costs. As the sustainability of organic rice farming relies significantly on hired labor, the key question is how can the large-, medium-, and small-scale farmers respond to such changes?

It is correct to say that the large-scale farmers have economies of scale due to having more land and capital, but they depend entirely on hired labor to work on their organic farms. The greater reliance on hired labor creates higher production costs, thereby reducing the economies of scale. For this reason, large-scale farmers have not been very willing to engage in export-oriented organic rice farming.

In contrast, the smallholders have less land and capital, but they are the group who participate most in export-oriented organic agriculture; they rely almost entirely on their family labor to carry out the work. The transition from conventional to organic agriculture means they squeeze themselves financially at the expense of higher profits and reduced labor costs. However, the smallholders are interested in participating in organic agriculture because they can potentially make more profit while at the same time improve the quality of their land. For the smallholders, the decision to participate in export-oriented organic agriculture does not stem from economic reasons alone, although these reasons are an important factor. The smallholders receive fewer economic benefits from participation in the niche markets than the large-scale farmers, due to their more limited access to land and capital. Ecological and social factors are important for the smallholders, since organic agriculture helps improve the fertility of their land, which also means an improvement in their longer term livelihood security.

23.6 Conclusion

Market integration has had a considerable impact on peasant livelihoods and the environment and agricultural practices in northeastern Thailand. The emergence of export-oriented organic jasmine rice production in northeastern Thailand has been a response to emerging niche markets. It has also been a response to Thailand's national agricultural restructuring policy which has moved in the direction of the production of high value foods for export. The export-oriented production of organic rice is a buyer-driven commodity chain in which large retailers, brand-named merchandisers, and trading companies play an important role in setting up decentralized production systems in exporting countries, typically located in the Third World.

These buyer-driven food chains have contributed to a change in agricultural production systems in northeastern Thailand, since they have re-arranged the conventional rice commodity chain to incorporate the demands of large retailers, consumer concerns about safety, health, the environment, and social justice, and at the same time the chain has been re-arranged in response to the forces of abstraction imposed by international regulations. The organic jasmine rice commodity chain is a vertical network, because contract farming is the principle means through which the local farmers are integrated into the fair trade and organic global markets. Contract farming allows retailers and other global actors to control quantities and standardize the quality of organic rice produced by individual farmers. Under these contracts, the livelihoods of farmers and their farming practices have been reshaped by the global actors and international regulations, even if these actors are largely absent from Thailand itself.

Competition within the organic jasmine rice commodity chain is created by re-arrangement of the agricultural practices and the environment of northeastern Thailand, in order to comply with the demands of retailers and international regulations. The socio-economic, ecological, and geographical conditions in northeastern Thailand have contributed to the competitive advantages created by the organic jasmine rice scheme. On the one hand, the socio-economic and environmental conditions in northeastern Thailand have set a limit on the effects of the Green Revolution. The ineffectiveness of the Green Revolution in northeastern Thailand left the environment relatively safe (free of chemicals) for organic farming and thus created a comparative advantage in terms of the conversion of the area to organic rice farming. The socio-economic conditions in the northeastern Thailand have also created a form of flexible specialization, as northeastern Thai farmers in the Northeast are typically regarded as flexible specialty laborers, diversifying their economic activities by combining farm activities with off-farm and non-farm activities as a strategy to make a living and to obtain a better livelihood.

On the other hand, the key challenge of the organic jasmine rice scheme is that it depends significantly on the ability of the farmers to re-arrange their time and space to comply with the relevant regulations. These international regulations have been enforced on the ground through the establishment of 20 local-based rules, which represent both codes of practices for on-farm management by the farmers and codes of practice for inspections by outside agents.These codes of practice are based on the ideals of both the fair trade and organic agriculture movement. However, the enforcement of these international regulations on the ground creates tensions for the farmers, as the transition to organic agriculture requires intensive labor inputs and increasing production and transaction costs. These costs adversely affect the capability of farmers to integrate into niche markets.

The debate about the disappearance or persistence of family farms is an ongoing one and relevant to the development of organic rice farming in this area. Many family farms have been transformed by capitalism because they have had to adapt to various innovations introduced by mechanization and by diversifying their enterprises and adopting pluri-activity (Brookfield 2008). Technological innovation has also made higher land productivity possible and less-contested and has created higher labor production rates (Kautsky 1988). Drawing on the arguments of Brookfield and Kautsky, I argue that the persistence of family farms in northeastern Thailand under post-Fordist capitalism has been made possible through the diversification of livelihood strategies and the use of technological innovations based on the transition to organic agriculture.

However, the capacity of farmers to integrate into the new niche markets is a different matter. Large-scale farmers have more land and capital and therefore have a greater capacity to hire labor and to access innovative technologies and activities which then help them to generate higher yields. However, the greater reliance on hired labor creates higher production costs, thereby reducing the economies of scale of the large-scale farmers. As a result, large-scale farmers have not been willing to engage in export-oriented organic rice farming.

In contrast the smallholders have less land and capital, but have pursued organic agriculture to the greatest extent. For them, the conversion to organic agriculture requires greater self-discipline, yet they are interested in making this change because they can gain a higher profit and more security from the improved quality of their land. Though the smallholders receive fewer economic benefits from their integration into the niche markets than the large-scale farmers, they prefer to practice organic agriculture because it helps them improve the fertility of their land and thus creates longer term livelihood security.

References

Brookfield H (2008) Family farms are still around: time to invert the old agrarian question. Geography Compass 2(1):108–126

Browne AW, Harris PJC, Hofny-Collins AH, Pasiecznik N, Wallace RR et al (2000) Organic production and ethical trade: definition, practice and links. J Food Policy 25:69–89

Darnhofer I (2005) Organic farming and rural development: some evidence from Austria. J Sociologia Ruralis 45(4):308–323

Dawe D (2002) The changing structure of the world rice market, 1950–2000. J Food Policy 27:355–370

Ellis F (1998) Household strategies and rural livelihood diversification. J Dev Stud 35(1):1–38

Gereffi G (1994) The organization of buyer-driven global commodity chains: how U.S. retailers shape overseas production networks. In: Gereffi G, Morzeniewicz M (eds) Commodity chains and global capitalism. Praeger, Westport, Connecticut, and London, pp 95–122

Guthman J (1998) Regulating meaning, appropriating nature: the condification of California organic agriculture. Antipode 30(2):135–154

Hart G (1997) Multiple trajectories of rural industrialisation: an agrarian critique of industrial restructuring and the new institutionalism. In: Goodman D, Watts MJ (eds) Globalizing food: agrarian questions and global restructuring. University of California Press, California, pp 56–73

Henson S, Loader R (2001) Barriers to agricultural exports from developing countries: the role of sanitary and phytosanitary requirements. J World Dev 29(1):85–102

Hughes A (2004) Accounting for ethical trade: global commodity networks, virtualism and the audit economy. In: Hughes A, Reimer S (eds) Geographies of commodity chains. Routledge, London and New York, pp 215–232

Hseith SC, Ruttan VW (1967) Environmental, technological, and institutional factors in the growth of rice production: Philippines, Thailand and Taiwan. Food Research Institute Studies, 7(3)

Isvilanonda S, Seiichi F (2005) Export potentials and constraints for development of jasmine rice production in Thailand. J Int Cooperation Stud 13(2):29–48

Kautsky K (1988) The agrarian question. Winchester, London

Kolseus H (1972) Modernization of paddy rice farming in Northeast Thailand with special reference to use of fertilizer. Dissertation for the Degree of Doctor of Philosophy of Agricultural Economics. University of Kentucky, Kentucky city

Manarangsan S, Suwanjindar S (1992) Contract farming and outgrower schemes in Thailand. In: Contract farming in Southeast Asia. Three country studies. Institute for Advanced Studies, University of Malaya, Kuala Lumpur

Mills MB (2001) Thai women in the global labor force: consuming desires, contested selves. Library of Congress Cataloging-in-Publication Data, New York

Panyakul W (2004) Organic farming: how to be certified? Earth Net Foundation, Bangkok

Phatharathananunth S (2002) The politics of the NGO movement in Northeast Thailand. J Asian Review 15:21–36

Pingali PL, Hossain M, Gerpacio RV et al (1997) Asian rice bowls: the returning crisis? Cabinternational and International Rice Research Insititute (IRRI), New York and Manila

Raynolds LT (1994) Institutionalizing flexibility: a comparative analysis of Fordist and post-Fordist models of third world agro-export production. In: Gereffi G, Morzeniewicz M (eds)

Commodity chains and global capitalism. Praeger, Westport, Connecticut, and London, pp 143–161

Rice BA (2001) Noble goals and challenging terrain: organic and fair trade coffee movements in the global market place. J Agric Environ Ethics 14:39–66

Rigg J (2001) More than the soil. Pearson, London

Rigg J (2006) Land, farming, livelihoods and poverty: rethinking the links in the rural south. J World Dev 34(1):180–202

Schoenberger E (1994) Competition, time, and space in industrial change. In: Gereffi G, Morzeniewicz M (eds) Commodity chains and global capitalism. Praeger, Westport, Connecticut and London, pp 51–66

Siamwala A (1989) The question for the development of Thai agricultural sector in the future. J Thai Parliament 37(9):32–52

Silcock TH (1970) The economic development of Thai agriculture. Australian National University Press, Canberra

Young N (2006) Distance as a hybrid actor in rural economies. J Rural Stud 22:253–266

Chapter 24
The Impact of Weather Variability on Rice and Aquaculture Production in the Mekong Delta

Dang Kieu Nhan, Nguyen Hieu Trung and Nguyen Van Sanh

Abstract Understanding the impact of weather variability on current agricultural production systems is of great importance for the development of strategies for adapting to and mitigating the potential impact of climate change on food security in the Mekong Delta of Vietnam. This essay assesses the impact of short-term weather variability on rice and aquaculture production, documents ways in which farmers have dealt with weather anomalies, and suggests strategies to adapt to weather and climate variability in the future. Statistical series data (1990–2008) and information collected from previous projects were analyzed. The probabilities of occurrences of weather anomalies were calculated and multiple regression analyses were conducted to identify and estimate significant effects of various weather variables on yields of rice and fish or shrimp in both irrigated and coastal regions. The results showed that temperature and rainfall were the key weather variables that strongly influenced rice and shrimp production. Vulnerability levels to weather variability differed by crops, crop-development stages, cropping seasons, and regions. Rice production was found to be more sensitive to weather variability than was aquaculture, and shrimp production was more sensitive than was *Pangasius* catfish culture. The impacts became more severe during the early vegetative, flowering, and ripening stages of rice crops. The wet-season rice crop and the coastal region were more vulnerable to weather anomalies than the dry-season rice crop and the irrigated region, respectively. Local farmers have coped with temperature and rainfall anomalies in the past by applying appropriate farming techniques. Nonetheless, further measures for adapting to weather and climate variability are essential.

D.K. Nhan (✉) and N.V. Sanh
Mekong Delta Development Research Institute, Can Tho University, Can Tho, Vietnam
e-mail: dknhan@ctu.edu.vn; nvsanh@ctu.edu.vn

N.H. Trung
College of Environment and Natural Resources, Can Tho University, Can Tho, Vietnam
e-mail: nhtrung@ctu.edu.vn

M.A. Stewart and P.A. Coclanis (eds.), *Environmental Change and Agricultural Sustainability in the Mekong Delta*, Advances in Global Change Research 45,
DOI 10.1007/978-94-007-0934-8_24,

Keywords Weather variability • Climate change • Rice production • Aquaculture • Mekong Delta

24.1 Introduction

In the Mekong Delta of Vietnam, agricultural production and people's livelihoods are highly susceptible to abnormal events in the Mekong River's flow and in the weather and/or climate. The delta is a low-lying region, where the livelihoods of about 13 million people, 70% of the total population of the area, mainly rely on agriculture and natural-resources exploitation. Hydrological conditions and soil characteristics, which are strongly linked, determine agricultural land uses in the delta (Nhan et al. 2007). The delta is prone to monsoon flooding in the upper delta in the wet season and to salinity intrusion in downstream parts in the dry season. Recent studies have suggested that climate change and sea-level rise would have strong negative impacts on agricultural production and households in the delta (Wassmann et al. 2004; Carew-Reid 2007; ADB 2009; MONRE 2009). Recognizing this challenge, the Vietnamese Prime Minister in 2008 approved the national target program that specifies three stages for coping with climate change up to 2015 (MONRE 2008). The program has been in effect for 2 years. However, the aspect of "adaptation to climate change" still seems to be at an early stage, and practical guidelines for the implementation of adaptation measures are also scarce (Heine 2009).

Rice culture and aquaculture are the principal farming activities of the agricultural sector in the delta. In 2008, around 1.8 million hectares of land in the delta were devoted to rice production and 0.8 million hectares of land were used for aquaculture. The delta accounted for 55% of the total amount of rice and 72% of total aquacultural production in Vietnam (GSO 2010). In the period 2000–2008, the amount of rice-land cultivated in the delta declined by an average of 1% annually, while the average annual growth rate of rice output in the region was 3%. The corresponding figures for the annual aquaculture growth rate were 10% and 23%, respectively (GSO 2010). The main reason for the decline in the land area devoted to rice in the delta is that farmers converted rice land into shrimp farms in coastal areas during the 2000–2008 period. The fact that rice production grew despite the fall in the amount of land in cultivation and the fact that aquacultural output grew more rapidly than the amount of land devoted to aquaculture suggests a trend toward greater intensity in these farming sectors. Coastal shrimp and freshwater *Pangasius* culture are the main drivers of development in aquaculture.

In order to address issues relating to global climate change and food insecurity, the Vietnamese government in 2009 issued a national strategic plan for agricultural land use and food security to 2020. In this plan, the Mekong Delta is recognized as playing a crucial role in national food security. Therefore, understanding the effects of weather variability and the potential impact of climate change on rice and aquaculture systems in the delta is important for developing strategies for adapting to climate change in the future.

The agronomic literature dealing with climatic effects on agricultural production is large and has a long history (Rosenberg 2010). According to data from

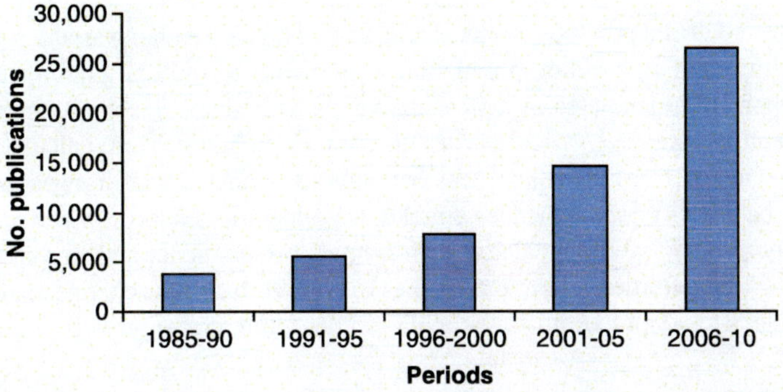

Fig. 24.1 The number of publications related to the linkage of climate change and agricultural and biological sciences by Science Direct (Based on data from http://www.sciencedirect.com, assessed on March 24, 2010)

ScienceDirect, literature on the impact of anthropogenic climate change on agricultural and biological systems dates from the early 1980s, and the number of publications on this subject has rapidly increased since the early 2000s (Fig. 24.1). The studies on this subject mostly rely on simulation models to provide the estimates of the impact of climate change on production (Rosenberg, 2010). For Vietnam in general and the Mekong Delta in particular, studies on the impact of potential climate change and sea-level rise are quite new (e.g., Wassmann et al. 2004; Carew-Reid 2007; ADB 2009; MONRE 2009). These studies, too, rely on simulation models to project the potential impact of future climate change and sea-level rise, based on pre-determined scenarios. In other words, there is still a lack of detailed, site-specific information regarding the climatic impact on agricultural production in the delta. We argue that in the past Mekong Delta farmers have dealt with climatic variability by modifying their farming practices. Understanding climatic effects on *current* agricultural production systems is necessary to identify possible solutions to improving food-production systems in the face of climate change and dealing with food insecurity in the future. By integrating a combination of census-data analysis and participatory community and household appraisals, this essay quantifies the effects of short-term weather variability on rice and aquaculture production, and it subsequently suggests possible ways to improve current agricultural production systems to effectively adapt to future weather variability and climate change.

24.2 Materials and Methods

24.2.1 Study Framework

To achieve the objective of the work, the present study integrated analyses of census data from the GSO (2010) and the CSO (2010) with participatory community-appraisal data collected from previous projects.[1] First, principal weather variables that have strongly impacted agricultural production were identified through participatory community appraisals conducted in representative research areas. Second, monthly values of the identified weather variables with different probabilities were calculated, based on data collected from provincial weather stations. Third, critical ranges and occurrence probabilities of the weather variables were identified and estimates of impacts on crop production were calculated with farmers' assessments and linear-regression methods. Finally, farmers' current measures to cope with climate change are described and strategies to adapt to climate variability in the future are suggested.

24.2.2 Data Collection

Participatory community appraisals were conducted in three districts: Vinh Thanh (Can Tho City), Tra Cu (Tra Vinh Province), and An Bien (Kien Giang Province). Vinh Thanh District is located in a flood-prone, freshwater, and irrigated zone, where intensive rice growing with two or three crops a year and freshwater fish culture are considered the major farming activities. The two other districts are located in coastal zones, where rice farming with two crops a year relies heavily on rainfall during the wet season, while black tiger (*Penaeus monodon*) shrimp farming totally depends on saline water from the sea in the dry season. For each district, two representative communes were identified, based on agro-ecological characteristics, from which one representative hamlet per commune was selected for data collection. For each hamlet, different dominant farming systems were determined. Focused group discussions were done with 7–10 people per farming system to get a general understanding of current farming practices, principal weather variables affecting crop production, and their occurrence time. Subsequently, household interviews were conducted with group members to estimate impacts on crop production.

Census data of An Giang, Can Tho, Hau Giang, Tra Vinh, Soc Trang, Bac Lieu, Cau Mau, and Kien Giang in the period of 1990–2008 were collected from yearly statistical books (CSO 2000, 2010; GSO 2010). Variables included monthly air temperature (mean, maximum, and minimum), monthly accumulative rainfall, and yearly yields of rice, fish, and shrimp.

24.2.3 Statistical Analysis

We applied statistical methods to analyze census data. We estimated monthly values of weather variables, which were considered to have significant impacts on crop

[1] (1) Hazard vulnerability and capacity assessments in Can Tho City of Vietnam in 2009, funded by Challenge to Change, and (2) improved agricultural water productivity for poverty reduction in coastal areas of the Vietnamese Mekong delta in 2008, funded by SEI-Asia.

production, and the probability that such values would be exceeded by 10%, 25%, 50% and 75%, assuming that data of the considered variables are log-normally distributed (Shahin et al. 1993). To confirm and further explain results obtained from farmers' assessments, multiple linear regression methods were applied to assess the impact of weather variables on rice, fish, and shrimp yields. To meet the assumptions for multiple linear regression analysis, we tested the normality, the variance homogeneity, the autocorrelation, and the multicollinearity of variables considered. We used the forward stepwise method to select variables. The validity of the results from the representative models was assessed using non-parametric bootstrapping, which creates a validation sample by sampling with replacement from the original sample (Hair et al., 1998). Combined with farmers' observations, the probability and multiple regression analysis allowed us to determine critical values and impacts of extreme temperature and rainfall on rice production.

24.3 Results and Discussions

24.3.1 Identification of Weather Variables Affecting Rice and Aquaculture

Results from participatory community appraisals show that anomalies in air temperature and rainfall, which usually occur in the onset of the dry and wet seasons, caused negative effects on rice and fish or shrimp production. Monthly distribution of air temperature and rainfall with probabilities of exceedence (*P*) is shown in Figs. 24.2 and 24.3. Colder days usually take place in December or January (the onset of the dry season) while hotter days are almost always observed in April (the early period of the wet season). In April, the probability of the maximum temperature exceeding 34.5°C, 35°C, 35.5°C, and 36°C is 75%, 50%, 25%, and 10%, respectively (Fig. 24.2b). Similarly, in January, the probability of the minimum temperature exceeding 19°C, 20°C, 21°C, and 21.5°C is 75%, 50%, 25%, and 10%, respectively (Fig. 24.2c).

During the dry season (December–April), rainfall is light and some areas experience droughts. Peak rainfall occurs between July and October. In the irrigated zone, in February the probability of rainfall exceeding 2, 5, and 11 mm is 50%, 25%, and 10% (Fig. 24.3a). In the coastal zone, the corresponding rainfall values at 50%, 25%, and 10% probabilities are 1.5, 2.5, and 12 mm, respectively (Fig. 24.3b). During the dry season with low flows of the Mekong River, salinity intrusion occurs in the lower part of the delta. During the wet months, in contrast, local rainfall, combined with the heavy discharge of water from upstream on the Mekong River, causes monsoon floods in the upper delta between August and November. A combination of local rainfall and the hydrological patterns of the Mekong River determines the seasonal cropping calendar of rice and shrimp culture in both the flood and coastal zones (Fig. 24.4). In irrigated and freshwater zones, rice culture and

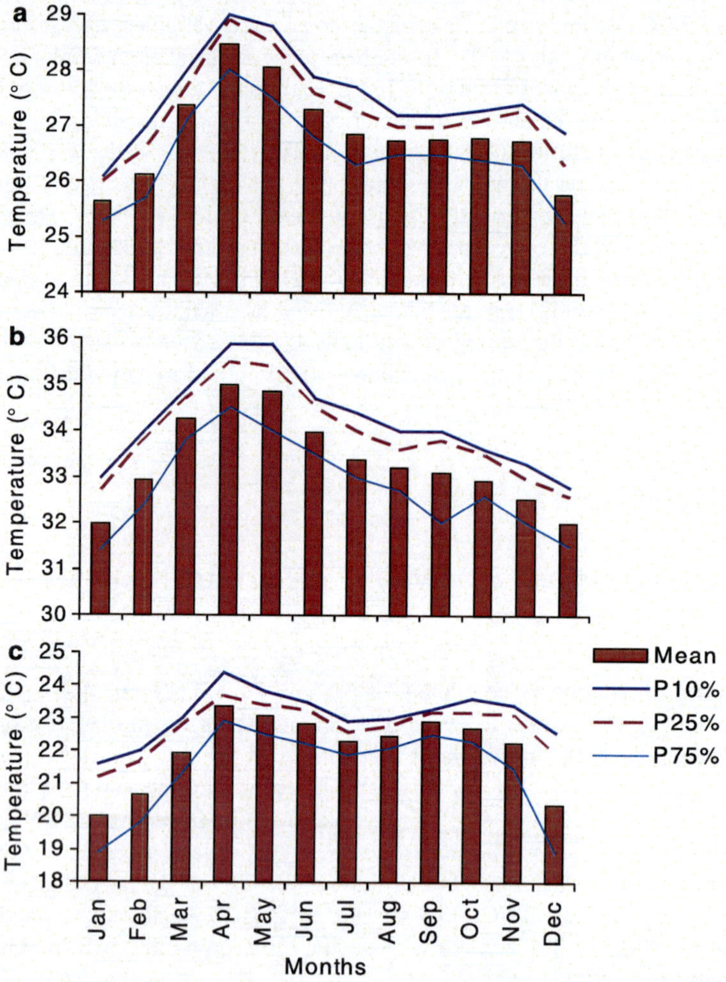

Fig. 24.2 Monthly air temperature with probability of exceedence (P) = 10, 25, 50 (mean) and 75%: (**a**) mean temperature, (**b**) maximum temperature, and (**c**) minimum temperature

aquaculture rely heavily on water from the Mekong River. During flood periods with high and prolonged inundation, however, rice cannot be grown. In the lower part of the delta along the coast, in contrast, rice production totally depends on rainfall during the wet season, while shrimp is grown during the dry season, using saline water from the sea.

In the irrigated zone, according to local people, extremely cold and/or rainy days occurring in January and February, when the rice crop is in flowering and ripening stages, would cause significant losses in the dry season (DS) crop. In addition, in the wet-season (WS) rice crop, abnormally hot days occurring during vegetative stages of the rice plant (April–May) would constrain rice production through limiting

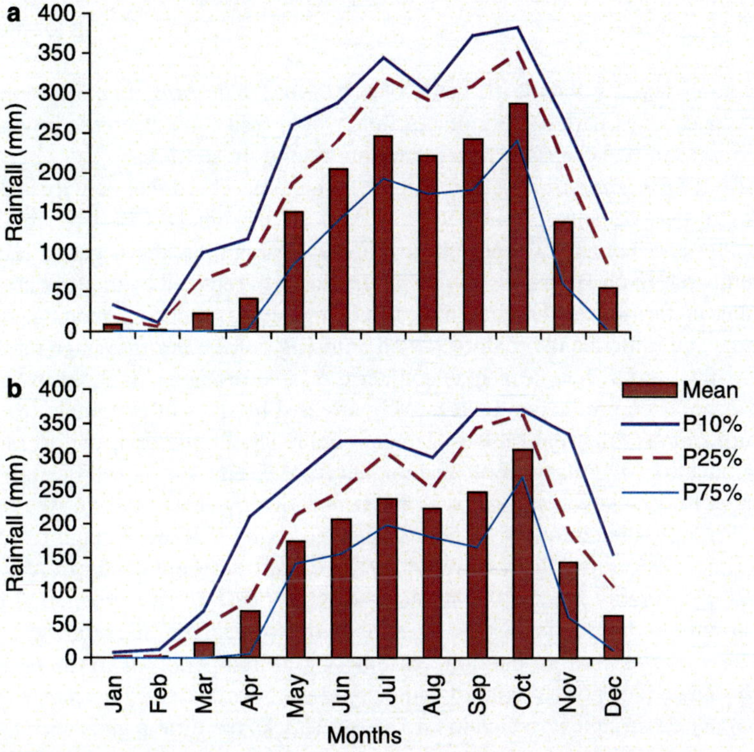

Fig. 24.3 Monthly rainfall with probability of exceedence (P) = 10, 25, 50 (mean) and 75% in the freshwater and irrigated zone (An Giang, Can Tho and Hau Giang provinces) (**a**) and the coastal zone (Tra Vinh, Soc Trang, Bac Lieu and Kien Giang provinces) (**b**)

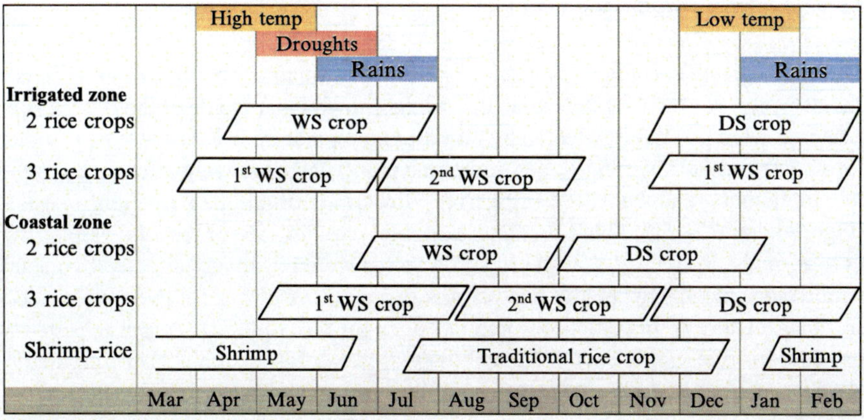

Fig. 24.4 The occurrence of abnormal weather events and seasonal calendar of rice and shrimp farming patterns in the irrigated zone (An Giang and Can Tho provinces) and in the coastal zone (Tra Vinh and Kien Giang provinces). *Pangasius* catfish culture is practiced year round in the irrigated zone

rice tillering activity and the incidence of rice thrips (*Baliothips bijomio*). Moreover, more cloudy days with heavy rains in June would result in a decrease in yields of the wet-season rice crops through increasing rice-grain sterility.

For intensive *Pangasius* catfish culture, farmers observed that activity and food intake of fish became problematic on days of extremely low temperature in December and January. According to farmers' evaluations, the negative effect of low temperature on *Pangasius* catfish production was reduced by appropriate interventions in farming techniques like deepening ponds, reducing feeding, adding vitamins and minerals into fish feed, and reducing pond water exchange rates.

In coastal areas, rice farmers encounter the same problems as those in the irrigated zone. Rice production is negatively affected by days of extremely low temperature and/or abnormal rains in January–February. In addition, due to rain-fed rice production, the occurrence of droughts in May and June would cause water shortages for rice in vegetative stages and hence lower yields of the wet-season rice crop. Through simulation, Tuan and Chinvanno found that the average maximum temperature in the dry season would increase and that rainfall during the early periods of the rainy season would decrease. In the case of rice production in the coastal zone, adaptive strategies to combat severe drought stress are therefore necessary. For shrimp production, extremely high temperatures in the period of March–May and abnormal rainfall in the dry season would cause big changes in the pond-water environment, which in turn causes shocks for shrimp and hence harvest losses.

24.3.2 Assessments of the Impact of Weather Variability

24.3.2.1 Rice Production

Multiple regression analysis confirmed and further explained findings from farmers' assessments. For the irrigated zone, the results showed that yields of the dry-season rice crop were positively affected by minimum temperature in January (i.e., panicle formation and/or flowering stages of the rice plant) and negatively affected by rainfall in January and February (i.e., from flowering, milking to ripening stages) (Figs. 24.4 and 24.5a). These predictor variables explain 15% of the total variability in rice yields. In the wet-season crop, rice yields were negatively affected by mean temperature in May, when the rice plant is in vegetative stages for the double rice-cropping pattern or in panicle-formation stages for the triple rice-cropping pattern. In addition, rice yields decreased with increased rainfall in June. These predictor variables explained 44% of the total variability in rice yields of the wet-season crop (Fig. 24.5b).

For the coastal zone, as in the irrigated zone, abnormal rains taking place in January and February resulted in a significant decrease in yields of the dry-season rice crop (Fig. 24.6a). This rainfall anomaly accounted for 24% of the total variability in rice yields. For the wet-season crop, rice yields tended to be significantly

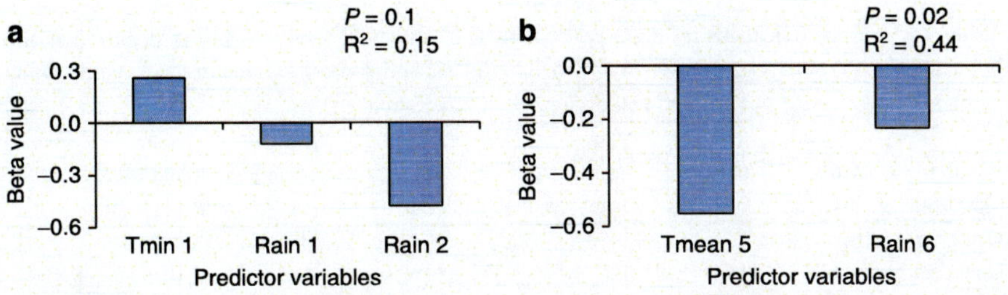

Fig. 24.5 Standardized partial regression coefficient (beta value) for predictor variables of rice yields in the irrigated zone: (**a**) dry-season crop and (**b**) wet-season crop. Model significance (*P*), adjusted coefficient of the determination (R^2), rainfall (Rain), minimum temperature (T_{min}), mean temperature (T_{mean}). For each predictor variable, the nearby number indicates month. The same explanation is also applied to Fig. 24.6

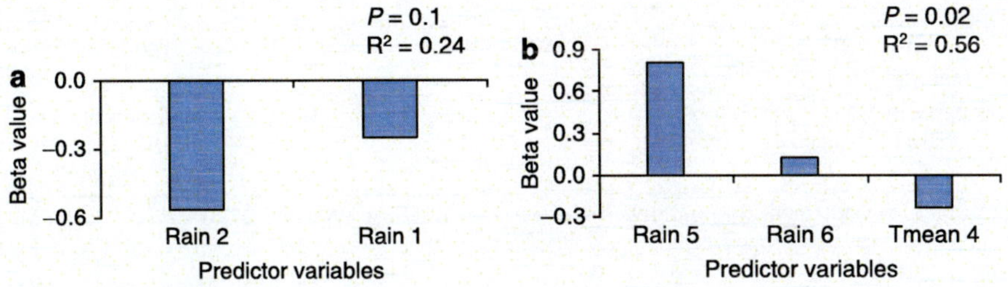

Fig. 24.6 Standardized partial regression coefficient (beta value) for predictor variables of rice yields in the coastal zone: (**a**) dry-season crop and (**b**) wet-season crop. Model significance (*P*) and adjusted coefficient of the determination (R^2)

greater with higher rainfall in May and June and with lower temperature in April (Fig. 24.6b). These predictor variables explain 56% of the total variability in rice yields. In the coastal zone, rice production relies totally on rain water during the rainy season, and rice production is highly susceptible to the occurrence of severe drought stress during early periods of the rainy season. In this zone, droughts lead not only to water shortages for the rice plant but also to soil acidification and salinization, due to acid sulphate soils and salinity intrusion from the estuary, which in turn adds more negative effects on rice production (Nhan et al. 2008). Looking at significance levels (*P*) and coefficients of determination (R^2), one can find that the negative impacts of temperature and rainfall anomalies on rice production were more significant during the wet season and for the coastal zone than for the dry-season crop and the irrigated zone, respectively. In adapting rice production in the delta to future weather and climate variability, we therefore need to pay more attention to the wet-season rice crop and to the coastal zone.

Farmers estimated the impact of temperature and rainfall variability on rice production and the frequency of occurrence of abnormal variation for both the dry season and the wet season in the irrigated zone as well as the coastal zone (Table 24.1).

Table 24.1 Estimates from farmers' perception and statistical analysis on the occurrence and impacts on rice yields of weather variability in irrigated and coastal regions in the Mekong Delta

Weather variability by crop	Farmers' estimation		Statistical calculation[a]		
	Return period (years)	Yield losses (tons paddy)	Value	Return period (years)	Yield losses (tons paddy)
Dry-season crop					
Low temperature in Jan	3–4	0.6	<19°C	4	0.12 for each 1°C decrease
Rainfall in Feb	3–4	0.6	>10 mm	4	0.3–0.4 for each 10 mm increase
Wet-season crop					
Extremely hot temperature in May (irrigated region)	3–4	0.3	>35°C	4	0.38 for each 1°C increase
Heavy rainfall in Jun (irrigated region)	2	0.6	>250 mm	2	0.2 for each 100 mm increase
Droughts in May (coastal region)	5–6	1.8	<50 mm	4	0.6 for each 100 mm decrease

[a]Estimates are obtained from multiple regression models (Figs. 24.5 and 24.6)

For the dry-season rice crop, abnormally low temperatures or abnormal rains in January–February were estimated to occur every 3–4 years, which caused yield loss of about 0.6 tons per hectare, accounting for about 10% of the normal yield. For the wet-season rice crop, in the irrigated region farmers perceived that extremely hot temperatures in May occurred every 3–4 years, causing yield loss of 5% of the normal yield. In addition, they estimated that high levels of rainfall in June – levels that would cause yield loss of about 10% of normal yield – occurred every 2 years (Table 24.1). In the coastal zone, in contrast, water scarcity in early rainy periods is an important problem for the wet-season rice. According to local farmers, an extreme drought occurs once every 5–6 years, which would reduce the rice crop by 1.8 tons per hectare, accounting for about 40% of the normal yield.

Based on farmers' estimates of abnormal weather variation, critical values and impacts of extreme temperature and rainfall were determined through both probability and multiple regression methods. As mentioned earlier, statistical results revealed that there is 25% probability that the lowest temperature in December or January is below 19°C. Through a multiple regression equation, it was calculated that in January in the range of 18–22°C, for each 1°C drop, yield decreases by 0.12 tons paddy per hectare in the irrigated region. Similarly, in April with 25% probability that the highest temperature is above 35.5°C, and if the temperature is greater than 34°C, for each 1°C increase the yield would decrease by 0.38 tons of paddy per hectare. Our results are consistent with those of Yoshida (1981) on the critical importance of temperature for the development of rice plants at different growth stages in tropical regions. According to him, extreme temperature at the productive stages of the rice plant, especially at anthesis (i.e., 9 days before and at heading), results in grain sterility in a high percentage of plants.

Similarly, we determined critical rainfall and quantified its impacts on rice production. In February with 25% probability the rainfall was observed to be greater than 10 mm. For each 10-mm increase, a harvest loss is estimated of 0.3 tons of paddy per hectare in the coastal region or 0.4 tons of paddy in the irrigated region. By the same method, we determined critical rainfall in May and June and quantified its impacts on rice production in both irrigated and coastal zones (Table 24.1).

24.3.2.2 Aquaculture Production

Regarding the effects of extreme temperature and rainfall on intensive *Pangasius* catfish production, statistical data analysis did not reveal any significant impact. Farmers' measures in coping with weather variability in *Pangasius* catfish production could be the reason. For shrimp production, the regression analysis shows the significantly negative correlation of mean air temperature in April and in May with shrimp yields (Fig. 24.7). According to this relationship, the variability of mean temperature in April and in May explains about 35% and 21% of the total variance of shrimp yields, respectively. With these correlations, shrimp yields would decrease by 160 kg per hectare or 90 kg per hectare for each 1°C increase (in a range of 27.5 and 29.5°C). The probability of mean air temperature exceeding 28°C is 75% in April and 50% in May (Fig. 24.2a). On hot days in April and May, the maximum temperature of the air usually exceeds 34°C (Fig. 24.2b), which is greater than the optimum temperature range for many shrimp species (Piyakarnchana et al. 1975). Pond water tends to follow the general trends of temperature in the surrounding air. Extremely high temperature would cause, both directly and indirectly, negative effects on shrimp as well as the pond-water environment and hence on growth and survival of shrimp (Delince 1984). Previous studies reported that coastal shrimp farming is economically risky (Preston and Clayton 2003; Joffre and Bossma 2009).

24.3.3 Measures of Coping and Adapting to Weather Variability

24.3.3.1 Coping Measures

Local farmers and extension workers have improved farming techniques and institutions to cope with extreme temperature and rainfall. Coping measures have been mostly applied at the farm level. For irrigated rice production, farmers have applied integrated nutrient management to enhance rice-plant health and allow the rice to better tolerate weather anomalies while reducing agro-chemical input costs. Earlier maturity cultivars with hard straw have been grown to minimize risks of extreme weather conditions in both dry and wet-season crops. For rain-fed rice

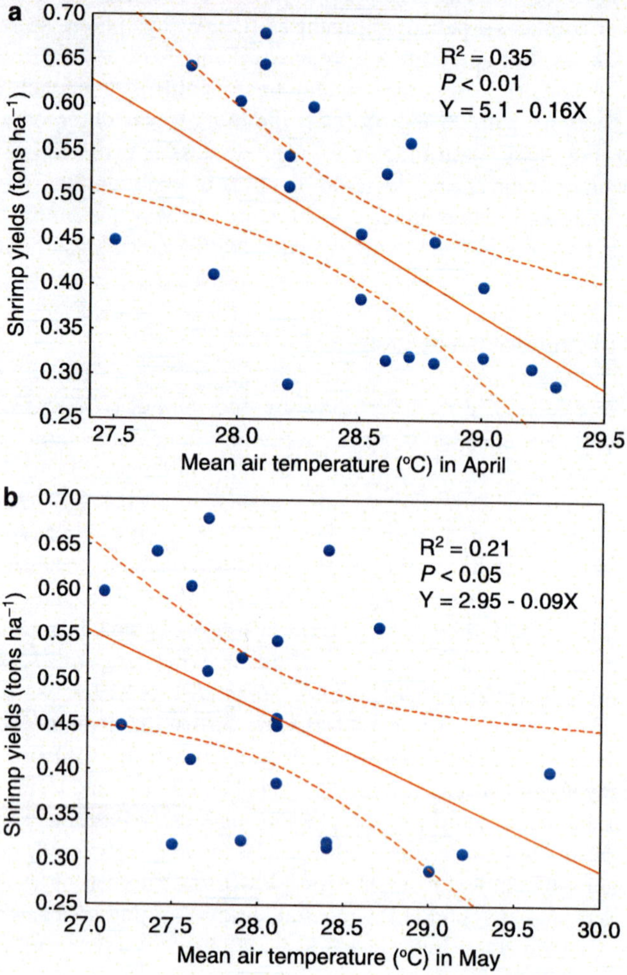

Fig. 24.7 The relationship of mean air temperature in April (**a**) and May (**b**) with shrimp yields. Regression line with the confident level at 95%, significance (P), the coefficient of determination (R^2), and the regression equation. Data collected from Tra Vinh, Soc Trang, Bac Lieu, and Ca Mau provinces were employed

production in the coastal zone, the rice crop is determined after consideration of rainfall distribution, the amount of freshwater available in irrigation canal systems, soil moisture content, and mid-range weather forecasts by local weather centers. In addition, farmers have modified their rice fields by digging small ditch systems in order to drain surplus rain water, to store rain water for maintaining soil moisture, and for preventing acidity from moving up from deep soils during soil-oxidation processes. Furthermore, farmers have extracted ground water to irrigate crops in the dry season. This measure, however, is not advocated by the government. The use by farmers of rice cultivars that can tolerate drought, salinity, and/or acidity has still been limited, due to the scarcity of such cultivars.

For intensive *Pangasius* catfish farming, farmers seem to deal effectively with weather variability using appropriate farming techniques as mentioned earlier. For shrimp production, farmers have grown *Scirpus littoralis* (Schrab) in shrimp farms at an appropriate density to stabilize the temperature and aquatic environment of shrimp ponds. In addition, farmers also deepen shrimp ponds or irrigate the pond with a greater amount of water to ensure that the temperature of the pond water does not become too hot. Furthermore, farmers have excavated deep peripheral trenches within the pond for the shrimp to seek shelter on hot days.

Nhan et al. (2008), analyzing the improvement of agricultural-water productivity in the Mekong delta's coastal region, reported that under the same agro-ecological system, water-use ability, and efficiency of resource-poor households are lower than in richer households, even though more support policies are given to poorer people by the government. This would suggest that the effectiveness of various coping measures is also influenced by the resources of households. Coping measures and interventions appear to be not enough to further improve the livelihoods of rural poor people.

24.3.3.2 Adaptation Measures

According to Maclver (1998), "adaptation is an important component of an integrated and balanced strategy to climatic variability. Adaptation is largely a time-dependent, location-specific learning process." A systems approach, which includes integral components regarding agricultural-production systems, food security, rural people's livelihoods, and combined "bottom-up" responses and "top-down" policy making, is needed. Such an approach is not new, but there is a gap between theoretical knowledge and its application in real situations. Successful climate-change adaptation and vulnerability reduction is rarely undertaken with respect to CC alone, and vulnerability reduction appears to be most effective if undertaken in combination with other strategies and plans at various levels (Smit and Wandel 2006).

In recent decades, the Vietnamese government has invested a great deal of money in developing irrigation systems in both irrigated and coastal zones. Such actions have facilitated coping measures by farmers, which seem to be successful at the farm level. For rice, appropriate seeding, proper cognizance of the length of the crop season, and selection and development of rice varieties with short-growth duration with different abilities to tolerate high temperature, drought, and salinity are necessary. For aquaculture, selection and development of fish/shrimp species that tolerate extreme temperature is of great importance in the future. In order to promote adaptation, technical measures need to be integrated to provide farmers a choice of technological packages appropriate for specific contexts rather than one generic package of technologies. A technological package would include relevant components such as adaptive cultivars, site-specific information regarding farming practices and integrated farming systems, and specific advice on how to manage agricultural water so as to adapt to weather and climate variability. In addition, crop-yield forecasting and simulation need to be applied to predict crop growth or production from the time of the forecast up to the time of harvest and to identify measures for minimizing crop vulnerability and risk due to weather fluctuations and climate variability (Bouman et al. 1997). Such risk-reduction measures are still new in the Mekong Delta. Moreover, the fact that at present weather forecasts are only made with lead times of a few days to a few weeks is still a constraint.

In addition, efforts to improve the agricultural sector's ability to adapt to weather and climate variability, which are usually made at farm scale, need to be reconsidered and placed in broader, more dynamic, and more heterogeneous contexts – the livelihoods of rural people and food security, for example – and taken from household, to regional, national, and even global scale (Ingram et al. 2008). In so doing, adaptation options developed will not only be made effective in terms of agricultural production but also more robust in terms of environmental and economic policy. Agriculture production in the delta has faced increasing challenges from weather and climate variability. Such challenges put more pressure on the government to develop appropriate policies to enhance the adaptability of farmers. It is therefore necessary to integrate the "top-down" scenario-based approach with the "bottom-up" vulnerability-perspective approach (Wall and Smit 2005). By assessing the impact of policies, analyzing the adaptability of local people, and identifying the factors that enhance or discourage adaptability, vulnerability assessments can help to improve agricultural policy. In Vietnam, however, the integration of the two approaches seems to be limited.

24.4 Conclusions

This essay has focused on the vulnerability of rice and aquaculture to temperature and rainfall variability in both irrigated and coastal zones in the Mekong Delta, and it has suggested strategies to improve the ability of current agricultural systems to effectively adapt to weather and climate variability in the future. Impacts of temperature and rainfall variability differ with different crops, the development stages of crops, and by season and region. The impacts on rice production are more clear-cut than on aquaculture production. Shrimp production is more directly sensitive to weather variability than is *Pangasius* catfish production. The impacts become more severe during the early vegetative, flowering, and ripening stages of rice crops. The wet-season rice crop and the coastal region are more vulnerable to weather anomalies than the dry-season rice crop and the irrigated region. So far, the measures farmers have used to cope with weather variability are mostly immediate ones. Sustainable rice and aquaculture for food security will require adaptation strategies to weather variability and climate change. Technical solutions for coping are available and should be further improved. Policy support for agricultural research and development is of great importance for the realization of such solutions. More attention should be paid to the coastal zone.

References

Asian Development Bank (ADB) (2009) The economics of climate change in Southeast Asia: a regional review. Asian Development Bank, Manila

Bouman BAM, Van Diepen CA, Vossen P, Van der Wal T (1997) Simulation and systems analysis tools for crop yield forecasting. In: Teng PS et al (eds) Applications of systems approaches at the farm and regional levels, vol 1. IRRI and ICASA/Kluwer, Dordrecht, pp 325–340

Carew-Reid J (2007) Rapid assessment of the extent and impact of sea level rise in Viet Nam. Climate Change Discussion Paper 1, ICEM –International Centre for Environmental Management, Brisbane

CSO (Can Tho Statistical Office) (2000, 2010) Yearly statistical book. Can Tho Publishing House, Can Tho

Delince G (1984) The ecology of the pond ecosystem: with special reference to Africa. Kluwer Academic Publishers

GSO (General Statistical Office) (2010) Yearly statistical book. GSO Publishing House, Hanoi

Hair JF, Anderson RE, Tatham RL, Black WC (1998) Multivariate data analysis, 5th edn. Prentice-Hall International, Upper Saddle River

Heine B (2009) Adaptation to climate change in coastal areas of the Mekong delta, Vietnam (unpublished paper)

Ingram JSI, Gregory PJ, Izac A-M (2008) The role of agronomic research in climate change and food security policy. Agric Ecosyst Environ 126:4–12

Joffre OM, Bosma RH (2009) Typology of shrimp farming in Bac Lieu province, Mekong Delta, using multivariate statistics. Agric Ecosyst Environ 132:153–159

Maclver DC (1998) Adaptation to climatic variability and change. In: Workshop summary, San Jose. Atmospheric Environment Service, Environment Canada, Toronto

Ministry of Natural Resources and Environment of Vietnam (MONRE) (2008) National target program to respond to climate change (a draft version). MONRE, Hanoi

Ministry of Natural Resources and Environment of Vietnam (MONRE) (2009) Climate change and sea level rise scenarios for Vietnam (internal report). MONRE, Hanoi

Nhan DK, Be NV, Trung NH (2007) Water use and competition in the Mekong Delta. In: Be TT, Sinh BT, Miller F (eds) Challenges to sustainable development in the Mekong delta – regional and national policy issues and research needs. SUMERNET, Bangkok, pp 143–188

Nhan DK, Son NN, Be TT (2008) Raising basin level agricultural water productivity for poverty reduction: Mekong delta case study. Mekong Basin Focal Project report

Piyakarnchana T, Hungspreugs M, Tamiyavanich S (1975) Some limiting factors on the growth and survival of the banana prawn, Penaeus merguiensis de Man cultivated in a tambak. In: Special symposium on marine sciences, Hong Kong, 7–16 Dec 1973. The Pacific Science Association, Hong Kong, pp. 71–74

Preston N, Clayton H (eds) (2003) Rice-shrimp farming in the Mekong Delta: biophysical and socioeconomic issues. ACIAR Technical Report No. 52e, ACIAR, Canberra

Rosenberg NJ (2010) Climate change, agriculture, water resources: what do we tell those that need to know? Clim Change 100:113–117

Shahin M, van Oorschot HJL, de Lange SJ (1993) Statistical analysis in water resources engineering. Balkema, Rotterdam

Smit B, Wandel J (2006) Adaptation, adaptive capacity and vulnerability. Glob Environ Change 16:282–292

Wall E, Smit B (2005) Climate change adaptation in light of sustainable agriculture. J Sustain Agric 27:113–123

Wassmann R, Hien NX, Hoanh CT, Tuong TP (2004) Sea level rise affecting the Vietnamese Mekong delta: water elevation in the flood season and implications for rice production. Clim Change 66:89–107

Yoshida S (1981) Fundamentals of rice crop science. IRRI, Los Baños

Index

M.A. Stewart and P.A. Coclanis (eds.), *Environmental Change and Agricultural Sustainability in the Mekong Delta*, Advances in Global Change Research 45, DOI 10.1007/978-94-007-0934-8,